Revit 2024 Architecture Basics: From the Ground Up

Elise Moss

Authorized Developer

SDC Publications
P.O. Box 1334
Mission, KS 66222
913-262-2664
www.SDCpublications.com
Publisher: Stephen Schroff

Copyright 2023 by Elise Moss

All rights reserved. This document may not be copied, photocopied, reproduced, transmitted, or translated in any form or for any purpose without the express written consent of the publisher, SDC Publications.

It is a violation of United States copyright laws to make copies in any form or media of the contents of this book for commercial or educational purposes without written permission.

Examination Copies
Books received as examination copies are for review purposes only and may not be made available for student use. Resale of examination copies is prohibited.

Electronic Files
Any electronic files associated with this book are licensed to the original user only. These files may not be transferred to any other party.

Trademarks
The following are registered trademarks of Autodesk, Inc.: AutoCAD, AutoCAD Architecture, Revit, Autodesk, AutoCAD Design Center, Autodesk Device Interface, VizRender, and HEIDI.
Microsoft, Windows, Word, and Excel are either registered trademarks or trademarks of Microsoft Corporation.
All other trademarks are trademarks of their respective holders.

The authors and publisher of this book have used their best efforts in preparing this book. These efforts include the development, research, and testing of material presented. The author and publisher shall not be held liable in any event for incidental or consequential damages with, or arising out of, the furnishing, performance, or use of the material herein.

ISBN-13: 978-1-63057-600-4
ISBN-10: 1-63057-600-X

Printed and bound in the United States of America.

Preface

Revit is a parametric 3D modeling software used primarily for architectural work. Traditionally, architects have been very happy working in 2D, first on paper, and then in 2D CAD, usually in AutoCAD.

The advantages of working in 3D are not initially apparent to most architectural users. The benefits come when you start creating your documentation and you realize that your views are automatically defined for you with your 3D model. Your schedules and views automatically update when you change features. You can explore your conceptual designs faster and in more depth.

Revit will not make you a better architect. However, it will allow you to communicate your ideas and designs faster, easier, and more beautifully. I wrote the first edition of this text more than ten years ago. Since that first edition, Revit has become the primary 3D CAD software used in the AEC industry. Revit knowledge has become a valuable skill in today's market and will continue to be in demand for at least another decade.

The book is geared towards users who have no experience in 3D modeling and very little or no experience with AutoCAD. Some experience with a computer and using the Internet is assumed. Autodesk is launching a browser-based version of Revit in 2023. This will allow users to "rent" the use of the software for a specific project instead of purchasing a full license.

This book uses the installed version of the software, but the interface is similar. The reason I opted for the non-browser-based software is 1) to minimize the consumption of data on my internet plan 2) to be able to work "off-line" when no internet is available and 3) to avoid data lag due to slow connection speeds. My expectation is that most schools and training centers will also opt for the non-browser installation for similar reasons.

I have endeavored to make this text as easy to understand and as error-free as possible…however, errors may be present. Please feel free to email me if you have any problems with any of the exercises or questions about Revit in general.

Acknowledgements

A special thanks to Gerry Ramsey, Scott Davis, James Balding, Rob Starz, and all the other Revit users out there who provided me with valuable insights into the way they use Revit.

Thanks to Stephen Schroff, Zach Werner and Karla Werner who work tirelessly to bring these manuscripts to you, the user, and provide the important moral support authors need.

My eternal gratitude to my life partner, Ari, my biggest cheerleader throughout our years together.

Elise Moss
Elise_moss@mossdesigns.com

TABLE OF CONTENTS

Lesson 10
Customizing Revit

Revit Hot Keys

About the Author

Class Files

To download the files that are required for this book, type the following in the address bar of your web browser:

SDCpublications.com/downloads/978-1-63057-600-4

Lesson 1
The Revit Interface

 Go to Start→Programs→Autodesk → Revit 2024.

You can also type Revit in the search field under Programs and it will be listed.

When you first start Revit, you will see this screen:

It is divided into three sections:

The Left section has two panels. The top panel is to Open or Start a new project. Revit calls the 3D building model a project. Some students find this confusing.

The bottom panel in the left section is used to open, create, or manage Revit families. Revit buildings are created using Revit families. Doors, windows, walls, floors, etc., are all families.

Recent Files show recent files which have been opened or modified as well as sample files.

If you select the Online Help link at the bottom of the left panel, a browser tab will open.

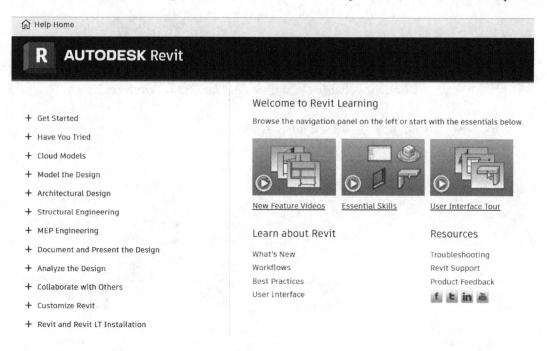

You should be able to identify the different areas of the user interface in order to easily navigate around the software.

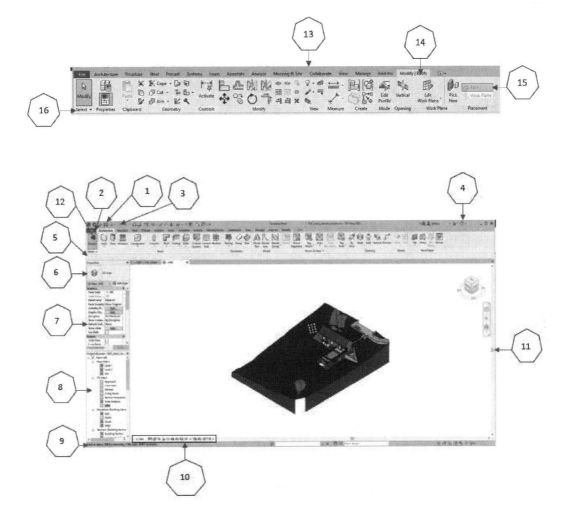

1	Revit Home	9	Status Bar
2	File tab	10	View Control Bar
3	Quick Access Toolbar	11	Drawing Area/Window
4	Info Center	12	Ribbon
5	Options Bar	13	Tabs on the ribbon
6	Type Selector	14	Contextual Ribbon – appears based on current selection
7	Properties Palette	15	Tools on the current tab of the ribbon
8	Project Browser	16	Panels – used to organize the ribbon Tools

The Revit Ribbon

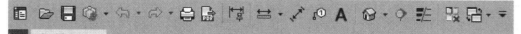

The Revit ribbon contains a list of panels containing tools. Each button, when clicked, reveals a set of related commands.

The Quick Access Toolbar (QAT)

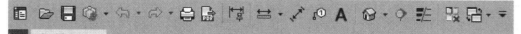

Most Windows users are familiar with the standard tools: New, Open, Save, Undo, and Redo.

 Opens the Home tab which has the Recent Files listed.

To return to the active project select the Back arrow.

	Synchronize to Central is used in team environments where users check in and check out worksets on a shared project. The Central location should be a shared drive or server that all team members can access. The **Synchronize to Central** tool is greyed out unless you have set up your project as a shared project with a central location.
	Exports your project to PDF. You can export the active window, a portion of the window or selected views and/or sheets.
	Controls the visibility of constraints.
Measure Between Two References / Measure Along An Element	Measure is used to measure distances.
	Places a permanent linear dimension.
	Tag by Category adds a label or symbol on doors, windows, equipment, etc.
A	Adds text to the current view.
Default 3D View / Camera / Walkthrough	Allows the user to switch to a default 3D isometric view, place a camera or create a walkthrough.
	Create a section view.
	When this is enabled, all lineweights display as a single width. This simplifies the view and uses less memory.
	Closes non-active windows. Reduces the number of tabs you see and uses less memory.
	Displays a list of open windows, so you can select a different window.
Customize Quick Access Toolbar / New / Open / Save / Synchronize with Central / Undo / Redo / Print / PDF / Activate Controls and Dimensions / Measure / Aligned Dimension / Tag by Category / Text / Default 3D View / Section / Thin Lines / Close Inactive Views / Switch Windows / Customize Quick Access Toolbar / Show Below the Ribbon	The down arrow allows users to customize which tools appear on the Quick Access toolbar.

Printing

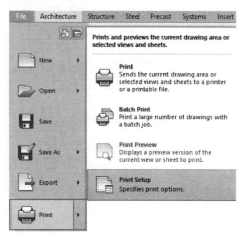

Print is located in the Application Menu as well as the Quick Access Toolbar.

The Print dialog is fairly straightforward.
Select the desired printer from the drop-down list, which shows installed printers.

You can set to 'Print to File' by enabling the check box next to Print to File.
The Print Range area of the dialog allows you to print the current window, a zoomed in portion of the window, and selected views/sheets.

Undo

The Undo tool allows the user to select multiple actions to undo. To do this, use the drop-down arrow next to the Undo button; you can select which recent action you want to undo. You cannot skip over actions (for example, you cannot undo 'Note' without undoing the two walls on top of it).
Ctrl-Z also acts as UNDO.

Redo

The Redo button also gives you a list of actions, which have recently been undone. Redo is only available immediately after an UNDO. For example, if you perform UNDO, then WALL, REDO will not be active.

Ctrl-Y is the shortcut for REDO.

Viewing Tools

A scroll wheel mouse can replace the use of the steering wheel. Click down on the scroll wheel to pan. Rotate the scroll wheel to zoom in and out.

The Rewind button on the steering wheel takes the user back to the previous view.

Different steering wheels are available depending on whether or not you are working in a Plan or 3D view.

The second tool has a flyout menu that allows the user to zoom to a selected window/region or zoom to fit (extents).

| ✓ Zoom in Region |
| Zoom Out(2x) |
| Zoom to Fit |
| Zoom All to Fit |
| Zoom Sheet Size |
| Previous Pan/Zoom |
| Next Pan/Zoom |

➤ Orient to a view allows the user to render an elevation straight on without perspective. Orient to a plane allows the user to create sweeps along non-orthogonal paths.

| Minimize to Tabs |
| Minimize to Panel Titles |
| Minimize to Panel Buttons |
| ✓ Cycle through All |

➤ If you right click on the Revit Ribbon, you can minimize the ribbon to gain a larger display window.

General
User Interface
Colors
Graphics
Hardware
File Locations
Rendering
Check Spelling
SteeringWheels
ViewCube
Macros
Cloud Model

Text visibility
- Show tool messages (always ON for basic wheels)
- Show tooltips (always ON for basic wheels)
- Show tool cursor text (always ON for basic wheels)

Big steering wheel appearance
Size: Normal   Opacity: 50%

Mini wheel appearance
Size: Normal   Opacity: 50%

Look tool behavior
- Invert vertical axis (Pull mouse back to look up)

Walk tool
- Move parallel to ground plane
  Speed Factor:
  0.1   50.0   3

Zoom tool
- Zoom in one increment with each mouse click (always ON for basic wheels)

Orbit tool
- Keep scene upright

Rewind history
- Save snapshots of current view (Turn OFF for better performance)

You can control the appearance of the steering wheels by right clicking on the steering wheel and selecting Options.

Some users have a 3D Connexion device – this is a mouse that is used by the left hand to zoom/pan/orbit while the right hand selects and edits.

Revit 2024 will detect if a 3D Connexion device is installed and add an interface to support the device.

It takes some practice to get used to using a 3D mouse, but it does boost your speed.

✓ Object Mode
Walk Mode
Fly Mode
2D Mode
✓ Keep Scene Upright
2D Zoom Direction
Center Tool
3Dconnexion Properties..

For more information on the 3D mouse, go to 3dconnexion.com.

Exercise 1-1:
Using the Steering Wheel & ViewCube

Drawing Name: *basic_project.rvt*
Estimated Time: 30 minutes

This exercise reinforces the following skills:

- ❑ ViewCube
- ❑ 2D Steering Wheel
- ❑ 3D Steering Wheel
- ❑ Project Browser
- ❑ Shortcut Menus
- ❑ Mouse

1. Select Home from the QAT or click **Ctl+D.**

2. The Home screen launches.

Sample Architecture Proj...

3. Select the Sample Architectural Project file. The file name is
rac_basic_sample_project.rvt.

   File name: basic_project   *If you do not see the Basic Sample Project listed, you can use the file basic_project included with the Class Files available for download from the publisher's website. To download the Class Files, type the following in the address bar of your web browser:* **SDCpublications.com/downloads/978-1-63057-600-4**

4. In the Project Browser:
   Expand the 3D Views.
   Double left click on the (3D) view.
   This activates the view.
   The active view is already in BOLD.

5.  If you have a mouse with a scroll wheel,
experiment with the following motions:
If you roll the wheel up and down, you can zoom in and out.
Hold down the SHIFT key and click down the scroll wheel at the
same time. This will rotate or orbit the model.
Release the SHIFT key. Click down the scroll wheel. This will pan
the model.

6.  When you are in a 3D view, a tool called the ViewCube is visible in
the upper right corner of the screen.

7.  Click on the top of the cube as shown.

8.  The display changes to a plan view.
Use the rotate arrows to rotate the view.

9. Click on the little house (home/default view tool) to return to
the default view.

*The little house disappears and reappears when your mouse
approaches the cube.*

10.  Select the Steering Wheel tool located on the View Control
toolbar.

11. A steering wheel pops up.

 Notice that as you mouse over sections of the steering wheel they highlight.
Mouse over the Zoom section and hold down the left mouse button. The display should zoom in and out.
Mouse over the Orbit section and hold down the left mouse button. The display should orbit.
Mouse over the Pan section and hold down the left mouse button. The display should pan.

12.

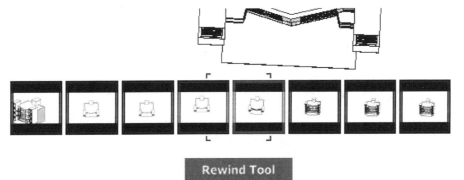

Select the Rewind tool and hold down the left mouse button.
A selection of previous views is displayed.
You no longer have to back through previous views. You can skip to the previous view you want.
Select a previous view to activate.
Close the steering wheel by selecting the X in the top right corner.

13.

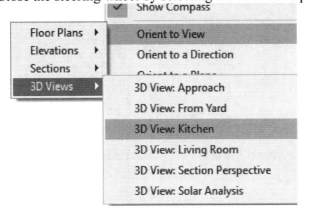 Place the mouse cursor over the View Cube.
Right click and a shortcut menu appears.
Select **Orient to View→ 3d Views→3D View: Kitchen**.

14. 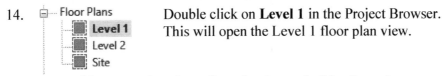 Double click on **Level 1** in the Project Browser.
This will open the Level 1 floor plan view.

Be careful not to select the ceiling plan instead of the floor plan.

15.

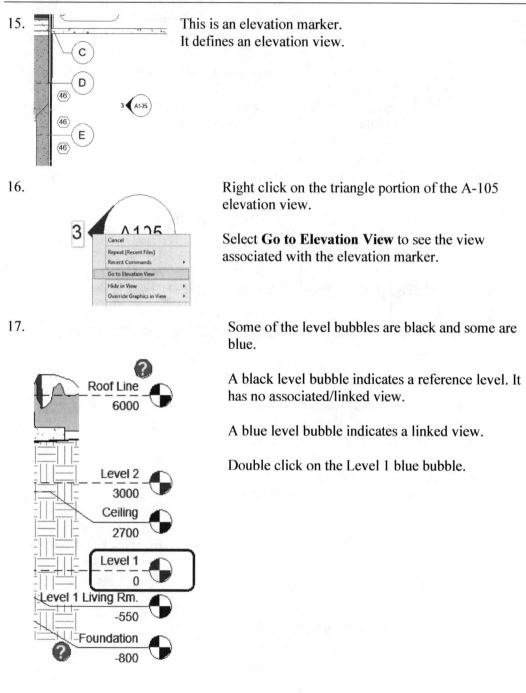

This is an elevation marker.
It defines an elevation view.

16.

Right click on the triangle portion of the A-105 elevation view.

Select **Go to Elevation View** to see the view associated with the elevation marker.

17.

Some of the level bubbles are black and some are blue.

A black level bubble indicates a reference level. It has no associated/linked view.

A blue level bubble indicates a linked view.

Double click on the Level 1 blue bubble.

18.

The Level 1 floor plan becomes active.

Notice that there are tabs at the top of the window for all of the open views.

19.

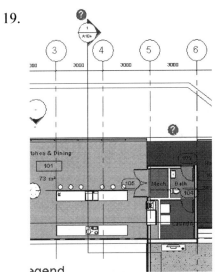

Locate the section line on Level 1 labeled A104.

20. The question mark symbols in the drawing link to help html pages. Click on one of the question marks.

21.

Help Button	
Select for Help Link	

Generic Annotations (1)   ⌄ 🔳 Edit Type

Identity Data
Order   12
Learning Level   01 - Basic
General
Summary   Create section views.
Location   Floor Plan Level 1
Learning Link   http://help.autodesk.com/view/RVT/2018/ENU/?contextId...
Other

Look in the Properties palette.
In the Learning Links field, a link to a webpage is shown. Click in that field to launch the webpage.

22.

A browser will open to the page pertaining to the question mark symbol.

Return to Revit.

23. Double left click on the arrow portion of the section line.

24. This view has a callout.

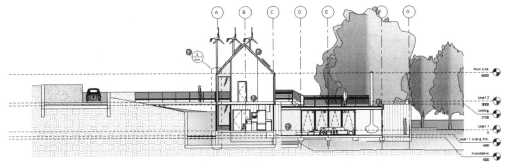

Double left click on the A104 callout bubble.

The bubble is located on the left.

25.

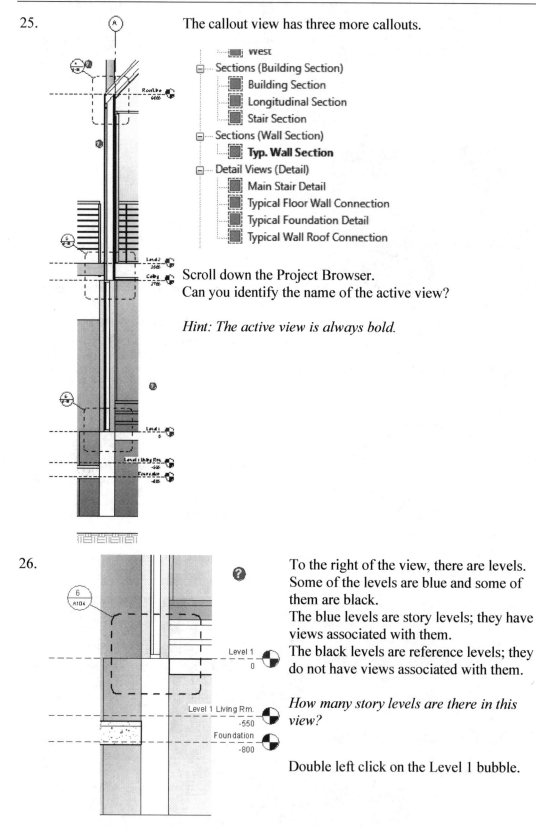

The callout view has three more callouts.

- ⬛ West
- ⊟ Sections (Building Section)
  - ⬛ Building Section
  - ⬛ Longitudinal Section
  - ⬛ Stair Section
- ⊟ Sections (Wall Section)
  - ⬛ **Typ. Wall Section**
- ⊟ Detail Views (Detail)
  - ⬛ Main Stair Detail
  - ⬛ Typical Floor Wall Connection
  - ⬛ Typical Foundation Detail
  - ⬛ Typical Wall Roof Connection

Scroll down the Project Browser.
Can you identify the name of the active view?

Hint: The active view is always bold.

26.

To the right of the view, there are levels.
Some of the levels are blue and some of them are black.
The blue levels are story levels; they have views associated with them.
The black levels are reference levels; they do not have views associated with them.

How many story levels are there in this view?

Double left click on the Level 1 bubble.

27. The Level 1 floor plan is opened. *We are back where we started!*

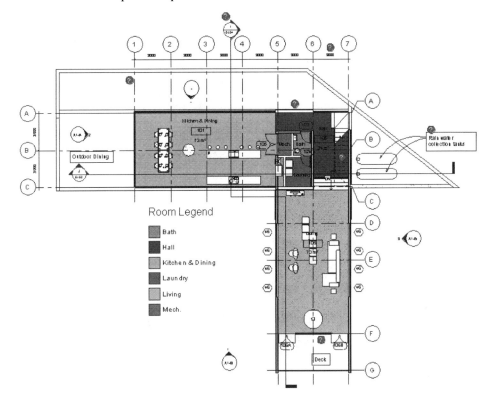

Room Legend

- Bath
- Hall
- Kitchen & Dining
- Laundry
- Living
- Mech.

Level 1 is also in BOLD in the project browser.

28.

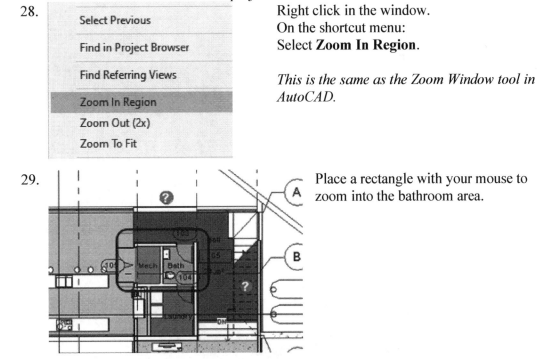

Select Previous

Find in Project Browser

Find Referring Views

Zoom In Region

Zoom Out (2x)

Zoom To Fit

Right click in the window.
On the shortcut menu:
Select **Zoom In Region**.

This is the same as the Zoom Window tool in AutoCAD.

29.

Place a rectangle with your mouse to zoom into the bathroom area.

30.

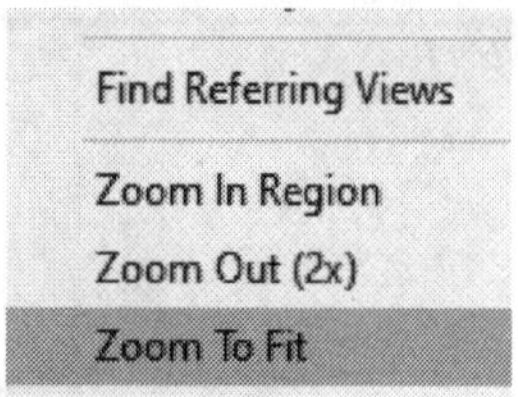

Find Referring Views

Zoom In Region

Zoom Out (2x)

Zoom To Fit

Right click in the window.
On the shortcut menu:
Select **Zoom To Fit**.
This is the same as the Zoom Extents tool in AutoCAD.
You can also double click on the scroll wheel to zoom to fit.

31.

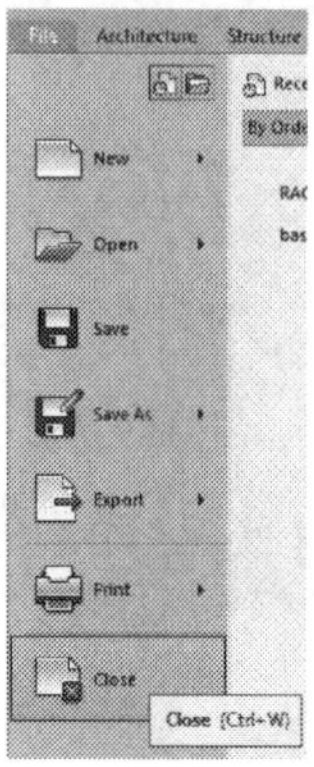

Close the file without saving.

Close by clicking **Ctl+W** or using
File→Close.

Exercise 1-2:
Changing the View Background

Drawing Name: *basic_project.rvt*
Estimated Time: 5 minutes

This exercise reinforces the following skills:
- Graphics mode
- Options

Many of my students come from an AutoCAD background and they want to change their background display to black.

1.   Go to **File→Open→Project**.

2.   Locate the file called *basic_project.rvt*.
This is included with the Class Files you downloaded from the publisher's website.

3. 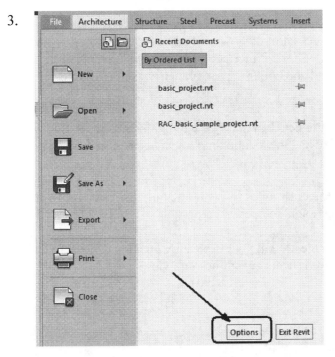   Select the drop-down on the Application Menu.
Select **Options**.

4.

Select **Colors**.
Locate the color tab next to Background.
Click on the color button.

Select the Black color.
Click **OK** twice.

5.
Activate the **View** ribbon.

6.

Locate **Canvas Theme** at the far right end of the ribbon.
Select **Canvas Theme.**

Notice you can toggle the background from white to black using Canvas Theme.

7.   Close the file without saving.

Revit's Project Browser

The Project Browser displays a hierarchy for all views, schedules, sheets, families and elements used in the current project. It has an interface similar to the Windows File Explorer, which allows you to expand and collapse each branch as well as organize the different categories.

You can change the location of the Project Browser by dragging the title bar to the desired location. Changes to the size and the location are saved and restored when Revit is launched.

You can search for entries in the Project Browser by right-clicking inside the browser and selecting Search from the right-click menu to open a dialog.

The Project Browser is used to navigate around your project by opening different views.
You can also drag and drop views onto sheets or families into views.

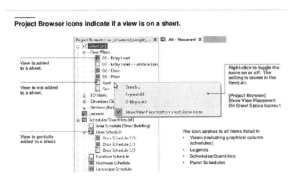

Project Browser icons indicate if a view is on a sheet.

Views can only be consumed (or placed on a sheet) once. The icons next to a view indicates whether it has been placed on a sheet.

If you require more than one version of a view (For example, you may have one view showing the fire ratings of the walls or electrical wiring and one view dimensioning the floor plan), you need to duplicate the view.

Revit's Properties Palette

The Properties Palette is a contextually based dialog. If nothing is selected, it will display the properties of the active view. If something is selected, the properties of the selected item(s) will be displayed.

By default, the Properties Palette is docked on the left side of the drawing area.

The Type Selector is a tool that identifies the selected family and provides a drop-down from which you can select a different type.

To make the Type Selector available when the Properties palette is closed, right-click within the Type Selector, and click Add to Quick Access Toolbar. To make the Type Selector available on the Modify tab, right-click within the Properties palette, and click Add to Ribbon Modify Tab. Each time you select an element, it will be reflected on the Modify tab.

Immediately below the Type Selector is the Properties filter; this may be used when more than one element is selected. When different types of elements are selected, such as walls and doors, only the instance properties common to the elements selected will display on the palette.

1. Type Selector
2. Properties filter
3. Edit Type button
4. Instance properties

The Edit Type button is used to modify or duplicate the element family type. Revit uses three different family classes to create a building model: system, loadable, and in-place masses. System families are walls, floors, roofs, and ceilings. Loadable families are doors, windows, columns, and furniture. In-place masses are basic shapes which can be used to define volumetric objects and are project-specific.

When you place an element into a building project, such as a wall, it has two types of properties: Type Properties and Instance Properties. The Type Properties are the structural materials defined for that wall and they do not change regardless of where the wall is placed. The Instance Properties are unique to that element, such as the length of the wall, the volume (which changes depending on how long the wall is), the location of the wall, and the wall height.

Exercise 1-3:
Closing and Opening the Project Browser and Properties Palette

Drawing Name: *basic_project.rvt*
Estimated Time: 5 minutes

This exercise reinforces the following skills:
- User Interface
- Ribbon
- Project Browser
- Properties panel

Many of my students will accidentally close the project browser and/or the properties palette and want to bring them back. Other students prefer to keep these items closed most of the time so they have a larger working area.

1. Go to **File→Open→Project**.

2. Locate the file called *basic_project.rvt*.
This is included with the Class Files you downloaded from the publisher's website.

3. Close the Properties palette by clicking on the x in the corner.

4. Close the Project Browser by clicking on the x in the corner.

5. Activate the View ribbon.
Go to the User Interface dropdown at the far right.
Place a check next to the Project Browser and Properties to make them visible.

6. Right click in the Browser.

Left click on **Show View Placement on Sheet Status Icons**.
Notice the Project Browser updates to hide the Sheet Status Icons.

7. Close without saving.

Revit's System Browser

The System Browser is useful in complex projects where you want to coordinate different building components, such as an HVAC system, structural framing, or plumbing.

The System Browser opens a separate window that displays a hierarchical list of all the components in each discipline in a project, either by the system type or by ones.

Customizing the View of the System Browser

The options in the View bar allow you to sort and customize the display of systems in the System Browser.

- **Systems**: displays components by major and minor systems created for each discipline.

- **Zones**: displays zones and spaces. Expand each zone to display the spaces assigned to the zone.

- **All Disciplines**: displays components in separate folders for each discipline (mechanical, piping, and electrical). Piping includes plumbing and fire protection.

- **Mechanical**: displays only components for the Mechanical discipline.

- **Piping**: displays only components for the Piping disciplines (Piping, Plumbing, and Fire Protection).

- **Electrical**: displays only components for the Electrical discipline.

- **AutoFit All Columns**: adjusts the width of all columns to fit the text in the headings.

  Note: You can also double-click a column heading to automatically adjust the width of a column.

- **Column Settings**: opens the Column Settings dialog where you specify the columnar information displayed for each discipline. Expand individual categories (General, Mechanical, Piping, Electrical) as desired, and select the properties that you want to appear as column headings. You can also select columns and click Hide or Show to select column headings that display in the table.

Exercise 1-4:
Using the System Browser

Drawing Name: *basic_project.rvt*
Estimated Time: 10 minutes

This exercise reinforces the following skills:
- ❑ User Interface
- ❑ Ribbon
- ❑ System Browser

This browser allows the user to locate and identify different elements used in lighting, mechanical, and plumbing inside the project.

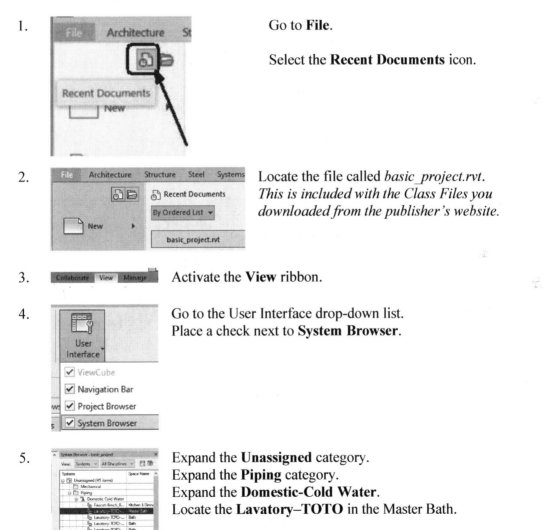

1.   Go to **File**.

    Select the **Recent Documents** icon.

2.   Locate the file called *basic_project.rvt.*
*This is included with the Class Files you
downloaded from the publisher's website.*

3.   Activate the **View** ribbon.

4.   Go to the User Interface drop-down list.
Place a check next to **System Browser**.

5.   Expand the **Unassigned** category.
Expand the **Piping** category.
Expand the **Domestic-Cold Water**.
Locate the **Lavatory–TOTO** in the Master Bath.

6. 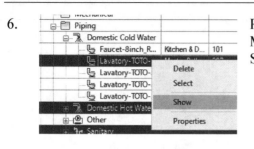 Right click on the **Lavatory –TOTO** in the Master Bath.
Select **Show**.

7. There is no open view that shows any of the highlighted elements. Searching through the closed views to find a good view could take a long time. Continue?

Click **OK**.

8. Select the **Show** button.

9. Click the **Show** button until you see this view.

Click **Close**.

10. Note that the active view is the **Level 2** floor plan.
This is bold in the Project Browser.

11. Zoom out to see the Master Bath.

To close the System Browser, select the x button in the upper right corner of the dialog.

12. Close the file without saving.

Revit's Ribbon

The ribbon displays when you create or open a project file. It provides all the tools necessary to create a project or family.

An arrow next to a panel title indicates that you can expand the panel to display related tools and controls.

By default, an expanded panel closes automatically when you click outside the panel. To keep a panel expanded while its ribbon tab is displayed, click the push pin icon in the bottom-left corner of the expanded panel.

Exercise 1-5:
Changing the Ribbon Display

Drawing Name: *basic_project.rvt*
Estimated Time: 15 minutes

This exercise reinforces the following skills:
- ❏ User Interface
- ❏ Ribbon

Many of my students will accidentally collapse the ribbon and want to bring it back. Other students prefer to keep the ribbon collapsed most of the time so they have a larger working area.

1.

Recent Files

Select **basic_project** from the Recent Files window.

MODELS

basic_project

2.

Project Browser - basic_project.rvt

Activate **Level 1** in the Project Browser.

If you don't see the icons next to the view names, you can restore them by right clicking in the Project Browser and enabling that option.

3.

Modify

On the ribbon: Locate the two small up and down arrows.

Room   Room   Tag
        Separator Room

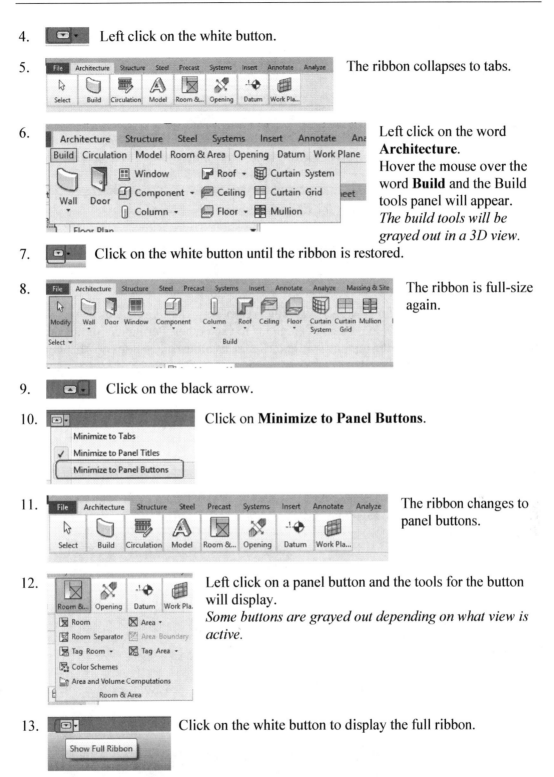

4. Left click on the white button.

5. The ribbon collapses to tabs.

6. Left click on the word **Architecture**.
Hover the mouse over the word **Build** and the Build tools panel will appear. *The build tools will be grayed out in a 3D view.*

7. Click on the white button until the ribbon is restored.

8. The ribbon is full-size again.

9. Click on the black arrow.

10. Click on **Minimize to Panel Buttons**.

11. The ribbon changes to panel buttons.

12. Left click on a panel button and the tools for the button will display. *Some buttons are grayed out depending on what view is active.*

13. Click on the white button to display the full ribbon.

14. Close without saving.

The Modify Ribbon

When you select an entity, you automatically switch to Modify mode. A Modify ribbon will appear with different options.

Select a wall in the view. Note how the ribbon changes.

Revit uses three types of dimensions: Listening, Temporary, and Permanent. Dimensions are placed using the relative positions of elements to each other. Permanent dimensions are driven by the value of the temporary dimension. Listening dimensions are displayed as you draw or place an element.

Exercise 1-6:
Temporary, Permanent and Listening Dimensions

Drawing Name: dimensions.rvt
Estimated Time: 30 minutes

This exercise reinforces the following skills:
- ❑ Project Browser
- ❑ Scale
- ❑ Graphical Scale
- ❑ Numerical Scale
- ❑ Temporary Dimensions
- ❑ Permanent Dimensions
- ❑ Listening Dimensions
- ❑ Type Selector

1. Browse to *dimensions.rvt* in the Class Files you downloaded from the publisher's website.
   Save the file to a folder.
   Open the file.

   Activate **Level 1** by double left clicking on Level 1 in the Project Browser.

The file has four walls.
The horizontal walls are 80′ in length.
The vertical walls are 52′ in length.

We want to change the vertical walls so that they are 60′ in length.

2. Select the House icon on the Quick Access toolbar.

3. 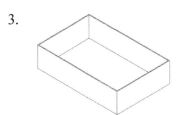 *This switches the display to a default 3D view.*

You see that the displayed elements are walls.

4. Floor Plans   Double left click on Level 1 under Floor Plans in the Project Browser.

The view display changes to the Level 1 floor plan.

5. 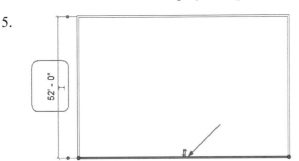 Select the bottom horizontal wall.

A temporary dimension appears showing the vertical distance between the selected wall and the wall above it.

6. A small dimension icon is next to the temporary dimension.

Left click on this icon to place a permanent dimension.

Left click anywhere in the window to release the selection.

7. Select the permanent dimension extension line, not the text.

A lock appears. If you left click on the lock, it would prevent you from modifying the dimension.

If you select the permanent dimension, you cannot edit the dimension value, only the dimension style.

8. 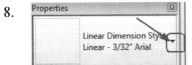 The Properties palette shows the dimension style for the dimension selected.

Select the small down arrow.
This is the Type Selector.

The Type Selector shows the dimension styles available in the project.
If you do not see the drop-down arrow on the right, expand the properties palette and it should become visible.

9.

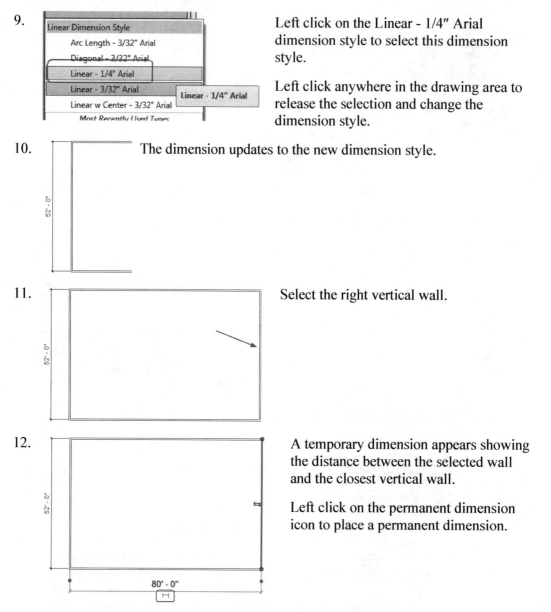

Left click on the Linear - 1/4″ Arial dimension style to select this dimension style.

Left click anywhere in the drawing area to release the selection and change the dimension style.

10. The dimension updates to the new dimension style.

11. Select the right vertical wall.

12. A temporary dimension appears showing the distance between the selected wall and the closest vertical wall.

Left click on the permanent dimension icon to place a permanent dimension.

13.   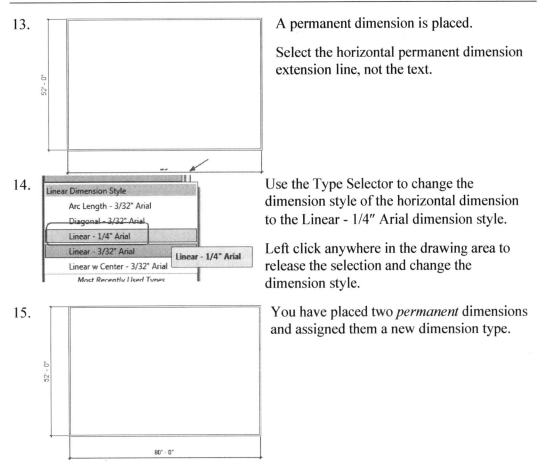   A permanent dimension is placed.

Select the horizontal permanent dimension extension line, not the text.

14.   Use the Type Selector to change the dimension style of the horizontal dimension to the Linear - 1/4″ Arial dimension style.

Left click anywhere in the drawing area to release the selection and change the dimension style.

15.   You have placed two *permanent* dimensions and assigned them a new dimension type.

16.   Hold down the Control key and select both **horizontal** (top and bottom side) walls so that they highlight.

NOTE: *We select the horizontal walls to change the vertical wall length, and we select the vertical walls to change the horizontal wall length.*

17.   Select the **Scale** tool located on the Modify Walls ribbon.

In the Options bar, you may select Graphical or Numerical methods for resizing the selected objects.

18.   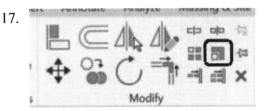   Enable **Graphical**.

The Graphical option requires three inputs.

> Input 1: Origin or Base Point
> Input 2: Original or Reference Length
> Input 3: Desired Length

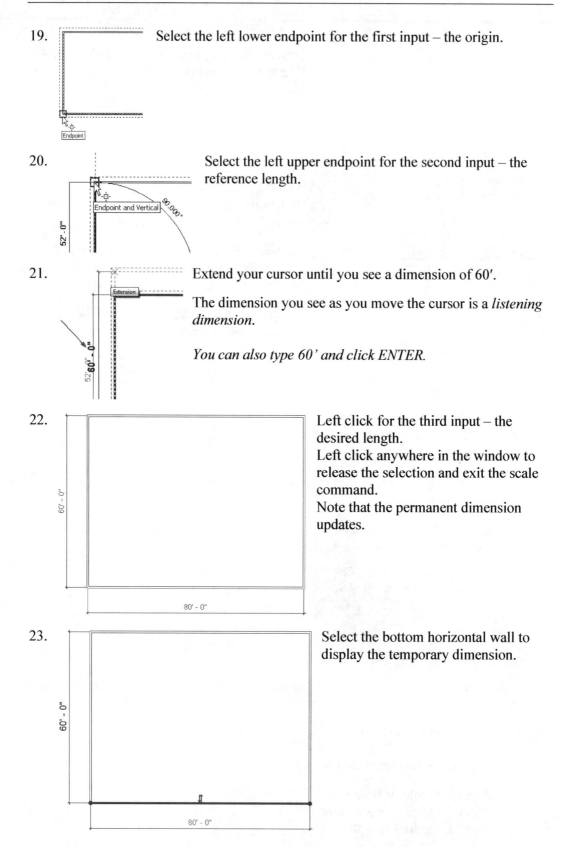

19.   Select the left lower endpoint for the first input – the origin.

20.   Select the left upper endpoint for the second input – the reference length.

21.   Extend your cursor until you see a dimension of 60′.

The dimension you see as you move the cursor is a *listening dimension.*

You can also type 60' and click ENTER.

22.   Left click for the third input – the desired length.
Left click anywhere in the window to release the selection and exit the scale command.
Note that the permanent dimension updates.

23.   Select the bottom horizontal wall to display the temporary dimension.

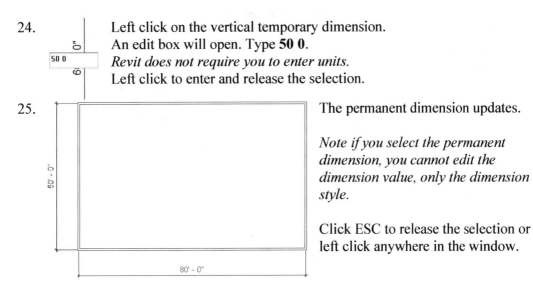

24.   Left click on the vertical temporary dimension.
        An edit box will open. Type **50 0**.
        Revit does not require you to enter units.
        Left click to enter and release the selection.

25.   The permanent dimension updates.

        *Note if you select the permanent
        dimension, you cannot edit the
        dimension value, only the dimension
        style.*

        Click ESC to release the selection or
        left click anywhere in the window.

26. Now, we will change the horizontal walls using the Numerical option of the Resize
    tool.
    Hold down the Control key and select the left and right vertical walls.

27.   Select the **Scale** tool.

28.   **Enable Numerical.**
        Set the Scale to **0.5**.
        This will change the wall length from **80'** to **40'**.
        The Numerical option requires only one input for the Origin or Base Point.

29.   Select the left lower endpoint for the first input – the origin.

30. The walls immediately adjust. Left click in the drawing
    area to release the selection.
    Note that the permanent dimension automatically
    updates.

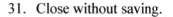

31. Close without saving.

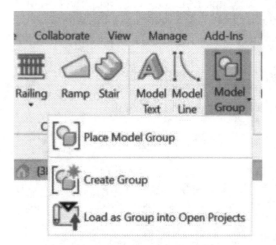

The Group tool works in a similar way as the AutoCAD GROUP. Basically, you are
creating a selection set of similar or dissimilar objects, which can then be selected as a
single unit. Once selected, you can move, copy, rotate, mirror, or delete them. Once you
create a Group, you can add or remove members of that group. Existing groups are listed
in your browser panel and can be dragged and dropped into views as required. Typical
details, office layouts, bathroom layouts, etc. can be grouped, and then saved out of the
project for use on other projects.

If you hold down the Control key as you drag a selected object or group,
Revit will automatically create a copy of the selected object or group.

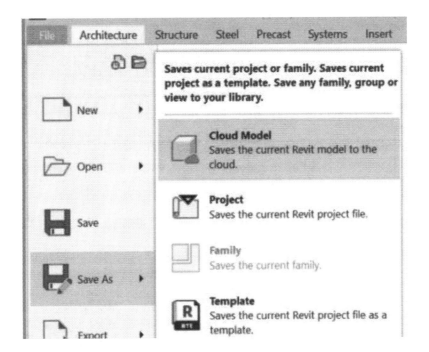

Revit has an option to save projects to the Cloud. In order to save to the cloud, the user needs to have a subscription for their software.

This option allows you to store your projects to a secure Autodesk server, so you can access your files regardless of your location. Users can share their project with each other and even view and comment on projects without using Revit. Instead you use a browser interface.

Autodesk hopes that if you like this interface you will opt for their more expensive option – BIM360, which is used by many AEC firms to collaborate on large projects.

The Collaborate Ribbon

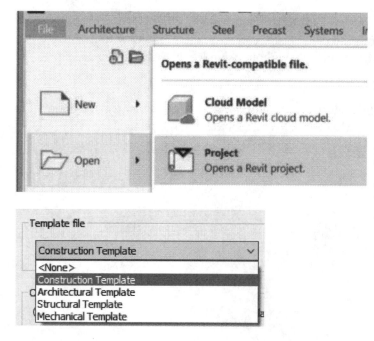

Collaborate tools are used if more than one user is working on the same project. The tools can also be used to check data sets coming in from other sources.

> ➤ If you plan to export to .dwg or import existing .dwg details, it is important to set your import/export setting. This allows you to control the layers that Revit will export to and the appearance of imported .dwg files.
> ➤ Clicking the ESC Key twice will always take you to the Modify command.
> ➤ As you get more proficient with Revit, you may want to create your own templates based on your favorite settings. A custom template based on your office standards (including line styles, rendering materials, common details, door, window and room schedules, commonly used wall types, door types and window types and typical sheets) will increase project production.

Revit's Menu

When you want to start a new project/building, you go to File→New→Project or use the hot key by clicking Control and 'N' at the same time.

When you start a new project, you use a default template (default.rte). This template creates two Levels (default floor heights) and sets the View Window to the Floor Plan view of Level 1.

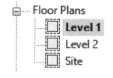

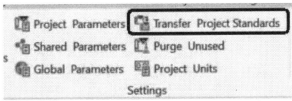

You can transfer the Project settings of an old project to a new one by opening both projects in one session of Revit, then with your new project active, select **Manage→Settings→Transfer Project Standards**.

Check the items you wish to transfer, and then click 'OK'.

Transfer Project Settings is useful if you have created a custom system family, such as a wall, floor, or ceiling, and want to use it in a different project.

The View Ribbon

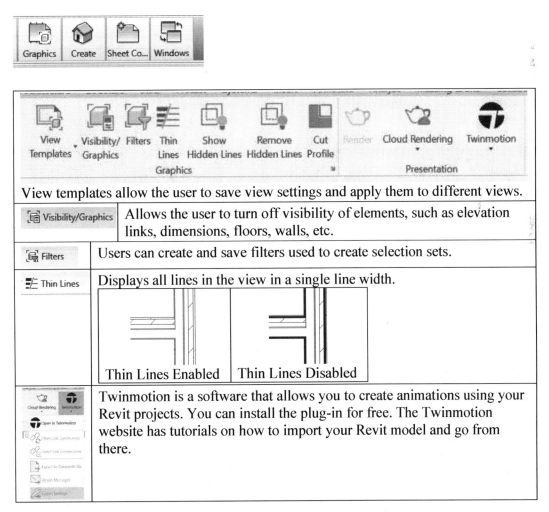

View templates allow the user to save view settings and apply them to different views.	
Visibility/Graphics	Allows the user to turn off visibility of elements, such as elevation links, dimensions, floors, walls, etc.
Filters	Users can create and save filters used to create selection sets.
Thin Lines	Displays all lines in the view in a single line width. Thin Lines Enabled Thin Lines Disabled
	Twinmotion is a software that allows you to create animations using your Revit projects. You can install the plug-in for free. The Twinmotion website has tutorials on how to import your Revit model and go from there.

	The 3D View tools allow the user to display an isometric view of the model, create a new view using a camera, or create a walkthrough animation.
	Creates a section view.
	Creates a detail view callout, useful for framing and foundation views.
	Creates a drafting view, useful in creating elevation details.
	Creates an elevation or framing elevation view.
	The Plan View tool creates a floor plan, reflected ceiling plan, plan region or area plan. A plan region can be used when you have a sunken or raised floor plan.
	Duplicate view can be used to create similar views for different phases or to display different elements, such as furniture or space layouts.

Legend / Keynote Legend	The Legend tool is used to place a legend on a sheet.
Schedules — Scope Box, Sheet, View / Schedule/Quantities / Graphical Column Schedule / Material Takeoff / Sheet List / Note Block / View List	The Schedules tool is used to create schedules.
Scope Box	The Scope box is used to control the display of levels or grids. It is also useful to create a section view of the 3D model.

Sheet View Title Block Revisions Guide Grid Matchline View Reference Viewports

Sheet Composition

	Places a new sheet in the project file.
View	Inserts or adds a view to a sheet.
Title Block	Adds a title block to a sheet.
	The Revisions tool adds revision clouds to a sheet.

	Places a matchline in a view. This is used when you have a view that spans more than one sheet and you want to be able to line them up.
	Guide Grid Used to line up views or notes on sheets.
View Reference	View Reference adds an annotation to a view with the sheet and detail view number.
Viewports ▾ Activate View Deactivate View	Viewports→Activate View/Deactivate View works similarly to Model/Paper Space on a layout tab. Allows the user to edit the elements within the view or change display settings.
Switch Windows Close Inactive Tab Views Tile Views User Interface Canvas Theme Windows	
Switch Windows	Switch Windows is used to switch from one open project or window to another. This is similar to the Window menu or using the tab key.
Close Hidden	Closes all open windows except the active window in the display.
	Opens a new window of the current window.
Tab Views	Creates a set of tabs for all open views and files, similar to the drawing tabs in AutoCAD.
	Tiles all the open windows. This is useful when working with masses or Revit families.

User Interface ☑ ViewCube ☑ Navigation Bar ☑ Project Browser ☐ System Browser ☑ Properties ☐ MEP Fabrication Parts ☐ P&ID Modeler ☑ Status Bar ☑ Status Bar - Worksets ☑ Status Bar - Design Options ☐ Recent Files Browser Organization Keyboard Shortcuts	Controls the visibility of different user navigation systems. Browser Organization allows the user to customize how views appear in the Project Browser. We do not use Worksets or Design Options in this text, so both those status bars can be disabled.
Canvas Theme	Allows you to toggle between a white or black background in the display window.

Creating standard view templates (exterior elevations, enlarged floor plans, etc.) in your project template or at the beginning of a project will save lots of time down the road.

The Manage Ribbon

⊛ Materials

Revit comes with a library of materials, which can be applied to walls, windows, doors, etc. You can also edit and create your own custom materials.

Any image file can be used to create a material.
My favorite sources for materials are fabric and paint websites.

Object Styles is used to control the line color, line weight, and line style applied to different Revit elements. It works similarly to layers in AutoCAD, except it ensures that all the similar elements, like doors, look the same.

Snaps in Revit are similar to object snaps in AutoCAD. You can set how your length and dimensions snap as well as which object types your cursor will snap to when making a selection.

Like AutoCAD, it is best not to make your snaps all-encompassing so that you snap to everything as that defeats the purpose and usefulness of snaps.

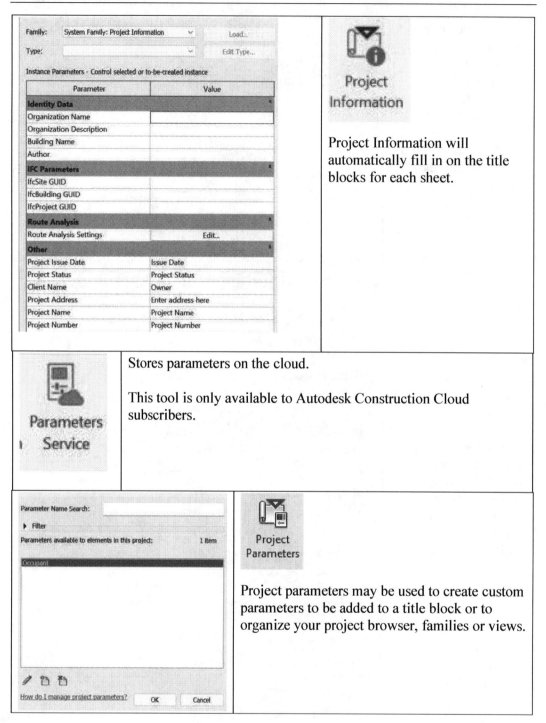

Project Information

Project Information will automatically fill in on the title blocks for each sheet.

Parameters Service

Stores parameters on the cloud.

This tool is only available to Autodesk Construction Cloud subscribers.

Project Parameters

Project parameters may be used to create custom parameters to be added to a title block or to organize your project browser, families or views.

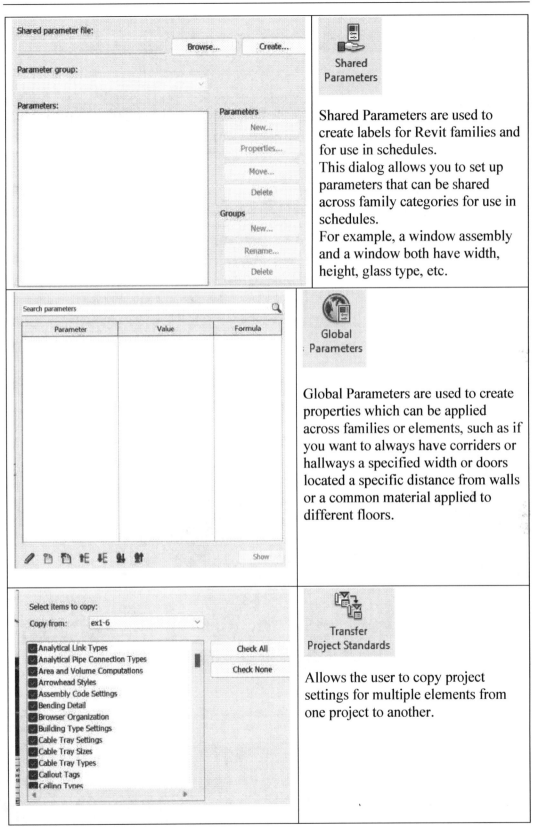

Shared Parameters

Shared Parameters are used to create labels for Revit families and for use in schedules.
This dialog allows you to set up parameters that can be shared across family categories for use in schedules.
For example, a window assembly and a window both have width, height, glass type, etc.

Global Parameters

Global Parameters are used to create properties which can be applied across families or elements, such as if you want to always have corridors or hallways a specified width or doors located a specific distance from walls or a common material applied to different floors.

Transfer Project Standards

Allows the user to copy project settings for multiple elements from one project to another.

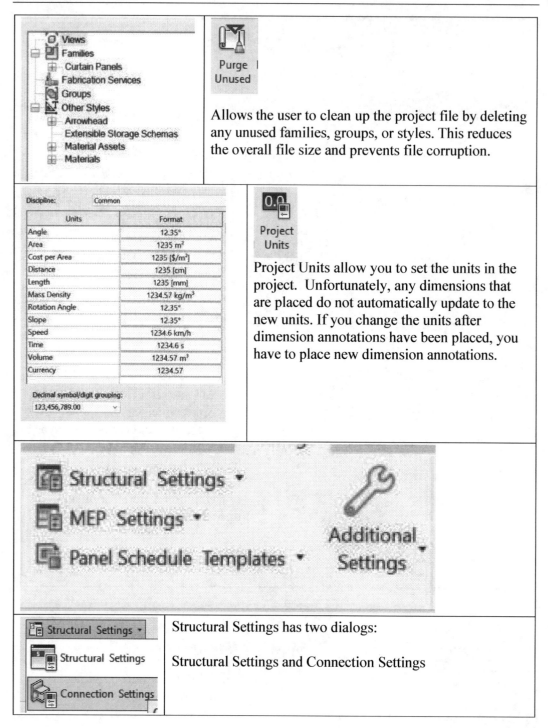

Allows the user to clean up the project file by deleting any unused families, groups, or styles. This reduces the overall file size and prevents file corruption.

Project Units allow you to set the units in the project. Unfortunately, any dimensions that are placed do not automatically update to the new units. If you change the units after dimension annotations have been placed, you have to place new dimension annotations.

Structural Settings has two dialogs:

Structural Settings and Connection Settings

Structural Settings ▾ The Structural Settings dialog allows you to control the symbols used for beams, joists, and columns as well as how to define loads, boundary conditions, and analysis to determine stress and load on structural assemblies.	Symbolic Representation Settings Load Cases Load Combinations Boundary Conditions Settings Loads Scaling Display Symbolic Cutback Distance Brace: 2.5000 mm Column: 1.5000 mm Beam/Truss: 2.5000 mm Brace Symbols Plan representation: Parallel Line Parallel line offset: 2.5000 mm ☐ Show brace above Symbol: ☐ Show brace below Symbol: Kicker brace symbol: Connection Symbols Display Symbols for: Beams and Braces Connection Type: Annotation Symbol: Moment Frame None Cantilever Moment None Load...
Connection Settings Allows the user to load different types of connections used in structural assemblies and apply them to a structural model.	Connections Parameters Connection Group: All Connections Available Connections: Apex haunch Base plate Base plate cut Base plate with traverse, bolted Base plate with traverse, welded Beam seat T Binding plates Bolting on gauge lines Bolting on gauge lines, 2 profiles Bracing I splice angle - additional object Bracing I splice angle double Bracing I splice angle single Bracing I splice plates - additional object All None Loaded Connections: Add --> <-- Remove

MEP Settings
Allow the user to control different parameters for MEP documentation, including the appearance of the specified elements.

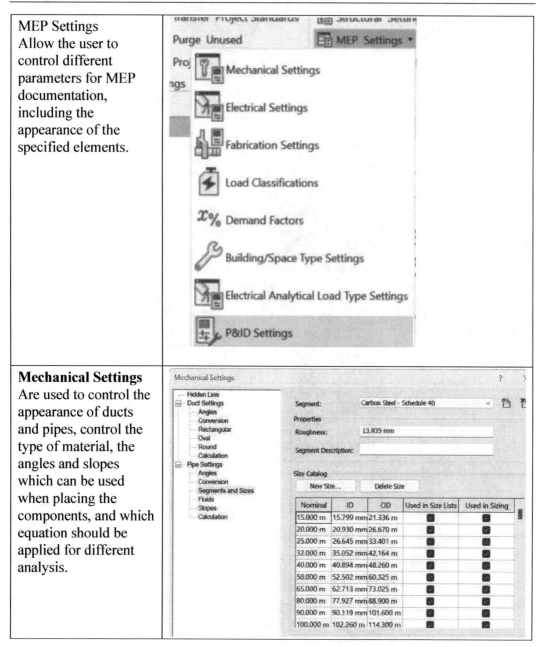

Mechanical Settings
Are used to control the appearance of ducts and pipes, control the type of material, the angles and slopes which can be used when placing the components, and which equation should be applied for different analysis.

Electrical Settings
Allow the user to assign allowable wire sizes, preferred angles to use when placing wiring, and how to calculate loads as well as format panel schedules.

Fabrication Settings
Determines which libraries are loaded and available for ductwork, electrical, and piping.

Load Classifications
Allows the user to select or create load classification types to be used in load calculations.

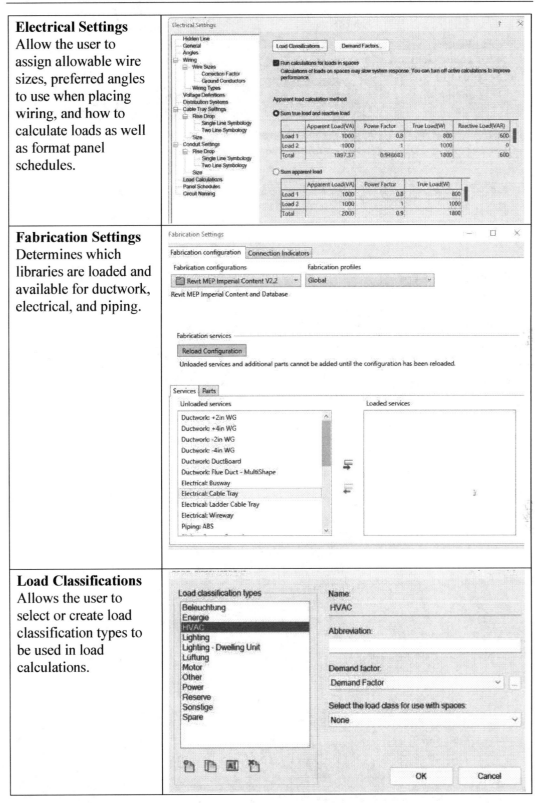

Demand Factors Allows the user to select or create demand factor types to be used in load calculations.	Demand Factors **Demand factor types** Default Demand Factor HVAC Lighting Lighting - Dwelling Unit Motor Other Power Reserve Spare Standard Vorgabe **Name:** HVAC **Calculation method:** Constant Constant By quantity By load **Demand factor:** 100.00% ☐ Add an additional load to the calculated result 0 VA
Building/Space Type Settings Are used for energy analysis. Users can assign different values to different building or space types or create their own custom types.	Building/Space Type Settings ? ✕ Filter: Enter Search Words 🔍 ◉ Building Type ○ Space Type Automotive Facility Convention Center Courthouse Dining Bar Lounge or Leisure Dining Cafeteria Fast Food Dining Family Dormitory Exercise Center Fire Station Gymnasium Hospital or Healthcare Hotel Library Manufacturing Motel Motion Picture Theatre Multi Family Museum Office Parking Garage Penitentiary Performing Arts Theater Police Station Post Office Religious Building Retail | Parameter	Value | |---|---| | **Energy Analysis** | | | Area per Person | 6.667 m² | | Sensible Heat Gain per person | 73.27 W | | Latent Heat Gain per person | 58.61 W | | Lighting Load Density | 9.69 W/m² | | Power Load Density | 16.15 W/m² | | Plenum Lighting Contribution | 20.0000% | | Occupancy Schedule | Warehouse Occupancy - 7 AM | | Lighting Schedule | Retail Lighting - 7 AM to 8 PM | | Power Schedule | Retail Lighting - 7 AM to 8 PM | | Outdoor Air per Person | 2.36 L/s | | Outdoor Air per Area | 0.30 L/(s-m²) | | Air Changes per Hour | 0.000000 | | Outdoor Air Method | by People and by Area | | Opening Time | 8:00 AM | | Closing Time | 6:00 PM | | Unoccupied Cooling Set Point | 27.78 °C | OK Cancel

Electrical Analytical Load Type Settings

• To create an area-based load type, click 🗎 (New). Enter a name for the area-based load type, and click OK.

• To duplicate an area-based load type, click 🗎 (Duplicate). Enter a name for the area-based load type, and click OK.

• To rename an area-based load type, click 🔠 (Rename). Enter a new name for the area-based load type, and click OK.

• To delete an area-based load type, click 🗎 (Delete). Click Yes to confirm. |
| To facilitate modeling an AutoCAD P&ID drawing in Revit, you'll need to map P&ID components and schematic lines to real-world 3D model components, such as a P&ID valve symbol with a Revit family, such as a ball valve. | |

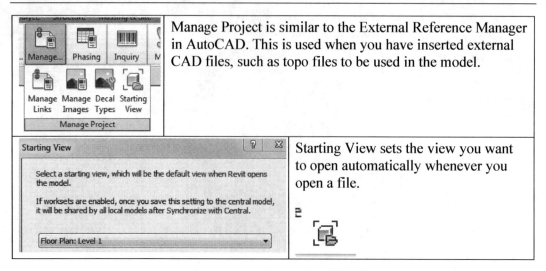

Manage Project is similar to the External Reference Manager in AutoCAD. This is used when you have inserted external CAD files, such as topo files to be used in the model.

Starting View

Select a starting view, which will be the default view when Revit opens the model.

If worksets are enabled, once you save this setting to the central model, it will be shared by all local models after Synchronize with Central.

Floor Plan: Level 1

Starting View sets the view you want to open automatically whenever you open a file.

The Additional Settings menu allows the user to customize the interface.

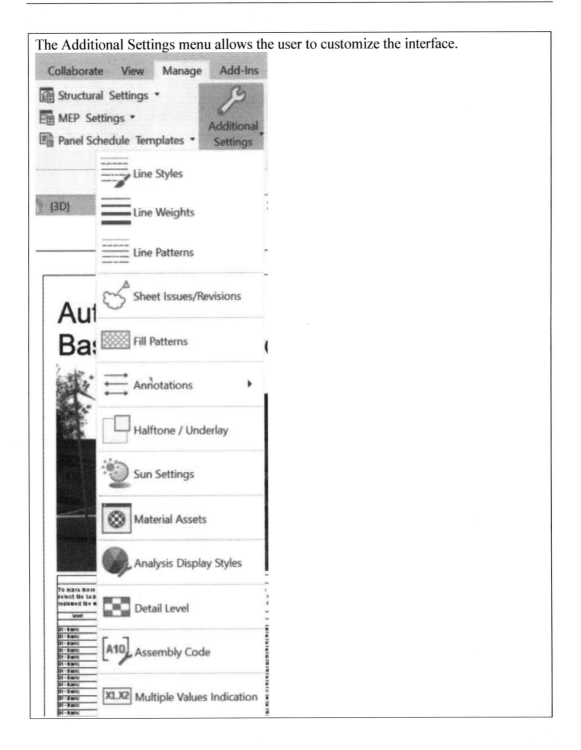

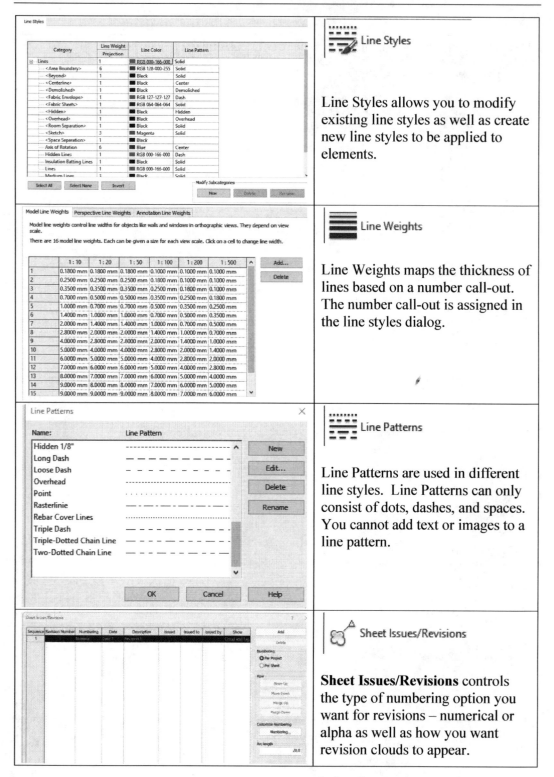

Line Styles

Line Styles allows you to modify existing line styles as well as create new line styles to be applied to elements.

Line Weights

Line Weights maps the thickness of lines based on a number call-out. The number call-out is assigned in the line styles dialog.

Line Patterns

Line Patterns are used in different line styles. Line Patterns can only consist of dots, dashes, and spaces. You cannot add text or images to a line pattern.

Sheet Issues/Revisions

Sheet Issues/Revisions controls the type of numbering option you want for revisions – numerical or alpha as well as how you want revision clouds to appear.

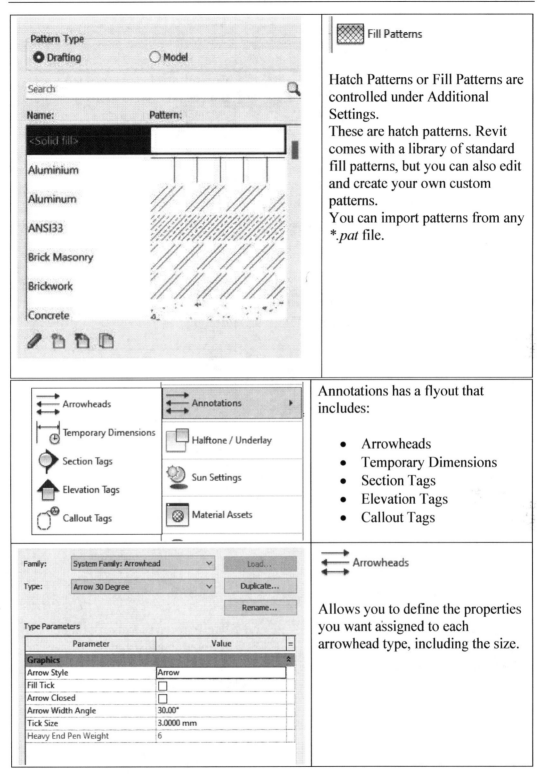

Fill Patterns

Hatch Patterns or Fill Patterns are controlled under Additional Settings.

These are hatch patterns. Revit comes with a library of standard fill patterns, but you can also edit and create your own custom patterns.

You can import patterns from any *.pat* file.

Annotations has a flyout that includes:

- Arrowheads
- Temporary Dimensions
- Section Tags
- Elevation Tags
- Callout Tags

Arrowheads

Allows you to define the properties you want assigned to each arrowhead type, including the size.

Temporary dimensions measure from: **Walls** ○ Centerlines ● Faces ○ Center of Core ○ Faces of Core **Doors and Windows** ○ Centerlines ● Openings	**Temporary Dimensions** Controls the behavior of temporary dimensions in terms of where the extension lines will land.
Family: System Family: Section Tag Load... Type: Section Head - Filled, Section Tail - Fillec Duplicate... Rename... Type Parameters <table><tr><td>Parameter</td><td>Value</td><td>=</td></tr><tr><td>Graphics</td><td></td><td>≫</td></tr><tr><td>Section Head</td><td>M_Section Head - Filled</td><td></td></tr><tr><td>Section Tail</td><td>M_Section Tail - Filled</td><td></td></tr><tr><td>Broken Section Display Style</td><td>Gapped</td><td></td></tr></table>	**Section Tags** Is used to set the default Section Tag family when creating a section. This is important if you have a custom section tag family you want to use.
Type Properties ✕ Family: System Family: Elevation Tag Load... Type: 12mm Circle Duplicate... Rename... Type Parameters <table><tr><td>Parameter</td><td>Value</td><td>=</td></tr><tr><td>Graphics</td><td></td><td>≫</td></tr><tr><td>Elevation Mark</td><td>M_Elevation Mark Body_Circle-1</td><td></td></tr></table>	**Elevation Tags** Is used to set the default Elevation Tag family when placing an elevation. This is important if you have a custom elevation family you want to use.
Type Properties ✕ Family: System Family: Callout Tag Load... Type: Callout Head w 3mm Corner Radius Duplicate... Rename... Type Parameters <table><tr><td>Parameter</td><td>Value</td><td>=</td></tr><tr><td>Graphics</td><td></td><td>≫</td></tr><tr><td>Callout Head</td><td>M_Callout Head : Metric Callout</td><td></td></tr><tr><td>Corner Radius</td><td>3.0000 mm</td><td></td></tr></table>	**Callout Tags** Is used to set the default Callout family when placing a callout. This is important if you have a custom callout family you want to use for your callouts.

	Halftone / Underlay **Halftone/Underlay** controls the appearance of levels which are below the active level (underlays) as well as linked models (halftone).
	Sun Settings **Sun Settings** is used when doing solar/energy analysis as well as to control the lighting for renderings.
	Material Assets Material Assets allows the user to define new materials using the Asset Browser. These materials can be used in the active document or saved to a custom library. Materials in the Asset Browser can not be modified. They are read-only files. This is why you see a lock next to the file name. Instead, you have to select the material, duplicate it, rename it, and redefine to create your own version of the material.

Analysis Display Styles

Analysis Display Styles

Styles
Enter Search Words

Settings Color Legend

Color mapping:
- Gradient
- Ranges

Preview

Value	Color

Use analysis display styles to visualize the results of an analysis generated by an add-in application.

Analysis display styles are permanently stored in Revit for use in project views. Within a view, you can select a different style and immediately apply it in the analysis result.

Detail Level

View scale-to-detail level correspondence

Use this table to control the detail level used for new views by scale.

Coarse		Medium		Fine
1 : 5000	>>	1 : 25	>>	1 : 5
1 : 2000		1 : 20		1 : 2
1 : 1000	<<	1 : 10	<<	1 : 1
1 : 500				
1 : 200				
1 : 100				
1 : 50				

Controls the detail level based on the view scale.

Assembly Code

File Location
C:\ProgramData\Autodesk\RVT 2024\Libraries\English-Imperial\UniformatClassifications

File Path (for local files)
- Absolute
- Relative
- At library locations

Browse...
View...
Reload

Determines which file to use when assigning assembly codes to Revit families.

When a set of elements contains a parameter with multiple values:

○ Display as <varies>
○ Display custom text:

X1,X2 Multiple Values Indication

You can specify the values displayed in properties, schedules, and tags when multiple elements are selected that contain different values for parameters.

When multiple elements are selected, scheduled, or tagged the parameters they have in common display. If the parameter values are the same for all elements selected, the actual value displays. If the parameter values are different across the multiple elements, the value set in the Multiple Values Indication setting displays.

For example, if multiple walls are selected and their unconnected height is set to 8000mm, 8000mm displays as the Unconnected Height in the Properties palette. If the multiple walls are all different heights, the value set in the Multiple Values Indication setting displays.

Location Coordinates ▾ Position ▾ Project Location	**Project Location Panel**

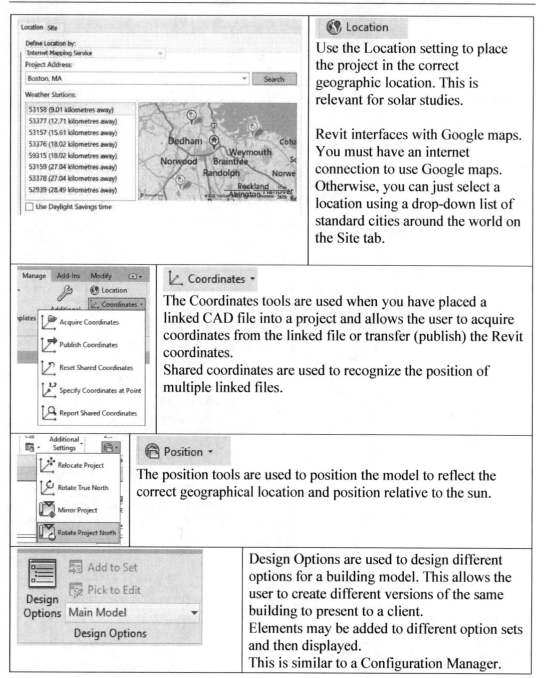

Location

Use the Location setting to place the project in the correct geographic location. This is relevant for solar studies.

Revit interfaces with Google maps. You must have an internet connection to use Google maps. Otherwise, you can just select a location using a drop-down list of standard cities around the world on the Site tab.

Coordinates ▾

The Coordinates tools are used when you have placed a linked CAD file into a project and allows the user to acquire coordinates from the linked file or transfer (publish) the Revit coordinates.
Shared coordinates are used to recognize the position of multiple linked files.

Position ▾

The position tools are used to position the model to reflect the correct geographical location and position relative to the sun.

Design Options are used to design different options for a building model. This allows the user to create different versions of the same building to present to a client.
Elements may be added to different option sets and then displayed.
This is similar to a Configuration Manager.

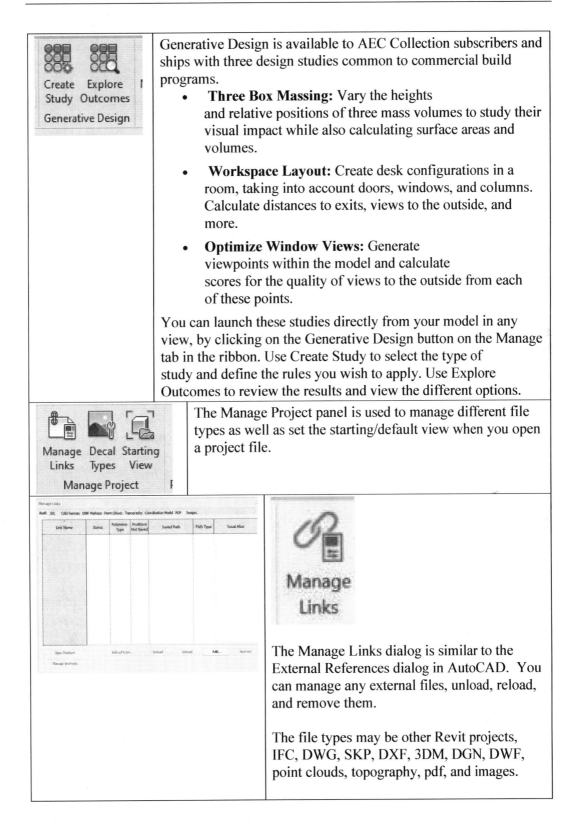

Create Study / Explore Outcomes — **Generative Design**	Generative Design is available to AEC Collection subscribers and ships with three design studies common to commercial build programs. • **Three Box Massing:** Vary the heights and relative positions of three mass volumes to study their visual impact while also calculating surface areas and volumes. • **Workspace Layout:** Create desk configurations in a room, taking into account doors, windows, and columns. Calculate distances to exits, views to the outside, and more. • **Optimize Window Views:** Generate viewpoints within the model and calculate scores for the quality of views to the outside from each of these points. You can launch these studies directly from your model in any view, by clicking on the Generative Design button on the Manage tab in the ribbon. Use Create Study to select the type of study and define the rules you wish to apply. Use Explore Outcomes to review the results and view the different options.
Manage Links / Decal Types / Starting View — **Manage Project**	The Manage Project panel is used to manage different file types as well as set the starting/default view when you open a project file.

The Manage Links dialog is similar to the External References dialog in AutoCAD. You can manage any external files, unload, reload, and remove them.

The file types may be other Revit projects, IFC, DWG, SKP, DXF, 3DM, DGN, DWF, point clouds, topography, pdf, and images.

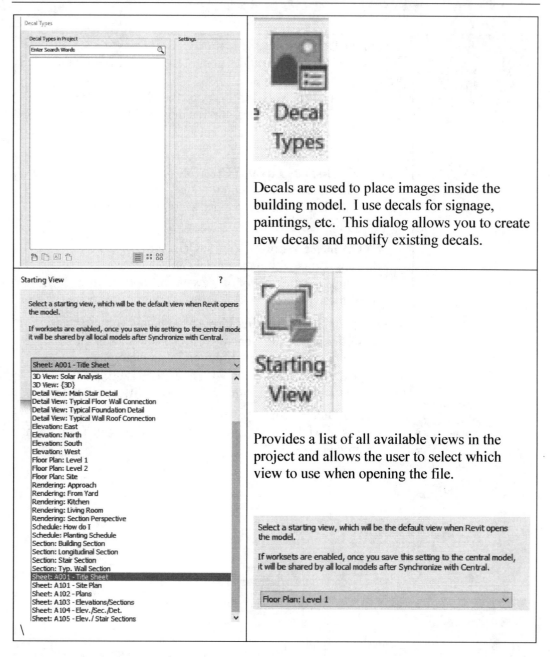

Decals are used to place images inside the building model. I use decals for signage, paintings, etc. This dialog allows you to create new decals and modify existing decals.

Provides a list of all available views in the project and allows the user to select which view to use when opening the file.

Select a starting view, which will be the default view when Revit opens the model.

If worksets are enabled, once you save this setting to the central model, it will be shared by all local models after Synchronize with Central.

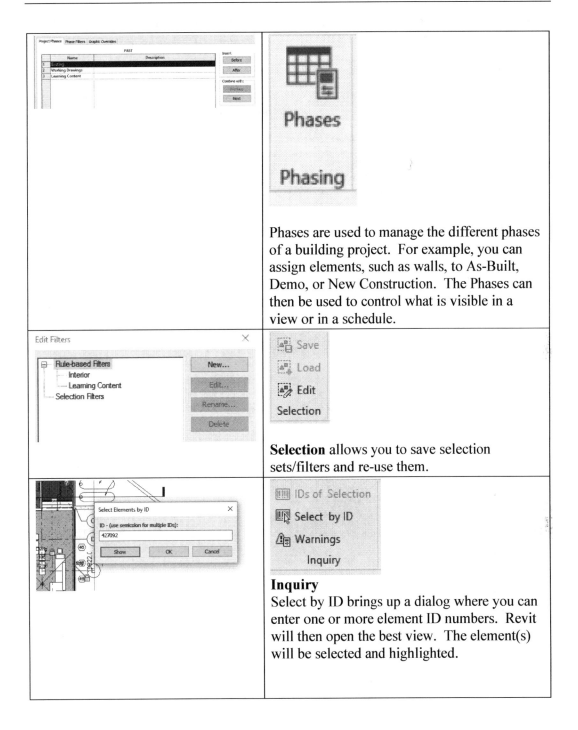

Phases are used to manage the different phases of a building project. For example, you can assign elements, such as walls, to As-Built, Demo, or New Construction. The Phases can then be used to control what is visible in a view or in a schedule.

Selection allows you to save selection sets/filters and re-use them.

Inquiry

Select by ID brings up a dialog where you can enter one or more element ID numbers. Revit will then open the best view. The element(s) will be selected and highlighted.

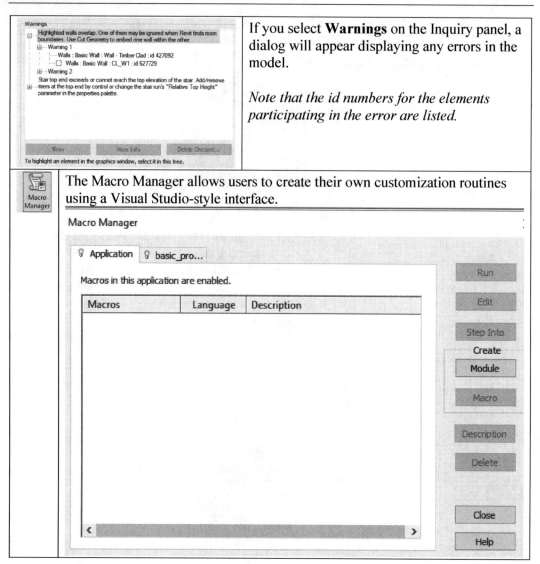

If you select **Warnings** on the Inquiry panel, a dialog will appear displaying any errors in the model.

Note that the id numbers for the elements participating in the error are listed.

The Macro Manager allows users to create their own customization routines using a Visual Studio-style interface.

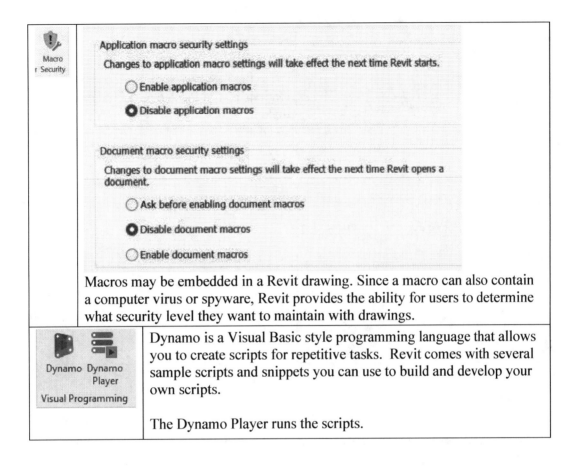

Macros may be embedded in a Revit drawing. Since a macro can also contain a computer virus or spyware, Revit provides the ability for users to determine what security level they want to maintain with drawings.

Dynamo is a Visual Basic style programming language that allows you to create scripts for repetitive tasks. Revit comes with several sample scripts and snippets you can use to build and develop your own scripts.

The Dynamo Player runs the scripts.

➤ By default, the Browser lists all sheets and all views.
➤ If you have multiple users of Revit in your office, place your family libraries and rendering libraries on a server and designate the path under the File Locations tab under Options; this allows multiple users to access the same libraries and materials.

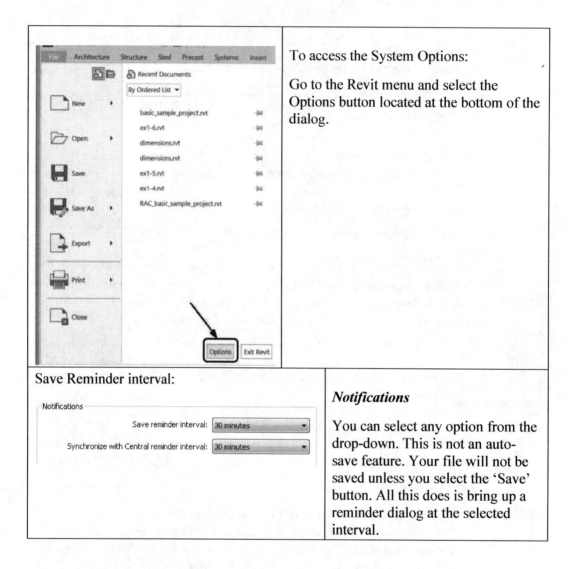

To access the System Options:

Go to the Revit menu and select the Options button located at the bottom of the dialog.

Save Reminder interval:

Notifications

You can select any option from the drop-down. This is not an auto-save feature. Your file will not be saved unless you select the 'Save' button. All this does is bring up a reminder dialog at the selected interval.

User Name	This indicates the username that checks in and checks out documents. You can only change the username when there are no open projects. The username can also be used as a property for sheets.
Username Elise	**Username** smoss@peralta.edu You are currently signed in. Your Autodesk ID is used as the username. If you need to change your username, you will need to sign out. Sign Out If you use Autodesk Docs and are signed in, then the Autodesk ID is used as the username. Signing into Autodesk allows you to save a backup of your project on the cloud as well as render in the cloud – using Autodesk's servers.

Journal File Cleanup	Determines how many past versions of a project's transcriptions can be stored. You can set the number of project versions to be saved and delete any number exceeding that value after a set number of days. Transcripts are used to recover a project file if it gets damaged or corrupted.
Journal File Cleanup When number of journals exceeds: 3 then Delete journals older than (days): 10 There is currently no option to place the journal file anywhere but the default location.	The transcripts give you the ability to clean up previously created journals, which are located in "C:\Program Files\<Revit product name and version>\Journals." They can be opened with a WordPad, NotePad, or any other text-based program.
Worksharing update frequency Less Frequent ⸻ More Frequent Every 5 seconds	If you are in a worksharing environment, where you and other team members are working on the same project, you can set how often the common project is updated.
View Options Default view discipline: Architectural ▾ Architectural Structural Mechanical Electrical Plumbing OK Coordination	The View options set the display style for views depending on discipline.

The User Interface Options

Configure

Tools and analyses:

- ☑ Architecture tab and tools
- ☑ Structure tab and tools
  - ☑ Steel tab and tools
  - ☑ Structural analysis and tools
- ☑ Systems tab: mechanical tools
  - ☑ Mechanical analysis tools
- ☑ Systems tab: electrical tools
  - ☑ Electrical analysis tools
- ☑ Systems tab: piping tools
  - ☑ Piping analysis tools

Enabling the Tools and analyses controls the visibility of the ribbons. You can hide the ribbons for those tools which you don't need or use by unchecking those tabs on the ribbons.

Search: Type a command name

Filter: All

Assignments:

Command	Shortcuts	Paths
Modify	MD	Create>Select; Insert>Select; Ann...
Select links		Create>Select; Insert>Select; Ann...
Select underlay elements		Create>Select; Insert>Select; Ann...
Select pinned elements		Create>Select; Insert>Select; Ann...
Select elements by face		Create>Select; Insert>Select; Ann...
Drag elements on select...		Create>Select; Insert>Select; Ann...
Type Properties		Create>Properties; Modify>Prope...
Properties	PP	Create>Properties; View>Windo...
	Ctrl+1	
	VP	
Family Category and Par...		Create>Properties; Modify>Prope...

Press new keys: ⊹ Assign — Remove

Import... Export... OK Cancel

Keyboard Shortcuts (also located under User Interface on the View ribbon) allow you to assign keyboard shortcuts to your favorite commands.

Customize Double-click Settings ☒	Double-click Options allow you to determine what happens when you double-left mouse click on different types of elements.

Select double-click behaviors:

Element Type:	Double-click action:
Family	Edit Family
Sketched Element	Edit Element
Inside Views / Schedules on Sheets	Activate View
Outside Views on Sheets	Deactivate View
Assemblies	Edit Element
Groups	Edit Element
Component Stairs	Edit Element

Tooltip Assistance

Tooltip assistance: Normal

None
Minimal
Normal
High

This setting controls the number of help messages you will see as you work.

Tooltip assistance:

☑ Enable Recent Files page at startup

You can enable/disable whether the Recent Files page is visible when you launch Revit.

Ribbon tab switching

After clearing a selection or after exiting...

Project Environment: Return to the previous tab

Family Editor: Stay on the Modify tab

☒ Display the contextual tab on selection

Users can also control what happens when they tab from one window to another.

View switching

Ctrl+(Shift)+Tab: Tab position order

Tab position order
History order

View switching allows you to quickly use the Ctl+Shift+Tab keys to navigate through views.

Colors tab	Allows you to customize the colors used in the interface.
UI active theme: Use system setting / Dark / Light / Use system setting Canvas colors Canvas color scheme: Light Background: ☐ White Selection: ■ RGB 000-059-189 ■ Semi-transparent Pre-selection: ■ RGB 000-059-189 Alert: ■ RGB 255-128-000 Calculating: ■ Cyan Rebar Editing: ■ RGB 128-255-064	
Background	Allows you to set the background color of the display window.
Selection Color	Set the color to be used when an object(s) is selected. You can also enable to be semi-transparent.
Pre-selection Color	Set the color to be used when the cursor hovers over an object.
Alert Color	Sets the color for elements that are selected when an error occurs.
Calculating	Sets the color for formulas.
Rebar Editing	Sets the color when modifying rebar elements.

The Graphics tab

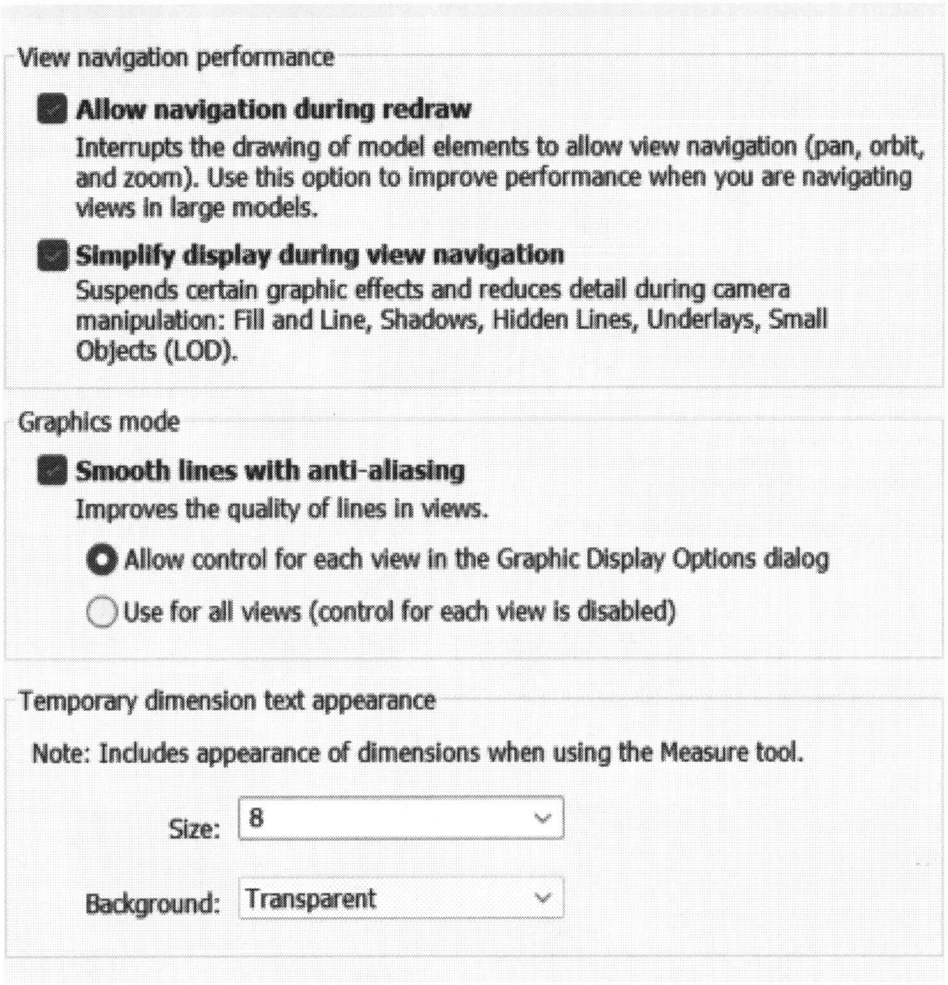

View navigation performance

☑ **Allow navigation during redraw**

Interrupts the drawing of model elements to allow view navigation (pan, orbit, and zoom). Use this option to improve performance when you are navigating views in large models.

☑ **Simplify display during view navigation**

Suspends certain graphic effects and reduces detail during camera manipulation: Fill and Line, Shadows, Hidden Lines, Underlays, Small Objects (LOD).

Graphics mode

☑ **Smooth lines with anti-aliasing**

Improves the quality of lines in views.

◉ Allow control for each view in the Graphic Display Options dialog

◯ Use for all views (control for each view is disabled)

Temporary dimension text appearance

Note: Includes appearance of dimensions when using the Measure tool.

Size: | 8 ⌄ |

Background: | Transparent ⌄ |

Most of this dialog is fairly straightforward.
Use this section to change the appearance of temporary dimensions. You can increase the font size or change the font background.

Hardware Setup

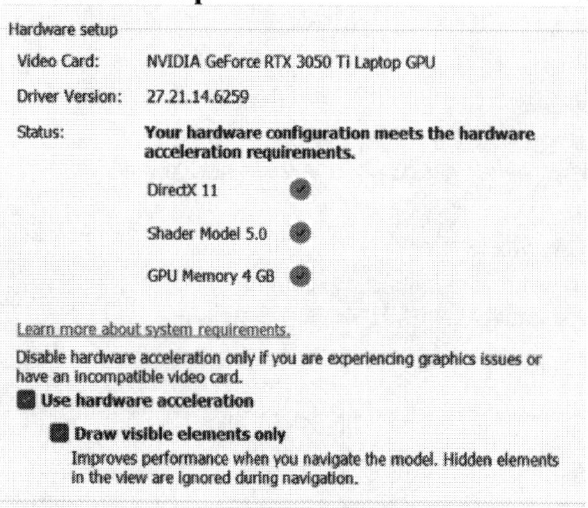

Hardware setup

Video Card: NVIDIA GeForce RTX 3050 Ti Laptop GPU

Driver Version: 27.21.14.6259

Status: **Your hardware configuration meets the hardware acceleration requirements.**

DirectX 11

Shader Model 5.0

GPU Memory 4 GB

Learn more about system requirements.

Disable hardware acceleration only if you are experiencing graphics issues or have an incompatible video card.

■ **Use hardware acceleration**

   ■ **Draw visible elements only**
   Improves performance when you navigate the model. Hidden elements in the view are ignored during navigation.

Displays the video card in use and provided driver status.

File Locations

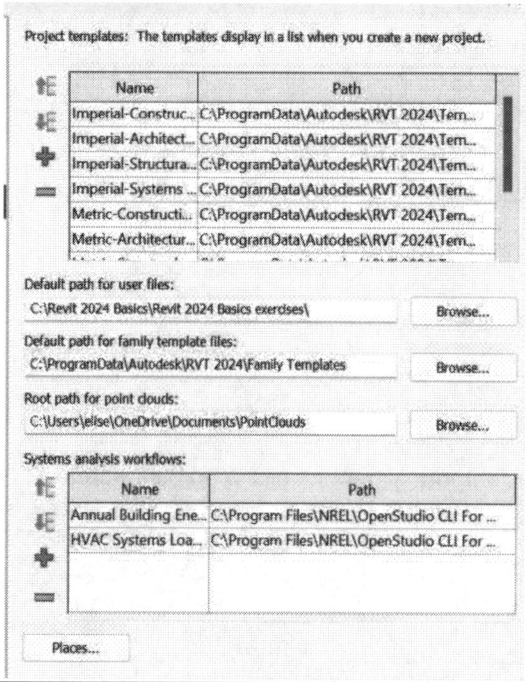

Project templates: The templates display in a list when you create a new project.

Name	Path
Imperial-Construc...	C:\ProgramData\Autodesk\RVT 2024\Tem...
Imperial-Architect...	C:\ProgramData\Autodesk\RVT 2024\Tem...
Imperial-Structura...	C:\ProgramData\Autodesk\RVT 2024\Tem...
Imperial-Systems ...	C:\ProgramData\Autodesk\RVT 2024\Tem...
Metric-Constructi...	C:\ProgramData\Autodesk\RVT 2024\Tem...
Metric-Architectur...	C:\ProgramData\Autodesk\RVT 2024\Tem...

Default path for user files:

C:\Revit 2024 Basics\Revit 2024 Basics exercises\   Browse...

Default path for family template files:

C:\ProgramData\Autodesk\RVT 2024\Family Templates   Browse...

Root path for point clouds:

C:\Users\elise\OneDrive\Documents\PointClouds   Browse...

Systems analysis workflows:

Name	Path
Annual Building Ene...	C:\Program Files\NREL\OpenStudio CLI For ...
HVAC Systems Loa...	C:\Program Files\NREL\OpenStudio CLI For ...

Places...

File Locations is used to set the default search locations for templates and families.

Places lists of all the places that will appear on the left pane of the Open dialog. Use for templates, project locations, and libraries.

| Rendering | Rendering controls where you are storing your AccuRender files and directories where you are storing your materials.

This allows you to set your paths so Revit can locate materials and files easily.

If you click the Get More RPC button, your browser will launch to the Archvision website. You must download and install a plug-in to manage your RPC downloads. There is a free exchange service for content, but you have to create a login account. |
|---|---|

| Check Spelling | Check Spelling allows you to use your Microsoft Office dictionary as well as any custom dictionaries you may have set up. These dictionaries are helpful if you have a lot of notes and specifications on your drawings.

To add a word to the Custom Dictionary, click the **Edit** button. |
|---|---|

| | Notepad will open and you can just type in any words you want to add to the dictionary.

Save the file. |
|---|---|

Steering Wheels

Text visibility
- Show tool messages (always ON for basic wheels)
- Show tooltips (always ON for basic wheels)
- Show tool cursor text (always ON for basic wheels)

Big steering wheel appearance
Size: Normal   Opacity: 50%

Mini wheel appearance
Size: Normal   Opacity: 50%

Look tool behavior
- Invert vertical axis (Pull mouse back to look up)

Walk tool
- Move parallel to ground plane
  Speed Factor:
  0.1 ————————————— 50.0   3

Zoom tool
- Zoom in one increment with each mouse click (always ON for basic wheels)

Orbit tool
- Keep scene upright

Rewind history
- Save snapshots of current view (Turn OFF for better performance)

Steering Wheels control the appearance and functionality of the steering wheels.

Steering Wheels are used to change the display.

ViewCube

ViewCube appearance
- Show the ViewCube

Show in: All 3D Views
On-screen position: Top Right
ViewCube size: Automatic
Inactive opacity: 50 %

When dragging the ViewCube
- Snap to closest view

When clicking on the ViewCube
- Fit-to-view on view change
- Use animated transition when switching views
- Keep scene upright

Compass
- Show the compass with the ViewCube (current project only)

ViewCube controls the appearance and location of the ViewCube.

The user can also determine how the display changes when the ViewCube is selected.

Macros
 	
Application macro security settings
Changes to application macro settings will take effect the next time Revit starts.
 	
    ○ Enable application macros
    ● Disable application macros
 	
Document macro security settings
Changes to document macro settings will take effect the next time Revit opens a document.
 	
    ○ Ask before enabling document macros
    ● Disable document macros
    ○ Enable document macros
	Macros is where the user sets the security level for Macros.

Cloud Model Cloud model cache Cache Location: [Browse...] C:\Users\elise\AppData\Local\Autodesk\Revit Choose a cache folder on a fast local hard drive that provides as much space as possible. Learn more	This tab allows you to select the location to store any back-ups for cloud documents.

To boost productivity, store all your custom templates on the network for easy access and set the pointer to the correct folder.

The Help Menu

Revit Help (also reached by function key F1) brings up the Help dialog.

What's New allows veteran users to quickly come up to speed on the latest release.

Essential Skills Videos is a small set of videos to help get you started. They are worth watching, especially if your computer skills are rusty.

Revit Community is an internet-based website with customer forums where you can search for solutions to your questions or post your questions. Most of the forums are monitored by Autodesk employees who are friendly and knowledgeable.

Additional Resources launches your browser and opens to a link on Autodesk's site. Autodesk Building Solutions take you to a YouTube channel where you can watch video tutorials.

> Support Knowledge Base
>
> 3rd Party Learning Content
>
> Autodesk.com/revit
>
> Autodesk Building Solutions

Privacy Settings
Takes you to a website where you can opt out of sharing your data and how you use Revit.

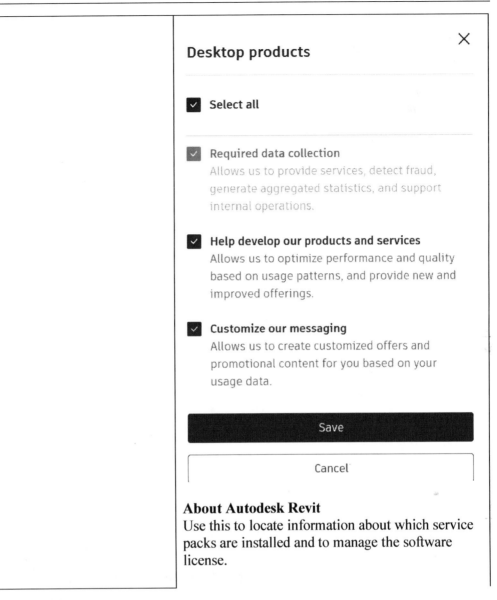

About Autodesk Revit
Use this to locate information about which service packs are installed and to manage the software license.

Exercise 1-7:
Setting File Locations

Drawing Name: Close all open files
Estimated Time: 5 minutes

This exercise reinforces the following skills:
- Options
- File Locations

1. Close all open files or projects.

2. Go to the **File Menu**.

   Select the **Options** button at the bottom of the window.

3. File Locations   Select the **File Locations** tab.

4. Default path for user files:

   C:\Users\Elise\Documents   Browse...

   In the **Default path for users** section, pick the **Browse** button.

5. **Default path for user files:**

   C:\Revit 2024 Basics\Revit 2024 Basics exercises

   Navigate to the local or network folder where you will save your files. When the correct folder is highlighted, pick **Open**. Your instructor or CAD manager can provide you with this file information.

I recommend to my students to bring a flash drive to class and back up each day's work onto the flash drive. That way you will never lose your valuable work. Some students forget their flash drive. For those students, make a habit of uploading your file to Google Drive, Dropbox, Autodesk Docs, or email your file to yourself.

Exercise 1-8:
Adding the Default Template to Recent Files

Drawing Name: Close all open files
Estimated Time: 5 minutes

This exercise reinforces the following skills:
- Options
- File Locations
- Project Templates
- Recent Files

1. 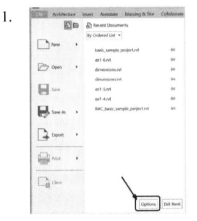 Select **Options** on the File Menu.

2. Project templates: The templates display in a list when you create a new project.

Name	Path
Imperial-Construc...	C:\ProgramData\Autodesk\RVT 2024\Tem...
Imperial-Architect...	C:\ProgramData\Autodesk\RVT 2024\Tem...
Imperial-Structura...	C:\ProgramData\Autodesk\RVT 2024\Tem...
Imperial-Systems ...	C:\ProgramData\Autodesk\RVT 2024\Tem...
Metric-Constructi...	C:\ProgramData\Autodesk\RVT 2024\Tem...
Metric-Architectur...	C:\ProgramData\Autodesk\RVT 2024\Tem...

Activate File Locations.

Click the **Plus** icon to add a project template.

3.

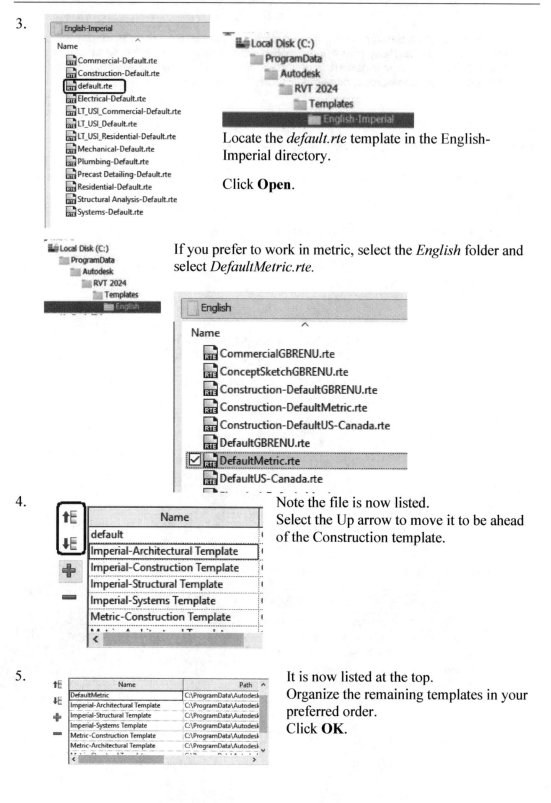

Locate the *default.rte* template in the English-Imperial directory.

Click **Open**.

If you prefer to work in metric, select the *English* folder and select *DefaultMetric.rte*.

4. Note the file is now listed.
Select the Up arrow to move it to be ahead of the Construction template.

5. It is now listed at the top.
Organize the remaining templates in your preferred order.
Click **OK**.

6.

The default template is now available in the drop-down list when you start a new project.

Exercise 1-9:
Turning Off the Visibility of Ribbons

Drawing Name: Close all open files
Estimated Time: 5 minutes

This exercise reinforces the following skills:
- ❏ Options
- ❏ User Interface
- ❏ Ribbon Tools

1.

Select **Options** on the Application Menu.

2.

Highlight **User Interface**.

Uncheck/disable Structure and tab tools, System tab: mechanical tools, Systems tab: electrical tools, Systems tab: piping tools and energy analysis and tools.

Click **OK**.

3. Notice that the available ribbons update to only display the tools you are interested in.

Notes:

Lesson 1 Quiz

True or False

1. Revit Warnings do not interrupt commands, so they can be ignored and later reviewed.
2. Deactivate View on the View menu is used to remove Views from Sheets.
3. The Visibility\Graphics dialog is used to control what elements are visible in a view.

Multiple Choice [Select the Best Answer]

4. Temporary dimensions:

    A. drive the values that appear in the permanent dimensions.
    B. display the relative position of the selected element.
    C. display the size of the selected element.
    D. display only when an element is selected.

5. When you select an object, you can access the element properties by:

    A. Right clicking and selecting Element Properties.
    B. Selecting Element Properties→Instance Properties on the Ribbon Bar.
    C. Going to File→Properties.
    D. Using the Properties pane located on the left of the screen.

6. The interface item that changes appearance constantly depending on entity selections and command status is the _____.

    A. Ribbon
    B. Project Browser
    C. Menu Bar
    D. Status Bar

7. Display Options available in Revit are _____.

    A. Wireframe
    B. Hidden Line
    C. Shaded
    D. Consistent Colors
    E. All of the Above

8. The shortcut key to bring up the Help menu is _____.

    A. F2
    B. F3
    C. VV
    D. F1

9. To add elements to a selection set, hold down the _____ key.

    A. CONTROL
    B. SHIFT
    C. ESCAPE
    D. TAB

10. The color blue on a level bubble indicates that the level:

    A. has an associated view.
    B. is a reference level.
    C. is a non-story level.
    D. was created using the line tool.

ANSWERS:

1) T; 2) F; 3) T; 4) A, B, & D 5) D; 6) A; 7) E; 8) D; 9) A; 10) A

Lesson 2
Mass Elements

Mass Elements are used to give you a conceptual idea of the space and shape of a building without having to take the time to put in a lot of detail. It allows you to create alternative designs quickly and easily and get approval before you put in a lot of effort.

Massing Tools

Show Mass	Show Mass controls the visibility of mass entities.
In-Place Mass	Creates a solid shape.
Place Mass	Inserts a mass group into the active project.
Curtain System Roof Wall Floor Model by Face	Model by Face: Converts a face into a Roof, Curtain Wall System, Wall, or Floor.

When creating a conceptual mass to be used in a project, follow these steps:

1. Create a sketch of the desired shape(s).
2. Create levels to control the height of the shapes.
3. Create reference planes to control the width and depth of the shapes.
4. Draw a sketch of the profile of the shape.
5. Use the Massing tools to create the shape.

Masses can be used to create a component that will be used in a project, such as a column, casework, or lighting fixture, or they can be used to create a conceptual building.

Exercise 2-1
Shapes

Drawing Name: shapes.rfa
Estimated Time: 5 minutes

This exercise reinforces the following skills:

- ❑ Creating basic shapes using massing tools
- ❑ Create an extrude
- ❑ Modify the extrude height
- ❑ Create a revolve
- ❑ Create a sweep
- ❑ Create a blend
- ❑ Modify a blend

1.

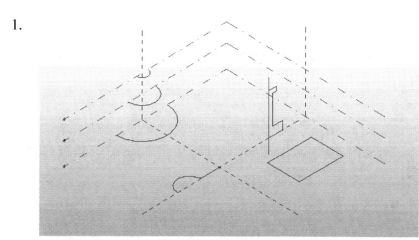

There are several sketches in the file.

Each set of sketches will be used to create a specific type of mass form.

Revit knows what type of shape you want to create based on what you select. It's all in the ingredients!

2.

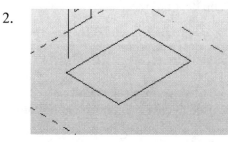

The most basic mass form is an extrude. This mass form requires a single closed polygonal sketch.

The sketch should have no gaps or self-intersecting lines.

Select the rectangle so it highlights.

3. 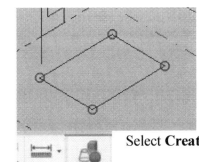 When the sketch is selected, you will see grips activated at the vertices.

4. 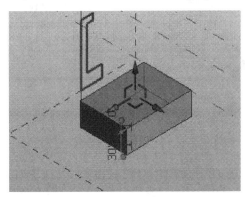 Select **Create Form→Solid Form** to create the extrude.

You won't see the Create Form tool on the ribbon unless a sketch is selected.

5. A preview of the extrude is displayed.

You can use the triad on the top of the extrude to extend the shape in any of the three directions.

There are two dimensions:

The bottom dimension controls the height of the extrude.
The top dimension controls the distance from top face of the extrude and the next level.

6. Activate the **East Elevation**.

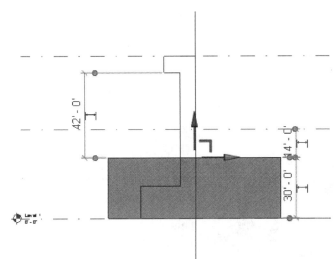

You can see how the dimensions are indicating the relative distances between the top face and the upper level, the bottom face, and another sketch.

7.

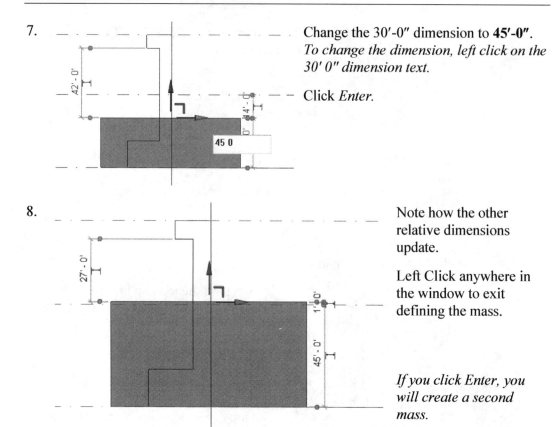

Change the 30'-0" dimension to **45'-0"**.
To change the dimension, left click on the 30' 0" dimension text.

Click *Enter.*

8.

Note how the other relative dimensions update.

Left Click anywhere in the window to exit defining the mass.

If you click Enter, you will create a second mass.

9. Switch to a 3D view.

10.

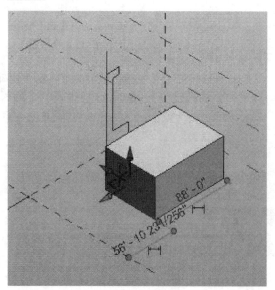

Select the Back/North face of the block. *Use the Viewcube to verify the orientation.*

To select the face, place the mouse over the face and click the tab key until the entire face highlights.

This activates the triad and also displays the temporary dimensions.

11.

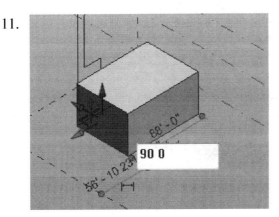

You will see two temporary dimensions. One dimension indicates the distance of the face to the opposite face (the length of the block). One dimension indicates the distance of the face to the closest work plane.

Change the 88' 0" dimension to **90'-0"**.

Left click in the window to release the selection and complete the change.

If you click Enter, you will create a second mass.

12.

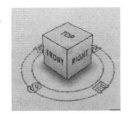

Use the View Cube to reorient the view so you can clearly see the next sketch.

This sketch will be used to create a Revolve.

A Revolve requires a closed polygonal shape PLUS a reference line which can be used as the axis of revolution.

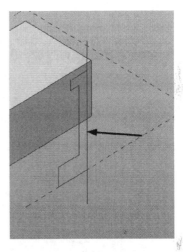

13.

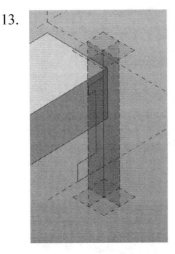

Hold down the CONTROL key and select the Axis Line (this is a reference line) and the sketch.

Note that a reference line automatically defines four reference planes. These reference planes can be used to place sketches.

14. 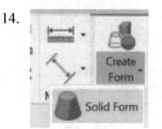 Select **Create Form→Solid Form**.

15. A Revolve will be created.

Our next mass form will be a SWEEP.

A sweep requires two sketches. One sketch must be a closed polygonal shape. This sketch is called the profile. The second sketch can be open or closed and is called the path. The profile travels along the path to create the sweep.

16. Hold down the CONTROL key and select the small circle and the sketch that looks like a question mark.

The two sketches will highlight.

17. 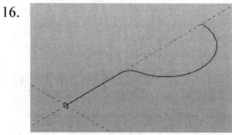 Select **Create Form→Solid Form**.

18. The sweep will be created.

The most common error when creating a sweep is to make the profile too big for the path. If the profile self-intersects as it travels along the path, you will get an error message. Try making the profile smaller to create a successful sweep. Sweeps are useful for creating gutters, railings, piping, and lighting fixtures.

19.

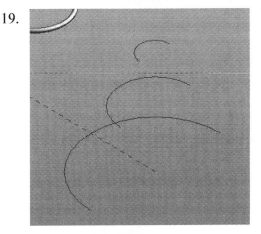

Our final shape will be a BLEND or LOFT. A blend is created using two or more open or closed sketches. Each sketch must be placed on a different reference plane or level.

Hold down the CONTROL key and select the three arcs.

20.

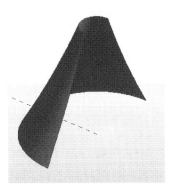

Select **Create Form→Solid Form**.

21.

A blend shape is created.

Blends are used for complex profiles and shapes.

Close without saving.

Challenge Task:

Can you create the four basic shapes from scratch?

Create an extrude, revolve, sweep and blend using New→Conceptual Mass from the Applications menu.

Steps to make an Extrude:

1. Set your active plane using the Set Work Plane tool.
2. Create a single polygonal sketch with no gaps or self-intersecting lines.
3. Select the sketch.
4. Create Form→Solid Form.
5. Green Check to finish the mass.

Steps to make a Revolve:

1. Set your active plane using the Set Work Plane tool.
2. Switch to a 3D view.
3. Draw a reference line to be used as the axis of revolution.
4. Pick one of the planes defined by the reference line as the active plane to place your sketch.
5. Create a single polygonal sketch – no gaps, overlapping or self-intersecting lines – for the revolve shape.
6. Hold down the CONTROL key and select BOTH the sketch and reference line. If the reference line is not selected, you will get an extrude.
7. Create Form→Solid Form.
8. Green Check to finish the mass.

Steps to make a Sweep:

1. Activate Level 1 floor plan view. You need to select one reference plane for the profile and one reference plane for the path. The reference planes must be perpendicular to each other. Set Level 1 for the path's reference plane.
2. Draw a path on Level 1. The path can be a closed or open sketch.
3. Create a reference plane to use for the profile. Draw a reference plane on Level 1 – name it profile plane. Make this the active plane for the profile sketch.
4. Switch to a 3D view. Create a single polygonal sketch – no gaps, overlapping or self-intersecting lines. The profile sketch should be close to the path or intersect it so it can follow the path easily. If it is too far away from the path, it will not sweep properly, or you will get an error.
5. Hold down the CONTROL key and select BOTH the path and the profile. If only one object is selected, you will get an extrude.
6. Create Form→Solid Form.
7. Green Check to finish the mass.

Steps to make a Blend:

1. Blends require two or more sketches. Each sketch should be on a parallel reference plane. You can add levels or reference planes for each sketch. If you want your blend to be vertical, use levels. If you want your blend to be horizontal, use reference planes.
   a. To add levels, switch to an elevation view and select the Level tool.
   b. To add reference planes, switch to a floor plan view and select the Reference Plane tool. Name the reference planes to make them easy to select.
2. Set the active reference plane using the Option Bar or the Set Reference Plane tool.
3. Create a single polygonal sketch – no gaps, overlapping or self-intersecting lines.
4. Select at least one more reference plane to create a second sketch. Make this the active plane.
5. Create a single polygonal sketch – no gaps, overlapping or self-intersecting lines.
6. Hold down the CONTROL key and select all the sketches created. If only one object is selected, you will get an extrude.
7. Create Form→Solid Form.
8. Green Check to finish the mass.

Exercise 2-2
Create a Conceptual Model

Drawing Name: default.rte [DefaultMetric.rte]
Estimated Time: 5 minutes

This exercise reinforces the following skills:

- ❑ Switching Elevation Views
- ❑ Setting Project Units
- ❑ Add a Level

This tutorial uses metric or Imperial units. Metric units will be designated in brackets.

Revit uses a level to define another floor or story in a building.

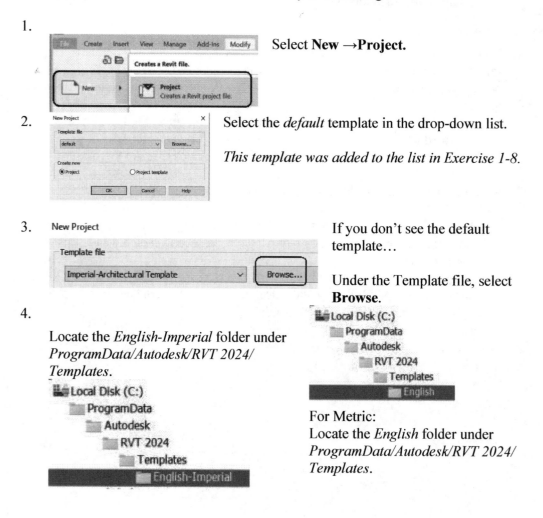

1. Select **New** →**Project.**

2. Select the *default* template in the drop-down list.

   This template was added to the list in Exercise 1-8.

3. If you don't see the default template…

   Under the Template file, select **Browse.**

4. Locate the *English-Imperial* folder under *ProgramData/Autodesk/RVT 2024/ Templates.*

   For Metric:
   Locate the *English* folder under *ProgramData/Autodesk/RVT 2024/ Templates.*

5.

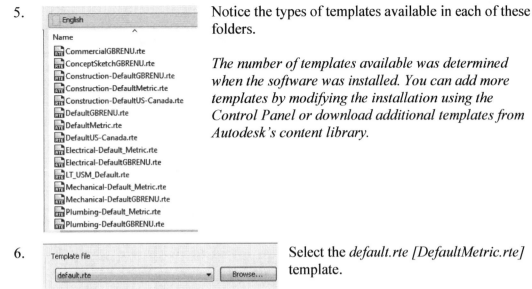

Notice the types of templates available in each of these folders.

The number of templates available was determined when the software was installed. You can add more templates by modifying the installation using the Control Panel or download additional templates from Autodesk's content library.

6.

Select the *default.rte [DefaultMetric.rte]* template.

Brackets indicate metric, which can be selected as an alternative.

Click **OK**.

If you accidentally picked Metric when you wanted Imperial or vice versa, you can change the units at any time.

To change Project Units, go to the **Manage** Ribbon.

Select **Settings→Project Units**.

Left click the Length button, then select the desired units from the drop-down list.

7. 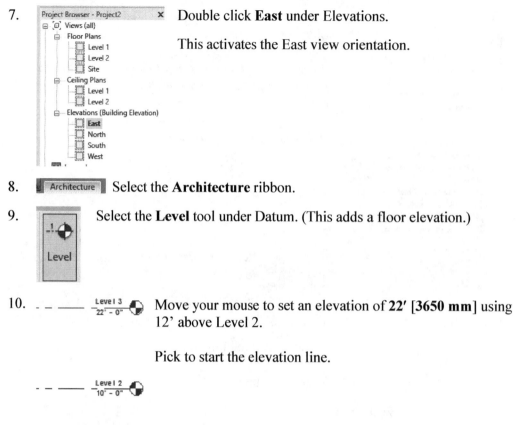 Double click **East** under Elevations.

   This activates the East view orientation.

8. Architecture   Select the **Architecture** ribbon.

9. Select the **Level** tool under Datum. (This adds a floor elevation.)

10. Level 3
    22' - 0"   Move your mouse to set an elevation of **22' [3650 mm]** using 12' above Level 2.

    Pick to start the elevation line.

    Level 2
    10' - 0"

11. ☑ Make Plan View   Plan View Types...   Offset: 0' 0"

    In the Options bar, enable **Make Plan View**.
    The Options bar can be located below the ribbon or at the bottom of the screen.

    If the Options bar is located below the ribbon and you would prefer it on the bottom of the screen, right click and select **Dock at Bottom**.

    If the Options bar is located at the bottom of the screen and you would prefer it to be below the ribbon, right click on the Options bar and select **Dock at Top**.

    Make Plan View should be enabled if you want Revit to automatically create a floor plan view of this level. If you forget to check this box, you can create the floor plan view later using the **View** Ribbon.

 Double click on the blue elevation symbol to automatically switch to the floor plan view for that elevation.

12. 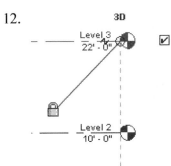 Pick to place the end point to position the level indicator above the other indicators.

13. Basically, you place a new level by picking two points at the desired height.

   Right click and select **Cancel** to exit the Level command.

   Revit is always looking for references even among annotations; you will notice that your level tags snap and lock together so when you move one to the right or left, all those in line with it will follow.

   The jogged line allows the user to create a jog if desired.

   If you need to adjust the position of the tag, just click on the line; 3 blue grips will appear. These can be clicked and dragged as needed. You can also right click on a level tag and select 'Hide annotation in view' and the tag and level line will disappear in that view only.
   Hide Annotation in View is only enabled if an object is selected first.

14. Save the file as a project as *ex2-2.rvt.*

Revit has an option where you can save the current file, whether it is a project file or a family, to a cloud server. In order for this option to be available, you have to pay a subscription fee to Autodesk for the cloud space.

Exercise 2-3
Adding an In-Place Mass

Drawing Name: in-place mass.rvt
Estimated Time: 10 minutes

This exercise reinforces the following skills:

- ❑   Switching Elevation Views
- ❑   Add Mass

1.   Open *in-place mass.rvt*.

2.   ▣ Views (all)
     └ Floor Plans
        ├ **Level 1**
        ├ Level 2
        ├ Level 3
        └ Site

     Activate the **Level 1** view.

3.   Massing & Site   Select the **Massing & Site** ribbon.

4.   In-Place
     Mass

     Select the **In-Place Mass** tool.

*Revit uses three different family categories. System families are families which are
defined inside the project, such as floors, walls, and ceilings. Loadable families are
external files which can be loaded into the project, such as doors, windows, and
furniture. The third family category is in-place masses, which are families created
on-the-fly. In-place masses are only available inside the project where they are
created and are usually unique to the project since they aren't loadable. However,
you can use the Copy and Paste function to copy in-place masses from one project
to another, if you decide to re-use it. That can be a hassle because you need to
remember which project is storing the desired in-place mass.*

5.   Massing - Show Mass Enabled   ✕

     Revit has enabled the Show Mass mode, so the
     newly created mass will be visible.

     To temporarily show or hide masses, select the Massing & Site
     ribbon tab and then click the Show Mass button on the
     Massing panel.

     Masses will not print or export unless you make the Mass
     category permanently visible in the View Visibility/Graphics
     dialog.

     ☐ Do not show me this message again   Close

     Masses, by default, are invisible. However, in
     order to create and edit masses you need to see
     what you are doing. Revit brings up a dialog to
     let you know that the software is switching the
     visibility of masses to ON, so you can work.

     Click **Close**.

*If you don't want to be bugged by this dialog, enable the **Don't show me this
message again** option.*

6. 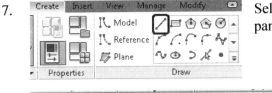 Enter **Mass 1** in the Name field.

Click **OK**.

Next, we create the boundary sketch to define our mass. This is the footprint for the conceptual building.

7. 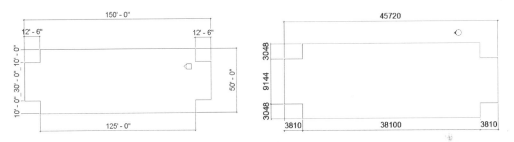 Select the **Line** tool located under the Draw panel.

8. Enable **Chain** in the Options bar.
This allows you to draw lines without always having to pick the start point.
If you hold down the SHIFT key while you draw, this puts you in orthogonal mode.

9. Create the shape shown.
The left figure shows the units in Imperial units (feet and inches).
The right figure shows the units in millimeters.

You can draw using listening dimensions or enter the dimension values as you draw.

For feet and inches, you do not need to enter the ' or " symbols. Just place a space between the numbers.

Revit doesn't have a CLOSE command for the LINE tool unlike AutoCAD, so you do have to draw that last line.

10. Exit out of drawing mode by right clicking and selecting Cancel twice, selecting ESC on the keyboard or by selecting the Modify button on the ribbon.

11. Switch to a 3D view.
Activate the **View** ribbon and select **3D View**.

You can also switch to a 3D view from the Quick Access toolbar by selecting the house icon.

12. Window around the entire sketch so it is highlighted.

13. 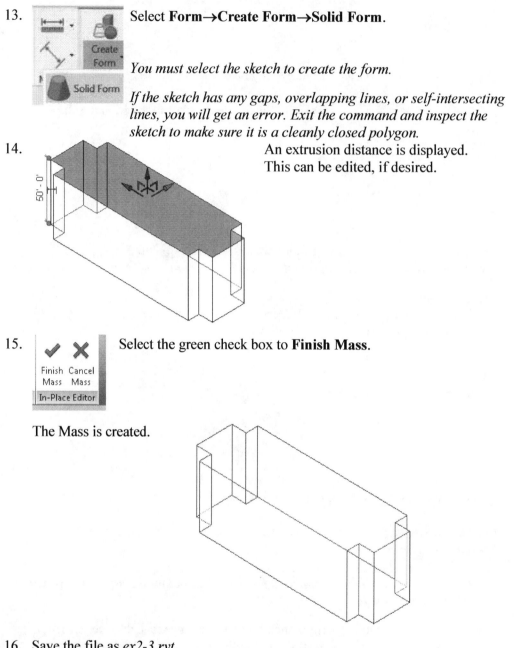 Select **Form→Create Form→Solid Form**.

You must select the sketch to create the form.

If the sketch has any gaps, overlapping lines, or self-intersecting lines, you will get an error. Exit the command and inspect the sketch to make sure it is a cleanly closed polygon.

14. An extrusion distance is displayed. This can be edited, if desired.

15. Select the green check box to **Finish Mass**.

The Mass is created.

16. Save the file as *ex2-3.rvt*.

Object tracking will only work if the sketch objects are active and available in the current sketch. You can use Pick to copy entities into the current sketch.

Exercise 2-4
Modifying Mass Elements

Drawing Name: modify mass.rvt
Estimated Time: 30 minutes

This exercise reinforces the following skills:

- ❑ Show Mass
- ❑ Align
- ❑ Modify Mass
- ❑ Mirror
- ❑ Create Form
- ❑ Save View

1.   Open *modify mass.rvt*.

2.  If you don't see the mass, use **Show Mass** on the Massing & Site ribbon to turn mass visibility ON.

Some students may experience this issue if they close the file and then re-open it for a later class.

3.   Elevations (Building Elevation
       ☐ **East**
       ☐ North
       ☐ South
       ☐ West

Activate the **East** Elevation.

*Remember if you don't see the mass, use **Show Mass** on the Massing & Site ribbon to turn mass visibility ON.*

4.   We see that top of the building does not align with Level 3.

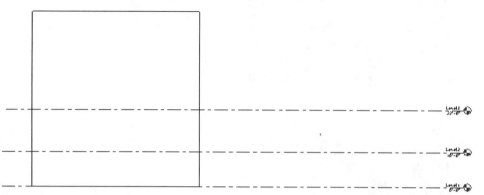

To adjust the horizontal position of the level lines, simply select the line and use the grip to extend or shorten it.

5.   [Modify]   Select the **Modify** Ribbon.

6.   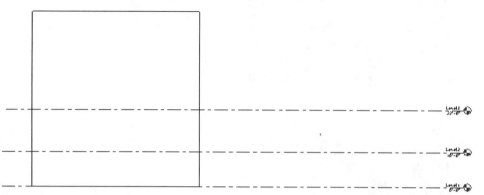   Select the **Align** tool.

When using Align, the first element selected acts as the source, and the second element selected shifts position to align with the first element.

7.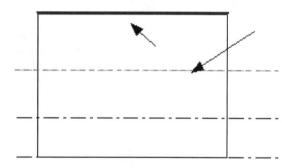

Select the top level line (Level 3) then select the top of the extrusion.

Right click and select **Cancel** twice to exit the Align command.

8.   The top of the extrusion now aligns to Level 3.

The lock would constrain or lock the top of the extrusion to the level. If the level elevation changes, then the extrusion will automatically update.

9.   ⊟ Floor Plans
        Level 1
        Level 2
        Level 3
        Site

Activate **Level 2** under Floor Plans.

10. 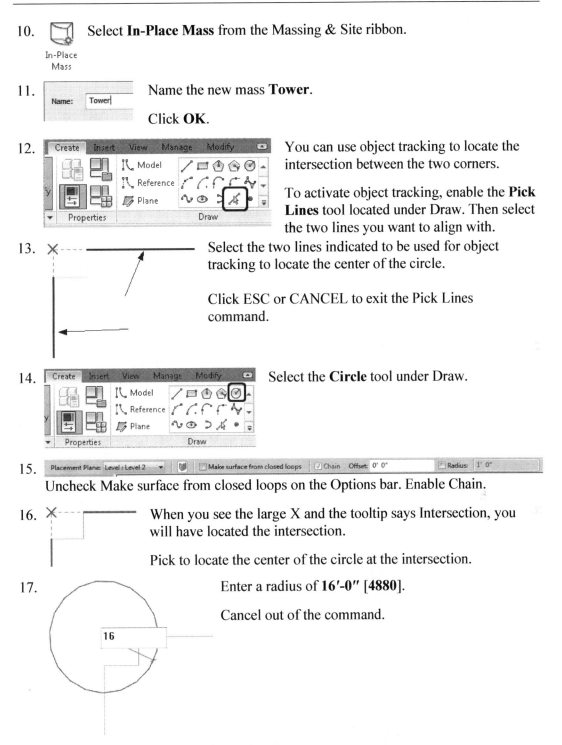 Select **In-Place Mass** from the Massing & Site ribbon.

In-Place
Mass

11. Name: Tower

Name the new mass **Tower**.

Click **OK**.

12. You can use object tracking to locate the intersection between the two corners.

To activate object tracking, enable the **Pick Lines** tool located under Draw. Then select the two lines you want to align with.

13. Select the two lines indicated to be used for object tracking to locate the center of the circle.

Click ESC or CANCEL to exit the Pick Lines command.

14. Select the **Circle** tool under Draw.

15. Placement Plane: Level : Level 2 | Make surface from closed loops | Chain | Offset: 0' 0" | Radius: 1' 0"

Uncheck Make surface from closed loops on the Options bar. Enable Chain.

16. When you see the large X and the tooltip says Intersection, you will have located the intersection.

Pick to locate the center of the circle at the intersection.

17. Enter a radius of **16'-0"** [4880].

Cancel out of the command.

16

When you used the Pick Line tool, you copied those lines into the current sketch. Once the lines were part of the current sketch, they could be used for object tracking.

18. Select the circle sketch so it is highlighted.

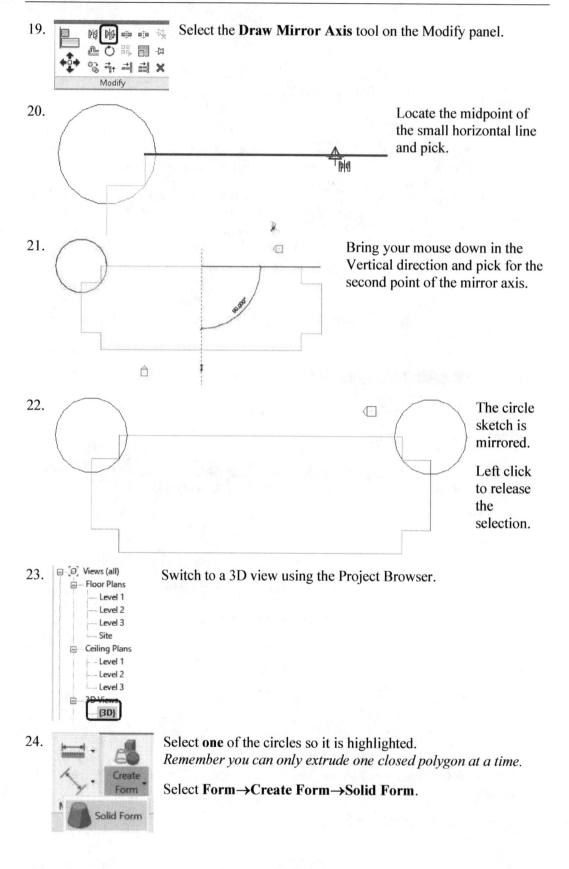

19. Select the **Draw Mirror Axis** tool on the Modify panel.

20. Locate the midpoint of the small horizontal line and pick.

21. Bring your mouse down in the Vertical direction and pick for the second point of the mirror axis.

22. The circle sketch is mirrored.

Left click to release the selection.

23. Switch to a 3D view using the Project Browser.

24. Select **one** of the circles so it is highlighted.
Remember you can only extrude one closed polygon at a time.

Select **Form→Create Form→Solid Form**.

25.

A small toolbar will appear with two options for extruding the circle.

Use the SPACE BAR to cycle through the different forms.

Click ENTER to select the option that looks like a cylinder or right click the mouse.

26.

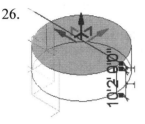

A preview of the extrusion will appear with the temporary dimension. You can edit the temporary dimension to modify the height of the extrusion.

Click ENTER or left click to accept the default height.

27.

If you click ENTER more than once, additional cylinders will be placed.

The circle is extruded.

28.

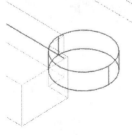

Select the remaining circle so it is highlighted.

Select **Form→Create Form→Solid Form**.

29.

A small toolbar will appear with two options for extruding the circle.

Use the SPACE BAR to cycle through the different forms.

Click ENTER to select the option that looks like a cylinder or right click the mouse.

30.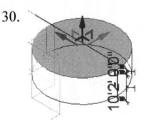

A preview of the extrusion will appear with the temporary dimension. You can edit the temporary dimension to modify the height of the extrusion.

Click ENTER or left click to accept the default height.

If you click ENTER more than once, you will keep adding cylinders.

Click ESC or right click and select CANCEL to exit the command.

Both circles are now extruded.

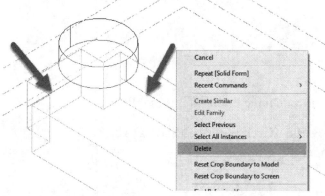

31.

Select the two lines used to locate the circle sketch.

Right click and select **Delete** from the shortcut menu to delete the lines.

You can also click the Delete key on the keyboard or use the Delete tool on the Modify panel.

32. Select **Finish Mass**.

If you do not delete the lines before you finish the mass, you will get an error message.

33. Activate the **South** Elevation.

34. Select each level line.

Maximize 3D Extents

Right click and select **Maximize 3D Extents**.
This will extend each level line so it covers the entire model.

35. Activate the Modify ribbon.

Select the **Align** tool from the Modify Panel.

36. ☑ Multiple Alignment Prefer:
    ☐ Lock Wall faces ▼

On the ribbon, enable **Multiple Alignment**.
Select **Wall faces** from the drop down list.

37. Select the Level 3 line as the source object.

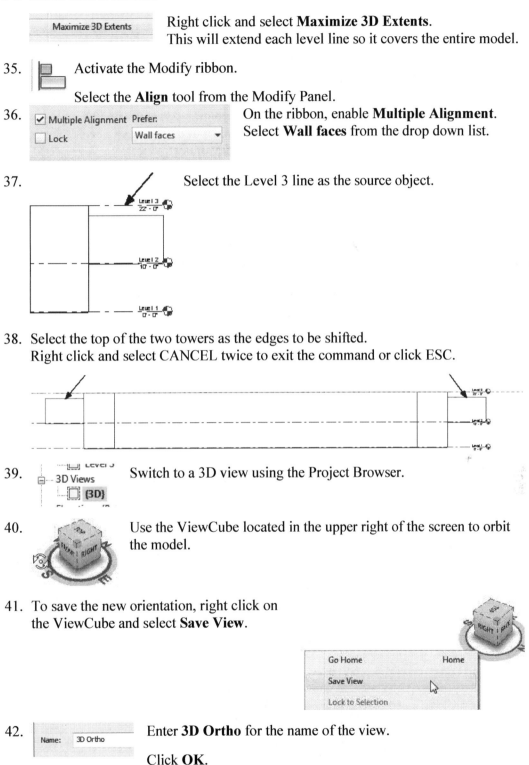

38. Select the top of the two towers as the edges to be shifted.
Right click and select CANCEL twice to exit the command or click ESC.

39. 3D Views
    {3D}

Switch to a 3D view using the Project Browser.

40. Use the ViewCube located in the upper right of the screen to orbit the model.

41. To save the new orientation, right click on the ViewCube and select **Save View**.

Go Home	Home
Save View	
Lock to Selection	

42. Name: 3D Ortho

Enter **3D Ortho** for the name of the view.

Click **OK**.

43. The **Saved** view is now listed in the Project browser under 3D Views.

44. Save the file as *ex2-4.rvt*.

> ➢ Pick a mass element to activate the element's grips. You can use the grips to change the element's shape, size, and location.
> ➢ You can only use the **View→Orient** menu to activate 3D views when you are already in 3D view mode.

Wall by Face

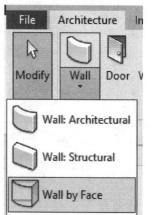

This tool can be used to place walls on non-horizontal faces of a mass instance or a generic model.

If you modify the mass, the wall will not update. To update the wall, use the **Update to Face** tool.

Exercise 2-5
Create Wall by Face

Drawing Name: wall by face.rvt
Estimated Time: 15 minutes

This exercise reinforces the following skills:

- Wall by Face
- Trim
- Show Mass

You can add doors and windows to your conceptual model to make it easier to visualize.

1. Open *wall by face.rvt*.

2. Activate the **3D Ortho** view under 3D Views.

3. Activate the **Massing & Site** ribbon.

4. Select **Model by Face→Wall**.

5. Note the wall type currently enabled in the Properties pane. A different wall type can be selected from the drop-down list available using the small down arrow.

   Imperial:
   Set the Default Wall Type to:
   Basic Wall: Generic- 8″.

   Metric:
   Set the Default Wall Type to:
   Basic Wall: Generic- 200 mm.

6. 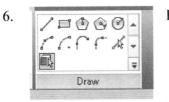 Enable **Pick Faces** from the Draw Panel on the ribbon.

7.

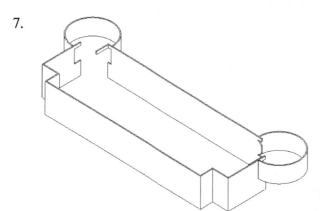

Select each wall and cylinder. The cylinder will be halved by the walls, so you will have to select each half.

You will have to do some cleanup work on the corners where the towers are.

Right click and select CANCEL to exit the command.

Some students will accidentally pick the same face more than once. You will see an error message that you have overlapping/duplicate walls. Simply delete the extra walls.

8.

Select any visible mass.
Right click and select **Hide in View→ Category**.

This will turn off the visibility of masses.

You can also disable Show Mass on the ribbon.

9.

Floor Plans
Level 1
Level 2
Level 3
Site

Activate **Level 1** floor plan.

10.

Window around all the walls to select.

11. Select the **Filter** tool from the ribbon.

Filter

12.

Category:
Walls

Uncheck all the boxes EXCEPT walls.
Click **OK**.

There are some duplicate walls in this selection.

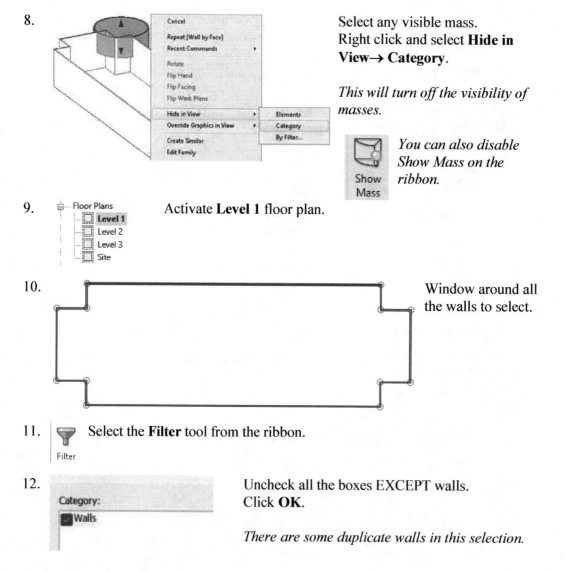

13.

Walls (12)	▼ 🔲 Edit Type
Constraints	⊠
Location Line	Finish Face: Exterior
Base Constraint	Level 1
Base Offset	0' 0"
Base is Attached	☐
Base Extension Distance	0' 0"
Top Constraint	Up to level: Level 3
Unconnected Height	22' 0"
Top Offset	0' 0"
Top is Attached	☐

In the Properties pane:

Set the Top Constraint to **up to Level 3**.

Right click and select Cancel to release the selection or left click in the display window to release the selection.

14.

Hold down the Ctrl Key.
Select the four walls indicated.

15.

Properties	✕
Basic Wall Generic - 8"	
Walls (4)	▼ 🔲 Edit Type
Constraints	⊠
Location Line	Finish Face: Exterior
Base Constraint	Level 1
Base Offset	0' 0"
Base is Attached	☐
Base Extension Distance	0' 0"
Top Constraint	Up to level: Level 2
Unconnected Height	10' 0"

In the Properties pane:

Set the Top Constraint to **Up to Level 2**.

Right click and select Cancel to release the selection or left click in the display window to release the selection.

16.

Floor Plans
　　Level 1
　　Level 2
　　Level 3
　　Site

Activate **Level 2** under Floor Plans.

17.

Underlay	⊠
Range: Base Level	Level 2
Range: Top Level	Level 3
Underlay Orientation	Look down

In the Properties Pane:
Go to the **Underlay** category.

Set the Range Base Level to **Level 2**.
Set the Range Top Level to **Level 3**.
Set the Underlay orientation to **Look down**.

This will turn off the visibility of all entities located below Level 2.

Each view has its own settings. We turned off the visibility of masses on Level 1, but we also need to turn off the visibility of masses on the Level 2 view.

18. 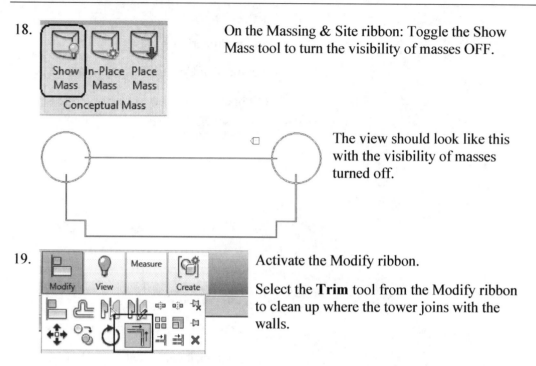 On the Massing & Site ribbon: Toggle the Show Mass tool to turn the visibility of masses OFF.

The view should look like this with the visibility of masses turned off.

19. Activate the Modify ribbon.

Select the **Trim** tool from the Modify ribbon to clean up where the tower joins with the walls.

20. When you select to trim, be sure to select the section you want to keep.

21. 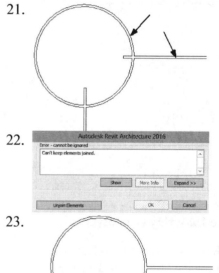 Select the arc and wall indicated.

22. If you see this dialog, click the **Unjoin Elements** button and proceed.

23. Select the arc and wall indicated.

Select the first line or wall to trim/extend. (Click on the part you want to keep)

Note that you have some instructions in the lower left of the screen to assist you.

24. 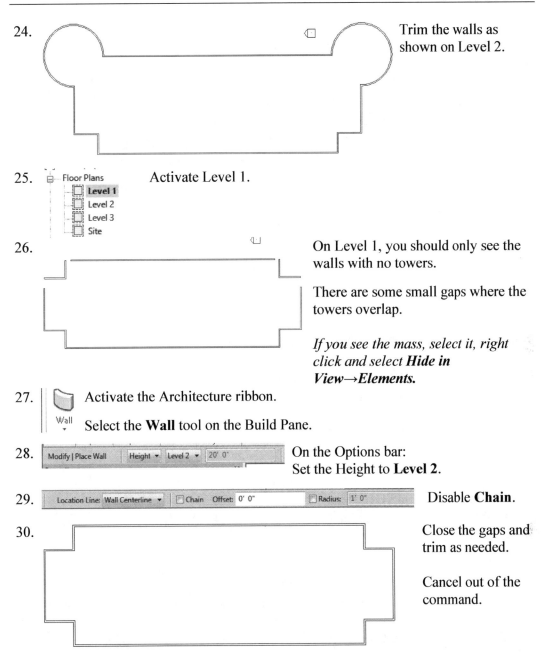 Trim the walls as shown on Level 2.

25. Floor Plans
    Level 1
    Level 2
    Level 3
    Site
    Activate Level 1.

26. On Level 1, you should only see the walls with no towers.

There are some small gaps where the towers overlap.

*If you see the mass, select it, right click and select **Hide in View→Elements**.*

27. Activate the Architecture ribbon.

Wall Select the **Wall** tool on the Build Pane.

28. Modify | Place Wall   Height ▾   Level 2 ▾   20' 0"   On the Options bar:
Set the Height to **Level 2**.

29. Location Line: Wall Centerline ▾   ☐ Chain   Offset: 0' 0"   ☐ Radius 1' 0"   Disable **Chain**.

30. Close the gaps and trim as needed.

Cancel out of the command.

31. Switch to a 3D view and orbit your model.

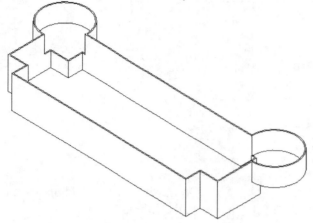

Check to make sure the walls and towers are adjusted to the correct levels.

Turn off the visibility of masses to see the walls.

32. Save as *ex2-5.rvt*.

Exercise 2-6
Adding Doors and Windows

Drawing Name: doors and windows.rvt
Estimated Time: 30 minutes

This exercise reinforces the following skills:

- Basics
- Door
- Load from Library
- Window
- Array
- Mirror
- Shading

You can add doors and windows to your conceptual model to make it easier to visualize.

1. Open *doors and windows.rvt*.

2. 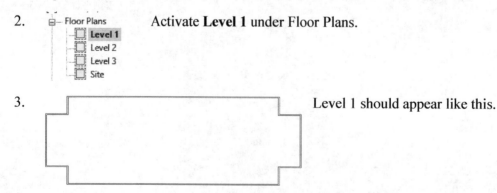 Activate **Level 1** under Floor Plans.

3. Level 1 should appear like this.

4. 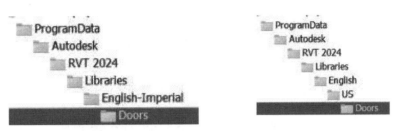 Activate the **Architecture** ribbon.

5. Select the **Door** tool under the Build panel.

Door

6. Select **Load Family** under the Mode panel.

Load
Family

Doors are loadable families.

7.

ProgramData   ProgramData
   Autodesk   Autodesk
      RVT 2024   RVT 2024
         Libraries   Libraries
           English-Imperial   English
               Doors   US
                                   Doors

Browse to the **Doors** folder under the Imperial or Metric library – use English-Imperial if you are using Imperial units or use English if you are using Metric units.

Preview

As you highlight each file in the folder, you can see a preview of the family.

Note that the files are in alphabetical order.

8.

Look in: Commercial
Name
Door-Exterior-Double
Door-Exterior-Double-Two_Lite
Door-Exterior-Revolving-Full Glass-M

Commercial
Name
M_Door-Exterior-Double

For Imperial Units: Go to the Commercial folder.
Locate the *Door-Exterior-Double.rfa* file.

For Metric Units: Go to the Commercial folder.
Locate the *M_Door-Exterior-Double.rfa* file.

Click **Open**.

9.

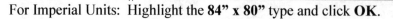

Family:	Types:		
Door-Exterior-Double.rfa	Type	Width	Height
		(all)	(all)
	72" x 80"	6' 0"	6' 8"
	72" x 82"	6' 0"	6' 10"
	72" x 84"	6' 0"	7' 0"
	84" x 80"	7' 0"	6' 8"
	84" x 84"	7' 0"	7' 0"
	96" x 80"	8' 0"	6' 8"
	96" x 84"	8' 0"	7' 0"

Select one or more types on the right for each family listed on the left   [OK]   [Cancel]

For Imperial Units: Highlight the **84" x 80"** type and click **OK**.

Family:	Types:		
M_Door-Exterior-Double.rfa	Type	Width	He
		(all)	(al
	1800 x 2000m	5' 10 111/128"	6' 6 189/256"
	1800 x 2050m	5' 10 111/128"	6' 8 181/256"
	1800 x 2100m	5' 10 111/128"	6' 10 173/256"
	2100 x 2000m	6' 10 173/256"	6' 6 189/256"
	2100 x 2100m	6' 10 173/256"	6' 10 173/256"
	2400 x 2000m	7' 10 125/256"	6' 6 189/256"
	2400 x 2100m	7' 10 125/256"	6' 10 173/256"

Select one or more types on the right for each family listed on the left   [OK]

For Metric Units: Highlight **2100 x 2000m** type and click **OK**.

10.

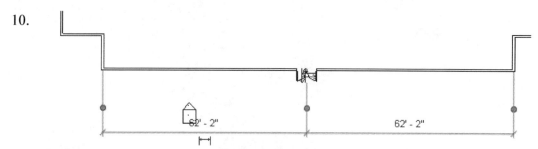

62' - 2"   62' - 2"

Place the door so it is centered on the wall as shown.
Doors are wall-hosted, so you will only see a door preview when you place your cursor over a wall.

Right Click and CANCEL to exit the command.

11.  If you click the space bar before you pick to place, you can control the orientation of the door.

After you have placed the door, you can flip the door by picking on it then pick on the vertical or horizontal arrows.

12. Pick the **Window** tool from the Build panel.

Windows are model families.

13. Select **Load Family** from the Mode panel.

14.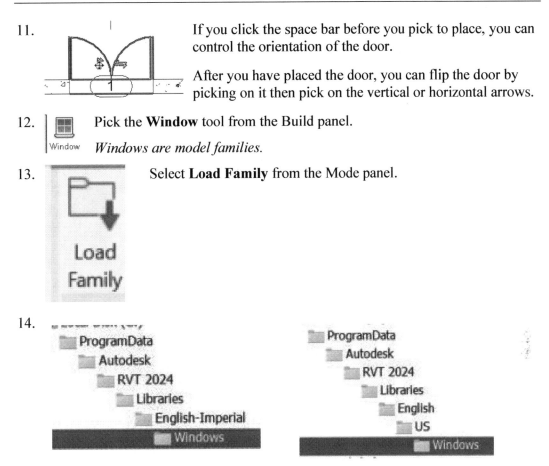

Browse to the **Windows** folder under the Imperial or Metric library – use English-Imperial if you are using Imperial units or use English if you are using Metric units.

15.

| File name: | Window-Casement-Double.rfa |
| Files of type: | All Supported Files (*.rfa, *.adsk) |

For Imperial Units:
Locate the *Window-Casement-Double.rfa* file.

| File name: | M_Window-Casement-Double.rfa |
| Files of type: | All Supported Files (*.rfa, *.adsk) |

For Metric Units:
Locate the *M_Window-Casement-Double.rfa* file.

Click **Open**.

16.

For Imperial Units:

Select the 48" x 24" type window to be loaded.

For Metric Units:

From the Type Selector drop-down list, select the 1400 x 600mm size for the M_Window-Casement-Double window.

17. The family is already loaded in the project but doesn't have this size. Select the second option to add the additional type.

You are trying to load the family Window-Casement-Double, which already exists in this project. What do you want to do?

→ Overwrite the existing version

→ Overwrite the existing version and its parameter values

18. Properties ×

Window-Casement-Double
48" x 24"

Verify that the correct window type is selected in the Type Selector on the Properties pane.

19. Place the window **6'-6"** [**3000 mm**] from the inner left wall.

Right click and select **Cancel** to exit the command.

20. The arrows appear on the exterior side of the window. If the window is not placed correctly, left click on the arrows to flip the orientation.

6' - 6"

21. Select the window so it highlights.

22. Select the **Array** tool under the Modify panel.

23. 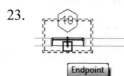 Select the midpoint of the window as the basepoint for the array.

Endpoint

24.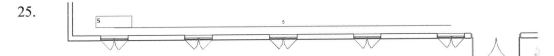

Enable Group and Associate. This assigns the windows placed to a group and allows you to edit the array.

Set the array quantity to **5** on the options bar located on the bottom of the screen. Enable **Last**.

Enable **Constrain.**

Enabling Constrain ensures that your elements are placed orthogonally.

Array has two options. One option allows you to place elements at a set distance apart. The second option allows you to fill a distance with equally spaced elements. We will fill a specified distance with five elements equally spaced.

25.

Pick a point **49'-0″ [14,935.20]** from the first selected point to the right.

26.

You will see a preview of how the windows will fill the space.

Click **ENTER** to accept.

Cancel out of the command.

27. Select the **Measure** tool on the Quick Access toolbar.

28. Check the distance between the windows and you will see that they are all spaced equally.

29.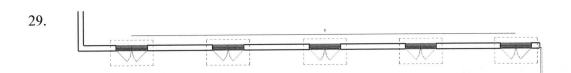

Window around the entire array to select all the windows.

The array count will display.

30. 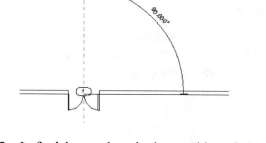 Use the **Mirror→Draw Mirror Axis** tool to mirror the windows to the other side of the wall opposite the door.

31. Select the center of the door as the start point of the mirror axis.

Move the cursor upwards at a 90 degree angle and pick a point above the door.

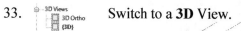

32. Left pick anywhere in the graphics window to complete the command.

You will get an error message, and the windows will not array properly if you do not have the angle set to 90 degrees or your walls are different lengths.

33. Switch to a **3D** View.

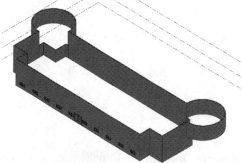

34. Set the Model Graphics Style to **Consistent Colors**.

We have created a conceptual model to show a client.

35. Save the file as *ex2-6.rvt*.

Exercise 2-7
Creating a Conceptual Mass

Drawing Name: New Conceptual Mass
Estimated Time: 60 minutes

This exercise reinforces the following skills:

- ❑ Masses
- ❑ Levels
- ❑ Load from Library
- ❑ Aligned Dimension
- ❑ Flip Orientation

1. Close any open files.

2. Use the Application Menu and go to **New→Conceptual Mass**.

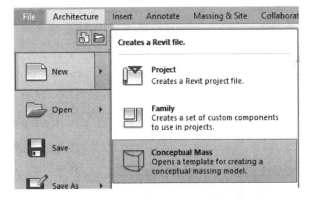

3.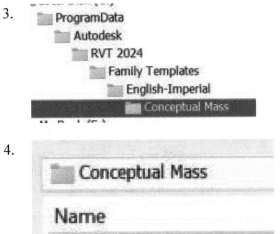

Browse to the *Conceptual Mass* folder under *Family Templates\English-Imperial.*

4. Select the **Mass** template.

   Click **Open**.

5. 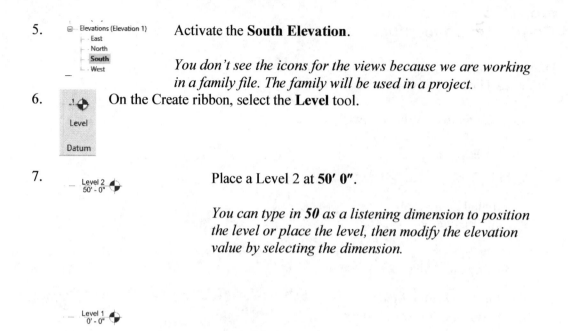   Activate the **South Elevation**.

You don't see the icons for the views because we are working in a family file. The family will be used in a project.

6.   On the Create ribbon, select the **Level** tool.

7.   Place a Level 2 at **50′ 0″**.

*You can type in **50** as a listening dimension to position the level or place the level, then modify the elevation value by selecting the dimension.*

8.   Activate the **Level 1** floor plan.

9. 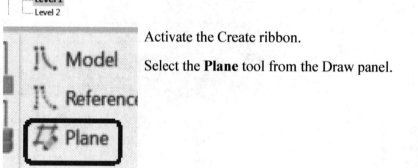   Activate the Create ribbon.

Select the **Plane** tool from the Draw panel.

10. Draw a vertical plane and a horizontal plane to form a box.

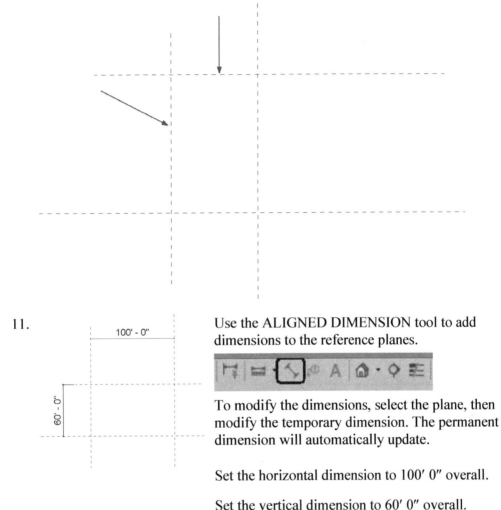

11. Use the ALIGNED DIMENSION tool to add dimensions to the reference planes.

To modify the dimensions, select the plane, then modify the temporary dimension. The permanent dimension will automatically update.

Set the horizontal dimension to 100′ 0″ overall.

Set the vertical dimension to 60′ 0″ overall.

Remember the permanent dimensions are driven by the values of the temporary dimensions. To set the temporary dimension values, select the reference planes. The planes you placed are unlocked – so their dimensions can be changed.

12. Activate the Insert ribbon.

Select Load Family from **Library→Load Family**.

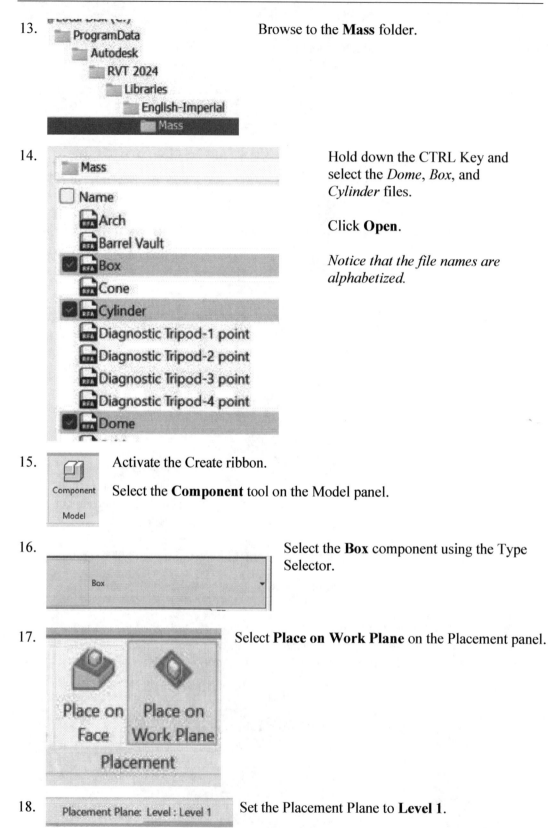

13. Browse to the **Mass** folder.

14. Hold down the CTRL Key and select the *Dome*, *Box*, and *Cylinder* files.

Click **Open**.

Notice that the file names are alphabetized.

15. Activate the Create ribbon.

Select the **Component** tool on the Model panel.

16. Select the **Box** component using the Type Selector.

17. Select **Place on Work Plane** on the Placement panel.

18. Set the Placement Plane to **Level 1**.

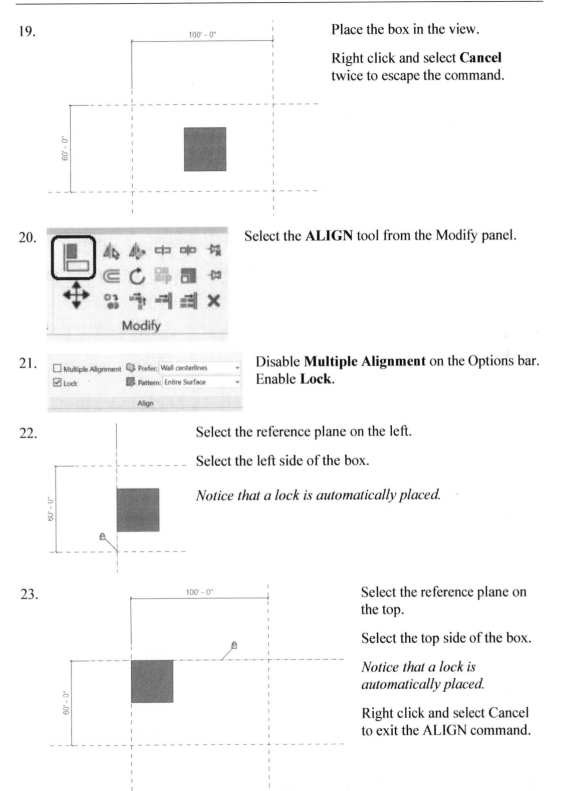

19.   Place the box in the view.

Right click and select **Cancel** twice to escape the command.

20.   Select the **ALIGN** tool from the Modify panel.

21.   Disable **Multiple Alignment** on the Options bar. Enable **Lock**.

22.   Select the reference plane on the left.

Select the left side of the box.

Notice that a lock is automatically placed.

23.   Select the reference plane on the top.

Select the top side of the box.

Notice that a lock is automatically placed.

Right click and select Cancel to exit the ALIGN command.

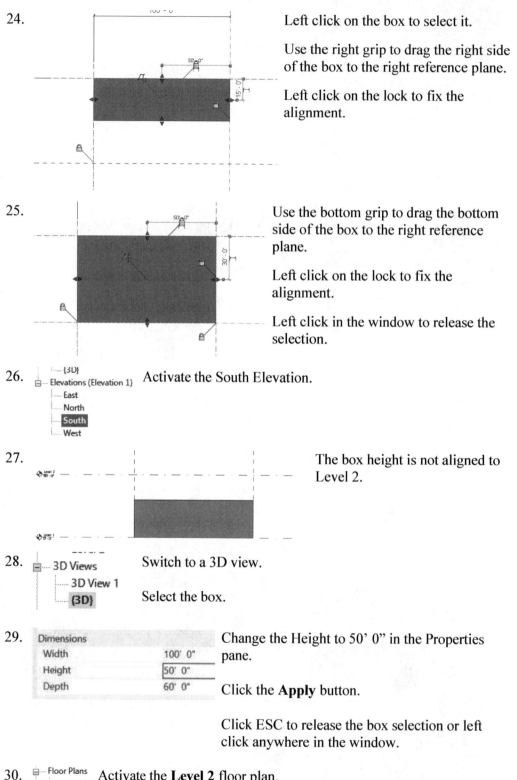

24.   Left click on the box to select it.

Use the right grip to drag the right side of the box to the right reference plane.

Left click on the lock to fix the alignment.

25.   Use the bottom grip to drag the bottom side of the box to the right reference plane.

Left click on the lock to fix the alignment.

Left click in the window to release the selection.

26.   Activate the South Elevation.

27.   The box height is not aligned to Level 2.

28.   Switch to a 3D view.

Select the box.

29.   Change the Height to 50' 0" in the Properties pane.

Dimensions	
Width	100' 0"
Height	50' 0"
Depth	60' 0"

Click the **Apply** button.

Click ESC to release the box selection or left click anywhere in the window.

30.   Activate the **Level 2** floor plan.

31. With Level 2 highlighted, scroll down the Properties pane to the Underlay category.

Set the Range: Base Level to **Level 1**.

Underlay	
Range: Base Level	Level 1
Range: Top Level	Level 2
Underlay Orientation	Look down

If you don't set the underlay correctly, you won't see the box that was placed on Level 1.

32. Activate the Create ribbon.

Select the **Component** tool on the Model panel.

33. Select the **Cylinder** component using the Type Selector.

34. In the Properties Pane:

Dimensions	
Radius	25' 0"
Height	5' 0"

Set the Radius to **25' 0"**.
Set the Height to **5' 0"**.

If you need to adjust the position of the cylinder: Set the Offset to -5' 0".

This aligns the top of the cylinder with the top of the box.

35. Select **Place on Work Plane** on the Placement panel.

36. Set the Placement Plane to **Level 2**.

Placement Plane: Level : Level 2

37.

Place a cylinder at the midpoint of the bottom edge of the box.

Right click and select **Cancel** twice to escape the command.

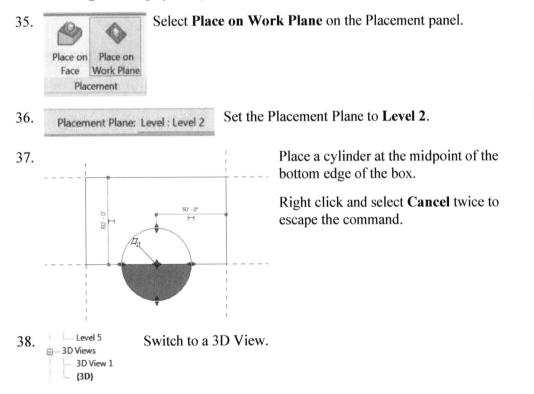

38. Switch to a 3D View.

Level 5
3D Views
  3D View 1
  {3D}

39. 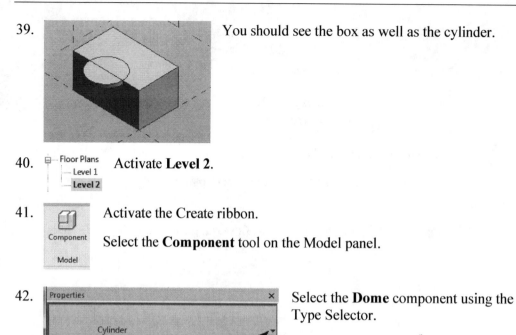 You should see the box as well as the cylinder.

40. Floor Plans   Activate **Level 2**.
    Level 1
    Level 2

41. Activate the Create ribbon.

    Component   Select the **Component** tool on the Model panel.

    Model

42.

    Properties   ×   Select the **Dome** component using the
                                                Type Selector.
                    Cylinder

    Box
        Box
    Cylinder
        Cylinder
    Dome
        Dome

43. In the Properties Pane:

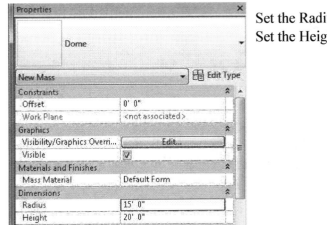

Set the Radius to **15′ 0″**.
Set the Height to **20′ 0″**.

44. Select **Place on Work Plane** on the Placement panel.

45. Set the Placement Plane to **Level 2**.

46. Place the dome so it is centered on the box.

Right click and select Cancel twice to exit the command.

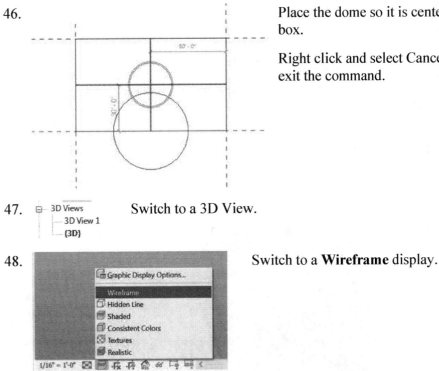

47. Switch to a 3D View.

3D Views
— 3D View 1
— {3D}

48. Switch to a **Wireframe** display.

49.

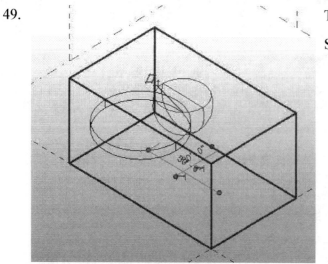

The dome is upside down.

Select the dome.

50.

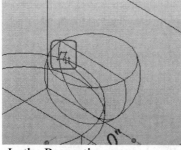

Left click on the orientation arrows to flip the dome.

51. In the Properties pane:

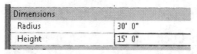

Change the Radius to **30′ 0″**.
Change the Height to **15′ 0″**.
Click **Apply**.

52.

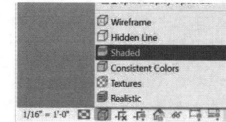

Change the display back to **Shaded**.

53. Save as *ex2-7.rfa*.

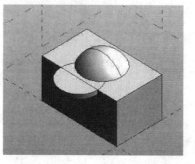

Using a Conceptual Mass in a Project

Drawing Name: New
Estimated Time: 30 minutes

This exercise reinforces the following skills:

- Masses
- Load from Library
- Edit Family
- Update to Face
- Visibility/Graphics

1. Close any open files.

   Use the Application Menu and go to
   New→Project.

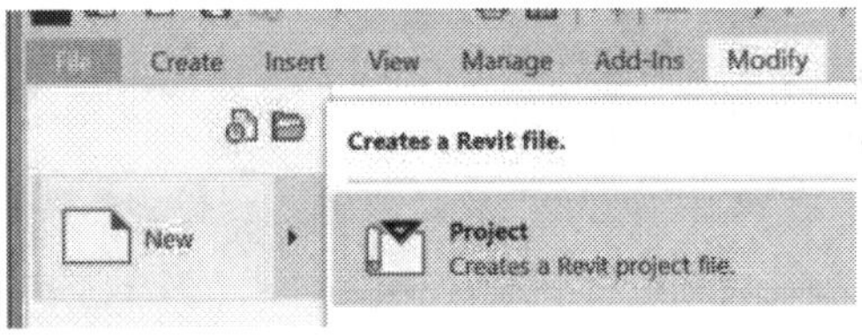

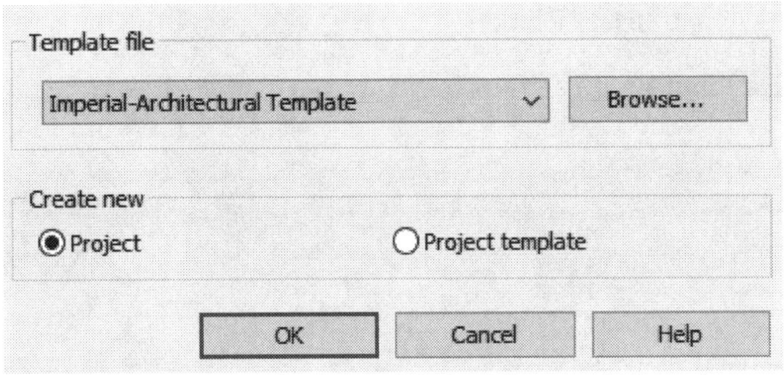

   Click **OK** to accept the Imperial-Architectural Template.

2. Type **VV** to launch the Visibility/Graphics dialog.

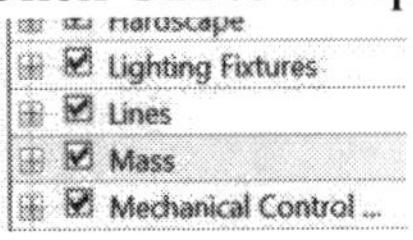

   *If this doesn't work, remember you can assign the shortcut
   using Keyboard Shortcuts under User Interface on the View
   ribbon.*

   Enable **Mass** visibility on the Model Categories tab.

   Click **OK**.

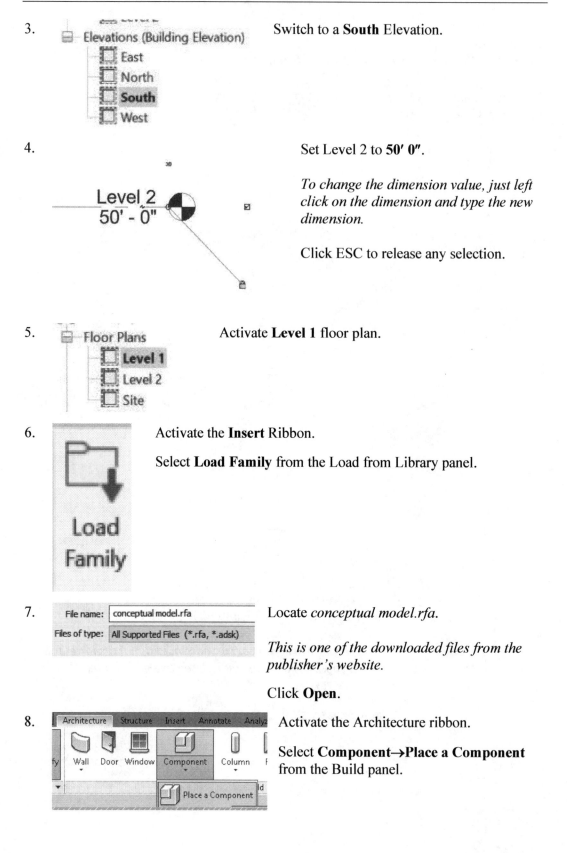

3. Switch to a **South** Elevation.

4. Set Level 2 to **50' 0"**.

   To change the dimension value, just left click on the dimension and type the new dimension.

   Click ESC to release any selection.

5. Activate **Level 1** floor plan.

6. Activate the **Insert** Ribbon.

   Select **Load Family** from the Load from Library panel.

7. Locate *conceptual model.rfa*.

   This is one of the downloaded files from the publisher's website.

   Click **Open**.

8. Activate the Architecture ribbon.

   Select **Component→Place a Component** from the Build panel.

9. 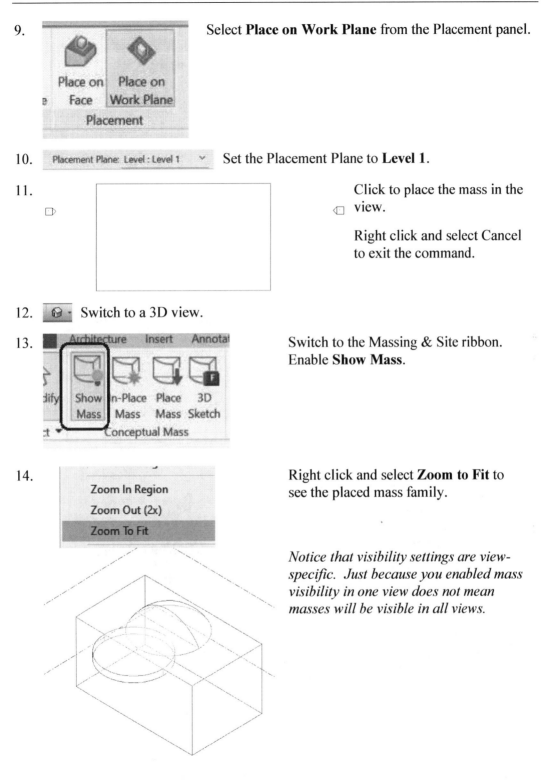 Select **Place on Work Plane** from the Placement panel.

10. Placement Plane: Level : Level 1 ⌄ Set the Placement Plane to **Level 1**.

11. Click to place the mass in the view.

Right click and select Cancel to exit the command.

12. Switch to a 3D view.

13. Switch to the Massing & Site ribbon. Enable **Show Mass**.

14. Right click and select **Zoom to Fit** to see the placed mass family.

Notice that visibility settings are view-specific. Just because you enabled mass visibility in one view does not mean masses will be visible in all views.

15.

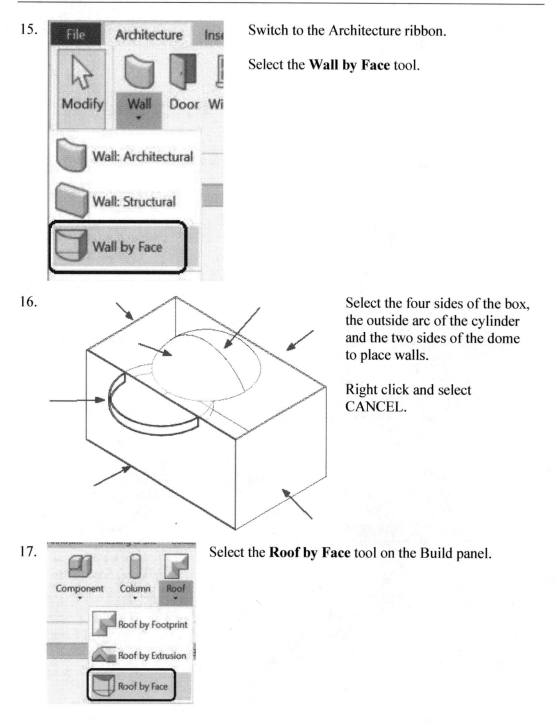

Switch to the Architecture ribbon.

Select the **Wall by Face** tool.

16.

Select the four sides of the box, the outside arc of the cylinder and the two sides of the dome to place walls.

Right click and select CANCEL.

17.

Select the **Roof by Face** tool on the Build panel.

18.

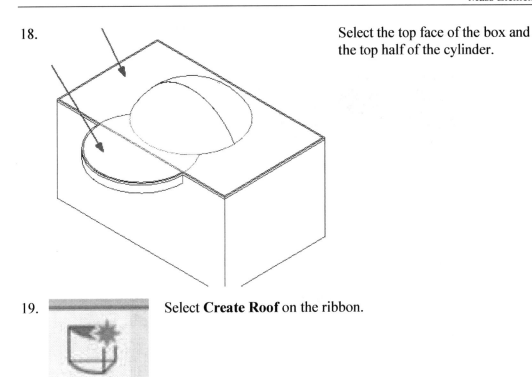

Select the top face of the box and the top half of the cylinder.

19.

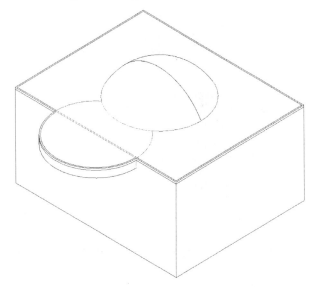

Select **Create Roof** on the ribbon.

20. The roof is placed. Right click and select **Cancel**.

21.

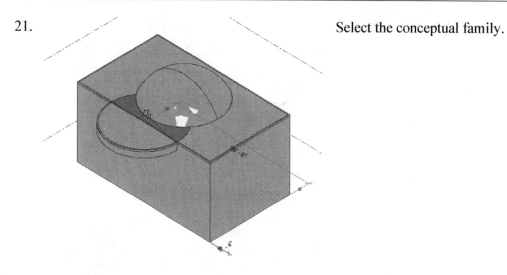

Select the conceptual family.

22.

Select **Edit Family** on the ribbon.

You will be modifying the external file.

23.

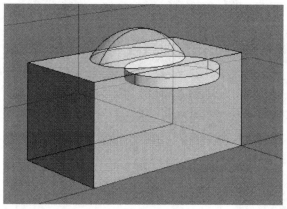

A window opens with the conceptual model.

Select the cylinder.

24.

Mass Material	Default Form	
Dimensions		
Radius	50' 0"	
Height	5' 0"	

Change the radius of the cylinder to **50'-0"**.

Left click in the window to release the selection.

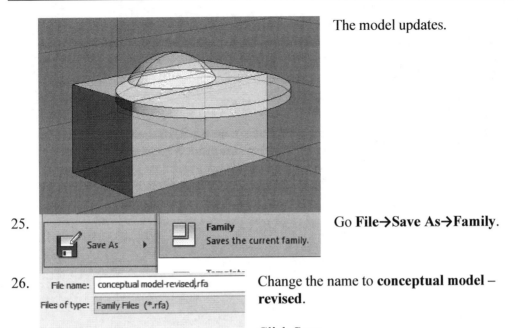

The model updates.

25.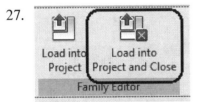

Go **File→Save As→Family**.

26. File name: conceptual model-revised.rfa

Files of type: Family Files (*.rfa)

Change the name to **conceptual model – revised**.

Click **Save.**
This ensures that the original file is not modified.

27.

Load into Project | Load into Project and Close

Family Editor

Select **Load into Project and Close**.

This loads the revised family into the existing project.

The family file is closed.

28. Click ESC to exit the insert command.

Revit automatically assumed you want to place the revised family.

29.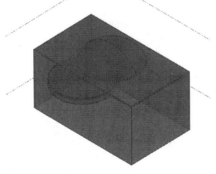

3D Views
    {3D}

Switch to a **3D** view.

You see the original conceptual model has not changed.

Select the conceptual model by windowing around it.

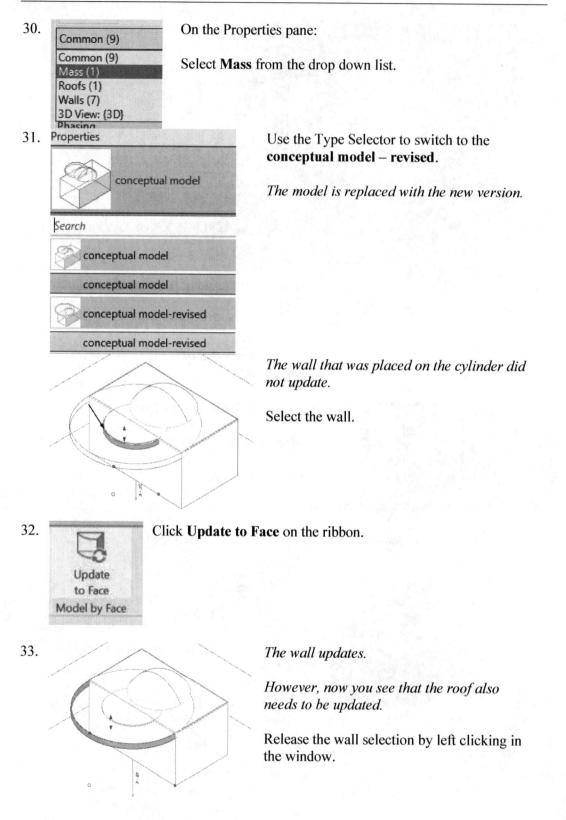

30. On the Properties pane:

Select **Mass** from the drop down list.

31. Use the Type Selector to switch to the **conceptual model – revised**.

The model is replaced with the new version.

The wall that was placed on the cylinder did not update.

Select the wall.

32. Click **Update to Face** on the ribbon.

33. *The wall updates.*

However, now you see that the roof also needs to be updated.

Release the wall selection by left clicking in the window.

34.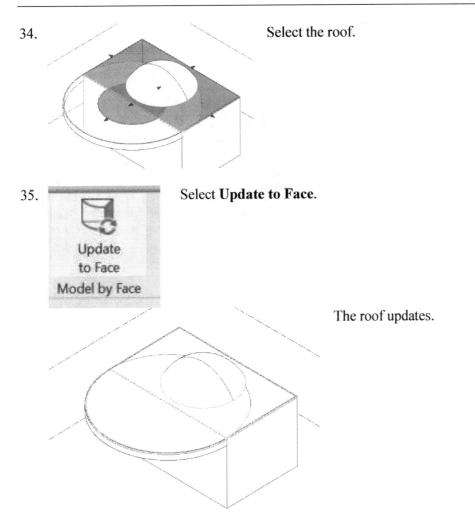

Select the roof.

35.

Update to Face
Model by Face

Select **Update to Face**.

The roof updates.

36.

Visibility
☑ Lines
☐ Mass
☑ Mechanical Equipm...
☑ Parking

Type **VV** to launch the Visibility/Graphics dialog.

Disable **Mass** visibility on the Model Categories tab.

Click **OK**.

You can also disable the Show Mass tool on the Massing & Site tab.

Architecture   Insert   Annotate   Massing & Site   Collabo

Show Mass | In-Place Mass | Place Mass | 3D Sketch Formit | Curtain System | Roof | Wall | Floor

Conceptual Mass   Model by Face

37. 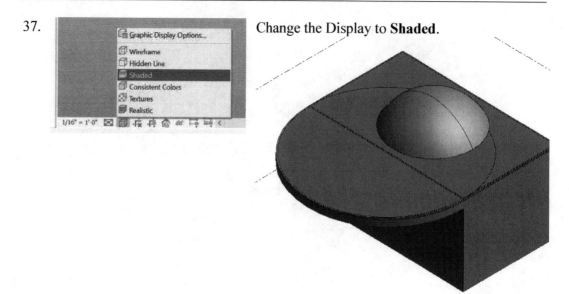 Change the Display to **Shaded**.

38. Save as *ex2-8.rvt*.

Additional Projects

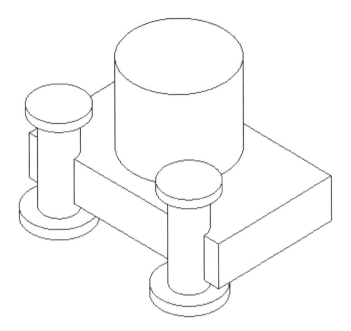

1) Create a conceptual mass family like the one shown.

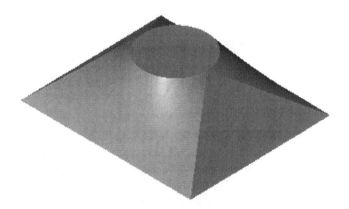

2) Make the shape shown using a Blend. The base is a rectangle and the top is a circle located at the center of the rectangle.

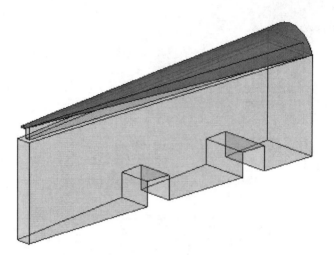

3) Make a conceptual mass family as shown.
   Use Solid Void to create the two openings.

4) Design a pergola using 4″ x 4″ posts and beams.

Lesson 2 Quiz

True or False

1. Masses can be created inside Projects or as a Conceptual Mass file.
2. Forms are always created by drawing a sketch, selecting the sketch, and clicking ⛃ Create Form.
3. In order to see masses, Show Mass must be enabled.
4. Masses are level-based.
5. You can modify the dimensions of conceptual masses.

Multiple Choice [Select the Best Answer]

6. Faces on masses can be converted to the following:

   A. Walls
   B. Roofs
   C. Floors
   D. Doors
   E. A, B, and C, but NOT D

7. You can adjust the distance a mass is extruded by:

   A. Editing the temporary dimension that appears before a solid form is created.
   B. Using the ALIGN tool.
   C. Using the 3D drag tool.
   D. Using the Properties pane located on the left of the screen.
   E. All of the above

8. To insert a conceptual mass family in a project:

   A. Use the INSERT tool.
   B. Use the PLACE COMPONENT tool.
   C. Use the BLOCK tool.
   D. Use the MASS tool.

9. Masses are hosted by:

   A. Projects
   B. Work Planes
   C. Families
   D. Files

10. Mass Visibility can be controlled using:

    A. The SHOW MASS tool
    B. Display Settings
    C. Object Settings
    D. View Properties

11. Each mass is defined by _____.

    A. A single profile sketch
    B. Multiple profile sketches
    C. Properties
    D. Materials

12. Revit comes with many pre-made mass shapes. Select the mass shape NOT available in the Revit mass library:

    A. BOX
    B. ARC
    C. CONE
    D. TRIANGLE-RIGHT

ANSWERS:

    1) T; 2) T; 3) T; 4) T; 5) T; 6) E; 7) E; 8) B; 9) B; 10) A; 11) A; 12) B

Lesson 3
Floor Plans

- ➤ Put a semi-colon between snap increments, not a comma.
- ➤ If you edit the first grid number to 'A' before you array, Revit will automatically increment them alphabetically for you: A, B, C, etc.
- ➤ You will need to 'Ungroup' an element in an array before you can modify any of the properties.
- ➤ To keep elements from accidentally moving, you can select the element and use **Pin Objects** or highlight the dimension string and lock the dimensions.
- ➤ You can purge the unused families and components in your project file in order to reduce the file space. Go to **File→Purge Unused**.
- ➤ Revit creates stairs from the center of the run, so it may be helpful to place some reference planes or detail lines defining the location of the starting points for the runs of any stairs.
- ➤ Floor plans should be oriented so that North is pointing up or to the right.
- ➤ The direction you draw your walls (clockwise or counterclockwise) controls the initial location of the exterior face of the wall. Drawing a wall from left to right places the exterior on the top. Drawing a wall from right to left places the exterior on the bottom. When you highlight a wall, the blue flip orientation arrows are always adjacent to the exterior side of the wall.

Throughout the rest of the text, we will be creating new types of families.
Here are the basic steps to creating a new family.

1. Select the element you want to define (wall, window, floor, stairs, etc.).
2. Select **Edit Type** from the Properties pane.
3. Select **Duplicate**.
4. Rename: Enter a new name for your family type.
5. Redefine: Edit the structure, assign new materials, and change the dimensions.
6. Reload or Reassign: Assign the new type to the element.

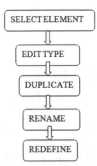

What is a Floor Plan?

A floor plan is a scaled diagram of a room or a building viewed from above. The floor plan may show an entire building, a single building level, or a single room. It may include dimensions, furniture, casework, appliances, electrical outlets, plumbing fixtures or anything else necessary to provide information. A floor plan is not a top view or a bird's eye view. It is drawn to scale. Most plan views represent a cut section of room, 4' above the floor.

Building projects typically start with a site plan. A site plan designates a building site or lot and involves grading the earth and placing the building foundation or pad. Some building projects may assume a pre-existing foundation or pad.

For most building projects, it is normal to start the project by laying out a grid. The grid lines allow references to be made in terms of position/location of various elements in the project. A surveyor will set out the grid lines on the site to help the construction crew determine the placement of structural columns, walls, doors, windows, etc.

Grids are system families. Grid lines are finite planes. Grids can be straight lines, arcs, or multi-segmented. In the User exam, you can expect at least one question about grids. It will probably be True/False.

Revit automatically numbers each grid. To change the grid number, click the number, enter the new value, and Click ENTER. You can use letters for grid line values. If you change the first grid number to a letter, all subsequent grid lines update appropriately. Each grid ID must be unique. If you have already assigned an ID, it cannot be used on another grid.

As you draw grid lines, the heads and tails of the lines can align to one another. If grid lines are aligned and you select a line, a lock appears to indicate the alignment. If you move the grid extents, all aligned grid lines move with it.

Grids are Annotation elements. But, unlike most annotation elements, they DO appear across different views. For example, you can draw a grid on your ground floor plan, and it would then appear on the subsequent floors (levels) of your model. You can control the display of grids on different levels using a scope box. Grids are datum elements.

A grid line consists of two main parts. The grid line itself and the Grid Header (i.e. the bubble at the end of the grid line). The default setting is for the grid line to have a grid header at one end only.

Exercise 3-1
Placing a Grid

Drawing Name: default.rte [DefaultMetric.rte]
Estimated Time: 20 minutes

This exercise reinforces the following skills:

- ❑ Units
- ❑ Snap Increments
- ❑ Grid
- ❑ Array
- ❑ Ungroup
- ❑ Dimension
- ❑ Dimension Settings

1.   Start a new project using the default template.

If you did not add the default template to the drop-down list, you can start a new project from the Applications Menu and then select default.rte from the Templates folder.

ProgramData
Autodesk
RVT 2024
Templates
English

File name: default.rte

Files of type: Template Files (*.rte)

File name: DefaultMetric.rte

Files of type: Template Files (*.rte)

2.   Views (all)
    Floor Plans
    Level 1
    Level 2
    Site
    Activate **Level 1**.

3.   View | Manage | Modify   Activate the **Manage** ribbon.

Under **Settings**, select **Snaps**.

Snaps

4.   ☑ Length dimension snap increments
    40;25;5
    Set the snap increments for length to **40′; 25′; 5′ [12200;7620;1520]**.

Use a semi-colon between the distance values.

◼ Length dimension snap increments
12200;7620;1520

Click **OK**.

5.   Architecture   Activate the **Architecture** ribbon.

6. Select **Grid** under the Datum panel.

7. [Click to enter grid start point] Pick on the left side of the graphics window to start the grid line.

8. Pick a point above the first point.

   Right mouse click and select Cancel twice to exit the Grid mode.

   Your first grid line is placed.

9. Select the Grid text to edit it.

10. Change the text to **A**.

    Left click in the graphics window to exit edit mode.

    If you edit the first grid letter before you array, Revit will automatically increment the arrayed grid elements.

11. Select the grid line so it is highlighted.

    Select the **Array** tool under the Modify panel.

12. Disable **Group and Associate**. This keeps each grid as an independent element.
    Set the Number to **8**.
    Enable Move To: **2nd**.
    Enable **Constrain**. This keeps the array constrained orthogonally.

13.

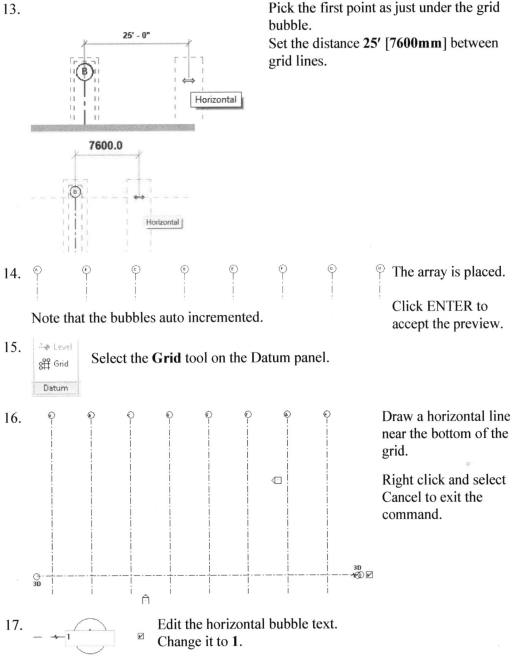

Pick the first point as just under the grid bubble.

Set the distance **25′ [7600mm]** between grid lines.

14.

Note that the bubbles auto incremented.

The array is placed.

Click ENTER to accept the preview.

15. Select the **Grid** tool on the Datum panel.

16.

Draw a horizontal line near the bottom of the grid.

Right click and select Cancel to exit the command.

17. Edit the horizontal bubble text.
Change it to **1**.

It might look like a 1, but it is really the letter I – the vertical grid lines stopped at H.

18. The square boxes that appear at the end of the grid lines control the appearance of the grid bubbles. Place a check in both ends of the horizontal grid line.

19.
Select the horizontal grid line so it is highlighted.

Select the **Array** tool under the Modify panel.

20.

Disable **Group and Associate**.
Set the Number to **3**.
Enable Move To: **2nd**.
Enable **Constrain**.

21.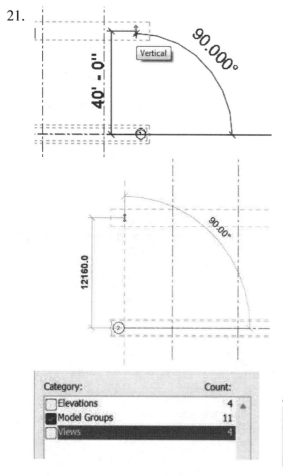

Pick the first point as just under the grid bubble.

Set a distance of **40' [12160]** between grid lines.

Window around the gridlines so they are all selected.

Filter

Use the FILTER tool, if necessary, to only select gridlines.

If you see Model Groups listed instead of Grids it means you forgot to disable the Group and Associate option when you created the array. Simply select the grid lines and select Ungroup from the ribbon and the lines will no longer be grouped.

22. 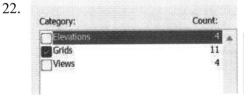 Window around the gridlines so they are all selected.

Filter

Use the FILTER tool, if necessary, to only select gridlines.

Click **OK**.

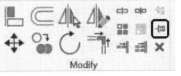 Select the PIN tool on the Modify panel. This will fix the gridlines in place so they can't accidentally be moved or deleted.

Modify

Tips Tricks

23. Save the file as *ex3-1.rvt[m_ex3-1].*

If you delete a grid line and then place another grid line, the numbering picks up where you left off. For example, if you delete grid line 3 and then add a new grid line, the new grid line will be labeled 4.

Exercise 3-2
Placing Walls

Drawing Name: walls1.rvt [m_walls1.rvt]
Estimated Time: 20 minutes

This exercise reinforces the following skills:

- ❑ Walls
- ❑ Mirror
- ❑ Filter
- ❑ Move

1. 📂 Open *walls1.rvt. [m_walls1.rvt]*

2. [Architecture] Activate the **Architecture** ribbon.

3. Type **VG** to launch the Visibility/Graphics dialog.

Select the **Annotations Categories** tab.

'4.

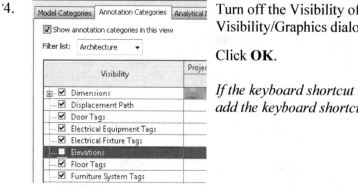

Turn off the Visibility of Elevations in the Visibility/Graphics dialog by unchecking the box.

Click **OK**.

If the keyboard shortcut key doesn't work, you can add the keyboard shortcut.

This turns off the visibility of the elevation tags in the display window. This does not DELETE the elevation tags or elevations.

Many users find the tags distracting.

5.

Select the **Wall** tool from the Build panel on the Architecture ribbon.

6.

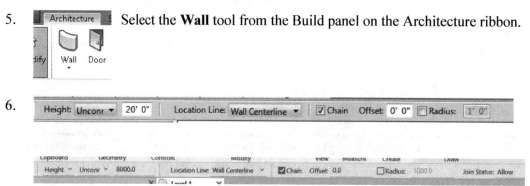

Enable **Chain**.

This eliminates the need to pick a start and end point.

7. Place walls as shown.

To enable the grid bubbles on the bottom, select the grid and place a check in the small square.

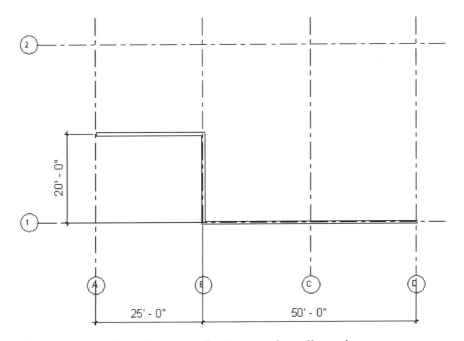

Dimensions are for reference only. Do not place dimensions.

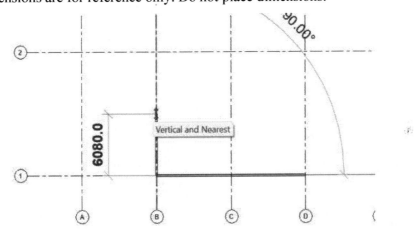

This layout shows units in millimeters.

8.   Window about the walls you just placed.

9.   Select the **Filter** tool.

10.   *You should just see Walls listed.*

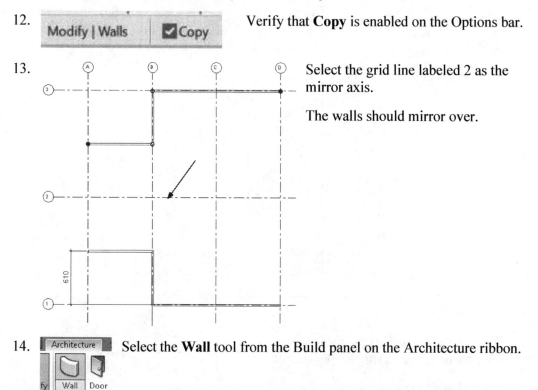

If you see other elements listed, uncheck them and this will select only the walls.

Click **OK**.

Many of my students use a crossing instead of a window or have difficulty selecting just the walls. Use the Filter tool to control your selections and potentially save a lot of time.

11. Select the **Mirror→Pick Axis** tool under the Modify Panel.

The Modify Walls ribbon will only be available if walls are selected.

12. Verify that **Copy** is enabled on the Options bar.

13. Select the grid line labeled 2 as the mirror axis.

The walls should mirror over.

14. Select the **Wall** tool from the Build panel on the Architecture ribbon.

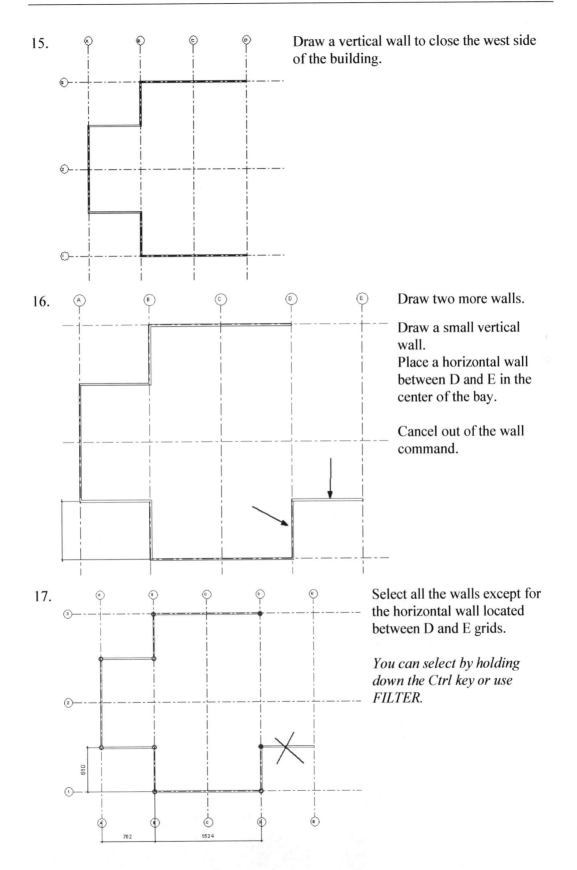

15. Draw a vertical wall to close the west side of the building.

16. Draw two more walls.

Draw a small vertical wall.
Place a horizontal wall between D and E in the center of the bay.

Cancel out of the wall command.

17. Select all the walls except for the horizontal wall located between D and E grids.

You can select by holding down the Ctrl key or use FILTER.

18. 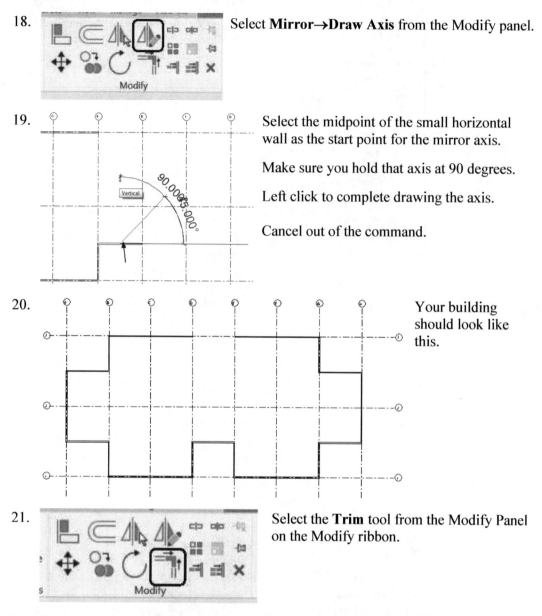 Select **Mirror→Draw Axis** from the Modify panel.

19. Select the midpoint of the small horizontal wall as the start point for the mirror axis.

Make sure you hold that axis at 90 degrees.

Left click to complete drawing the axis.

Cancel out of the command.

20. Your building should look like this.

21. Select the **Trim** tool from the Modify Panel on the Modify ribbon.

22. Select the two upper horizontal walls. The trim tool will join them together. Cancel out of the command.

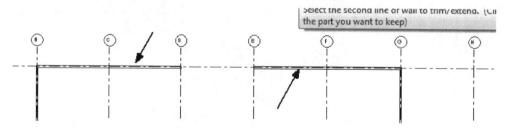

23. Activate the **Architecture** ribbon.

Select the **Grid** tool from the Datum panel on the Architecture ribbon.

24. Select the **Pick** mode.

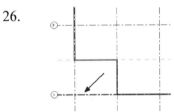

25. Set the Offset to **20'** on the Options bar.

For metric units: set the Offset to **6080** on the Options bar.

Offset: 6080.0

26. Select Grid 1.

You should see a preview of the new grid line.
Add a grid line along the center line of the walls between Grids 1 and 2.

Cancel out of the command.

27.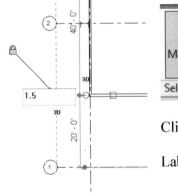

Click the Modify button.

Label the grid **1.5**.

28.

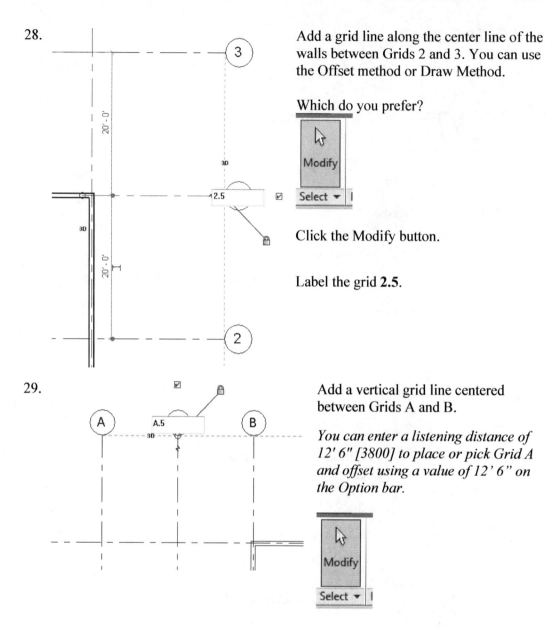

Add a grid line along the center line of the walls between Grids 2 and 3. You can use the Offset method or Draw Method.

Which do you prefer?

Click the Modify button.

Label the grid **2.5**.

29.

Add a vertical grid line centered between Grids A and B.

You can enter a listening distance of 12' 6" [3800] to place or pick Grid A and offset using a value of 12' 6" on the Option bar.

Click the Modify button.

Label the grid **A.5**.

30.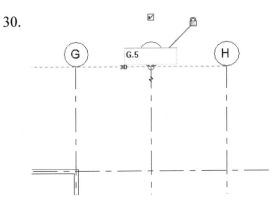

Add a vertical grid line centered between Grids G and H.

You can enter a listening distance of 12' 6" [3800] to place.

Click the Modify button.

Label the grid **G.5**.

31.

Activate the Modify Ribbon.
Select the **Align** tool on the Modify Panel.

32.

Enable **Multiple Alignment**.
Enable **Lock**.

Set **Wall centerlines** in the drop-down list.

33.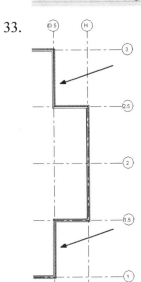

Shift the walls located on the G grid to the G.5 grid using ALIGN.

Select the G.5 grid first, then select the center line of the walls.

Right click and select CANCEL.

Hint: *You can click the TAB key to switch from the wall faces to the wall centerline.*

34.

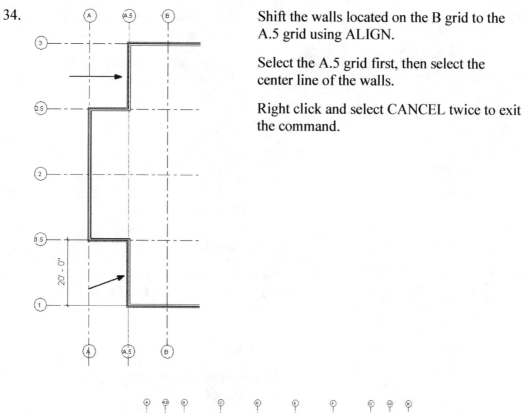

Shift the walls located on the B grid to the A.5 grid using ALIGN.

Select the A.5 grid first, then select the center line of the walls.

Right click and select CANCEL twice to exit the command.

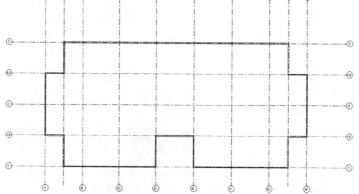

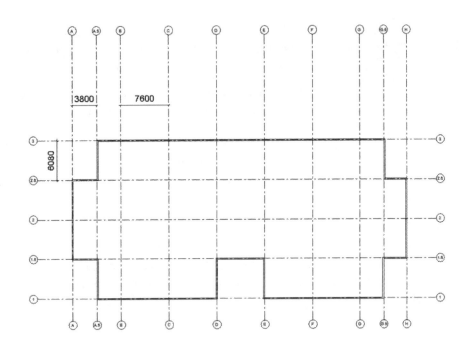

35.   Save as *ex3-2.rvt.[m_ex3-2.rvt]*

When drawing walls, it is important to understand the ***location line***. This is the layer from which the wall is drawn and controlled. The location line controls the position of the wall for flipping. It also sets the wall's position when wall thickness is modified. If you change the wall type, say from brick to wood stud, the stud location will be maintained. You can always select the walls and use 'Element Properties' to change the location line. The new user should experiment with this function to fully understand it.

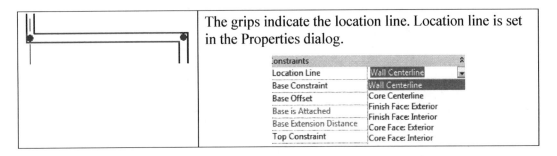

	The grips indicate the location line. Location line is set in the Properties dialog.

	Walls have an exterior side and an interior side. When a wall is selected the orientation/flip arrows will appear on the exterior side of the wall. When you draw walls clockwise, the exterior sides of the wall will be oriented towards the outside. When you draw walls counterclockwise, the exterior sides of the wall will be oriented towards the inside.

Starting from an AutoCAD file

On some projects, you will receive an AutoCAD file, either from the architect or a sub-contractor. In the next exercise, you learn how to bring in an AutoCAD file and use it to create a Revit 3D model.

Exercise 3-3
Converting an AutoCAD Floor plan

Drawing Name: autocad_floorplan.dwg
Estimated Time: 30 minutes

This exercise reinforces the following skills:

❑   Import CAD
❑   Duplicate Wall Type
❑   Wall Properties
❑   Trim
❑   Orbit

Metric units and designations are indicated in brackets [ ].
There is a video for this lesson on my Moss Designs YouTube channel:
https://www.youtube.com/user/MossDesigns

1. 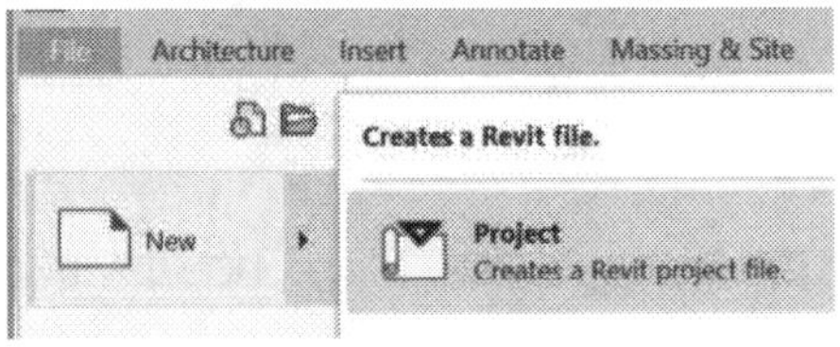   Go to **New→Project**.

2.    Click **OK** to use the default template.

3. 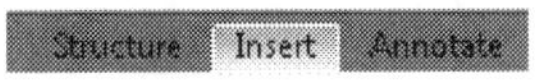   Activate the **Insert** ribbon.

4.    Select the **Import CAD** tool from the Import Panel.

5. 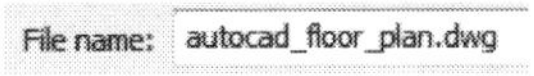   Locate the *autocad_floor_plan.dwg*.
     This file is included in the downloaded Class Files.

6.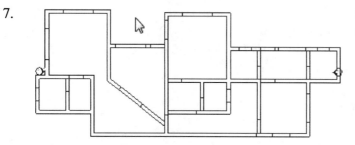

Set Colors to **Preserve**.
Set Layers to **All**.
Set Import Units to **Auto-Detect**.
Set Positioning to: **Auto-Center to Center**.
Click **Open**.

7.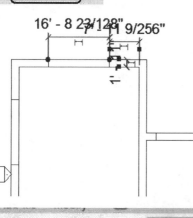

Select the imported CAD data so it highlights.

8.

Select the **Partial Explode** tool under the Import Instance panel.

9. 16' - 8 23/128" 9/256"
   7/128"

Select a line so it is highlighted.

⌐ Line Style:
0

Line Style

Note the Line Style listed in the Properties pane and on the ribbon.

10. Transfer Project Standards   Purge Unused   Project Units   Additional Settings

   Line Styles

Activate the **Manage** ribbon.

Under Settings→Additional Settings:

Select **Line Styles**.

11.

Category	Line Weight Projection	Line Color	Line Pat
⊟ Lines	1	■ RGB 000-166-000	Solid
0	1	■ Black	Solid
0wall	1	■ Black	Solid
0wallthick	4	■ Black	Solid
<Area Boundary>	6	■ RGB 128-000-255	Solid

Left click on the + symbol next to Lines to expand the list.
Locate the Line Styles named **0**, **0wall**, and **0wallthick**.

12.

	Projection		
⊟ Lines	1	■ RGB 000-166-000	Solid
0	1	■ Cyan	Solid
0wall	1	■ Cyan	Solid
0wallthick	4	■ Cyan	Solid

Change the color of **0**, **0wall**, and **0wallthick** line styles to **Cyan**.

You can use the CTL key to select all three line styles and change the color for all three at the same time.

Click on the color and then select **Cyan** from the dialog box.
Click **OK**.

13.

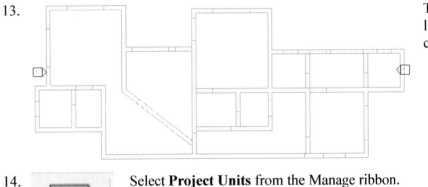

The imported lines will change color to Cyan.

14.

Project Units

Select **Project Units** from the Manage ribbon.

15.

Discipline: Common

Units	Format
Angle	12.35°
Area	1235 SF
Cost per Area	1235 [$/ft²]
Distance	1235 ft
Length	1' - 5 11/32"
Mass Density	1234.57 lb/ft³
Rotation Angle	12.35°
Slope	17 1/2" / 12"

Left click on the **Length** button.

16. Set the Rounding **To the nearest ¼"**.

Click **OK** twice to close the dialogs.

17. Select the **Measure** tool from the Quick Access toolbar.

18. Measure the wall thickness.

It should read 1'-11".

19. Activate the **Architecture** ribbon.

20. Select the **Wall** tool under the Build panel.

We need to create a wall style that is 1'-11" thick.

21. Select **Edit Type** from the Properties pane.

22. Select **Duplicate**.

23. Change the Name to **Generic – 1' 11"**.

24. Select **Edit** next to Structure.

25.

Change the thickness to **1' 11"**.

Click **OK**.

26. Click **OK** to exit the Properties dialog.

27. Select the **Pick Lines** tool under the Draw panel.

28.  On the Options bar, select **Finish Face: Exterior** for the Location Line.

This means you should select on the exterior side of the wall to place the wall correctly.

29. When you select the line, a preview will appear to indicate the wall orientation. To switch the wall orientation, click the SPACE bar or move the cursor to make a different selection. Select so the dashed line appears inside the wall.

30. Move around the model and select a section of each wall. Do not try to select the entire section of wall or you will get an error message about overlapping walls.

Warning
Highlighted walls overlap. One of them may be ignored when Revit finds room boundaries. Use Cut Geometry to embed one wall within the other.

Delete any overlapping walls.

31.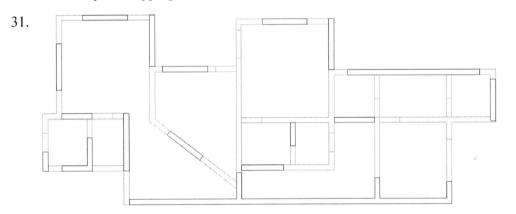

Your model should look similar to this. Note only partial walls have been placed on many of the sections.

32. Activate the View ribbon.

Select **3D View** from the Create panel.

33. The model shows the walls placed.

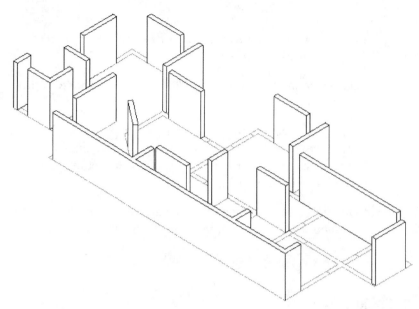

34. 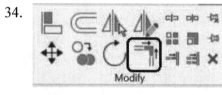 Use the Trim tool to extend and trim the walls to create the floor model.

35. 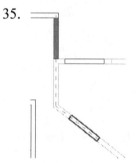 When you mouse over the second selection on the TRIM command, you will see a preview of how the walls will connect. If the preview does not look proper, shift your mouse slightly until the preview appears correct.

36.

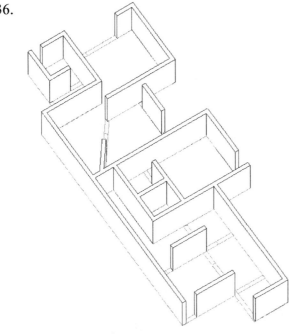

Not all walls can be connected using TRIM.

Some will require use of the EXTEND tool.

37.

Select the **Extend Single Element** tool from the Modify panel.

To use this tool, select the face of the wall you want to extend TO first, then the wall you want to extend.

38. Orbit the model so you can see where to trim and extend walls.

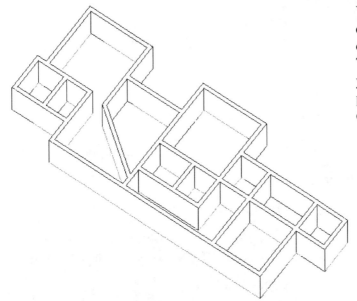

You will still see the colored lines from the original AutoCAD file. They are there to help you check that you placed the walls correctly.

39. Save as *ex3-3.rvt*.

Exercise 3-4
Wall Properties

Drawing Name: wall properties.rvt [m_wall properties.rvt]
Estimated Time: 40 minutes

This exercise reinforces the following skills:

- ❑ Walls
- ❑ Filter
- ❑ Wall Properties
- ❑ Flip Orientation
- ❑ Join

    [ ] brackets are used to indicate metric units or designations.

1.   Open *wall properties.rvt*.

2.   *Select the exterior walls using **Filter**.*
    Use crossing to select the floor plan.

3.   Select the **Filter** tool.

4.   Uncheck all the objects except for Walls.

    Click **OK**.
    This way, only walls are selected.

5.   Select **Edit Type** in the Properties panel.

6.   Select **Generic – 8″ [Generic – 200 mm]** under Type.

    Select **Duplicate**.

7.   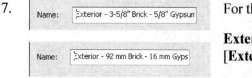   For the name, enter:

    Exterior – 3-5/8″ Brick – 5/8″ Gypsum
    [Exterior- 92mm Brick -16mm Gypsum].

    Click **OK**.

8.   Select **Edit** for Structure.

9. `<< Preview` Enable the Preview button to expand the dialog so you can see the preview of the wall structure.

10. `Insert` Select the **Insert** Button.

In the dialog, notice that the bottom of the list indicates toward the interior side and the top of the list indicates toward the exterior side of the wall.

11.

	Function	M:
1	Core Boundary	Layers /
2	Finish 1 [4] ▼	<By Cate
3	Structure [1]	fault V
4	Substrate [2]	yers
	Thermal/Air Layer [3]	
	Finish 1 [4]	
	Finish 2 [5]	
	Membrane Layer	

Select the second line in the Function column.

Select **Finish 1 [4]** from the drop-down list.

12.

Layers

	Function
1	Finish 1 [4]
2	Thermal/Air Layer [
3	Substrate [2]
4	**Core Boundary**
5	Structure [1]
6	**Core Boundary**
7	Finish 2 [5] ▼

Click the **Insert** button until you have seven layers total.

Arrange the layers as shown.

Assign the Functions as shown:
Layer 1: Finish 1 [4]   Layer 5: Structure [1]
Layer 2: Thermal/Air Layer   Layer 6: Core Boundary
Layer 3: Substrate [2]   Layer 7: Finish 2 [5]
Layer 4: Core Boundary

13. Select the **Material** column for Layer 1.

Layers   EXTERIOR SIDE

	Function	Material	
1	Finish 1 [4]	<By Category> ⋯	0
2	Thermal/Air Lay	<By Category>	0'
3	Substrate [2]	<By Category>	0'

14. Materials

`bri` ✕

In the Materials search box, type **bri**.

15.

Any materials with 'bri' as part of the name or description will be listed.

Highlight **Brick, Common** from the Name list.

16.

☑ Use Render Appearance

Color RGB 85 47 55

)arency

Note that we can set how we want this material to appear when we render and shade our model.

Click the Graphics tab.

Enable **Use Render Appearance**.
We only need to enable this for the outside materials. The inner materials are not visible during rendering.

17.

Identity | Graphics | **Appearance** | Physical | Thermal

Non-Uniform Running - Burgundy

▶ **Information**

▼ **Masonry**

Type Masonry

Image

Brick_Non_Uniform_Running_Burgundy.png

Finish Unfinished

▶ ☑ **Relief Pattern**

▶ ☐ **Tint**

Select the **Appearance** tab.

Note the image of how the brick will appear when the wall is rendered.

Click **OK**.

18.

| 1 | Finish 1 [4 | Brick, Common | | 0' 3 5/8" |

Change the Thickness to **3 5/8" [92mm]**.

	Function	Material	Thickness
1	Finish 1 [4]	Brick, Common	92.0

19.

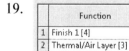

	Function
1	Finish 1 [4]
2	Thermal/Air Layer [3]

Locate Layer 2 and verify that it is set to Thermal/Air Layer. Select the **Material** column.

20.

On the Materials tab:

Type **air** in the search field.

Select **Air Infiltration Barrier**.

Click **OK**.

21.

	Function	Material	Thickness
1	Finish 1 [4]	Brick, Common	0' 3 5/8"
2	Thermal/Air Layer [3]	Air Infiltration Barrier	0' 1"

Set the Thickness of **Layer 2** to **1″ [25.4]**.

EXTERIOR SIDE

	Function	Material	Thickness
1	Finish 1 [4]	Brick, Common	92.0
2	Thermal/Air Layer [3]	Air Infiltration Barrier	25.4

22.

	Function	Material	Thickness
1	Finish 1 [4]	Brick, Common	0' 3 5/8"
2	Thermal/Air Layer [3]	Air Infiltration Barrier	0' 1"
3	Substrate [2]	Plywood, Sheathing	0' 0 1/2"

Set Layer 3 to Substrate [2], **Plywood, Sheathing**; Thickness to **½″ [17.2]**.

	Function	Material	Thickness
1	Finish 1 [4]	Brick, Common	92.0
2	Thermal/Air Layer [3]	Air Infiltration Barrier	25.4
3	Substrate [2]	Plywood, Sheathing	17.2

23.

Set the Material for Structure [1] to **Metal Stud Layer**.

Click **OK**.

24.

	Function	Material	Thickness
1	Finish 1 [4]	Brick, Common	0' 3 5/8"
2	Thermal/Air Layer [3]	Air Infiltration Barrier	0' 1"
3	Substrate [2]	Plywood, Sheathing	0' 0 1/2"
4	Core Boundary	Layers Above Wrap	0' 0"
5	Structure [1]	Metal Stud Layer	0' 6"

Set the Thickness for Layer 4 Structure [1] to **6″ [152.4]**.

The Core Boundary Layers may not be modified or deleted.

These layers control the location of the wrap when walls intersect.

	Function	Material	Thickness
1	Finish 1 [4]	Brick, Common	92.0
2	Thermal/Air Layer [3]	Air Infiltration Barrier	25.4
3	Substrate [2]	Plywood, Sheathing	17.2
4	Core Boundary	Layers Above Wrap	0.0
5	Structure [1]	Metal Stud Layer	152.4

25.

	Function
1	Finish 1 [4]
2	Thermal/Air Layer [3]
3	Substrate [2]
4	**Core Boundary**
5	Structure [1]
6	**Core Boundary**
7	Finish 2 [5]

Select the Material column for Layer 7: Finish 2 [5].

26.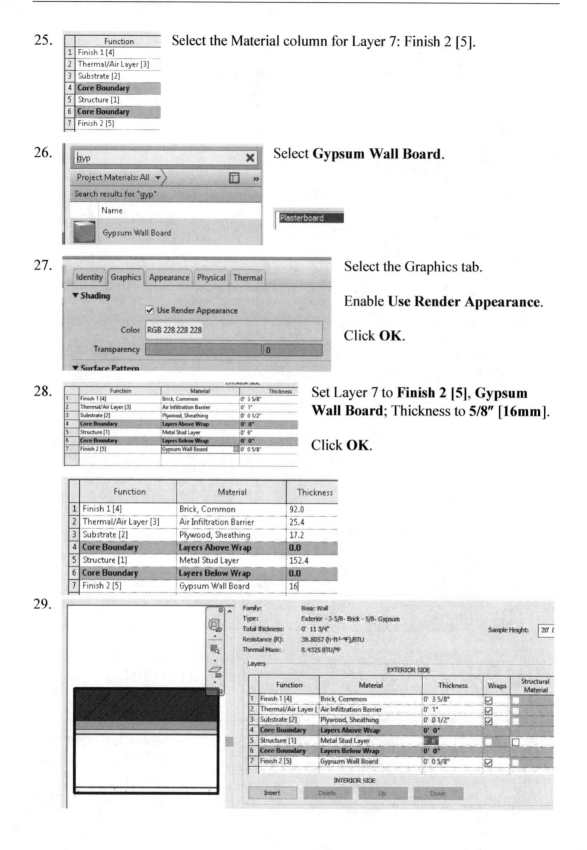

gyp ✕

Project Materials: All ▼ 〉

Search results for "gyp"

Name

Gypsum Wall Board

Plasterboard

Select **Gypsum Wall Board**.

27.

| Identity | Graphics | Appearance | Physical | Thermal |

▼ Shading

☑ Use Render Appearance

Color RGB 228 228 228

Transparency | 0

▼ Surface Pattern

Select the Graphics tab.

Enable **Use Render Appearance**.

Click **OK**.

28.

EXTERIOR SIDE

	Function	Material	Thickness
1	Finish 1 [4]	Brick, Common	0' 3 5/8"
2	Thermal/Air Layer [3]	Air Infiltration Barrier	0' 1"
3	Substrate [2]	Plywood, Sheathing	0' 0 1/2"
4	Core Boundary	Layers Above Wrap	0' 0"
5	Structure [1]	Metal Stud Layer	0' 6"
6	Core Boundary	Layers Below Wrap	0' 0"
7	Finish 2 [5]	Gypsum Wall Board	0' 0 5/8"

Set Layer 7 to **Finish 2 [5]**, **Gypsum Wall Board**; Thickness to **5/8" [16mm]**.

Click **OK**.

	Function	Material	Thickness
1	Finish 1 [4]	Brick, Common	92.0
2	Thermal/Air Layer [3]	Air Infiltration Barrier	25.4
3	Substrate [2]	Plywood, Sheathing	17.2
4	**Core Boundary**	**Layers Above Wrap**	**0.0**
5	Structure [1]	Metal Stud Layer	152.4
6	**Core Boundary**	**Layers Below Wrap**	**0.0**
7	Finish 2 [5]	Gypsum Wall Board	16

29.

Family: Basic Wall
Type: Exterior - 3-5/8- Brick - 5/8- Gypsum
Total thickness: 0' 11 3/4"   Sample Height: 20' 0
Resistance (R): 38.8057 (h·ft²·°F)/BTU
Thermal Mass: 8.4325 BTU/°F

Layers

EXTERIOR SIDE

	Function	Material	Thickness	Wraps	Structural Material
1	Finish 1 [4]	Brick, Common	0' 3 5/8"	☑	☐
2	Thermal/Air Layer [	Air Infiltration Barrier	0' 1"	☑	☐
3	Substrate [2]	Plywood, Sheathing	0' 0 1/2"	☑	☐
4	Core Boundary	Layers Above Wrap	0' 0"		
5	Structure [1]	Metal Stud Layer	0' 6	☐	☐
6	Core Boundary	Layers Below Wrap	0' 0"		
7	Finish 2 [5]	Gypsum Wall Board	0' 0 5/8"	☑	☐

INTERIOR SIDE

| Insert | Delete | Up | Down |

Layers					
		EXTERIOR SIDE			
	Function	Material	Thickness	Wraps	Structural Material
1	Finish 1 [4]	Brick, Common	92.0	☑	
2	Thermal/Air Layer [3]	Air Infiltration Barrier	25.4	☑	
3	Substrate [2]	Plywood, Sheathing	17.2	☑	
4	Core Boundary	Layers Above Wrap	0.0		
5	Structure [1]	Metal Stud Layer	152.4		☑
6	Core Boundary	Layers Below Wrap	0.0		
7	Finish 2 [5]	Gypsum Wall Board	16.0	☑	

INTERIOR SIDE

| Insert | Delete | Up | Down |

Set Layer 1 to Finish 1 [4], Brick, Common, Thickness: **3-5/8″ [92]**.
Set Layer 2 to Thermal/Air Layer [3], Air Infiltration Barrier, Thickness: **1″ [25.4]**.
Set Layer 3 to Substrate [2], Plywood, Sheathing, Thickness to **½″ [17.2]**.
Set Layer 4 to Core Boundary, Layers Above Wrap, Thickness to **0″**.
Set Layer 5 to Structure [1], Metal Stud Layer, Thickness to **6″ [152.4]**.
Set Layer 6 to Core Boundary, Layers Below Wrap, Thickness to **0″**.
Set Layer 7 to Finish 2 [5], Gypsum Wall Board, Thickness to **5/8″ [16]**.
Click **OK** to exit the Edit Assembly dialog.

30.

Graphics		☆
Coarse Scale Fill Pattern	Diagonal up	…
Coarse Scale Fill Color	■ Blue	
Materials and Finishes		☆

Set the Coarse Scale Fill Pattern to **Diagonal up**.

Set the Coarse Fill Color to **Blue**.

31. Click **OK** to exit the Properties dialog.
Because your walls were selected when you started defining the new wall style, your walls now appear with the new wall style.

32. If you zoom in, you will see that the coarse fill pattern is Diagonal Up and is Color Blue.

This only appears if the Detail Level is set to Coarse.

33.

☐ Coarse
◉ Medium
▩ Fine

-0″ ☐ ⬜ ⬚ ✕ ☒ ⬚ ⬚

Set the Detail Level to **Medium**.

You will not see the hatching or any change in line width unless the Detail Level is set to Medium or Fine.

34. We now see the wall details, but the brick side is toward the interior and the gypsum board is towards the exterior for some walls. In other words, the walls need to be reversed or flipped.

How do we know the walls need to be flipped?

35. When we pick the walls, the orientation arrows are located adjacent to the exterior side of the wall.

The orientation arrows are the blue arrows that are activated when an object is selected.

36. **Change wall's orientation**   Select the wall, right click and select **Change wall's orientation** or just click the blue arrows.

37. Go around the building and flip the orientation of the walls so that the brick is on the outside.

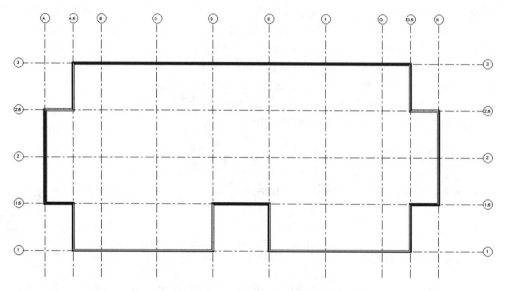

If you have problems selecting a wall instead of a grid line, use the TAB key to cycle through the selection.

38. Save the file as *ex3-4.rvt*.

Exercise 3-5
Add Level 1 Interior Walls

Drawing Name: walls2.rvt [m_walls2.rvt]
Estimated Time: 10 minutes

This exercise reinforces the following skills:
A video is available for this exercise on my Moss Designs YouTube channel:
https://www.youtube.com/user/MossDesigns

- ❑ Wall
- ❑ 3D View
- ❑ Visibility
- ❑ Wall Properties

1. Open *walls2.rvt.[m_walls2.rvt]*

2. 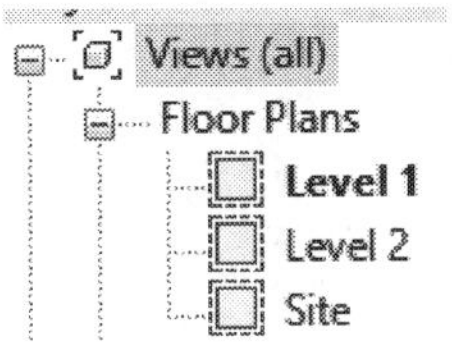 Activate **Level 1**.

3.  Select the **Wall** tool on the Build panel under the Architecture ribbon.

Wall

4. 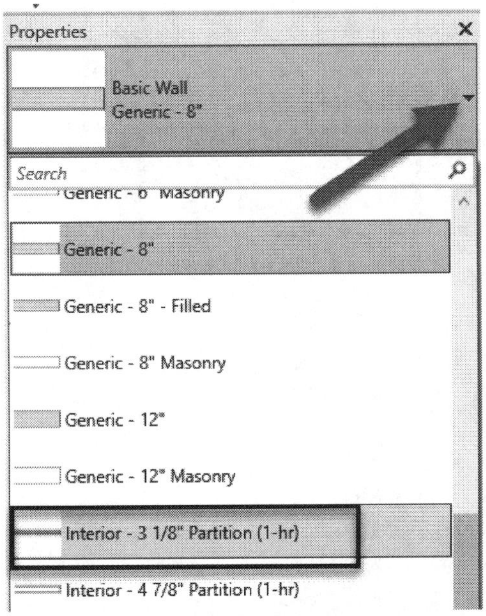 Scroll down the list of wall styles available under the Type Selector.

Locate the **Interior- 3 1/8" Partition (1-hr)** style.

For metric, look for the **Interior-138mm Partition (1-hr)** style.

5.

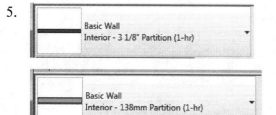

In the Properties panel, select
Interior: 3 1/8″ Partition (1-hr)
[Basic Wall: Interior - 138mm Partition (1-hr)].

6.   Place interior walls as shown.

Imperial Units

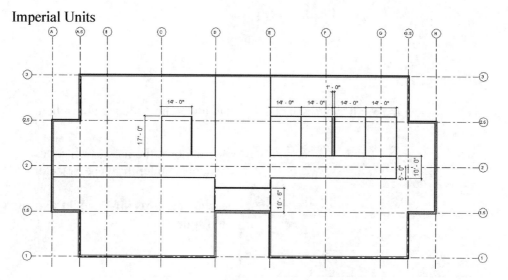

Metric Units (using millimeters)

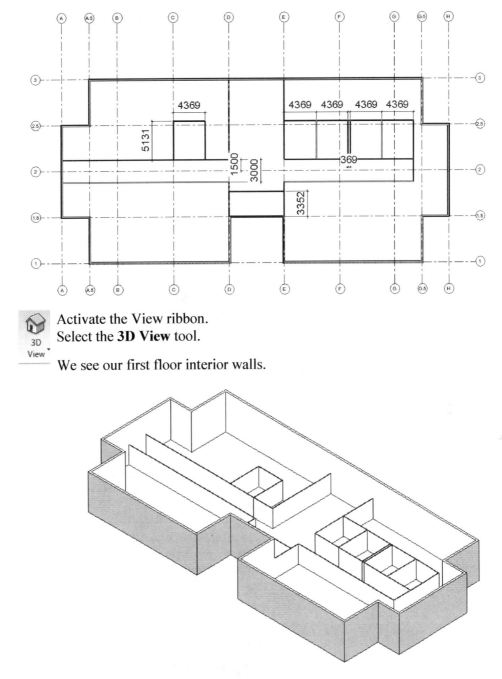

7.　Activate the View ribbon.
　　Select the **3D View** tool.

　　We see our first floor interior walls.

8.　Save the file as *ex3-5.rvt.[m_ex3-5.rvt]*

Exercise 3-6
Add Level 2 Interior Walls

Drawing Name: walls3.rvt [m_walls3.rvt]
Estimated Time: 20 minutes

This exercise reinforces the following skills:

- View Properties
- Wall
- Wall Properties
- 3D View

1.   Open *walls3.rvt. [m_walls3.rvt]*

2.   Activate **3D View**.

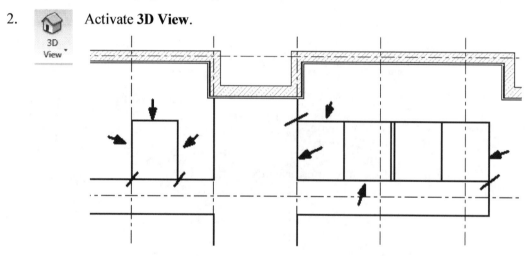

Some of our interior walls will be continuous from Level 1 to Level 2.
The walls that will be extended are indicated by arrows. A diagonal line indicates
where the walls need to be split or divided as a portion of the wall will be
continuous and a portion will be a single story.

3.   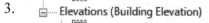   Switch view to a **South Elevation** view.

Double left click in the browser window on
the South Elevation view.

4. 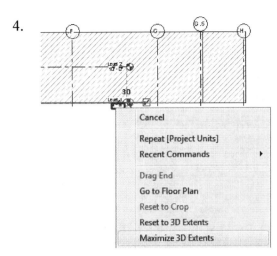 Select the level line.

Right click and select Maximize 3D Extents.

This will stretch the level to cover the entire building.

Repeat for the other level.

5. 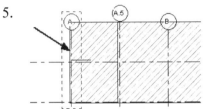 Select the first grid line labeled **A**.

6. Lift click on the pin to unpin the gridline.

You have to unpin the element before it can be modified.

7. 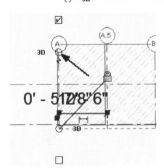 Use the grip indicated to drag the grid line above the building.

8. All the grid lines which were created as part of the array adjust. If you forgot to uncheck Group and Associate when you created the array, you will not be able to adjust the grids without either UNGROUPING or editing the group.

 To ungroup, simply select one of the gridlines and select Ungroup from the ribbon.

Ungroup

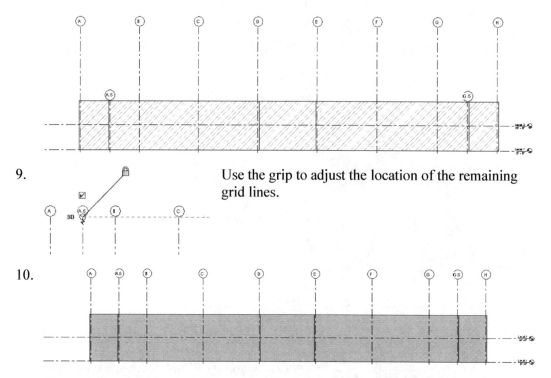

9. 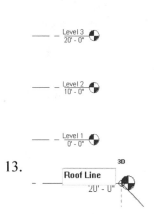   Use the grip to adjust the location of the remaining grid lines.

10.

The current model has level lines for the ground plane and second floor but not for the roof. We will add a level line for the roof to constrain the building's second level and establish a roof line.

11.   Select the **Level** tool on the Datum panel on the Architecture ribbon.

Level

12. Draw a level line **10′ [4000** mm] above Level 2.

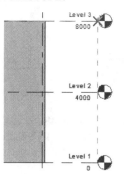

13.   Rename Level 3 by double left clicking on the level name. Rename to **Roof Line.**

14. Would you like to rename corresponding views?

☐ Do not show me this message again [Yes] [No]

Click **Yes**.

This renames the views in the Project Browser to Roof Line.

15. Switch back to the Level 1 Floor plan view.

16. Activate the **Modify** ribbon.

Select the **Split** tool.

17. ☐ Delete Inner Segment Uncheck **Delete Inner Segment** on the Options bar.

18. Split the walls where portions will remain only on level 1 and portions will be continuous up to the roof line.

19. Split the walls at the intersections indicated.

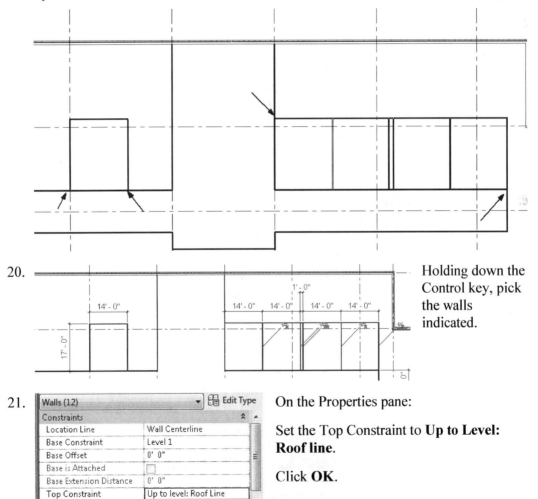

20. Holding down the Control key, pick the walls indicated.

21. On the Properties pane:

Walls (12)	▼ 🔲 Edit Type
Constraints	
Location Line	Wall Centerline
Base Constraint	Level 1
Base Offset	0' 0"
Base is Attached	☐
Base Extension Distance	0' 0"
Top Constraint	Up to level: Roof Line
Unconnected Height	20' 0"

Set the Top Constraint to **Up to Level: Roof line**.

Click **OK**.

By constraining walls to levels, it is easier to control their heights. Simply change the level dimension and all the walls constrained to that level will automatically update.

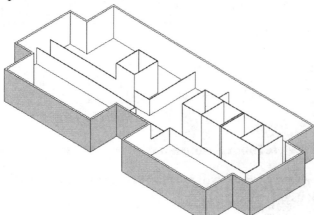

Walls that extend through multiple levels should be created out of just one wall, not walls stacked on top of each other. This improves model performance and minimizes mistakes from floor to floor. For a stair tower, create the wall on the first floor and set its top constraint to the highest level.

22.

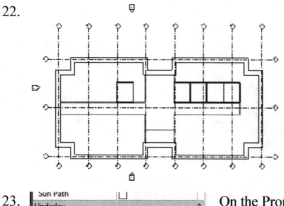

Activate the **Level 2** view.

It can be disconcerting to try to create elements on the second level (Level 2) when you can see the interior walls in the first level (Level 1).

23.

Sun Path	
Underlay	⌃
Range: Base Level	Level 2
Range: Top Level	Roof Line
Underlay Orientation	Look down
Extents	⌃

On the Properties Pane:
Locate the Underlay category.
Set the Range: Base Level to **Level 2.**

This means that the user will only see what is visible on Level 2.

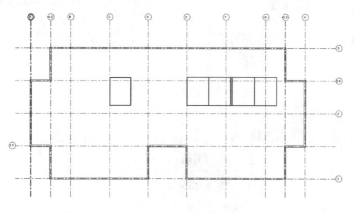

24. Save the file as *ex3-6.rvt*. *[m_ex3-6.rvt]*

Global Parameters

Global Parameters are project-based. They can be used to control material settings or used to define dimensions. For example, if you want to ensure that your project complies with building codes so that all hallways are a minimum of 8' wide, you can define a global parameter and apply it to any hallway dimensions that need to meet that code.

What makes this especially powerful is that if the building code changes or the building specification changes, the dimensions can update simply by changing the single global parameter.

The Reveal Constraints tool will show dimensions that have had global parameters assigned to them. It will display an additional red dimension with a label below the original dimension.

I collaborate with a large group using BIM 360 on building layouts. The CAD Manager will create a sheet in the project which lists all the "rules" or global parameters in use in the project. He then sets the project up so that the sheet is the first view displayed for anybody opening the project.

You can specify the view that Revit displays by default when a project is opened.

The default setting for a starting view is <Last Viewed>, that is, whichever view was active the last time the model was closed.

When a project is shared across a collaboration team, the specified starting view is applied to all local models. This view is opened when the central or any local model is opened, and when any team member uses the Open dialog to detach from the central or create a local model.

Best practices for starting views:

- You can reduce the amount of time required to open a model by choosing a simple view as the starting view.

- You might create a special view that contains important project information and notices that you want to share with team members. Use this view as the starting view so that team members always see this information when they open the project.

To specify a starting view:

1.   Open the model.

2.   Click Manage tab ➤ Manage Project panel ➤ 🖫 Starting View.

3.   In the Starting View dialog, specify the starting view, and click OK.

When Revit loads this model, the specified view is opened.

Adding Doors

Doors are loadable families. This means doors are external files which are inserted into a building project. Revit comes with a library of door families which you can load and insert into a building project. Each door family may have several different size combinations (Height and Width). You may also download door families from various websites, such as bimobject.com or revitcity.com. If you do a search for Revit families or Revit library, a long list of possible sources for content will be displayed.

Exercise 3-7
Add Doors

Drawing Name: doors1.rvt [m_doors1.rvt]
Estimated Time: 10 minutes

This exercise reinforces the following skills:

- ❏ Door
- ❏ Load From Library
- ❏ Global Parameters

1.   Open *doors1.rvt. [m_doors1.rvt]*

2.   Activate the **Level 1** view.

3.      Activate the **Architecture** ribbon.

4.   Select the **Door** tool from the Build panel.

5.   Select **Single-Flush 36″ x 84″ [M_Single-Flush: 0915 × 2134mm]** from the drop-down.

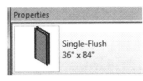 

6.   Enable **Tag on Placement** on the ribbon.

    This will automatically add door tags to the doors as they are placed.

7.   Place the doors as shown. **Verify that you are on Level 1.**
    Remember if you click the space bar that will flip the door orientation before you click to place.

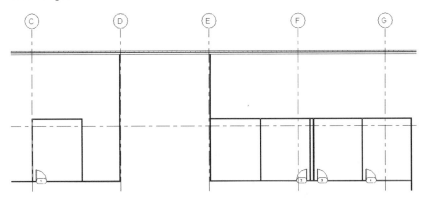

8. Activate the Manage ribbon.

   Select **Global Parameters**.

9. Select the **New Parameter** tool at the bottom of the dialog.

10.
   Name:

   Door Distance|

   Discipline:

   Common

   Type of parameter:

   Length

   Group parameter under:

   Dimensions

   Type **Door Distance** as the Name for the parameter.

   Click **OK**.

11.

Parameter	Value
Dimensions	
Door Distance	2' 6"

Parameter	Value
Dimensions	
Door Distance	800.0

   Set the value to **2' 6"**.

   Click **OK**.

   For metric:

   Set the value to **800**.

   Click **OK**.

12. Use the arrows on the doors to flip their orientation, if needed.

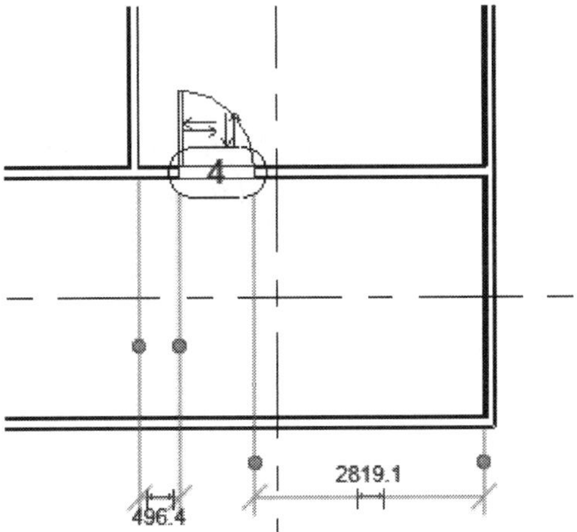

By default, the temporary dimensions for doors display the door opening.

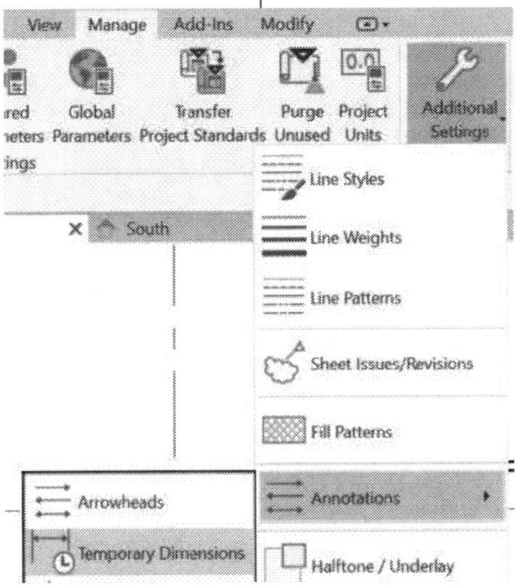

2819.1

496.4

If you are working in a metric project:

Go to the Manage ribbon.

Under **Additional Settings**:

Click **Annotations→Temporary Dimensions**.

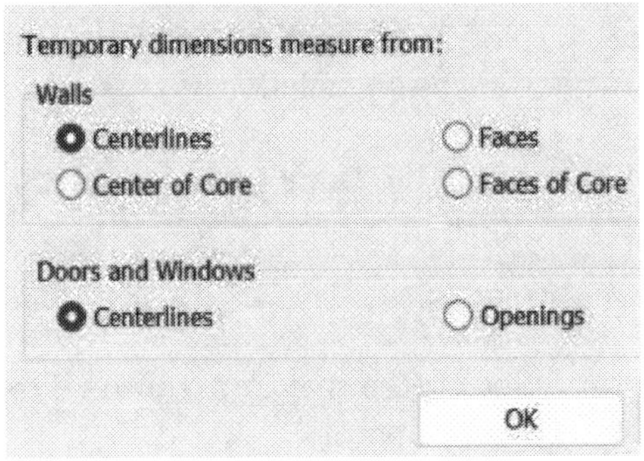

Enable **Centerlines** for Walls and Doors.

Click **OK**.

13.

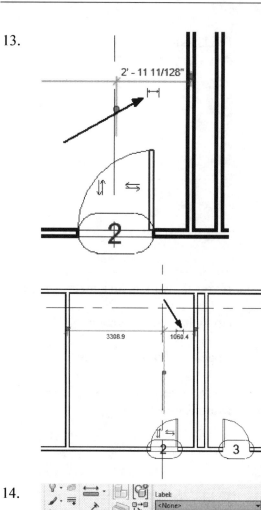

Select the door to activate the temporary dimension.

Select the small dimension symbol to add a permanent dimension.

14.

Select the permanent dimension.

Notice on the ribbon there is a label dimension panel.

Select the **Door Distance** from the drop-down.

15.

Select **Remove Constraints** if this error dialog appears.

16.   Repeat for all the doors.

Door numbers are often keyed to room numbers – 101A, etc. To edit the door tag, just click on the number and it will change to blue and you can renumber it. You can also change the door number from the Properties dialog box of the door or from the door schedule. A change in any of those locations will update automatically on all other views.

17. Select the **Door** tool on the Architecture ribbon.

Door

18. 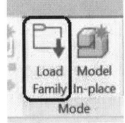 Select **Load Family** from the Mode panel.

19. Browse to the *Doors* folder.

20.

File name:	Door-Double-Glass
Files of type:	All Supported Files (*.rfa, *.adsk)

Locate the *Door - Double-Glass. Rfa [M_Double-Glass .rfa]* file under the *Doors* folder and **Open**.

File name:	M_Door-Double-Glass.rfa
Files of type:	Family Files (*.rfa)

21.

Types:

Type	Width	Heig
	(all)	(all)
68" x 80"	5' 8"	6' 8"
68" x 82"	5' 8"	6' 10"
68" x 84"	5' 8"	7' 0"
72" x 78"	6' 0"	6' 6"
72" x 80"	6' 0"	6' 8"
72" x 82"	6' 0"	6' 10"
72" x 84"	6' 0"	7' 0"

right for each family listed on the left [OK]

Select size **72" x 78" [1800 x 2000mm]** type to load.

Click **OK**.

Types:

Type	Width	
	(all)	
1700 x 2000m	1700.0	2000.0
1700 x 2050m	1700.0	2050.0
1700 x 2100m	1700.0	2100.0
1800 x 1950m	1800.0	1950.0
1800 x 2000m	1800.0	2000.0
1800 x 2050m	1800.0	2050.0
1800 x 2100m	1800.0	2100.0

22. 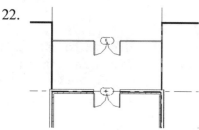 Place two double glass entry doors at the middle position of the entry walls.

The double doors should swing out in the path of egress for accuracy. The interior doors should typically swing into the room space not the hall.

23. Activate **Level 2**.

24. | Home | Activate the **Architecture** ribbon.

25.  Select the **Door** tool from the Build panel.

Door

26. Select **Single-Flush 36″ x 84″ [M_Single-Flush: 0915 × 2134mm]** from the drop-down.

27. Place the doors as shown.

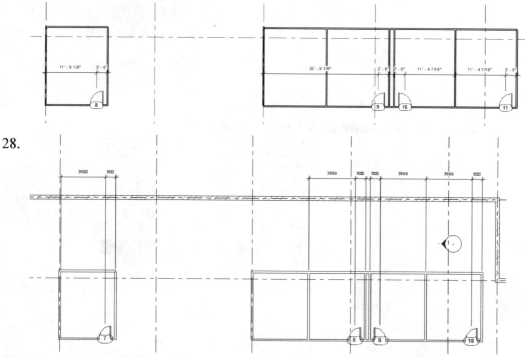

28.

29. Use the Door Distance global parameter to position the doors in each room.

We will be adjusting the room sizes in a later exercise.

30. Save the file as *ex3-7.rvt*. *[m_ex3-7.rvt]*

Exercise 3-8
Define a Starting View

Drawing Name: starting view.rvt [m_starting view.rvt]
Estimated Time: 20 minutes

This exercise reinforces the following skills:

- ❑ Add a Sheet
- ❑ Load a Sheet template
- ❑ Add Notes
- ❑ Global Parameters
- ❑ Define a Starting View

1.   Open *starting view.rvt*. [m_starting view.rvt]

2.   In the Project browser:
Highlight the Sheets
category.
Right click and select **New
Sheet.**

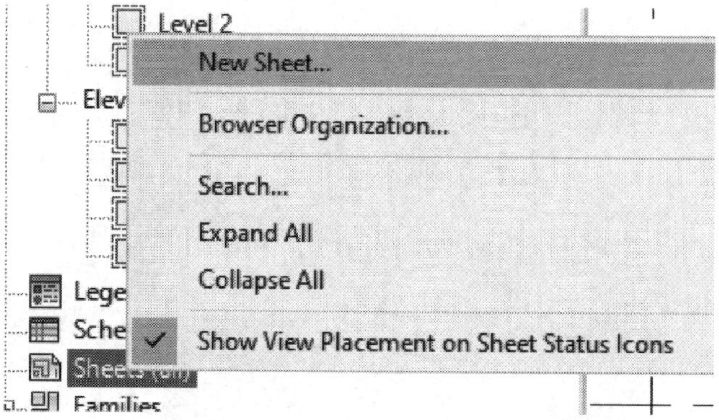

3.    Select the **Load** button.

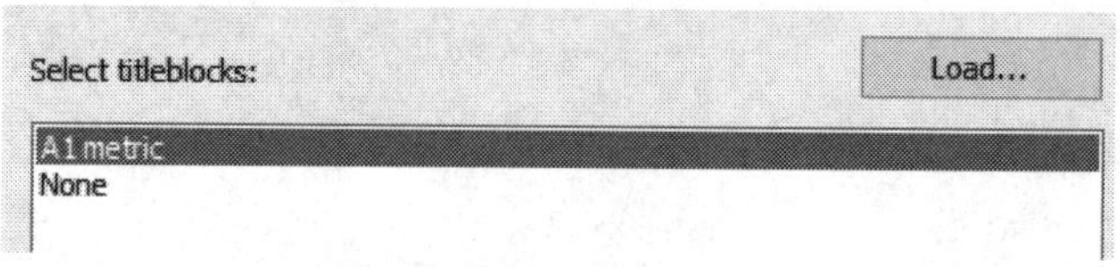

    Metric titleblocks are located in the
English/US/Titleblocks folder.

There are additional metric titleblocks in the
English/UK/Titleblocks folder.

4. 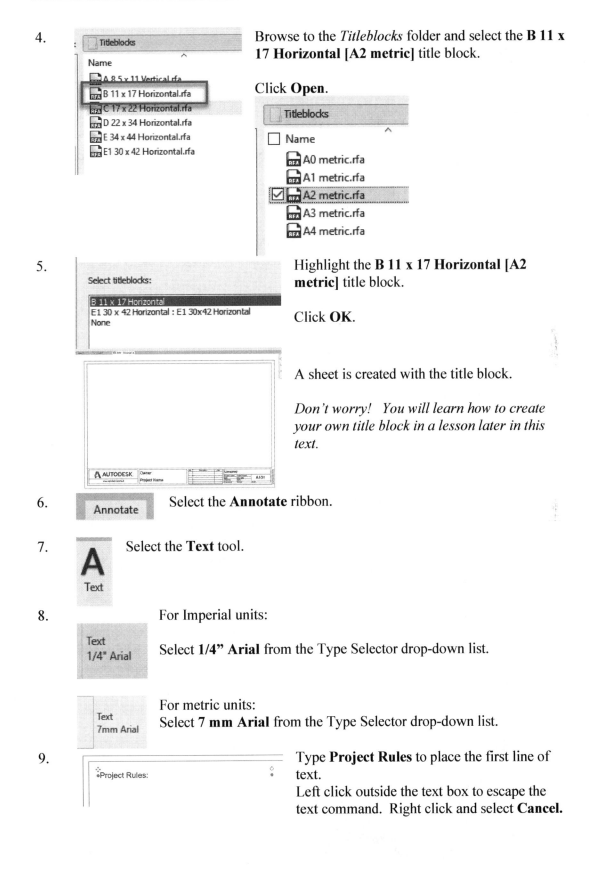 Browse to the *Titleblocks* folder and select the **B 11 x 17 Horizontal [A2 metric]** title block.

Click **Open**.

5. Highlight the **B 11 x 17 Horizontal [A2 metric]** title block.

Click **OK**.

A sheet is created with the title block.

Don't worry! You will learn how to create your own title block in a lesson later in this text.

6. Select the **Annotate** ribbon.

7. Select the **Text** tool.

8. For Imperial units:

Select **1/4" Arial** from the Type Selector drop-down list.

For metric units:
Select **7 mm Arial** from the Type Selector drop-down list.

9. Type **Project Rules** to place the first line of text.
Left click outside the text box to escape the text command. Right click and select **Cancel.**

| 10. | | Select the **Text** tool. |

A

Text

| 11. | | For Imperial units: |

Text
3/32" Arial

Select **3/32" Arial** from the Type Selector drop-down list.

Text
5mm Arial

For metric units:

Select **5mm Arial** from the Type Selector drop-down list.

| 12. | | Type **Door Distance – 2' 6"** below the Project Rules: header. |

Project Rules:

Door Distance - 2' 6"

Project Rules:

Door Distance - 800mm

Metric:

Type **Door Distance – 800mm** below the Project Rules: header.

| 13. | | In the Properties panel: |

Drawn By	Author
Sheet Number	A101
Sheet Name	Landing Page
Sheet Issue Date	01/03/23
Appears In Sheet List	☐

Change the Sheet Name to **Landing Page.**
Uncheck **Appears in Sheet List**.

Click **Apply**.

| 14. | | Activate the **Manage** ribbon. |

w   Manage

| 15. | | Select **Starting View** on the Manage Project panel on the ribbon. |

Starting
View

16.

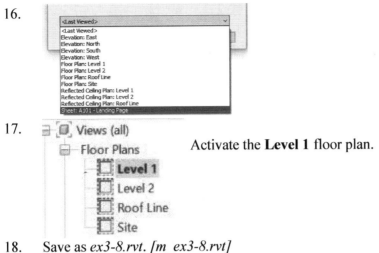

17. **Views (all)**

   Floor Plans

   Level 1

   Level 2

   Roof Line

   Site

Activate the **Level 1** floor plan.

18. Save as *ex3-8.rvt. [m_ex3-8.rvt]*
    Close the file and then re-open it.

19. Notice it re-opens to the Landing Page.

If you want the project to open to the last active view, change the Starting View to <Last Viewed>, which is the default.

Exercise 3-9
Exploring Door Families

Drawing Name: doors2.rvt [m_doors2.rvt]
Estimated Time: 60 minutes

This exercise reinforces the following skills:

- ❑ Door
- ❑ Hardware

1. Open *doors2.rvt. [m_doors2.rvt]*

2. Activate the **Level 1** view.

3. Zoom into the exterior double glass door.
   Select the door to activate the Properties pane.

4.

Properties	
Door-Double-Glass 72" x 78"	
Doors (1)	
Constraints	
Level	Level 1
Sill Height	0' 0"
Construction	
Swing Angle	45.00°
Frame Type	

Note that you can adjust the Swing Angle for the door.

Change it to 45 degrees and see how it updates in the display window.

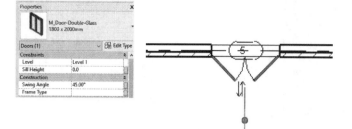

Note that the other double glass door doesn't change. The swing angle is controlled by each instance, not by type.

5. Select the **Door** tool on the Architecture ribbon.

Door

6. Select **Load Family** from the Mode panel.

7. Browse to the *Doors/Commercial* folder.

8.

File name:	Door-Exterior-Double-Two_Lite
Files of type:	All Supported Files (*.rfa, *.adsk)

Locate the *Door-Exterior-Double-Two_Lite. Rfa [M_ the Door-Exterior-Double-Two_Lite.rfa]* file under the *Doors* folder and **Open**.

File name:	M_Door-Exterior-Double-Two_Lite.rfa
Files of type:	All Supported Files (*.rfa, *.adsk)

9.

Types:

Type	Width (all)	
72" x 80"	6' 0"	6' 8"
72" x 82"	6' 0"	6' 10"
72" x 84"	6' 0"	7' 0"
84" x 80"	7' 0"	6' 8"
84" x 84"	7' 0"	7' 0"
96" x 80"	8' 0"	6' 8"
96" x 84"	8' 0"	7' 0"

ight for each family listed on the left   OK

Types:

Type	Width (all)	
1800 x 2000m	1800.0	2000.0
1800 x 2050m	1800.0	2050.0
1800 x 2100m	1800.0	2100.0
2100 x 2000m	2100.0	2000.0

Select the 72" x 82" [1800 x 2100] type.

Click **OK**.
Click ESC to exit the command.

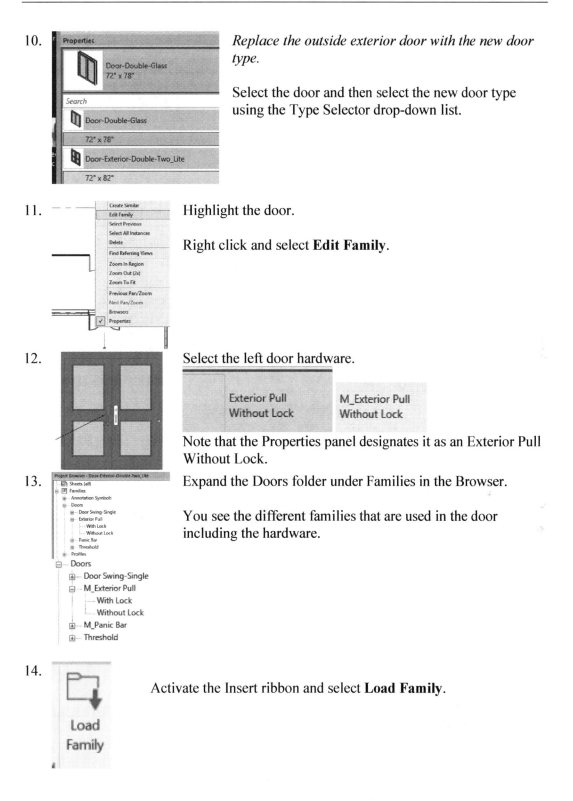

10. *Replace the outside exterior door with the new door type.*

Select the door and then select the new door type using the Type Selector drop-down list.

11. Highlight the door.

Right click and select **Edit Family**.

12. Select the left door hardware.

Note that the Properties panel designates it as an Exterior Pull Without Lock.

13. Expand the Doors folder under Families in the Browser.

You see the different families that are used in the door including the hardware.

14. Activate the Insert ribbon and select **Load Family**.

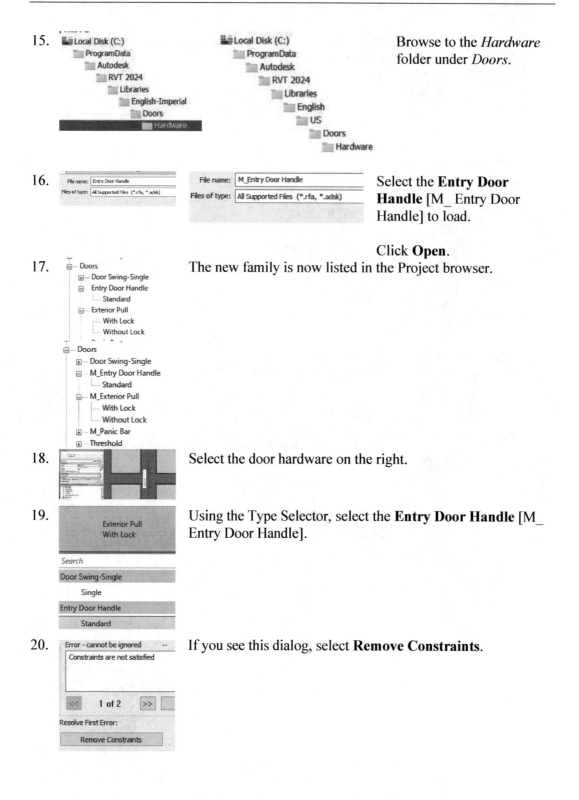

15. Browse to the *Hardware* folder under *Doors*.

16. Select the **Entry Door Handle** [M_ Entry Door Handle] to load.

Click **Open**.

17. The new family is now listed in the Project browser.

18. Select the door hardware on the right.

19. Using the Type Selector, select the **Entry Door Handle** [M_ Entry Door Handle].

20. If you see this dialog, select **Remove Constraints**.

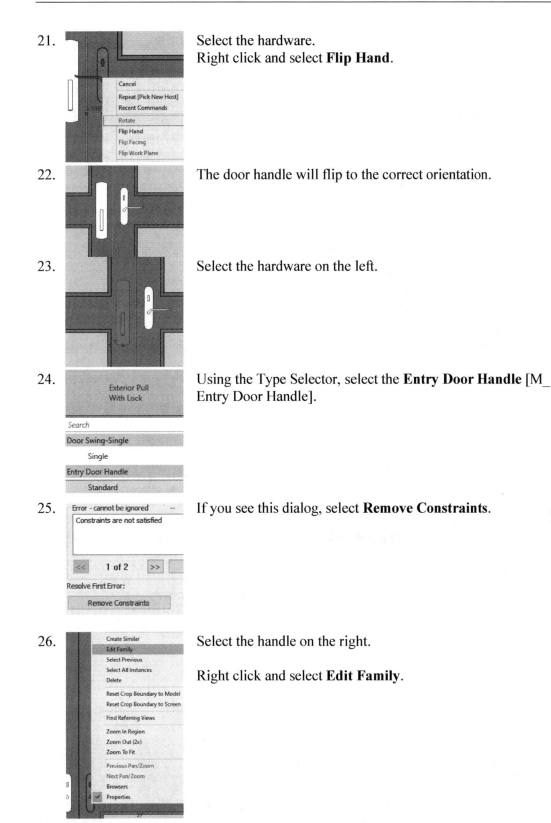

21. Select the hardware.
Right click and select **Flip Hand**.

22. The door handle will flip to the correct orientation.

23. Select the hardware on the left.

24. Using the Type Selector, select the **Entry Door Handle** [M_
Entry Door Handle].

25. If you see this dialog, select **Remove Constraints**.

26. Select the handle on the right.

Right click and select **Edit Family**.

27.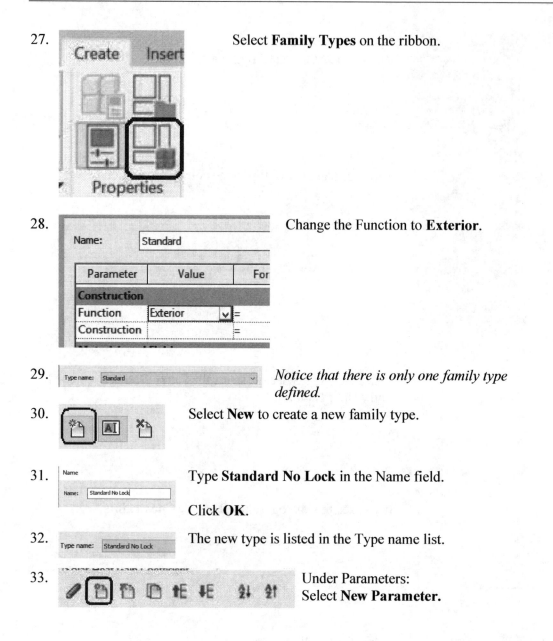

Select **Family Types** on the ribbon.

28. Change the Function to **Exterior**.

29. *Notice that there is only one family type defined.*

30. Select **New** to create a new family type.

31. Type **Standard No Lock** in the Name field.

Click **OK**.

32. The new type is listed in the Type name list.

33. Under Parameters:
Select **New Parameter.**

34.

Enable **Family Parameter**.
Type **Lock** in the Name field.
Enable **Type**.
Set Discipline to **Common**.
Set Type of Parameter to **Yes/No**.
Set Group Parameter under to **Construction**.
Click **OK**.

Click **OK** to close the dialog box.

35.

Select the cylinder representing the lock.

36.

Select the small gray box to the right of the Visible field.

37.

Highlight **Lock** and click **OK**.

38.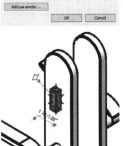

Orbit the model around.

Select the rectangle representing the lock.

39.

Select the small gray box to the right of the Visible field.

40.

Highlight **Lock** and click **OK**.

41.

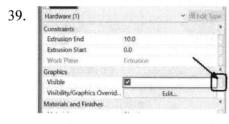

Launch the Family Types dialog again.

42.

Select the **Standard** type selected under the Type Name:

43.

Enable **Lock**.

44.

With **Standard No Lock** type selected:
Uncheck **Lock**.

Click **OK**.

45.

File name: Entry Door Handle-2
Files of type: Family Files (*.rfa)

Save As *Entry Door Handle-2.rfa [m_Entry Door Handle-2.rfa]* in your exercise folder.

File name:	M_Entry Door Handle 2.rfa
Files of type:	Family Files (*.rfa)

46.

Load into Project and Close

Select **Load into Project and Close** from the ribbon.

47.

Check the open Projects/Families you want to load the edited Family into
☑ Door-Exterior-Double-Two_Lite.rfa
☐ ex3-9.rvt

Load into the door family.

Check the open Projects/Families you want to load the edited Family into

☐ ex3-9.rvt
☑ M_Door-Exterior-Double-Two_Lite.rfa
☐ m_ex3-9.rvt

48.

Click ESC to exit the command which is active.

49.

Use the Type Selector to set the hardware on the right to *Entry Door Handle-2-Standard.*
Use the Type Selector to set the hardware on the left to *Entry Door Handle-2-Standard No Lock.*

50.

File name: Door-Exterior-Double-Two_Lite 2
Files of type: Family Files (*.rfa)

Save as *Door-Exterior-Double-Two_Lite 2.rfa [M_Door-Exterior- Double- Two_Lite 2.rfa]*

File name:	M_Door-Exterior-Double-Two_Lite 2.rfa
Files of type:	Family Files (*.rfa)

51.

Load into Project and Close

Load into the project and close.

Check the open Projects/Families you want to load the edited Family into

☐ ex3-9.rvt
☑ m_ex3-9.rvt

52.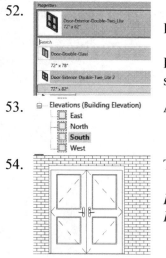

Escape out of the active command.

Replace the exterior door with the new version using the type selector.

53.

Activate the South elevation.

54.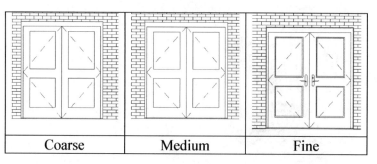

The door is displayed with the new hardware.

Hardware is only displayed when the detail level is set to Fine.

55.

Change the Display detail to see how the door display changes.

Coarse	Medium	Fine

56.

Save as *ex3-9.rvt.[m_ex3-9.rvt]*

Stairs

There are two basic methods to create a set of stairs in Revit. You can use a "run" which involves defining the stair type and then drawing a line by selecting two points. Revit will automatically generate all the stair geometry based on the location and distance between the two points. In the second method, you use a sketch. In the sketch, you must define the boundaries of the stairs – that is the inside and outside edges – and the riser lines. The distance between the risers determines the width of the tread.

The most common mistakes I see users make when sketching stairs are:

- Creating a closed boundary – the boundary lines should be open. The boundaries define where the stringers are for the stairs.

- Placing risers on top of each other. If you place a riser on top of a riser, you will get an error.

- Making the riser lines too short or too long. The riser line endpoints should be coincident to the inside and outside boundary lines.

You can assign materials to the stairs based on the stair type. Stairs are system families, which means they are unique to each project.

Exercise 3-10
Adding Stairs

Drawing Name: stairs1.rvt [m_stairs1.rvt]
Estimated Time: 15 minutes

This exercise reinforces the following skills:

❑   Stairs

1.   Open *stairs1.rvt. [m_stairs1.rvt]*

2.   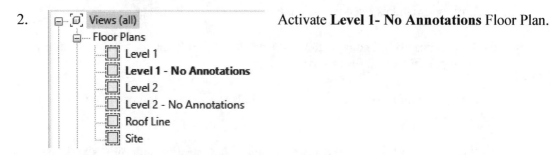 Activate **Level 1- No Annotations** Floor Plan.

3.   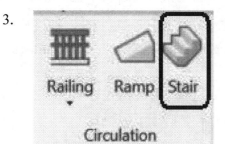 Select the **Stair** tool under the Circulation panel on the Architecture ribbon.

4.   Highlight **Run**.

Select the **Straight Run** option.

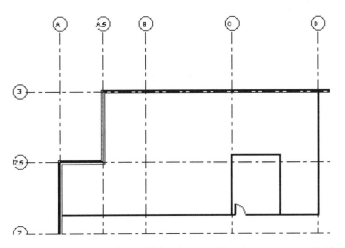

Our first set of stairs will be located in the room at Grids C-2.5.

5.

Location Line: Exterior Support: Left   ∨   Offset: 0' 0"   Actual Run Width: 3' 0"   ☑ Automatic Landing

On the Options bar:

Set the Location line to **Exterior Support: Left.**
Set the Offset to **0' 0".**
Set the Actual Run Width to **3' 0".**
Enable **Automatic Landing.**

Location Line: Exterior Support: Left   ∨   Offset: 0.0   Actual Run Width: 1000.0   ☑ Automatic Landing

For Metric:
On the Options bar:

Set the Location line to **Exterior Support: Left.**
Set the Offset to **0.0.**
Set the Actual Run Width to **1000.**
Enable **Automatic Landing.**

6.

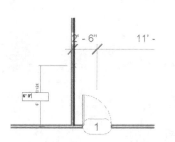

Start the stairs 6' 0" [1520mm] above the door by selecting the inside wall face and typing in the dimension.

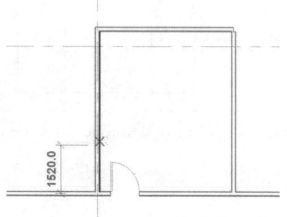

7.

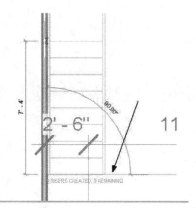

Drag the mouse straight up.

Notice at the start point where you selected Revit generates a riser count.

When you see a riser count of 9, left click to complete the first run of stairs.

8.

For Metric:

Drag the mouse straight up.

Notice at the start point where you selected Revit generates a riser count.

When you see a riser count of 11, left click to complete the first run of stairs.

9.

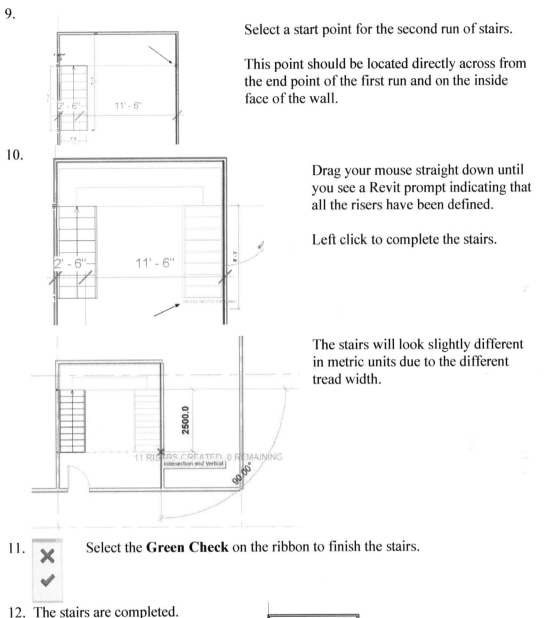

Select a start point for the second run of stairs.

This point should be located directly across from the end point of the first run and on the inside face of the wall.

10.

Drag your mouse straight down until you see a Revit prompt indicating that all the risers have been defined.

Left click to complete the stairs.

The stairs will look slightly different in metric units due to the different tread width.

11. Select the **Green Check** on the ribbon to finish the stairs.

12. The stairs are completed.

13. 3D Views [3D] Activate the **3D View**.

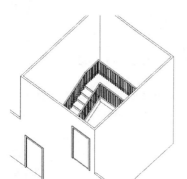

Railing was added on both sides of the stairs.

We will correct this in the next exercise.

14. Click F8 to access the orbit tool.

15. Save as *ex3-10.rvt.[m_ex3-10.rvt]*

Exercise 3-11
Creating a Handrail on a Wall

Drawing Name: railing1.rvt [m_railing1.rvt]
Estimated Time: 20 minutes

This exercise reinforces the following skills:

- ❑ Railings
- ❑ Railing Types

1.   Open *railing1.rvt. [m_railing1.rvt]*

2.   Views (all)
     Floor Plans
        Level 1
        Level 1 - No Annotations
        Level 2
        Level 2 - No Annotations
        Roof Line
        Site

     Activate **Level 1- No Annotations** Floor Plan.

3.   Railings : Railing : Handrail - Rectangular

     Hover your mouse over the Railing on the left side of the stairs.

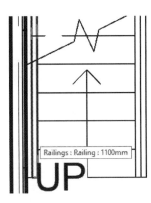

Railings : Railing : 1100mm

Left click to select the railing.

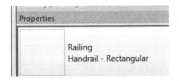

Properties

Railing
Handrail - Rectangular

If you have difficulty selecting the railing, use the TAB key to cycle the selection or use the FILTER tool. You can check that you have selected the railing because it will be listed in the Properties pane.

4. 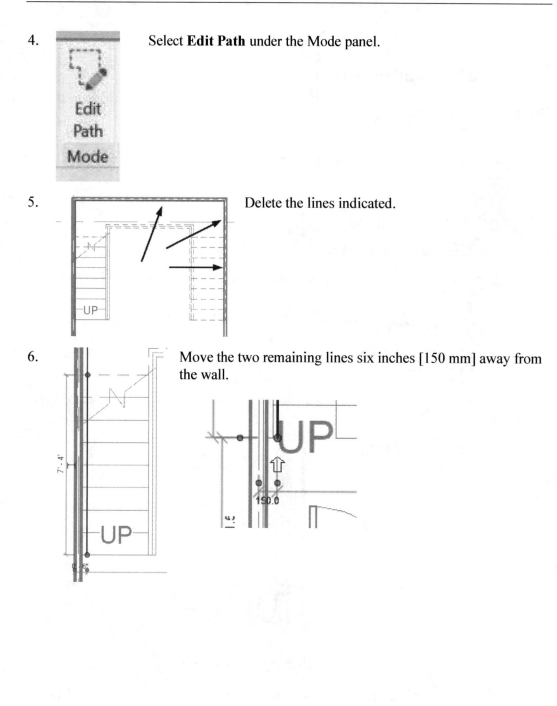   Select **Edit Path** under the Mode panel.

5.   Delete the lines indicated.

6.   Move the two remaining lines six inches [150 mm] away from the wall.

7.

Adjust the top vertical line so it is 1' 6" [460 mm] above the top riser.

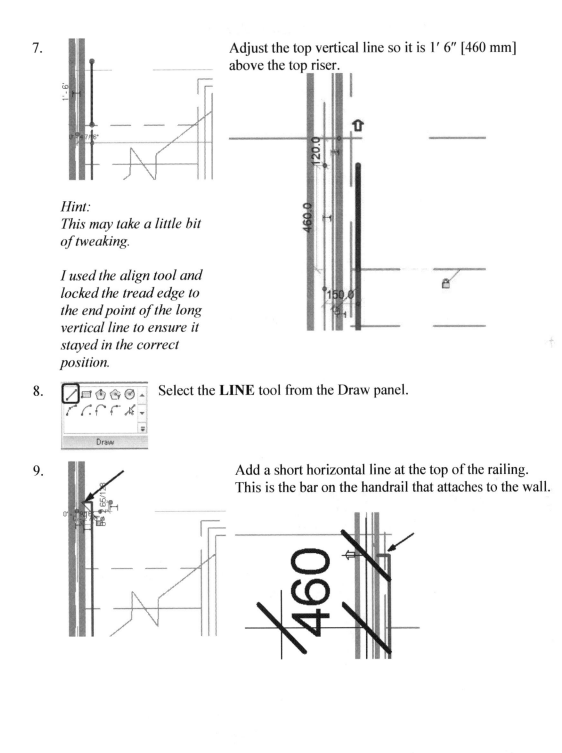

Hint:
This may take a little bit of tweaking.

I used the align tool and locked the tread edge to the end point of the long vertical line to ensure it stayed in the correct position.

8. Select the **LINE** tool from the Draw panel.

Draw

9.

Add a short horizontal line at the top of the railing. This is the bar on the handrail that attaches to the wall.

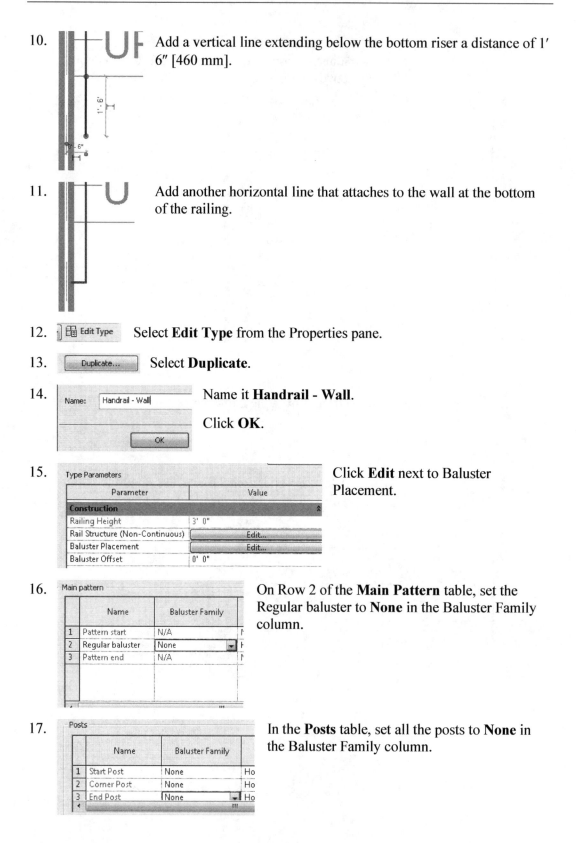

10. Add a vertical line extending below the bottom riser a distance of 1' 6" [460 mm].

11. Add another horizontal line that attaches to the wall at the bottom of the railing.

12. Select **Edit Type** from the Properties pane.

13. Select **Duplicate**.

14. Name it **Handrail - Wall**.

    Click **OK**.

15. Click **Edit** next to Baluster Placement.

16. On Row 2 of the **Main Pattern** table, set the Regular baluster to **None** in the Baluster Family column.

17. In the **Posts** table, set all the posts to **None** in the Baluster Family column.

18.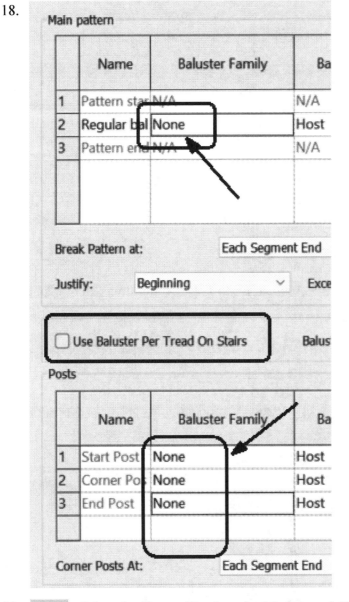

Verify that you have set the balusters to **None** in both areas.

Many of my students will set it to None in one area and not the other.

Click **OK** twice to exit all dialogs.

19. Select the **Green Check** under Mode to exit Edit Mode.

Warning	
The rail is not continuous. Breaks in the rail usually occur at sharply-angled transitions. To fix the problem, try: - Changing the transition style in the rail type properties, or - Modifying the railing path at the transition.	

If you get an error message, check the sketch. It must be a single open path. All line segments must be connected with no gaps and no intersecting lines or overlying lines.

20. 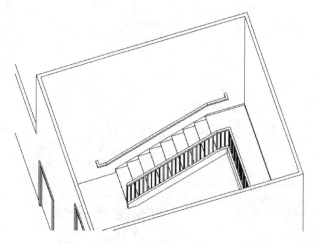 Switch to a 3D view.

21. Inspect the railing.

22.
Floor Plans
Level 1
Level 1 - No Annotations
Level 2
Level 2 - No Annotations
Roof Line
Site

Activate **Level 1- No Annotations** Floor Plan.

23.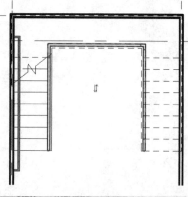

Select the handrail you just created.

24.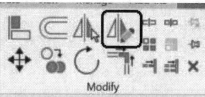

Select **Mirror→Draw Axis** from the Modify panel.

25. 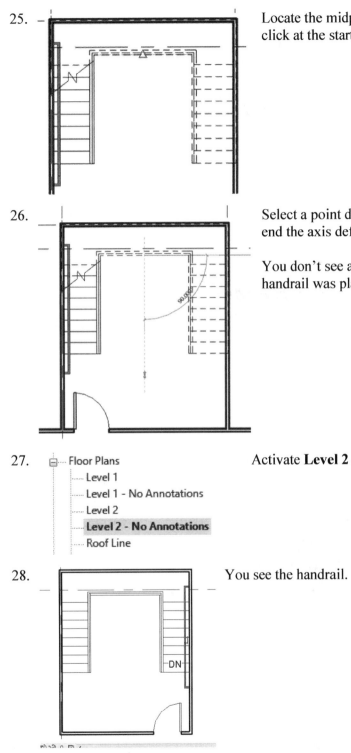 Locate the midpoint on the landing and left click at the start of the axis.

26. Select a point directly below the midpoint to end the axis definition.

You don't see anything – that's because the handrail was placed on the second level.

27.
- Floor Plans
  - Level 1
  - Level 1 - No Annotations
  - Level 2
  - **Level 2 - No Annotations**
  - Roof Line

Activate **Level 2 – No Annotations**.

28. You see the handrail.

DN

29. Switch to a 3D view.

30. Inspect the railing.

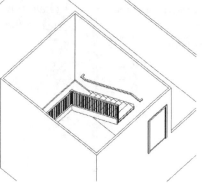

31. Save as *ex3-11.rvt.[m_ex3-11.rvt]*

Exercise 3-12
Modifying the Floor Plan – Skills Review

Drawing Name: copy1.rvt [m_copy1.rvt]
Estimated Time: 30 minutes

This exercise reinforces the following skills:

- ❑ 3D View
- ❑ Copy
- ❑ Align
- ❑ Floor
- ❑ Railing

1. Open *copy1.rvt. [m_copy1.rvt]*

2. Activate the **North** Elevation.

3. Set View Properties to **Wireframe**.

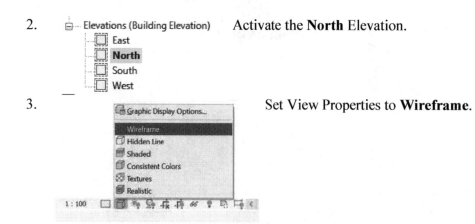

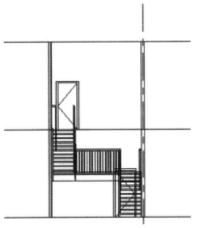

We see that our stairs in elevation look pretty good.

4. Activate **Level 2 – No Annotations**.

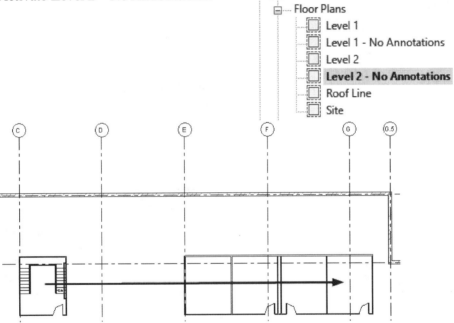

We want to copy the stairs, railings, and landing to the stairwell indicated.

5. Window around the walls, stair and railings.

6. Select the **Filter** tool.

7.

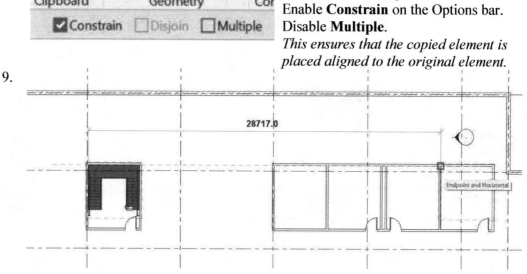

Category:	Count:
Railings	3
Railings: Top Rails	2
Stair Paths	1
Stairs	1
Stairs: Landings	1
Stairs: Runs	2
Stairs: Supports	8
Walls	1

Use **Filter** to ensure that only the Stairs and Railings are selected.

Click **OK**.

8.

Select **Copy** from the Modify panel.

The Modify→Copy can only be used to copy on the same work plane/level.

☑Constrain ☐Disjoin ☐Multiple

Enable **Constrain** on the Options bar.
Disable **Multiple**.
This ensures that the copied element is placed aligned to the original element.

9.

28717.0

Endpoint and Horizontal

Select the top left corner of the existing stairwell as the basepoint.

Select the top left corner of the target stairwell as the target location.
Left click to release.

(Instead of copying the stair, you can also repeat the process we used before to create a stair from scratch.)

10. 3D Views
       (3D) Activate **3D View**.

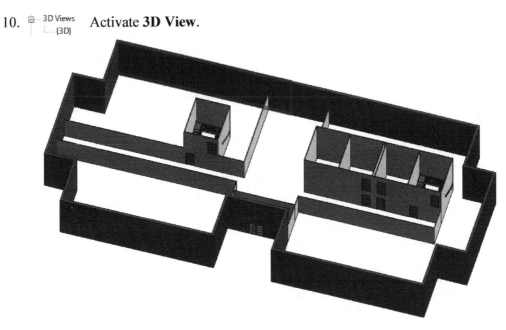

You should see a complete stair with landing and railing in both stairwells.

11. Save as *ex3-12.rvt*.[*m_ex3-12.rvt*]

Exercise 3-13
Defining a 2-hr Wall

Drawing Name: walls4.rvt [m_walls4.rvt]
Estimated Time: 10 minutes

This exercise reinforces the following skills:

- Split
- Wall Properties

Stairwell walls are usually 2-hour walls. The walls, as currently defined, are 1-hour walls. We need to load 2-hr walls into the project from the Project Browser.

1. Open *walls4.rvt. [m_walls4.rvt]*

2. ⊟ Views (all)
   ⊟ Floor Plans
      Level 1
      Level 1 - No Annotations
      Level 2
      Level 2 - No Annotations
      Roof Line
      Site

   Activate the **Level 2 – No Annotations** Floor Plan.

3.  Select the **Split** tool from the Modify panel on the Modify ribbon.

4. Split the walls at the intersections indicated.

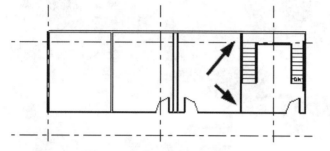

5. Select the stairwell walls indicated.

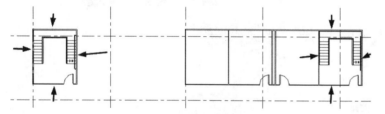

These walls will be defined as 2-hr walls.

6. You can window around both areas while holding down the CTRL key.

Then select the **Filter** tool from the ribbon.

7. Uncheck all the elements EXCEPT walls.

Click **OK**.

Category:	Count:
Doors	3
Railings	6
Railings: Top Rails	4
Stair Paths	2
Stairs	2
Stairs: Landings	2
Stairs: Runs	4
Stairs: Supports	16
Walls	8

8.

On the Properties pane, select the **Interior - 5″ Partition (2-hr)** [**Basic Wall: Interior - 135mm Partition (2-hr)**] from the Type Selector drop-down list.

Left click in the window to clear the selection.

9.   Save the file as *ex3-13.rvt. [m_ex3-13.rvt]*

Exercise 3-14
Adding an Elevator

Drawing Name: elevator1.rvt [m_elevator1.rvt]
Estimated Time: 40 minutes

This exercise reinforces the following skills:

- ❑ 3D View
- ❑ Delete
- ❑ Wall Properties
- ❑ Families

1.   Open *elevator1.rvt. [m_elevator1.rvt]*

2.   Activate the **Level 1 - No Annotations** Floor Plan.

3.   Activate the **Insert** ribbon.

      Select **Load Family** from the Load from Library panel.

4.   Locate the *Elevator* and the *Door-Elevator* in the downloaded Class Files.

      Click **Open**.

5. The elevator is now available in the Project browser for your current project.

     To locate it, look for the folder called Specialty Equipment and expand that folder.

6. Select the *6' 10" x 5' 2" family.*
    Hold down the left mouse button and drag into the display window.

    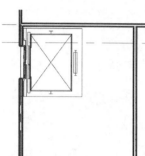

7.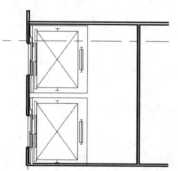

    The elevator will appear on the cursor when you hover over a wall.

    The elevator is wall-hosted, so needs a wall to be placed.

    Click the SPACE bar to rotate the elevator to orient it properly.

    Left pick to place.

8. Place two elevators.

    Right click and select CANCEL to exit the command.

9. Add an **Interior – 3 1/8" Partition (1-hr) [Interior-138mm Partition (1-hr)]** wall behind and to the right of the two elevators as shown.

    Add a wall between the two elevator shafts. **Use the midpoint of the wall behind the elevators to locate the horizontal wall.**

10.

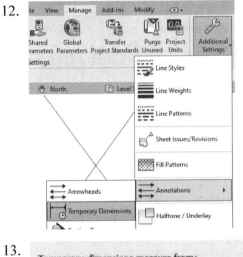

Constrain both walls to the Roof Line level.

11. **Manage** Modify Activate the **Manage** ribbon.

12. Go to **Additional Settings→Annotations→Temporary Dimensions**.

13. Enable **Centerlines** for Walls.

Enable **Centerlines** for Doors and Windows.

Click **OK**.

14. 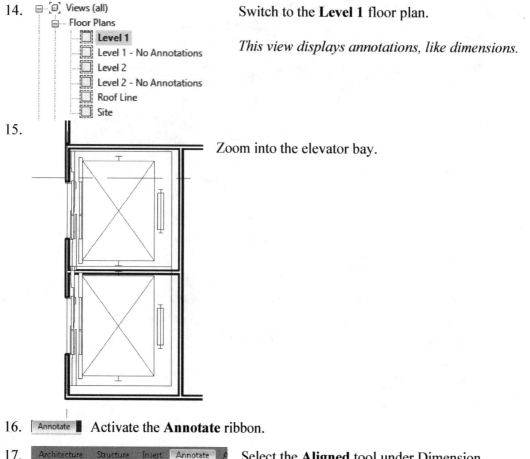 Switch to the **Level 1** floor plan.

This view displays annotations, like dimensions.

15. Zoom into the elevator bay.

16. Annotate ▌ Activate the **Annotate** ribbon.

17. Select the **Aligned** tool under Dimension.

Make sure visibility of dimensions is turned on in the Visibility/Graphics dialog if you don't see dimensions.

18. 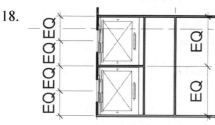 Select the center of the wall.
Select the center line of the elevator.
Select the center line of the wall.
Left pick to place the continuous dimension.

Repeat for the other elevator.

19. Select the dimension.
Left click on the EQ symbol to set the dimensions equal and position the elevator centered in the space.

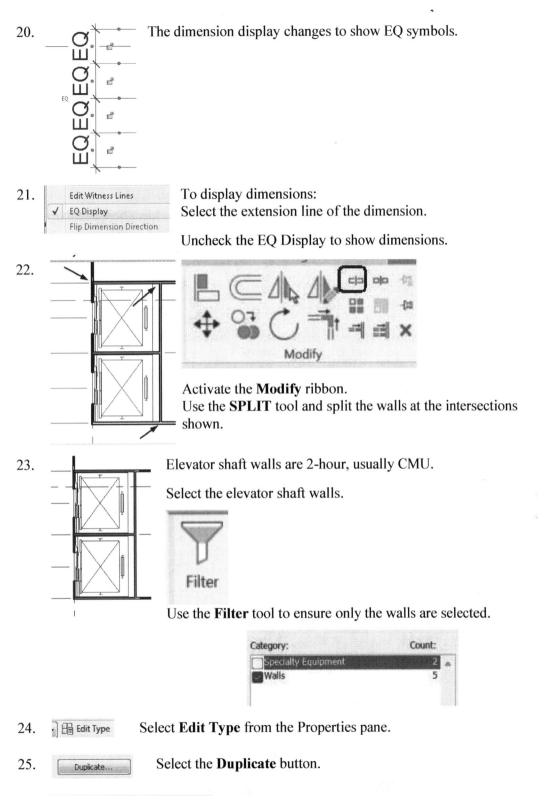

20. The dimension display changes to show EQ symbols.

21. Edit Witness Lines
    ✓ EQ Display
    Flip Dimension Direction

    To display dimensions:
    Select the extension line of the dimension.

    Uncheck the EQ Display to show dimensions.

22. Activate the **Modify** ribbon.
    Use the **SPLIT** tool and split the walls at the intersections shown.

23. Elevator shaft walls are 2-hour, usually CMU.

    Select the elevator shaft walls.

    Use the **Filter** tool to ensure only the walls are selected.

Category:	Count:
Specialty Equipment	2
Walls	5

24. Select **Edit Type** from the Properties pane.

25. Select the **Duplicate** button.

26. Name: Interior -4-3/8" CMU (2-hr)

    Name the new wall style **Interior - 4 3/8" CMU (2-hr) [Interior - 100 mm CMU (2-hr)]**.

Name: Interior -100mm|CMU (2-hr)   Click **OK**.

27.

Type Parameters	
Parameter	Value
Construction	
Structure	Edit...

Select **Edit** next to Structure.

28.

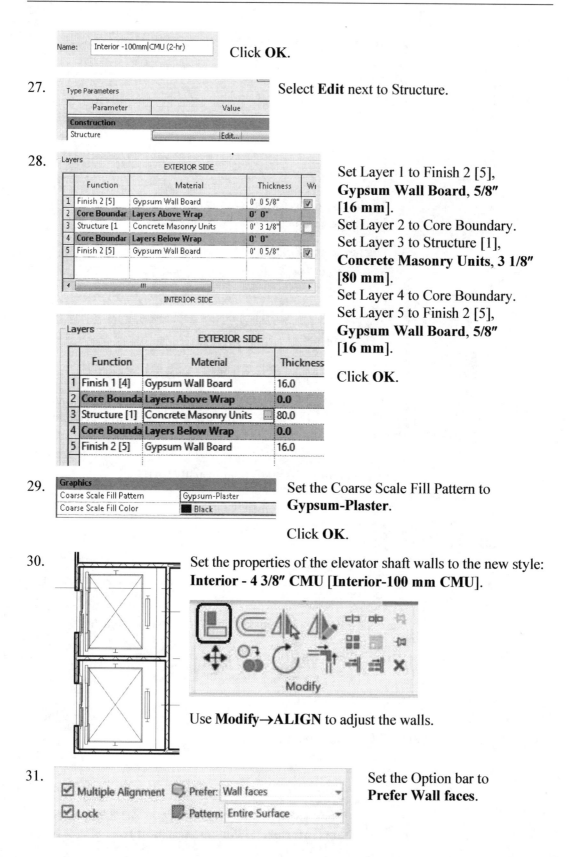

Layers

EXTERIOR SIDE

	Function	Material	Thickness	W
1	Finish 2 [5]	Gypsum Wall Board	0' 0 5/8"	✓
2	**Core Boundar**	**Layers Above Wrap**	**0' 0"**	
3	Structure [1	Concrete Masonry Units	0' 3 1/8"	
4	**Core Boundar**	**Layers Below Wrap**	**0' 0"**	
5	Finish 2 [5]	Gypsum Wall Board	0' 0 5/8"	✓

INTERIOR SIDE

Layers

EXTERIOR SIDE

	Function	Material	Thickness
1	Finish 1 [4]	Gypsum Wall Board	16.0
2	**Core Bounda**	**Layers Above Wrap**	**0.0**
3	Structure [1]	Concrete Masonry Units	80.0
4	**Core Bounda**	**Layers Below Wrap**	**0.0**
5	Finish 2 [5]	Gypsum Wall Board	16.0

Set Layer 1 to Finish 2 [5],
**Gypsum Wall Board, 5/8″
[16 mm]**.
Set Layer 2 to Core Boundary.
Set Layer 3 to Structure [1],
**Concrete Masonry Units, 3 1/8″
[80 mm]**.
Set Layer 4 to Core Boundary.
Set Layer 5 to Finish 2 [5],
**Gypsum Wall Board, 5/8″
[16 mm]**.

Click **OK**.

29.

Graphics	
Coarse Scale Fill Pattern	Gypsum-Plaster
Coarse Scale Fill Color	Black

Set the Coarse Scale Fill Pattern to
Gypsum-Plaster.

Click **OK**.

30.

Set the properties of the elevator shaft walls to the new style:
Interior - 4 3/8″ CMU [Interior-100 mm CMU].

Modify

Use **Modify→ALIGN** to adjust the walls.

31.

☑ Multiple Alignment Prefer: Wall faces

☑ Lock   Pattern: Entire Surface

Set the Option bar to
Prefer Wall faces.

32. 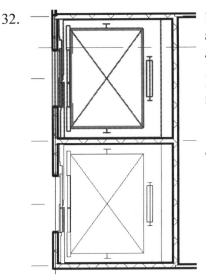 Use the ALIGN tool to adjust the walls so the faces are flush. *You can use the TAB key to tab through the selections.*

Re-adjust the position of the elevators to ensure they remain centered in the shaft.

You can click on the EQ toggle on the aligned dimension to update.

33.  Switch to a 3D view.

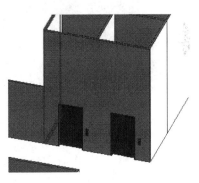

Verify that all the walls appear properly, with walls for the second floor constrained as needed.

Next, we add the elevator doors for the second floor.

34. Activate **Level 2**.

- ⊟ 🗇 Views (all)
  - ⊟ Floor Plans
    - 🗀 Level 1
    - 🗀 Level 1 - No Annotations
    - 🗀 **Level 2**
    - 🗀 Level 2 - No Annotations
    - 🗀 Roof Line
    - 🗀 Site

35. Activate the **Architecture** ribbon.

Door Select the **Door** tool.

36.  Verify that the Door-Elevator is selected on the Properties palette.

Door-Elevator
12000 x 2290 Opening

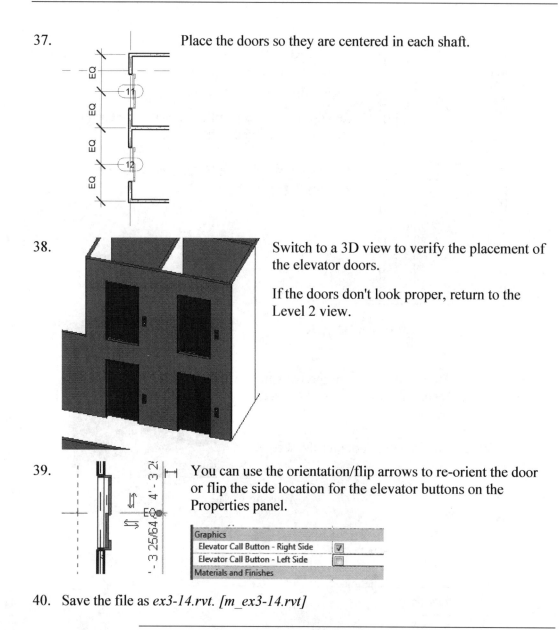

37. Place the doors so they are centered in each shaft.

38. Switch to a 3D view to verify the placement of the elevator doors.

If the doors don't look proper, return to the Level 2 view.

39. You can use the orientation/flip arrows to re-orient the door or flip the side location for the elevator buttons on the Properties panel.

Graphics	
Elevator Call Button - Right Side	☑
Elevator Call Button - Left Side	☐
Materials and Finishes	

40. Save the file as *ex3-14.rvt. [m_ex3-14.rvt]*

> ➤ The Snap Settings you use are based on your Zoom factor. In order to use the smaller increment snap distances, zoom in. To use the larger increment snap distances, zoom out.
> ➤ When placing a component, use the spacebar to rotate the component before placing.
> ➤ You can purge any unused families from your project by using Manage→ Settings→Purge Unused.

Exercise 3-15
Load Family

Drawing Name: load family.rvt [m_load family.rvt]

Estimated Time: 15 minutes

This exercise reinforces the following skills:

- ❑ Load Family
- ❑ Space Planning

   All of the content required in this exercise is included in the downloaded Class Files.

1.   Open *load family.rvt. [m_load family.rvt]*

2.   Activate **Level 1**.

```
⊟ ⊙  Views (all)
   ⊟   Floor Plans
         ☐ Level 1
         ☐ Level 1 - No Annotations
         ☐ Level 2
         ☐ Level 2 - No Annotations
         ☐ Roof Line
         ☐ Site
```

3.

Architecture	Insert	Annotate	Mass

Wall Door Window Component

Place a Component

Select the **Component→Place a Component** tool from the Build panel on the Architecture ribbon.

4.   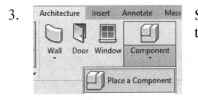   Select **Load Family** from the Mode panel.

Load Family Model In-place

Mode

5.

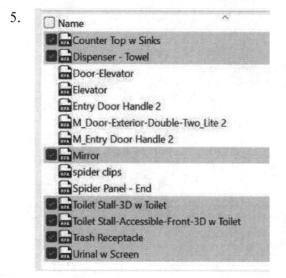

Locate the following files in the downloaded exercises for this text.

- Urinal w Screen
- Counter Top w Sinks
- Dispenser-Towel
  Toilet Stall-3D w Toilet
- Toilet Stall-Accessible-Front-3D w Toilet
- Mirror
- Trash Receptacle

You can select more than one file by holding down the CTRL key.
Click **Open**.

6. Under the Properties Pane:

Change Element Type:

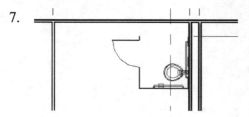

Select the **Toilet Stall-Accessible-Front-3D w Toilet**.

You need to select the 60" x 60" Clear family.

7.

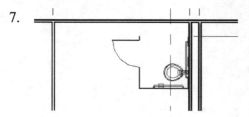

The toilet will be visible when the cursor is over a wall.
Click to place. Do not try to place it exactly.
Simply place it on the wall.

Use the SPACE bar to orient the family.

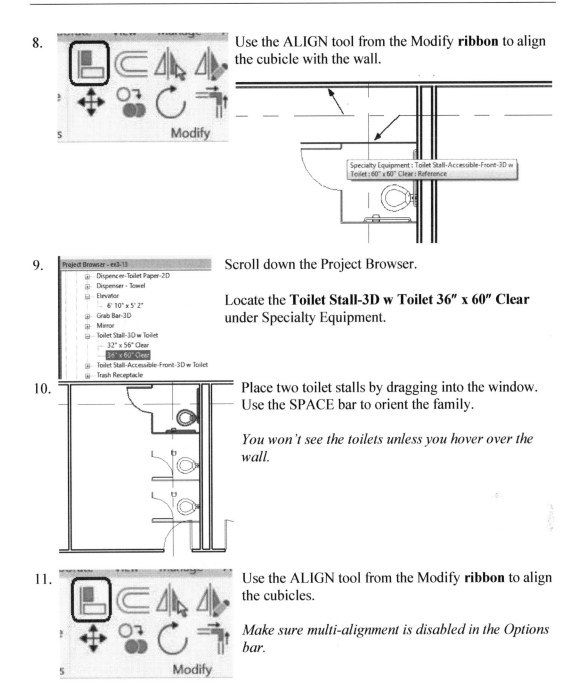

8. Use the ALIGN tool from the Modify **ribbon** to align the cubicle with the wall.

9. Scroll down the Project Browser.

   Locate the **Toilet Stall-3D w Toilet 36″ x 60″ Clear** under Specialty Equipment.

10. Place two toilet stalls by dragging into the window. Use the SPACE bar to orient the family.

    You won't see the toilets unless you hover over the wall.

11. Use the ALIGN tool from the Modify **ribbon** to align the cubicles.

    Make sure multi-alignment is disabled in the Options bar.

12.

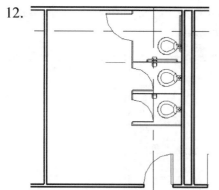

The toilet stalls should all be nestled against each other.

13.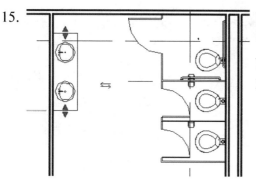

Select the **Component→Place a Component** tool from the Build panel on the Architecture ribbon.

14.

Select the **Counter Top w Sinks 600 mm x 1800 mm Long** from the Type drop-down list on the Properties pane.

15.

Place the **Counter Top w Sinks 600 mm x 1800 mm Long** in the lavatory.

Notice if you pull the grips on this family you can add additional sinks.

16.

Select the **Component→Place a Component** tool from the Build panel on the Architecture ribbon.

17.

Select the **Dispenser - Towel** from the Type drop-down list on the Properties pane.

18.

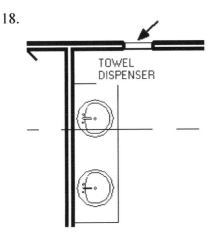

Place the **Dispenser – Towel.**

19. Select the **Component→Place a Component** tool from the Build panel on the Architecture ribbon.

20. Select the **Trash Receptacle** from the Type drop-down list on the Properties pane.

21.

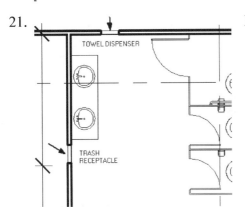

Place the **Trash Receptacle next to the sink.**

22. Select the **Component→Place a Component** tool from the Build panel on the Architecture ribbon.

23. Select the **Mirror: 72″ x 48″** from the Type drop-down list on the Properties pane.

24.

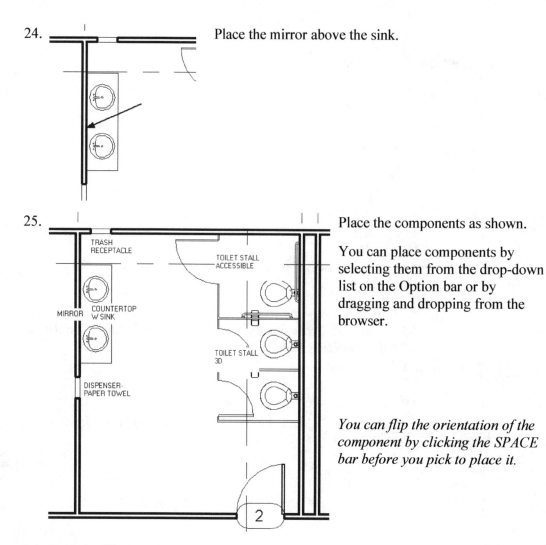

Place the mirror above the sink.

25. Place the components as shown.

You can place components by selecting them from the drop-down list on the Option bar or by dragging and dropping from the browser.

You can flip the orientation of the component by clicking the SPACE bar before you pick to place it.

26. Save the file as *ex3-15.rvt. [m_ex3-15.rvt]*

When you duplicate a view it uses the copied view name as the prefix for the new view name and appends Copy #; i.e. '[View Name] Copy 1' instead of naming the Duplicate View 'Copy of [View Name]'. This makes it easier to sort and locate duplicated views.

Exercise 3-16
Adjust Room Sizes

Drawing Name: dimensions2.rvt [m_dimensions2.rvt]

Estimated Time: 15 minutes

This exercise reinforces the following skills:

-   Pin
-   Global Parameters
-   Floor Plan
-   Dimensions

Verify that the rooms are the same size and that the plumbing space is 1'-0"[305mm] wide.
Pin the walls into position to ensure they don't move around on you.

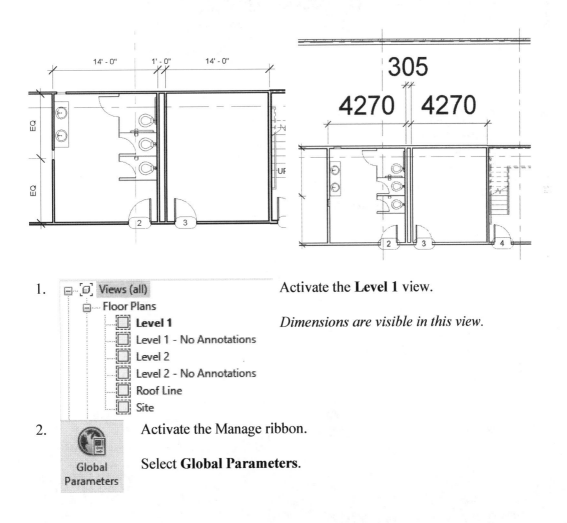

1. Views (all)
   Floor Plans
   - **Level 1**
   - Level 1 - No Annotations
   - Level 2
   - Level 2 - No Annotations
   - Roof Line
   - Site

   Activate the **Level 1** view.

   Dimensions are visible in this view.

2. Global Parameters

   Activate the Manage ribbon.

   Select **Global Parameters**.

3.   Select the **New Parameter** tool.

4.

Name:
Lavatory Room Width

☐ Reporting Parameter

(Can be used to extract value from a geometric condition and report it in a formula)

Discipline:
Common

Type of parameter:
Length

Group parameter under:
Dimensions

Tooltip description:
<No tooltip description. Edit this parameter to write a custom toolti…

Edit Tooltip…

Type **Lavatory Room Width** for the Name.

Click **OK**.

5.

Parameter	Value
Dimensions	
Door Distance	2' 6"
Lavatory Room Width	14' 0"

Parameter	Val
Dimensions	
Door Distance	800.0
Lavatory Width	4270.0

Set the Width to **14' 0"**.
[4270mm]

Click **OK**.

6.

Activate the Modify ribbon.

Enable Visibility for all constraints in the view by toggling on the **Activate** button.

7.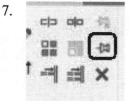

Select the **Pin** tool.

8.

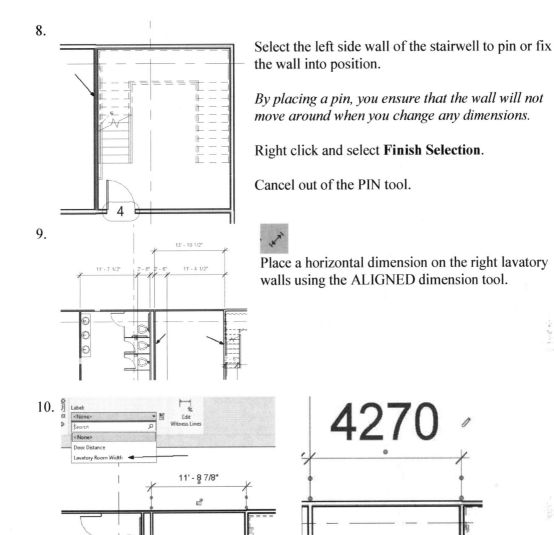

Select the left side wall of the stairwell to pin or fix the wall into position.

By placing a pin, you ensure that the wall will not move around when you change any dimensions.

Right click and select **Finish Selection**.

Cancel out of the PIN tool.

9.

Place a horizontal dimension on the right lavatory walls using the ALIGNED dimension tool.

10.

4270

Select the dimension.

Select the **Lavatory Room Width** label from the ribbon.

The walls will adjust.

11.

Select the ALIGNED dimension tool.

12.

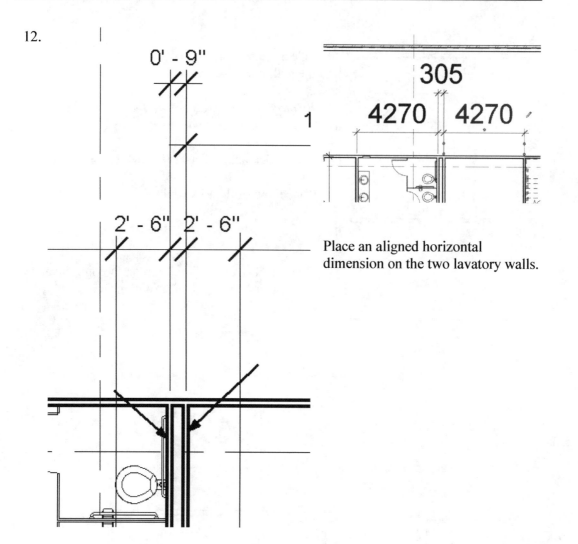

Place an aligned horizontal
dimension on the two lavatory walls.

13.

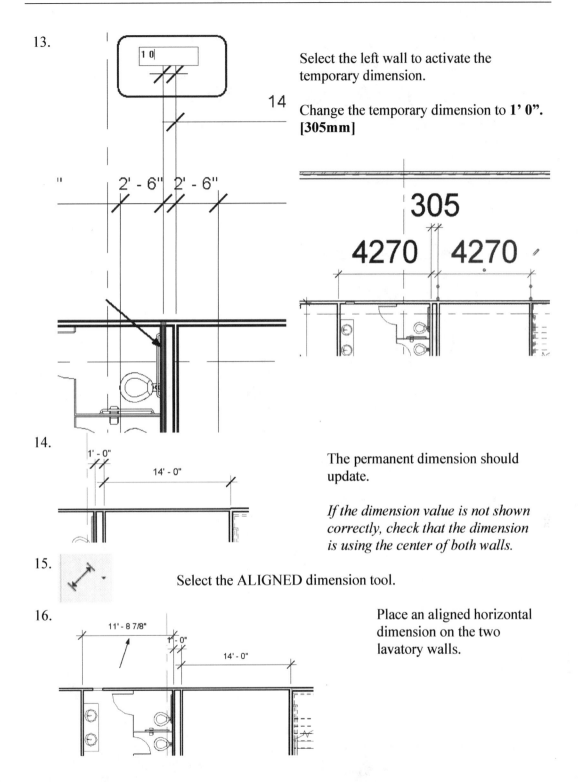

Select the left wall to activate the temporary dimension.

14

Change the temporary dimension to **1' 0"**. **[305mm]**

14.

The permanent dimension should update.

If the dimension value is not shown correctly, check that the dimension is using the center of both walls.

15.

Select the ALIGNED dimension tool.

16.

Place an aligned horizontal dimension on the two lavatory walls.

17.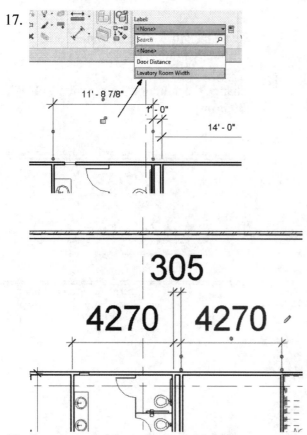

Select the dimension.

Select the **Lavatory Room Width** label from the ribbon.

The walls will adjust.

18. Save as *ex3-16.rvt. [m_ex3-16.rvt]*.

Exercise 3-17
Mirror Components

Drawing Name: mirror.rvt [m_mirror.rvt]
Estimated Time: 15 minutes

This exercise reinforces the following skills:

- ❑ Mirror
- ❑ Align
- ❑ Filter
- ❑ Wall Properties

In order for the Mirror to work properly, the rooms need to be identical sizes.
This is a challenging exercise for many students. It takes practice to draw a mirror axis
six inches from the end point of the wall. If you have difficulty, draw a detail line six
inches between the 1' gap used for plumbing, and use the Pick Axis method for Mirror.

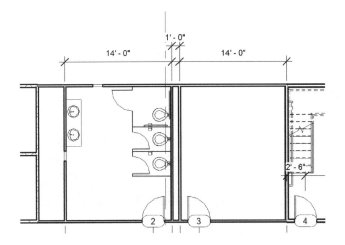

1.   Open *mirror.rvt [m_mirror.rvt]*.

2.   Activate **Level 1 – No Annotations**.

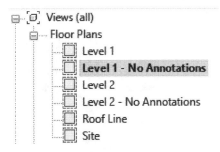

3.

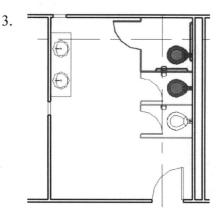

Hold down the Control key and select the top two stalls or just window around them.

They should highlight in red.
Select the countertop with sinks, the trash receptacle, the mirror and the towel dispenser.

Everything is selected EXCEPT one toilet stall.

4.

Select the **Mirror→Draw Mirror Axis** tool under the Modify panel.

5.

Start a line at the midpoint into the wall space.

Hint: Type 0 6 for 6" to start the line.
For metric, type =305/2 to start the line.

Then pick a point directly below the first point selected.

6.

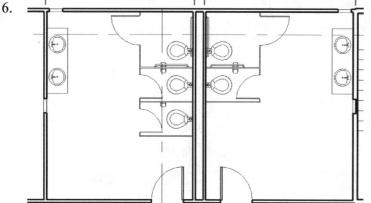

The stalls and other elements are mirrored.

7.

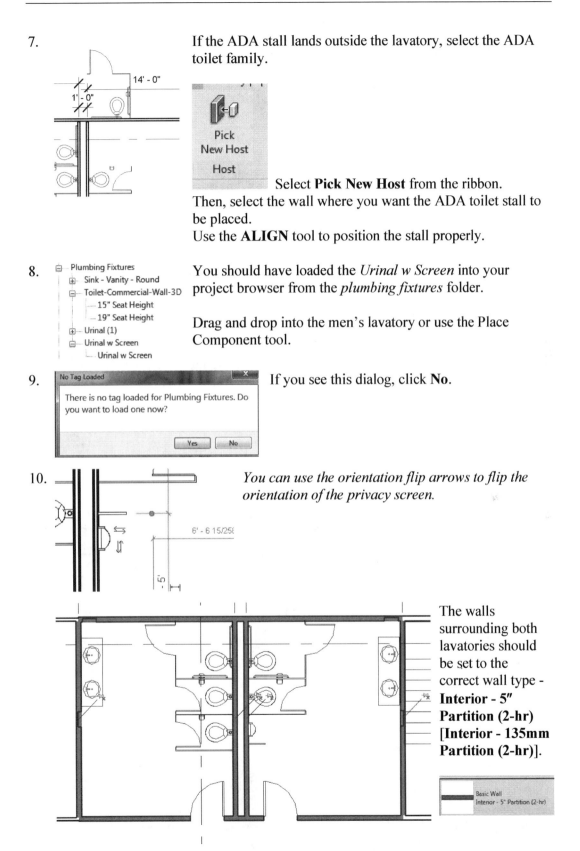

If the ADA stall lands outside the lavatory, select the ADA toilet family.

Select **Pick New Host** from the ribbon.
Then, select the wall where you want the ADA toilet stall to be placed.
Use the **ALIGN** tool to position the stall properly.

8.

You should have loaded the *Urinal w Screen* into your project browser from the *plumbing fixtures* folder.

Drag and drop into the men's lavatory or use the Place Component tool.

9.

If you see this dialog, click **No**.

10.

You can use the orientation flip arrows to flip the orientation of the privacy screen.

The walls surrounding both lavatories should be set to the correct wall type - **Interior - 5″ Partition (2-hr) [Interior - 135mm Partition (2-hr)]**.

11. Use **FILTER** to only select walls.

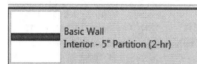

Category:	Count:
Casework	2
Doors	3
Plumbing Fixtures	1
Railings	2
Railings: Top Rails	2
Specialty Equipment	11
Stairs	1
Stairs: Landings	1
Stairs: Runs	2
Stairs: Supports	8
Walls	9

12. Set the walls to the correct type:

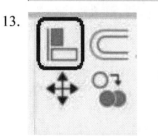

Basic Wall
Interior - 5" Partition (2-hr)

Interior - 5″ Partition (2-hr) [Interior - 135mm Partition (2-hr)] using the Properties pane type selector.

13.

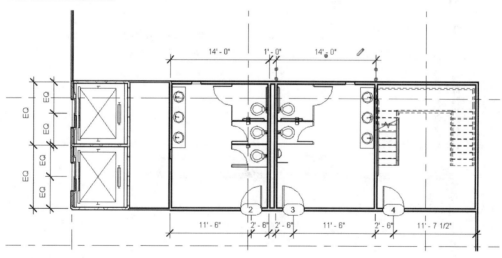

Use **ALIGN** from the Modify panel on the Modify ribbon to align the lavatory walls.

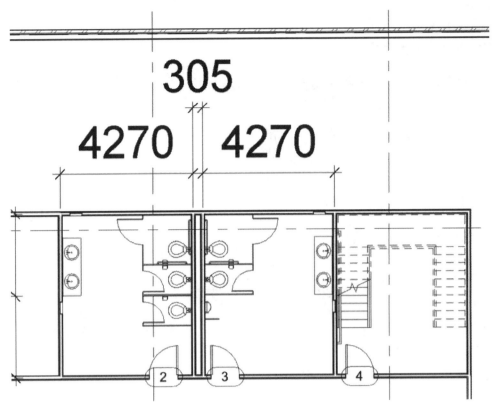

14. Save the file as *ex3-17.rvt [m_ex3-17.rvt]*.

To re-select the previous selection set, hit the left arrow key.

Exercise 3-18
Create a 3D View

Drawing Name: 3D section1.rvt [m_3D section1.rvt]
Estimated Time: 10 minutes

This exercise reinforces the following skills:

- 3D view
- Section Box
- Properties

1. Open *3D section1.rvt [m_3D section1.rvt]*.

2.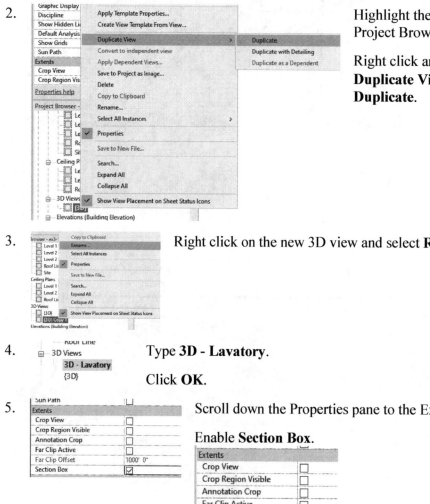
Highlight the 3D view in the Project Browser.

Right click and select
**Duplicate View→
Duplicate**.

3. Right click on the new 3D view and select **Rename**.

4. Type **3D - Lavatory**.

Click **OK**.

5. Scroll down the Properties pane to the Extents section.

Enable **Section Box**.

6.

A clear box appears around the model. Left click to select the box.

There are grips indicated by arrows on the box.

7.

Select the grips and reduce the size of the box, so only the lavatories are shown.

Use the **SPIN** tool to rotate your view so you can inspect the lavatories.

If you hold down the SHIFT key and middle mouse button at the same time, you can orbit the model.

8.

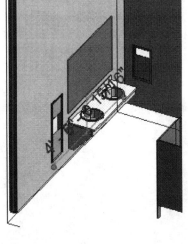

If you need to adjust the elevation/height of the mirror:

Select the mirror.

9.

In the Properties pane:

Set the elevation to **4′ 0″ [1600mm]**.

Click **Apply**.

Repeat for the other mirror.

10. Save as *ex3-18.rvt [m_ex3-18.rvt]*.

Exercise 3-19
Copying Lavatory Layouts

Drawing Name: copy2.rvt [m_copy2.rvt]
Estimated Time: 10 minutes

This exercise reinforces the following skills:

-   Filter
-   Group
-   Copy Aligned

1. Open *copy2.rvt [m_copy2.rvt]*.

2. Activate **Level 1 – No Annotations**.

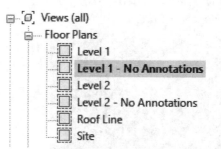

3. Select the components used in the lavatory layout.

4.

Hint: Use the Filter tool to select only the Casework, Plumbing Fixtures and Specialty Equipment.

5. Select the **Create Group** tool from the Create panel.

6. In the Name field, type **Lavatory**.

Click **OK**.

7. Select **Copy to Clipboard** from the Clipboard panel.

8. Select **Paste→Aligned to Selected Levels** from the Clipboard panel.

9. Select **Level 2**.

    Click **OK**.

10.

    Select **Fix Groups**, if you see this dialog.

11. How do you want to fix the inconsistent groups?

    → Ungroup the inconsistent groups
    Ungroups instances that cannot remain consistent with their group type.

    Select **Ungroup the inconsistent groups**.

    → Create new group types
    Creates new group types for instances that cannot remain consistent with their group type.

12. Switch to **Level 2 – No Annotations**.

If the ADA toilet stalls are acting up, select each stall and use the Pick New Host option to locate the stalls correctly.
Then use the ALIGN tool to position them properly in each lavatory.

The lavatory
should be placed.

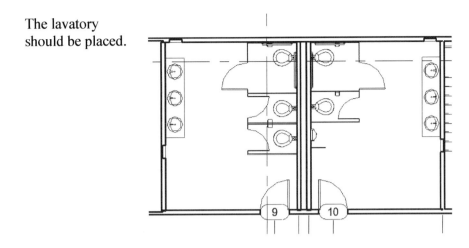

13. Save the file as *ex3-19.rvt.* *[m_ex3-19.rvt].*

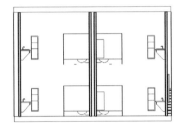

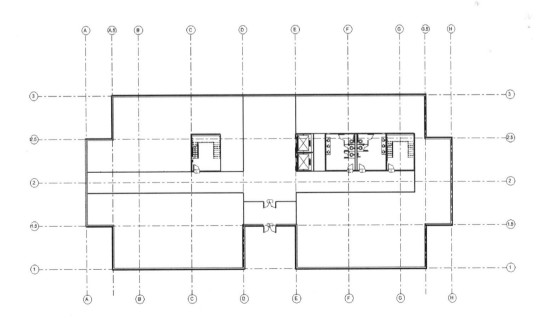

Level 1

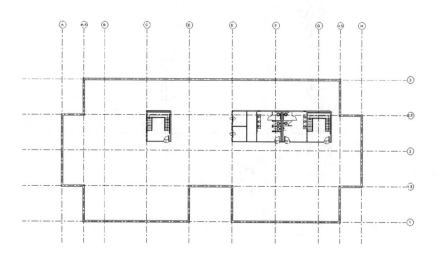

Level 2

Exercise 3-20
Add a Door to a Curtain Wall

Drawing Name: curtain wall1.rvt [m_curtain wall1.rvt]

Estimated Time: 30 minutes

This exercise reinforces the following skills:

- ☐ Curtain Wall
- ☐ Modify Wall Curtain Wall
- ☐ Elevations
- ☐ Grid Lines
- ☐ Load Family
- ☐ Properties
- ☐ Type Selector

1. Open *curtain wall1.rvt. [m_curtain wall1.rvt]*

2.  Activate **Level 1 – No Annotations** Floor Plan.

3. Select the East Wall.

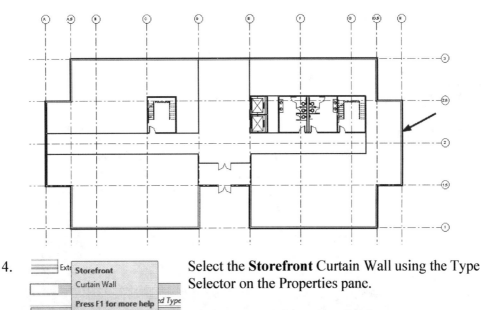

4. Select the **Storefront** Curtain Wall using the Type Selector on the Properties pane.

Click **ESC** to release the selection.

5.

Visibility	
☑ Elevations	
☑ Floor Tags	
☑ Furniture System Tags	

Type **VV** to launch the Visibility/Graphics dialog.
Switch to the Annotations tab.
Enable Elevations.
Click **OK**.

6.

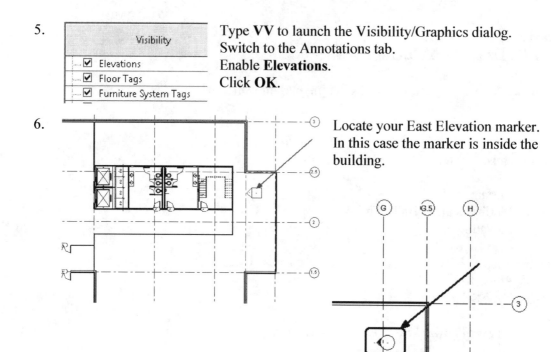

Locate your East Elevation marker.
In this case the marker is inside the
building.

7.

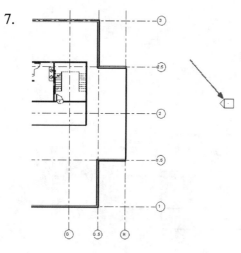

Place your cursor over the square part of the
elevation marker. Hold down your left mouse
button. Drag the marker so it is outside the
building.

The metric elevation marker may be pinned.
Left click on the pin icon to unpin and then
move the elevation marker outside the
building.

8.

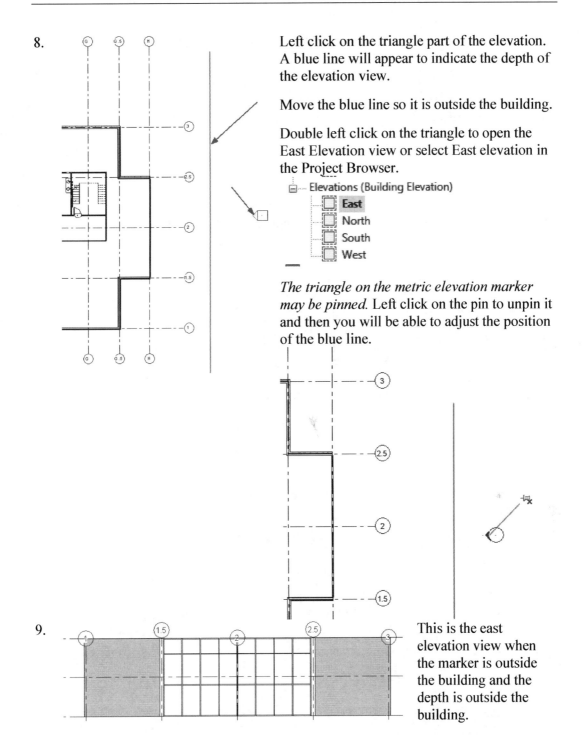

Left click on the triangle part of the elevation. A blue line will appear to indicate the depth of the elevation view.

Move the blue line so it is outside the building.

Double left click on the triangle to open the East Elevation view or select East elevation in the Project Browser.

Elevations (Building Elevation)
- East
- North
- South
- West

The triangle on the metric elevation marker may be pinned. Left click on the pin to unpin it and then you will be able to adjust the position of the blue line.

9.

This is the east elevation view when the marker is outside the building and the depth is outside the building.

10. 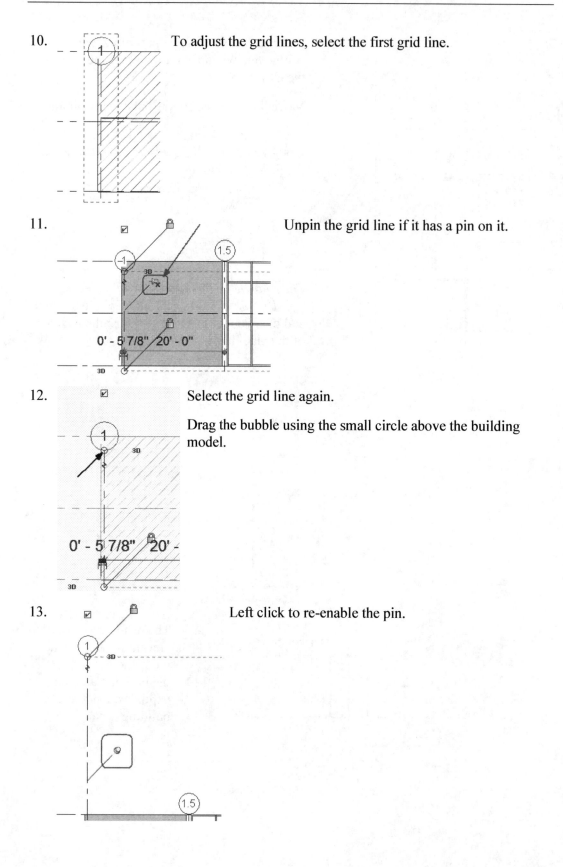 To adjust the grid lines, select the first grid line.

11. Unpin the grid line if it has a pin on it.

12. Select the grid line again.

Drag the bubble using the small circle above the building model.

13. Left click to re-enable the pin.

14.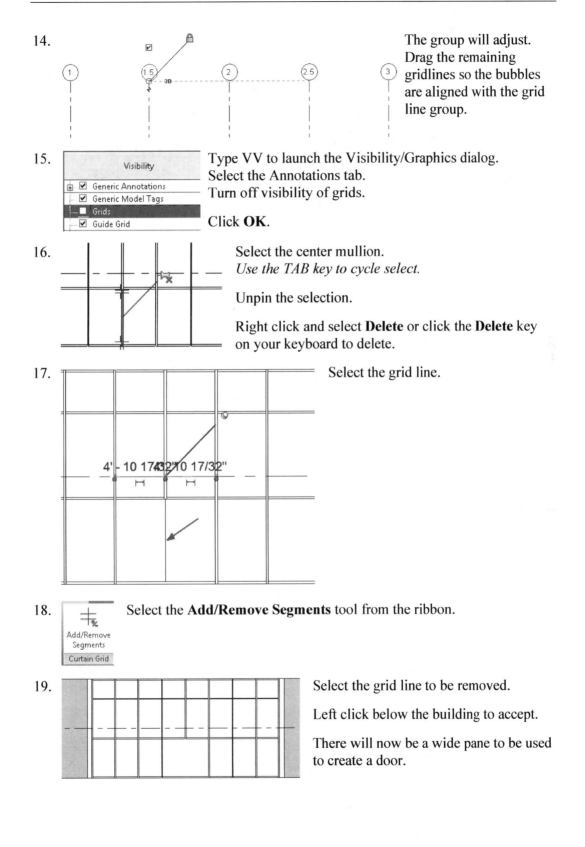

The group will adjust. Drag the remaining gridlines so the bubbles are aligned with the grid line group.

15.

Visibility

☑ Generic Annotations
☑ Generic Model Tags
☐ Grids
☑ Guide Grid

Type VV to launch the Visibility/Graphics dialog.
Select the Annotations tab.
Turn off visibility of grids.

Click **OK**.

16.

Select the center mullion.
Use the TAB key to cycle select.

Unpin the selection.

Right click and select **Delete** or click the **Delete** key on your keyboard to delete.

17.

4'- 10 17/32" 10 17/32"

Select the grid line.

18.

Add/Remove
Segments

Curtain Grid

Select the **Add/Remove Segments** tool from the ribbon.

19.

Select the grid line to be removed.

Left click below the building to accept.

There will now be a wide pane to be used to create a door.

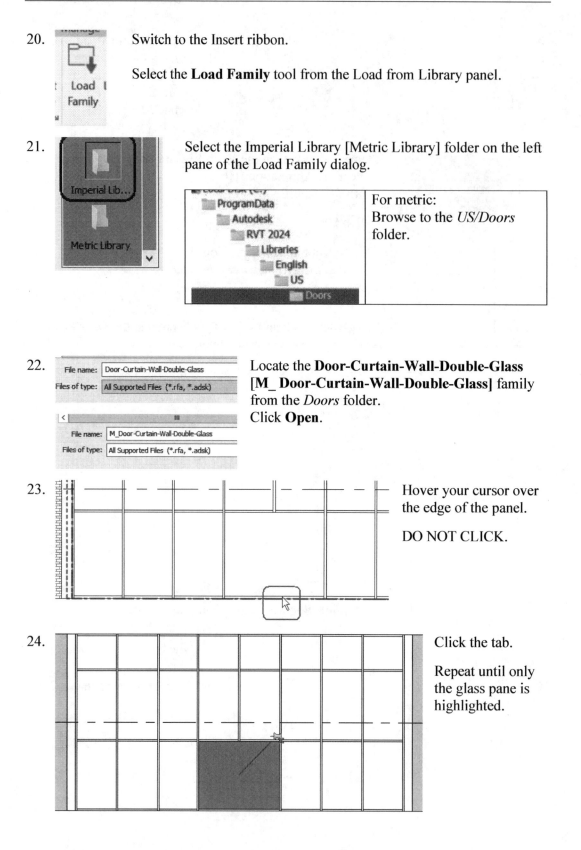

20. Switch to the Insert ribbon.

    Select the **Load Family** tool from the Load from Library panel.

21. Select the Imperial Library [Metric Library] folder on the left pane of the Load Family dialog.

    For metric:
    Browse to the *US/Doors* folder.

22. Locate the **Door-Curtain-Wall-Double-Glass [M_ Door-Curtain-Wall-Double-Glass]** family from the *Doors* folder.
    Click **Open**.

23. Hover your cursor over the edge of the panel.

    DO NOT CLICK.

24. Click the tab.

    Repeat until only the glass pane is highlighted.

25. 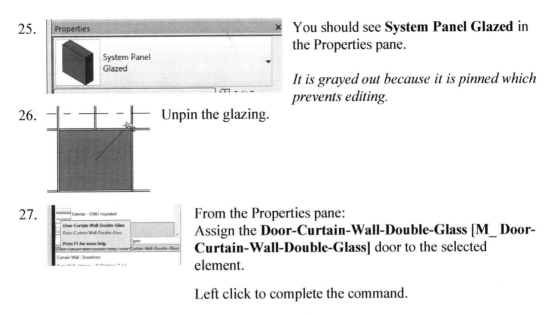 You should see **System Panel Glazed** in the Properties pane.

It is grayed out because it is pinned which prevents editing.

26. 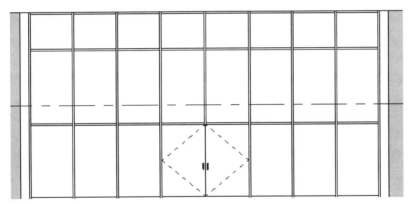 Unpin the glazing.

27. From the Properties pane:
Assign the **Door-Curtain-Wall-Double-Glass [M_ Door-Curtain-Wall-Double-Glass]** door to the selected element.

Left click to complete the command.

28. You will now see a door in your curtain wall.
Remember you won't see the hardware unless the Detail Level is set to Fine.

29. Save as *ex3-20.rvt [m_ex3-20.rvt]*.

Exercise 3-21
Modifying a Curtain Wall

Drawing Name: curtain wall2.rvt [m_curtain wall2.rvt]
Estimated Time: 30 minutes

This exercise reinforces the following skills:

- Use Match to change an Element Type
- Curtain Wall
- Curtain Wall Grids
- Curtain Wall Doors

Several students complained that the door was too big and asked how to make it a smaller size.

1.   Open *curtain wall2.rvt [m_curtain wall2.rvt].*

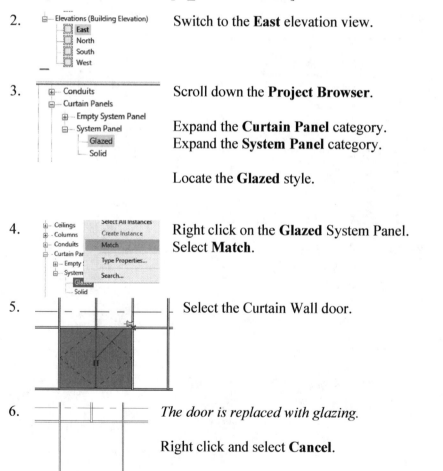

2.   Switch to the **East** elevation view.

3.   Scroll down the **Project Browser**.

    Expand the **Curtain Panel** category.
    Expand the **System Panel** category.

    Locate the **Glazed** style.

4.   Right click on the **Glazed** System Panel.
    Select **Match**.

5.   Select the Curtain Wall door.

6.   *The door is replaced with glazing.*

    Right click and select **Cancel**.

7. 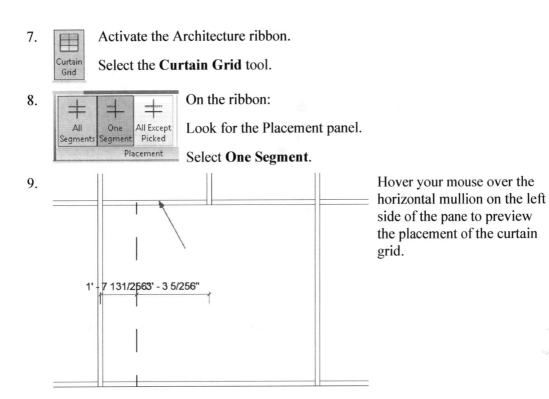   Activate the Architecture ribbon.

Select the **Curtain Grid** tool.

8.   On the ribbon:

Look for the Placement panel.

Select **One Segment**.

9.   Hover your mouse over the horizontal mullion on the left side of the pane to preview the placement of the curtain grid.

1' - 7 131/256" 3' - 3 5/256"

Note that you use the horizontal mullion to place a vertical grid and a vertical mullion to place a horizontal grid.

10. 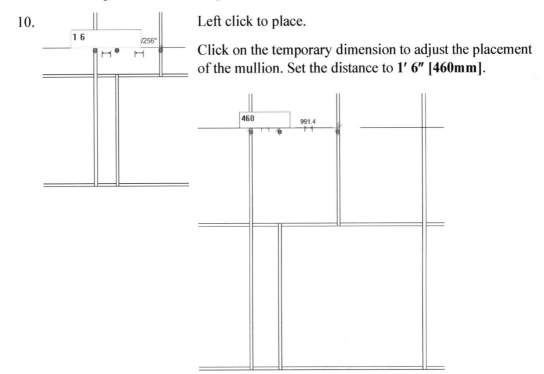   Left click to place.

Click on the temporary dimension to adjust the placement of the mullion. Set the distance to **1' 6" [460mm]**.

460   991.4

11.

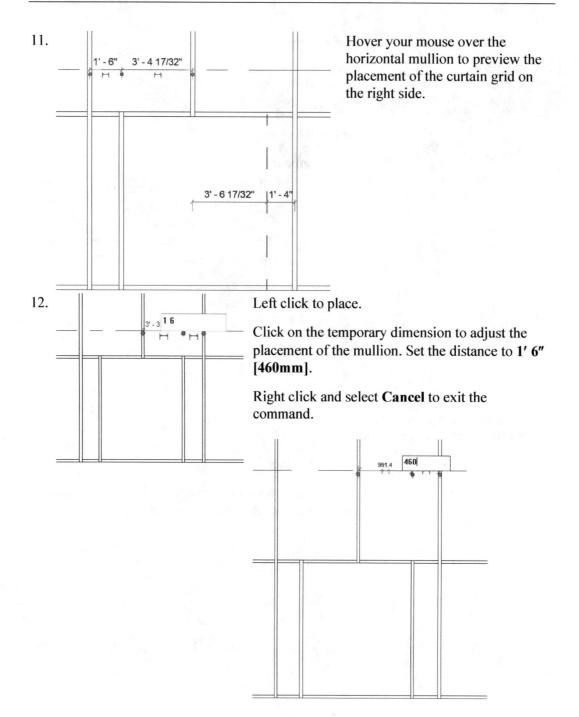

Hover your mouse over the horizontal mullion to preview the placement of the curtain grid on the right side.

12.

Left click to place.

Click on the temporary dimension to adjust the placement of the mullion. Set the distance to **1′ 6″ [460mm]**.

Right click and select **Cancel** to exit the command.

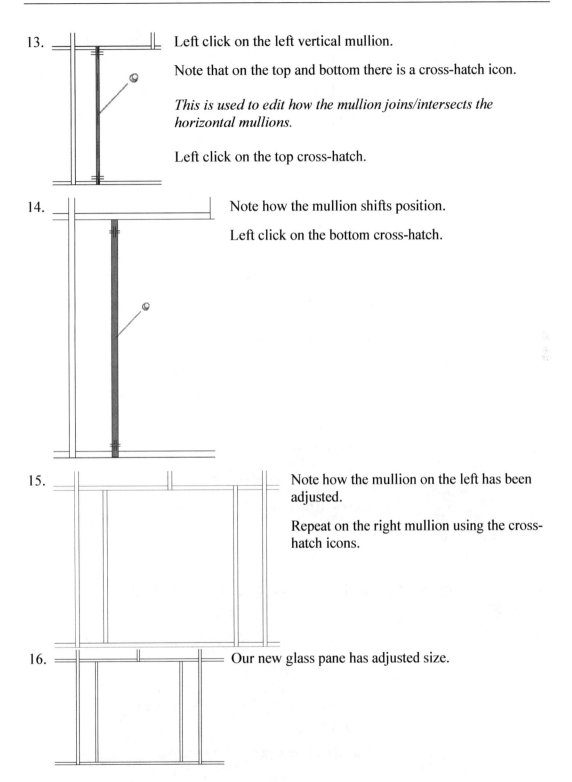

13. Left click on the left vertical mullion.

Note that on the top and bottom there is a cross-hatch icon.

This is used to edit how the mullion joins/intersects the horizontal mullions.

Left click on the top cross-hatch.

14. Note how the mullion shifts position.

Left click on the bottom cross-hatch.

15. Note how the mullion on the left has been adjusted.

Repeat on the right mullion using the cross-hatch icons.

16. Our new glass pane has adjusted size.

17. 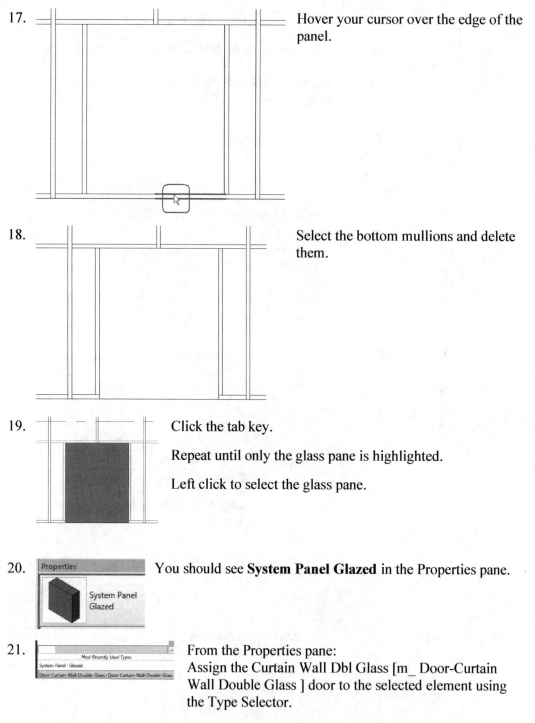 Hover your cursor over the edge of the panel.

18. Select the bottom mullions and delete them.

19. Click the tab key.

Repeat until only the glass pane is highlighted.

Left click to select the glass pane.

20. You should see **System Panel Glazed** in the Properties pane.

21. From the Properties pane:
Assign the Curtain Wall Dbl Glass [m_ Door-Curtain Wall Double Glass] door to the selected element using the Type Selector.

Left click to complete the command.

22. You will now see a door in your curtain wall.
    *The hardware is only visible if the Detail Level is set to **Fine**.*

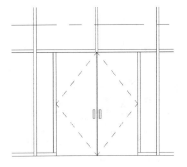

23. Save as *ex3-21.rvt [m_ex3-21.rvt]*.

Exercise 3-22
Curtain Wall with Spiders

Drawing Name: curtain wall3.rvt [m_curtain wall3.rvt]

Estimated Time: 15 minutes

This exercise reinforces the following skills:

- ❑ Curtain Walls
- ❑ System Families
- ❑ Properties

1. Open *curtain wall3.rvt [m_curtain wall3.rvt]*.

2. 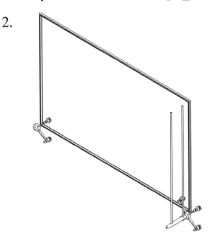 We are going to replace the existing panels in the curtain wall with a panel using spiders.

In order to use the panel with spiders, we need to load the family into the project.

3.   Go to the **Insert** ribbon.

    Select **Load Family**.

4. File name: spider clips
    Files of type: All Supported Files (*.rfa, *.adsk)

    Browse to where you are storing the exercise files.
    Select the *spider clips.rfa.*
    Click **Open**.

5.

    Select the curtain wall so that it is highlighted.

6. Edit Type   Select **Edit Type** from the Properties panel.

7. Duplicate...   Select **Duplicate**.

8. Name: Curtain Wall with Spiders

    Rename *Curtain Wall with Spiders.*
    Click **OK**.

9.
Parameter	Value
Construction	⚹
Function	Exterior
Automatically Embed	☑
Curtain Panel	System Panel : Glazed
Join Condition	Curtain Wall Dbl Glass : Curtain Wall Dbl Glass
Materials and Finishes	Empty System Panel : Empty
	Stacked Wall : Exterior - Brick Over CMU w Metal Stud
Structural Material	System Panel : Glazed
Vertical Grid	System Panel : Solid
Layout	spider clips : Panel - Center

    Locate the Curtain Panel parameter.

    Select the drop-down arrow.

    Select the spider clips: Panel - Center family.

10.
Parameter	Value
Construction	⚹
Function	Exterior
Automatically Embed	☑
Curtain Panel	spider clips : Panel - Center
Join Condition	Vertical Grid Continuous
Materials and Finishes	⚹

    You should see the spider clips panel family listed as assigned to the Curtain Panel.

11.

Vertical Mullions	
Interior Type	Rectangular Mullion : 2.5" x 5" rec
Border 1 Type	Rectangular Mullion : 2.5" x 5" rec
Border 2 Type	Rectangular Mullion : 2.5" x 5" rec
Horizontal Mullions	
Interior Type	Rectangular Mullion : 2.5" x 5" rec
Border 1 Type	Rectangular Mullion : 2.5" x 5" rec
Border 2 Type	Rectangular Mullion : 2.5" x 5" rec

Scroll down the dialog box and locate the Vertical and Horizontal Mullions.

Rectangular Mullion : 50 x 150mm
Rectangular Mullion : 50 x 150mm
Rectangular Mullion : 50 x 150mm

Rectangular Mullion : 50 x 150mm
Rectangular Mullion : 50 x 150mm
Rectangular Mullion : 50 x 150mm

12.

Adjust for Mullion Size	
Vertical Mullions	
Interior Type	Rectangular Mullion : 2.5" x 5" rectangular
Border 1 Type	None
Border 2 Type	Circular Mullion : 2.5" Circular
Horizontal Mullions	Rectangular Mullion : 1" Square
Interior Type	Rectangular Mullion : 1.5" x 2.5" rectangular
Border 1 Type	Rectangular Mullion : 2.5" x 5" rectangular
Border 2 Type	Rectangular Mullion : 2.5" x 5" rectangular
Identity Data	Rectangular Mullion : 2.5" x 5" rectangular
Keynote	

Use the drop-down selector to set all the Mullions to **None**.

13.

Vertical Mullions	
Interior Type	None
Border 1 Type	None
Border 2 Type	None
Horizontal Mullions	
Interior Type	None
Border 1 Type	None
Border 2 Type	None

Verify that all the mullions are set to **None**.

Click **OK**.

14.

Warning - can be ignored

Due to the changes in parameters of their hosts some mullions became non-type driven.

Show   More Info   Expand >>

Delete Mullions   OK   Cancel

If you see this dialog, click **Delete Mullions**.

15.

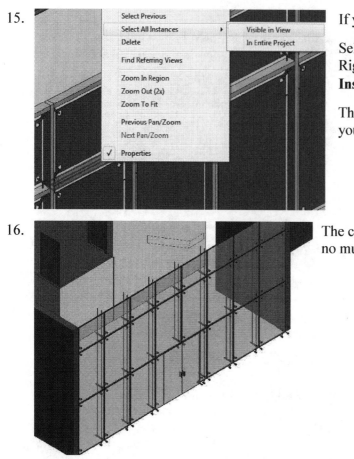

If you still see mullions…

Select one of the mullions. Right click and use **Select All Instances→Visible in View**.

Then click the Delete key on your keyboard.

16.

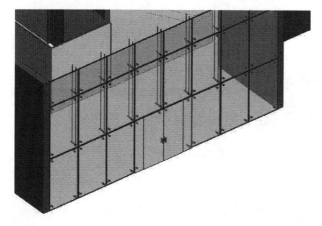

The curtain wall with spiders and no mullions is displayed.

17. Save as *ex3-22.rvt.* *[m_ex3-22.rvt]*

Extra: Replace the bottom panels with the Spider Panel – End Panel family to create a cleaner looking curtain wall.

Hint: *Load the Spider Panel – End Panel family and replace each bottom panel with the new family using the Type Selector (similar to placing the curtain wall door). You will need to unpin before you can replace the panel.*

Exercise 3-23
Adding Windows

Drawing Name: windows1.rvt [m_windows1.rvt]

Estimated Time: 60 minutes

This exercise reinforces the following skills:

- ❑ Window
- ❑ Window Properties
- ❑ Array
- ❑ Mirror
- ❑ Copy-Move

1. Open *windows1.rvt [m_windows1.rvt]*.

2. 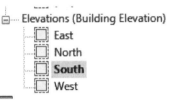 Switch to a **South** elevation view.

3. 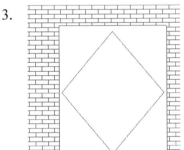 Zoom into the entrance area of the building.

    Display is set to Coarse, Hidden Line.

4. Select the **Architecture** ribbon.

    Window Select the **Window** tool from the Build panel.

5.

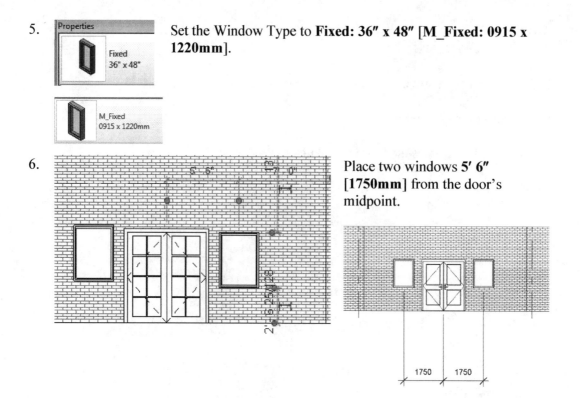

Set the Window Type to **Fixed: 36″ x 48″** [**M_Fixed: 0915 x 1220mm**].

6.

Place two windows **5′ 6″ [1750mm]** from the door's midpoint.

7. In plan view, the windows will be located as shown.

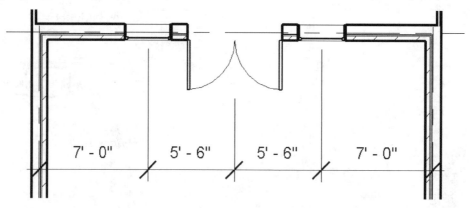

7' - 0"   5' - 6"   5' - 6"   7' - 0"

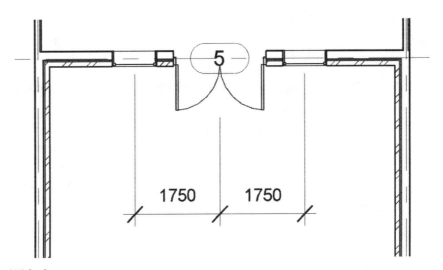

8.  Activate the **South** Elevation.

9. Zoom into the entrance area.

10. 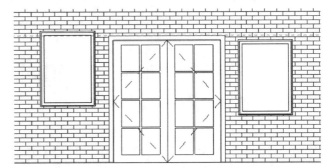 Use the **ALIGN** tool to align the top of the windows with the top of the door.
Select the **Align** tool on the Modify panel on the Modify ribbon.

11. Enable **Multiple Alignment** on the Options bar.

12. Select the top of the door, then the top of the left window. Then, select the top of the second window.

Right click and select **Cancel** twice to exit the command.

13. Activate **Level 1 – No Annotations**.

14. 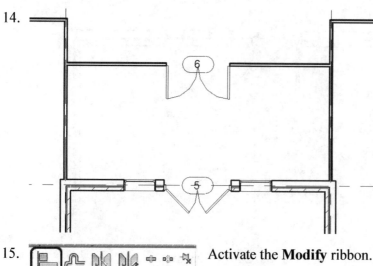 Zoom into the entry area.

We want the exterior walls to be centered on the grid lines.

We want the interior walls aligned to the interior face of the exterior walls.

15. Activate the **Modify** ribbon.

Select the **ALIGN** tool.

16. On the Options bar:
Disable **Multiple Alignment**.
Set Prefer to **Wall faces**.

17. Select the interior face on the exterior wall.

Then select the corresponding face on the interior wall.

Repeat for the other side.

18. Elevations (Building Elevation)
East
North
South
West

Activate the South elevation.

19. Place a **Fixed: 36″ x 48″ [M_Fixed: 0915 x 1220mm]** window on the Level 2 wall.

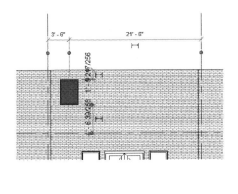

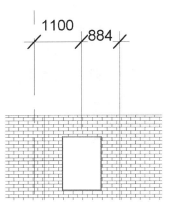

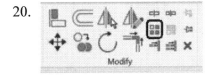

Locate the window 3′ 6″ [1100mm] to the right of grid D.

20.   Select the window.
       Select the **Array** tool from the Modify panel.

21. Enable **Group and Associate**.

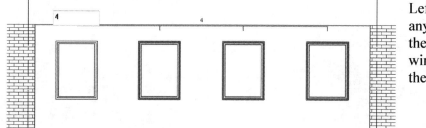

Set the Number to **4**.
Enable Move to: **Last**.
Enable **Constrain**. *This forces your cursor to move orthogonally.*

22.

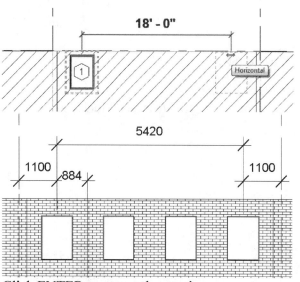

Pick the midpoint of the top of the window as the base points.

Move the cursor 18′ [5420 mm] towards the right.
Pick to place.

23. Click ENTER to accept the preview.

Left click anywhere in the display window to exit the command.

24.

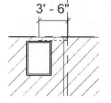

Use the **Measure** tool from the Quick Access tool bar to verify that the last window is 3′ 6″ [1100 mm] from grid E.

Verify that the four windows are equally spaced using the Measure tool.

25. Select the first window.
Select the window group.

Select **Edit Group** from the Group panel.

26. Select the first window.

27. In the Properties pane:
Set the Sill Height to **4' 0" [1200mm]**.

Windows (1)	
Constraints	
Level	Level 2
Sill Height	1200.0

28. Select **Finish** from the Edit group toolbar.

29. *The entire group of windows adjusts.*

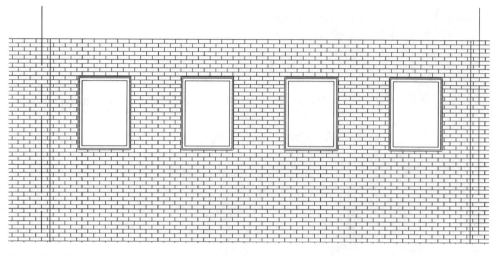

30. Save the file as *ex3-23.rvt [m_ex3-23.rvt]*.

Challenge Exercise:

Add additional windows to complete the project.
Use the MIRROR and ARRAY tools to place windows.

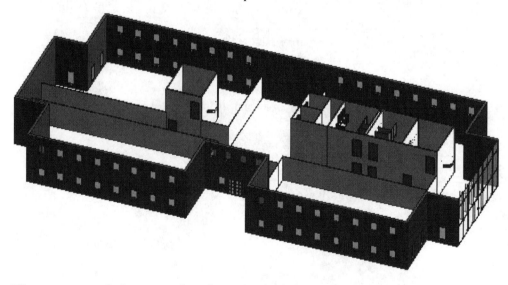

Elements can only be arrayed on the same work face.

Exercise 3-24:
Floor Plan Layout

Drawing Name: floor plan1.rvt [m_floor plan1.rvt]
Estimated Time: 20 minutes

This exercise reinforces the following skills:

- ❑ Loading a Title block
- ❑ Add a New Sheet
- ❑ Adding a view to a sheet
- ❑ Adjusting Gridlines and Labels
- ❑ Hiding the Viewport

1. Open *floor plan1.rvt [m_floor plan1.rvt]*.

2.  In the browser, highlight **Sheets**.
   Right click and select **New Sheet**.

3.  Select **Load**.

4.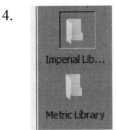

   Select the *Imperial Library [Metric Library]* folder located on the left pane.

   For Metric:

   Browse to *Libraries/English/US/Titleblocks*.

5.

Titleblocks		Titleblocks
Name		Name
A 8.5 x 11 Vertical		A0 metric
B 11 x 17 Horizontal		A1 metric
C 17 x 22 Horizontal		A2 metric
D 22 x 34 Horizontal		A3 metric
E 34 x 44 Horizontal		A4 metric
E1 30 x 42 Horizontal		

Browse to the *Titleblocks* folder.

Select the **D 22 x 34 Horizontal [A3 metric]** Titleblock.

Click **Open**.

6.

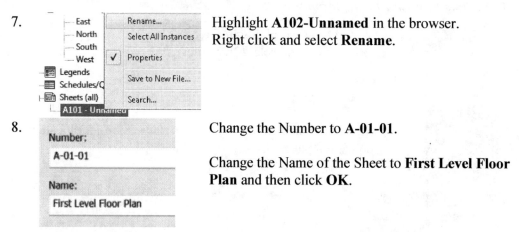

Highlight the **D 22 x 34 Horizontal [A3 metric]** Titleblock.

Click **OK**.

7. Highlight **A102-Unnamed** in the browser.
Right click and select **Rename**.

8. Change the Number to **A-01-01**.

Change the Name of the Sheet to **First Level Floor Plan** and then click **OK**.

9. Drag and drop the Level 1 Floor Plan from the browser into the sheet.

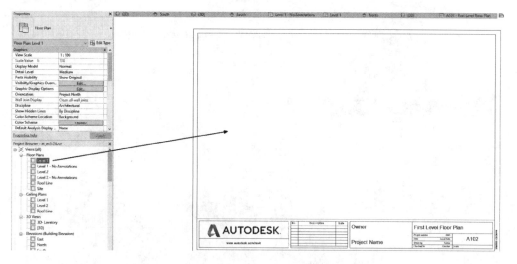

10.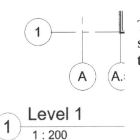

For metric:

Select the view on the sheet.
Change the View Scale to **1:200** in the Properties pane.

11.

Level 1
1/8" = 1'-0"

If you zoom into the View label, you see the scale is set to *1/8" = 1'-0" [1:200]*.

To adjust the view label line, click on the view, then select the line and use the grip on the end point to adjust the length.

Level 1
1 : 200

You can adjust the position of the label by picking on the label to activate the blue grip. Then hold the grip down with the left mouse button and drag to the desired location.

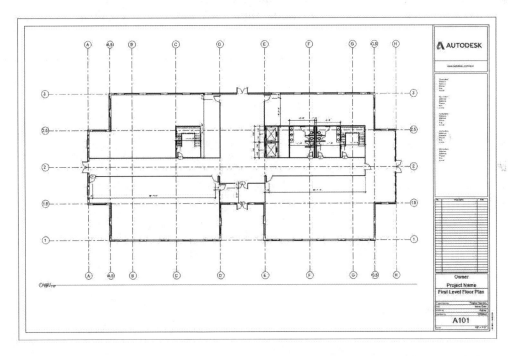

12.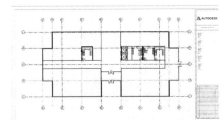

Adjust the position of the grids as needed.

Unpin one grid and then use the grip next to the bubble head to re-position.

Notice that the icon for the view updates in the Project Browser to indicate that it has been placed on a sheet.

Views can only be placed on a sheet once. If you need to place another view of the floor plan, you need to duplicate the view.

13. In the Properties Pane:
Change **Author** next to Drawn By to your name.

The title block updates with your name.

14. Save as *ex3-24.rvt [m_ex3-24.rvt]*.

Extra: *Place a second sheet with the Level 2 Floor Plan.*

Turn off the elevation marker visibility in the views.
To do this: Select the view. Right click and select Activate View. Then use VV/VG to disable the elevations. Once you are done, right click and select Deactivate View to return to the sheet.

Additional Projects

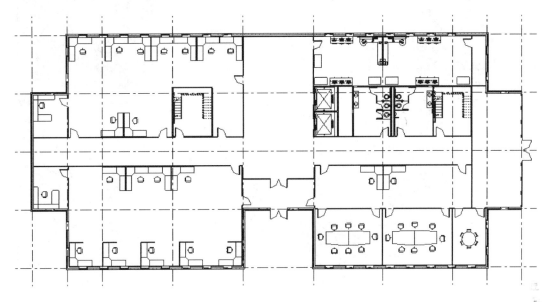

1) Duplicate the Level 1 Floor plan and create an office layout similar to the one shown. If you have access to the Generative Model tool (this is only available to users with a subscription), see if you can generate several conceptual layouts.

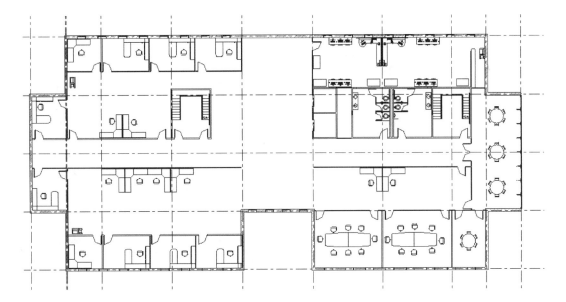

2) Duplicate the Level 2 Floor plan and create an office layout similar to the one shown.

3) On Level 2, in the cafeteria area, add furniture groups using a table and chairs.
   Add vending machines. (Families can be located from the publisher's website.)

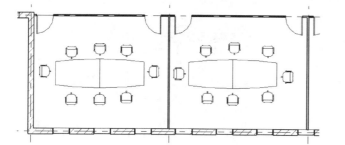

Add tables and chairs to the two conference room areas.

Use Group to create furniture combinations that you can easily copy and move.

There are conference table group families available on the publisher's website.

4) Furnish the two conference rooms on Level 2.

5) Furnish the office cubicles. Families can be found on the publisher's website.

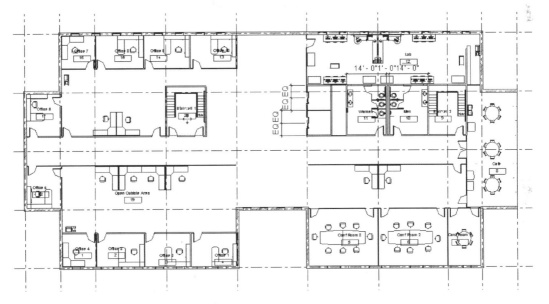

6) Add room tags to the rooms on Level 2.
   Create a room schedule.

<Room Schedule>		
A	**B**	**C**
Number	Name	Area
1	Office 4	136 SF
2	Office 3	187 SF
3	Office 2	181 SF
4	Office 1	183 SF
5	Conf Room 3	470 SF
6	Conf Room 2	479 SF
7	Conf Room 1	233 SF
8	Cafe	708 SF
9	Stairwell 2	229 SF
10	Men	228 SF
11	Women	228 SF
12	Lab	1058 SF
13	Office 10	183 SF
14	Office 9	181 SF
15	Office 8	187 SF
16	Office 7	136 SF
17	Office 6	176 SF
18	Office 5	176 SF
19	Open Cubicle A	6168 SF
20	Stairwell 1	225 SF

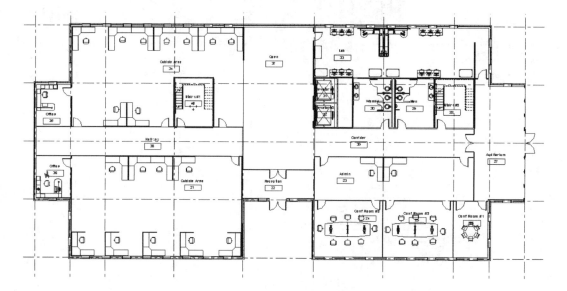

7) Use the Room Separator tool and the Room Tag tool to add rooms and room tags to Level 1.

Lesson 3 Quiz

True or False

1. Changes in the settings for one view do not affect other views.

Multiple Choice

2. When placing doors, which key is pressed to flip orientation?
   Choose one answer.

   A. Ctrl+S
   B. H
   C. Spacebar
   D. F
   E. L or R

3. Which of the following is NOT an example of bidirectional associativity?
   Choose one answer.

   A. Draw a wall in plan and it appears in all other views
   B. Add an annotation in a view
   C. Change a door type in a schedule and all the views update
   D. Flip a section line and all views update

4. Curtain Grids can be defined using all of the following except:
   Choose one answer.

   A. Vertical line
   B. Angled line
   C. Horizontal line
   D. Partial line

5. A stair can consist of all the following EXCEPT:

   A. Runs

   B. Landings

   C. Railings

   D. Treads

6. Which command is used to place a free-standing element, such as furniture?
    Choose one answer.

    A. Detail Component
    B. Load Family
    C. Repeating Detail
    D. Model In Place
    E. Place a Component

7. Select the TWO that are type properties of a wall:
    Choose at least two answers.

    A. FUNCTION
    B. COARSE FILL PATTERN
    C. BASE CONSTRAINT
    D. TOP CONSTRAINT
    E. LOCATION LINE

8. Which is NOT true about placing windows?
    Choose one answer.

    A. Windows require a wall as a host
    B. Windows cut an opening in the wall when placed
    C. The location of the exterior side of the window can be selected
    D. Sill height is adjustable in plan view

9. If you highlight Level 1, right click and select Duplicate View→Duplicate, what
    will the new view name be by default?

    A. Copy of Level 1
    B. Level 1 Copy 1
    C. Level 1 (2)
    D. None of the above

ANSWERS:

   1) T; 2) C; 3) B; 4) B; 5) D; 6) E; 7) A & B; 8) D; 9) B

Lesson 4
Materials

Materials are used:

- For Rendering
- For Scheduling
- To control hatch patterns
- To control appearance of objects

Revit divides materials into TWO locations:

- Materials in the Project
- Autodesk Asset Library

You can also create one or more custom libraries for materials you want to use in more than one project. You can only edit materials in the project or in the custom library – Autodesk Materials are read-only.

In the Revit 2024, users have the ability to store materials on the cloud. This makes it easier for users to share their custom materials and access them. This release only allows you to access and store the materials using the Autodesk Desktop Connector and you need to have a subscription for Autodesk Docs.

| Identity | Graphics | Appearance | Physical | Thermal |

Materials have FIVE asset definitions – Identity, Graphics, Appearance, Physical and Thermal. Materials do not have to be fully defined to be used – YOU DON'T HAVE TO DEFINE ALL FIVE ASSETS. Each asset contributes to the definition of a single material.

You can define Appearance, Physical, and Thermal Assets independently without assigning/associating it to a material. You can assign the same assets to different materials. An asset can exist without being assigned to a material, BUT a material must have at least THREE assets (Identity, Graphics, and Appearance). Assets can be deleted from a material in a project or in a user library, but you cannot remove or delete assets from materials in locked material libraries, such as the Autodesk Materials library, the AEC Materials library, or any locked user library. You can delete assets from a user library, BUT be careful because that asset might be used somewhere!

You can only duplicate, replace or delete an asset if it is in the project or in a user library. The Autodesk Asset Library is "read-only" and can only be used to check-out or create materials.

The Material Browser is divided into three distinct panels.
The Project Materials panel lists all the materials available in the project.
The Material Libraries list any libraries available.
The Material Editor allows you to modify the definition of materials in the project or in custom libraries.

The search field allows you to search materials using keywords, description or comments.

The button indicated toggles the display of the Material Libraries panel.

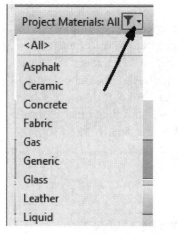

The Filter icon allows you to filter the display of project materials by material type.

Open Existing Library Create New Library Remove Library Create Category Delete Category Rename	Creates, opens, and edits user libraries. Create categories to help organize materials.
Create New Material Duplicate Material and Assets Duplicate Using Shared Assets	Create a new material or duplicate, rename, and redefine an existing material.
	Open or close the Asset Browser.
« »	Open or close the Material Editor.
	Create a custom parameter which can be used in a schedule.

Exercise 4-1
Modifying the Material Browser Interface

Drawing Name: basic_project.rvt
Estimated Time: 10 minutes

This exercise reinforces the following skills:

❑ Navigating the Material Browser and Asset Browser

1. Activate the **Manage** ribbon.

    Materials   Select the **Materials** tool.

2. Make the Material Browser appear as shown.

    Toggle the Materials Library pane OFF by selecting

    the button next to Materials Libraries.

    Collapse the Material Editor by selecting the >> Arrow on the bottom of the dialog.

3.

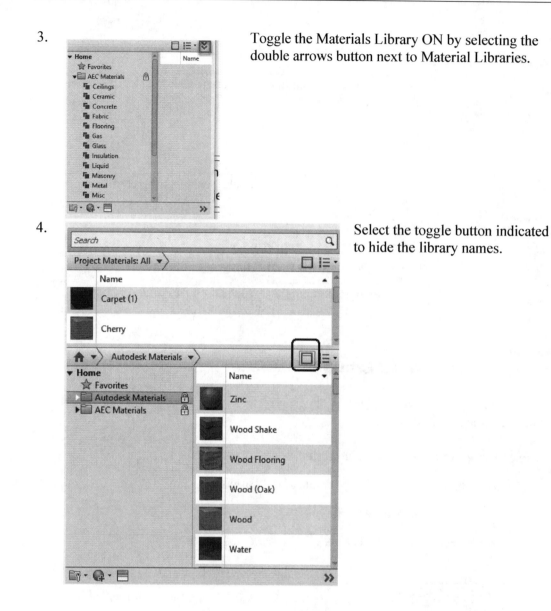

Toggle the Materials Library ON by selecting the double arrows button next to Material Libraries.

4.

Select the toggle button indicated to hide the library names.

5.

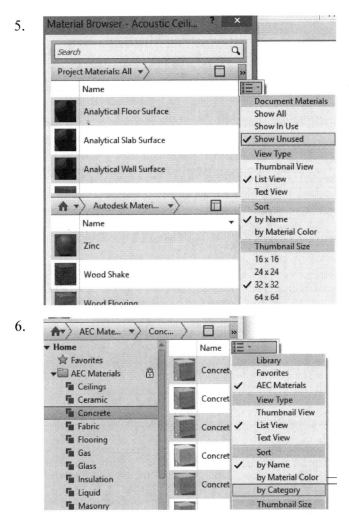

In the Project Materials panel, set the view to **Show Unused**.

Scroll through the list to see what materials are not being used in your project.

6.

In the Material Library pane, change the view to **sort by Category**.

7. Close the Material dialog.

The Material Browser has up to five tabs.

The Identity Tab

The value in the Description field can be used in Material tags.

Using comments and keywords can be helpful for searches.

Product Information is useful for material take-off scheduling.

Assigning the keynote and a mark for the legend makes keynote creation easier.

The Graphics Tab

Shading –
Determines the color and appearance of a material when view mode is SHADED.

If you have Use Render Appearance unchecked, then the material will show a color of the material when the display is set to Shaded.

If Use Render Appearance is checked, then the material uses the Render Appearance Color when the display is set to Shaded.

Comparing the appearance of the model when Render Appearance is disabled or enabled:

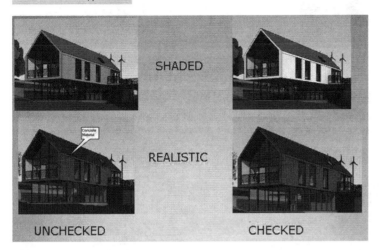

Surface Pattern will be displayed in all the following Display settings: Wireframe/Hidden/Shaded/Realistic/Consistent Colors.

Cut Patterns appear in section views.

The Appearance Tab

Appearance controls how material is displayed in a rendered view, Realistic view, or Ray Trace view.

A zero next to the hand means that only the active/selected material uses the appearance definition.

Different material types can have different appearance parameters.

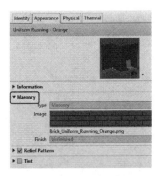

The assigned class controls what parameters are assigned to the material. If you create a custom class, it will use generic parameters. The images below illustrate how different classes use different appearance parameters.

The Physical Tab

This information may be used for structural and thermal analysis of the model.

You can only modify physical properties of materials in the project or in a user library.

The Thermal Tab

The Thermal tab is only available in Revit. (The Autodesk Material Library is shared by any installed Autodesk software, such as AutoCAD.)

This information is used for thermal analysis and energy consumption.

Material libraries organize materials by category or class.

Material structural classes are organized into the following types:
- Basic
- Generic
- Metal
- Concrete
- Wood
- Liquid
- Gas
- Plastic

This family parameter controls the hidden view display of structural elements. If the Structural Material Class of an element is set to Concrete or Precast, then it will display as hidden. If it is set to Steel or Wood, it will be visible when another element is in front of it. If it is set to Unassigned, the element will not display if hidden by another element.

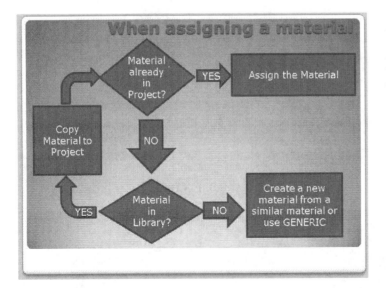

Use this flow chart as a guide to help you decide whether or not you need to create a new material in your project.

Exercise 4-2
Copy a Material from a Library to a Project

Drawing Name: copy material.rvt
Estimated Time: 10 minutes

In order to use or assign a material, it must exist in the project. You can search for a material and, if one exists in one of the libraries, copy it to your active project.

This exercise reinforces the following skills:

- ❑ Copying materials from the Materials Library
- ❑ Modifying materials in the project

1.   Go to the Manage ribbon.

2.   Select the **Materials** tool.

3.   Type **oak** in the search field.

4.   If you look in the Materials Library pane, you see that more oak materials are listed.

5.   Select the Up arrow to copy the **Oak, Red** material from the library to the active project.

6.   The material is now available for use in the project.

7.   Type **brick** in the search field.

8. There are several more brick materials in the library.

9. Locate Brick, Adobe in the project materials.
Highlight and right click.
Select **Duplicate Material and Assets**.

This allows you to copy the material to the project and edit it.

10. Activate the Identity tab.

Change the Name to **Brick, Adobe Unfinished**.

11. Select the button next to **Keynote**.

12. Highlight **Clay Unit Masonry**.
Click **OK**.

13. The keynote number auto-fills the text box.

14. Activate the Graphics tab.
Enable **Use Render Appearance**.

15.

Left click on <**none**> in the Pattern box under Surface Pattern Foreground and select **Masonry – Brick**.

Click **OK**.

16.

You can use the search field to help you locate the desired fill pattern.

17.

The fill pattern now appears in the Pattern box.

18.

Activate the **Appearance** tab.

Left click on the image name.

19.

Locate the *adobe brick* image file included in the exercise files.
Click **Open**.

20. 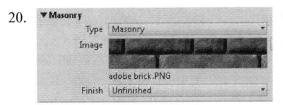 Change the Finish to **Unfinished**.

Click **OK**.

21. Close the Materials dialog.

22. Save as *ex4-2.rvt*.

You have modified the material definition in the project ONLY and not in the Materials library.

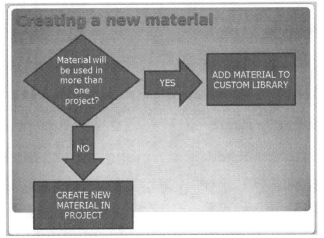

If you need to create a new material, you should first consider whether the material might be used in more than one project.

If the material may be used again, add the material to your custom library.

Steps for creating a new material:

1. Locate a material similar to the one you want, if you can (otherwise, use Generic)
2. Duplicate the material
3. Rename
4. Modify the Identity, Graphics, and Appearance tabs
5. Apply to the object
6. Preview using Realistic or Raytrace
7. Adjust the definition as needed

Create a Custom Material Library

Drawing Name: custom library.rvt
Estimated Time: 5 minutes

This exercise reinforces the following skills:

❑   Materials Library

1.   Go to the Manage ribbon.

2.   Select the **Materials** tool.
     Materials

3.   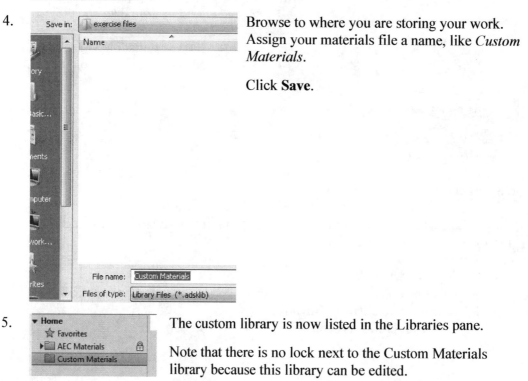 At the bottom of the dialog:

     Select **Create New Library**.

4.   Browse to where you are storing your work. Assign your materials file a name, like *Custom Materials*.

     Click **Save**.

5.   The custom library is now listed in the Libraries pane.

     Note that there is no lock next to the Custom Materials library because this library can be edited.

     Click **OK** to close the Materials dialog.

6.   Save the project as *ex4-3.rvt*.

Exercise 4-4
Create Paint Materials

Drawing Name: paint materials.rvt
Estimated Time: 30 minutes

This exercise reinforces the following skills:

❑ Materials

1. Go to the Manage ribbon.

2.  Select the **Materials** tool.

   Materials

3.  Type **Paint** in the search text box.

   Select the Create **New Material** tool at the bottom of the dialog.

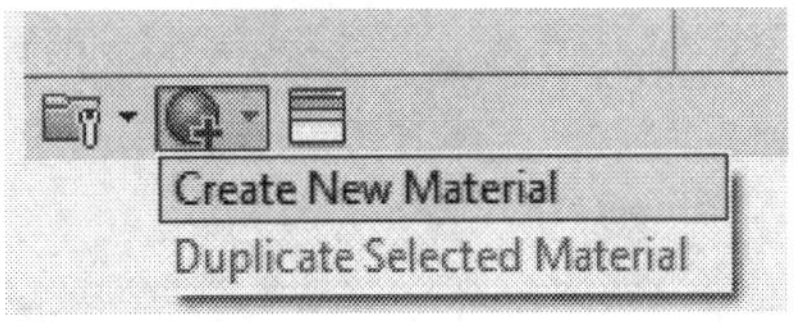

4. 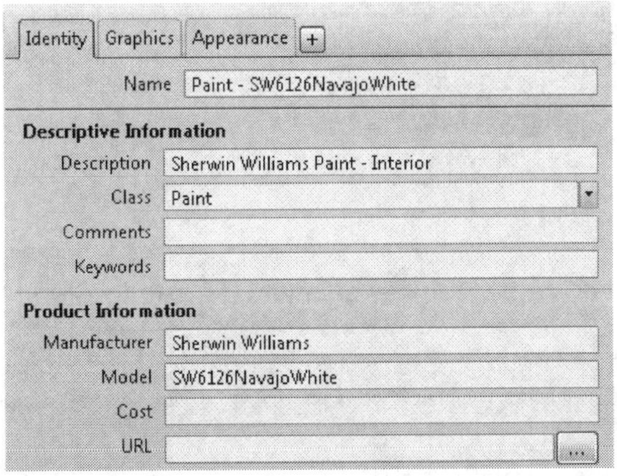 Activate the **Identity** tab.
   In the Name field, type **Paint - SW6126NavajoWhite**.
   In the Description field, type **Sherwin Williams Paint – Interior**.
   Select **Paint** for the Class under the drop-down list.
   In the Manufacturer field, type **Sherwin Williams**.
   In the Model field, type **SW6126NavajoWhite**.

5.
Revit Annotation Information

Keynote [ ] [...]

Mark [ ]

Select the **... browse** button next to Keynote.

6.

Division 09 — Finishes
09 22 00 — Supports for Plaster and Gypsum Board
09 23 00 — Gypsum Plastering
09 24 00 — Portland Cement Plastering
09 29 00 — Gypsum Board
09 30 00 — Tiling
09 51 00 — Acoustical Ceilings
09 64 00 — Wood Flooring
09 65 00 — Resilient Flooring
09 68 00 — Carpeting
09 72 00 — Wall Coverings
09 73 00 — Wall Carpeting
09 81 00 — Acoustic Insulation
09 84 00 — Acoustic Room Components
09 91 00 — Painting
09 91 00.A1 — Paint Finish
09 91 00.A2 — Semi-Gloss Paint Finish

Select the **09 91 00.A2 Semi-Gloss Paint Finish** in the list.

Click **OK**.

7.
Revit Annotation Information

Keynote [09 91 00.A2] [...]

Mark [ ]

Note the keynote information auto-fills based on the selection.

8.

Identity | Graphics | Appearance | [+]

▼ **Shading**

☑ Use Render Appearance

Color [RGB 120 120 120]

Transparency [0]

Activate the Graphics tab.

Place a check on **Use Render Appearance**.

9.

Identity | Graphics | Appearance | [+]

0 Generic(13)

▶ **Information**

▼ **Generic**

Color [RGB 80 80 80]

Image [ ]

(no image selected)

Select the Appearance tab.

Left click on the Color button.

10.

Custom colors:

Select an empty color swatch under Custom colors.

11. In the Red field, enter **234**.
In the Green field, enter **223**.
In the Blue field, enter **203**.

Click **Add**.

Hue:	26	Red:	234
Sat:	102	Green:	223
Lum:	206	Blue:	203

This matches Sherwin William's Navajo White, SW6126.

12. Select the color swatch.

Click **Add**.

The color swatch will be added to the pallet.

Click **OK**.

Custom colors:

Name:
RGB 234-223-203

Original   New

Hue:	26	Red:	234
Sat:	102	Green:	223
Lum:	206	Blue:	203

Add   PANTONE...

OK   Cancel

13. Change the Scene to display **Walls**.

This way you can see what the color will look like on a wall.

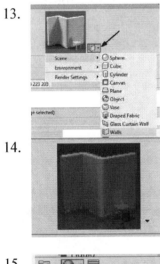

Scene
Environment
Render Settings

Sphere
Cube
Cylinder
Canvas
Plane
Object
Vase
Draped Fabric
Glass Curtain Wall
Walls

14. The preview will update to display the new color.

Click **Apply**.

15. Click on the bottom of the Material Browser to add a material.

Create New Material
Duplicate Selected Material

16. On the Identity tab:

On the Identity tab:

Type **Paint - Interior-SW0068 Heron Blue** for the name of the material.

Identity | Graphics | Appearance | +
Name | Paint - Interior SW0068 Heron Blue

17.

Enter the Description in the field.

You can use Copy and Paste to copy the Name to the Description field.

Set the Class to **Paint**.
Enter the Manufacturer and Model information.

18.

Select the **...** browse button next to Keynote.

19.

09 91 00	Painting	
09 91 00.A1	Paint Finish	
09 91 00.A2	Semi-Gloss Paint Finish	

Locate the **Semi-Gloss Paint Finish**.
Click **OK**.

20.

The Identity tab should appear as shown.

21.

Activate the **Graphics** tab.

Enable **Use Render Appearance**.

22.

Activate the Appearance tab.

Select the down arrow next to Color to edit the color.

23. Select an empty slot under Custom Colors.

Set Red to 173.
Set Green to 216.
Set Blue to 228.

Click **Add**.

24. Select the custom color.

Click **OK**.

25. The color is previewed.

26.

If you return to the Graphics tab, you see that the Color has been updated.

27. Click **OK** to close the Materials dialog.

28. Save as *ex4-4.rvt*.

Exercise 4-5
Add Categories and Materials to a Custom Library

Drawing Name: material library.rvt
Estimated Time: 10 minutes

This exercise reinforces the following skills:

- ❑ Material Libraries
- ❑ Create Library Categories
- ❑ Add Materials to Custom Libraries

1.   Go to the Manage ribbon.

2.   Select the **Materials** tool.

3.   In the Library Pane:

    Highlight the Custom Materials library.
    Right click and select **Create Category**.

4.   Type over the category name and rename it to **Paint**.

5.   Locate the Navajo White paint in the Project Materials list.

    Right click and select **Add to → Custom Materials → Paint**.

6.   The Paint material is now listed in your Custom Materials library.

The material will now be available for use in other projects.

7. Locate Heron Blue paint in the Project Materials list.

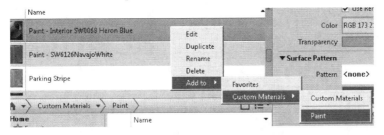

Right click and select **Add to → Custom Materials → Paint**.

8.

There are now two paint materials available in the Custom Materials library.

9. Close the Materials Manager.

10. Save as *ex4-5.rvt*.

Exercise 4-6
Defining Wallpaper Materials

Drawing Name: wallpaper.rvt
Estimated Time: 15 minutes

This exercise reinforces the following skills:

- ❑ Materials
- ❑ Defining a new class
- ❑ Using the Asset Manager

1. Go to **Manage→Materials**.

2. Type **Wallpaper** for the name of the material.

3. There are no wallpaper materials in the project or the library.

4. Clear the search field.

   Locate the **Default** material in the project.

5. Highlight the **Generic** material.
   Right click and select **Rename**.

6. Rename to **Wallpaper - Striped**.

7. Activate the Identity tab.

   In the Description field: type
   Wallpaper – Striped.

   In the Class field, type **Wallpaper**.

This adds Wallpaper to the Class drop-down list automatically.

8. Browse to assign a keynote.

9. Locate the **Vinyl Wallcovering** keynote.
   Click **OK**.

10.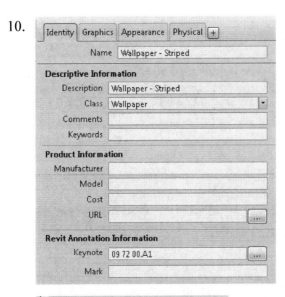

The Identity tab should appear as shown.

11. Launch the **Asset Browser** by clicking the button on the dialog.

12. Type **stripes** in the search field.

13. Highlight **Wall Covering**.
    There are several materials available.

Asset Name	Aspect	Type	Category
Vertical Stripes - Blue-Gray	Appe...	Generic	Wall Covering
Vertical Stripes - Multi-colored	Appe...	Generic	Wall Covering
Vertical Stripes - Pink-Beige	Appe...	Generic	Wall Covering
Wallpaper - Stripes Splatter	Appe...	Generic	Wall Covering

Highlight the **Vertical Stripes – Blue-Gray**.

14. Select the double arrow on the far right to copy the material properties to the Wallpaper-Striped definition.

15. Activate the Appearance tab.

Note how the appearance has updated.

16.

Activate the Graphics tab.

Place a check on Use Render Appearance.

Note that the color has updated for the Render Appearance.

17. Close the Material Browser.

18. Save as *ex4-6.rvt*.

Extra: *Create a Wallpaper – Floral material using the Asset Browser.*

Note: The materials you defined are local only to this project. To reuse these materials in a different project, they should be added to a custom library or use Transfer Project Standards on the Manage ribbon.

Exercise 4-7
Defining Vinyl Composition Tile (VCT)

Drawing Name: vct.rvt
Estimated Time: 30 minutes

This exercise reinforces the following skills:

- Materials
- Copy a material
- Rename a material
- Use the Asset Browser

1.   ⬨   Go to **Manage→Materials**.

   Materials

2. | laminate | Type **laminate** in the search field.

3.   Highlight the **Laminate - Ivory, Matte** material.

   | Identity | Graphics | Appearance | + |

   Name: Laminate - Ivory, Matte

   Descriptive Information
   Description: Laminate- Ivory, Matte
   Class: Laminate
   Comments:
   Keywords:

   Product Information
   Manufacturer:
   Model:
   Cost:
   URL:

   Revit Annotation Information
   Keynote:
   Mark:

   On the Identity tab:

   Type in the Description field:
   Laminate - Ivory, Matte

   Type in the Class field:
   Laminate.

   Browse for the keynote.

4.
09 64 00	Wood Flooring
09 65 00	Resilient Flooring
09 65 00.A1	Resilient Flooring
09 65 00.A2	Vinyl Composition Tile
09 65 00.A3	Rubber Flooring

   Select the keynote for **Vinyl Composition Tile**.

   Click **OK**.

5.

The Identity tab should appear as shown.

6.

On the Graphics tab:
Enable **Use Render Appearance**.

7.

Select the Pattern button under Surface Pattern Foreground to assign a hatch pattern.

8.

Scroll down and select the **Plastic** fill pattern.

You can also use the Search field to help you locate the fill pattern.

Click **OK**.

9.

Select the Appearance tab.

In the Preview window:

Select the down arrow.
Highlight Canvas to switch the preview to display a canvas.

10.

The Appearance tab should look as shown.

11.

There are multiple Ivory, Matte laminate materials in the Project.

Highlight the ones that were not modified.

Right click and click **Delete**.

12.

Verify that there is only one Ivory, Matte laminate material in the project by doing a search on ivory.

13. `tile`   Type **tile** in the search box.

14. 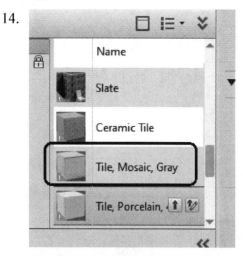   Locate the **Tile, Mosaic, Gray** in the library panel.

Copy into the project materials.

15. On the Identity tab:

Change the Description.

Change the Class to **Laminate**.

16. Add **VCT, gray** to the Keywords list.

17. Assign **Vinyl Composition Tile** to the keynote.

18.

Name: Tile, Mosaic, Gray

Descriptive Information

Description: Mosiac tile, gray

Class: Laminate

Comments:

Keywords: VCT, gray

Product Information

Manufacturer:

Model:

Cost:

URL:

Revit Annotation Information

Keynote: 09 65 00.A2

Mark:

The Identity tab should appear as shown.

19.

▼ Shading

☑ Use Render Appearance

Color: RGB 186 186 186

Transparency: 0

On the Graphics tab:

Enable **Use Render Appearance**.

20.

▼ Shading

☑ Use Render Appearance

Color: RGB 198 189 176

Transparency:

▼ Surface Pattern

▼ Foreground

Pattern: <none>

Color: RGB 0 0 0

Alignment: Texture Alignment...

Select the Pattern button under Surface Pattern Foreground to assign a hatch pattern.

21.

Name: | Pattern:

Aluminum

Concrete

Crosshatch

Crosshatch-small

Scroll down and select the **Crosshatch** fill pattern.

Click **OK**.

22.

The Graphics tab should appear as shown.

23. | TILE | Type **TILE** in the search field.

24.

Scroll down the Document materials and locate the Vinyl Composition Tile.

Right click and select **Rename**.

25. | Vinyl Composition Tile - Diamond Pattern |

Name the new material **Vinyl Composition Tile - Diamond Pattern**.

26.

Activate the Identity tab.

In the Description field: Type **VCT-Diamond Pattern**.
Set the Class to **Laminate**.
Enter **VCT, Diamond** for Keywords.

Assign **09 65 00.A2** as the Keynote.

27.

Activate the Graphics tab.

Enable **Use Render Appearance** under Shading.

28.

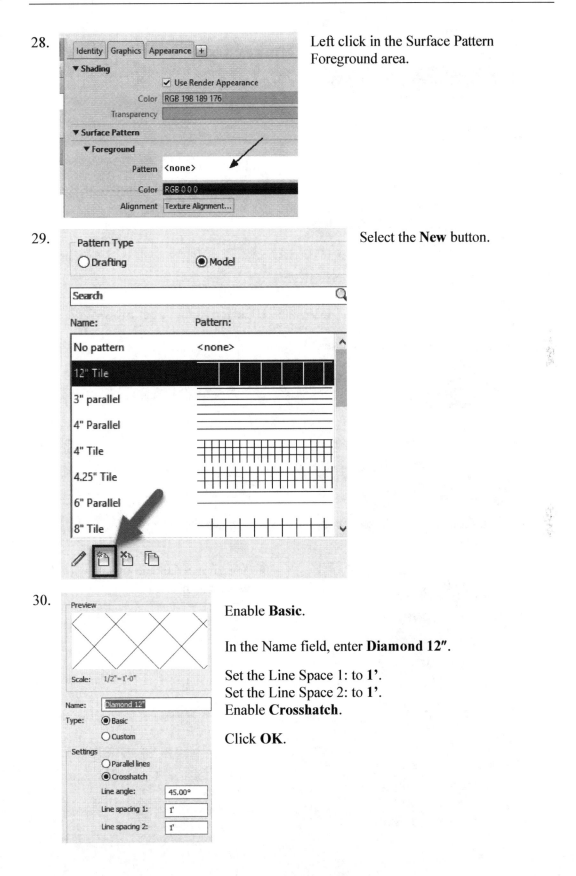

Left click in the Surface Pattern Foreground area.

29.

Select the **New** button.

30.

Enable **Basic**.

In the Name field, enter **Diamond 12″**.

Set the Line Space 1: to **1'**.
Set the Line Space 2: to **1'**.
Enable **Crosshatch**.

Click **OK**.

31.

Ceiling 24x48

Diamond 12"

HerringBone 2x6

Pattern Type

○ Drafting   ◉ Model

No Pattern   OK   Cancel

The new hatch pattern is listed.

Highlight to select and click **OK**.

32.

▼ Surface Pattern

Pattern Diamond 12"

Color RGB 0 0 0

Alignment Texture Alignment...

The new hatch pattern is displayed.

33. Launch the Asset Browser.

34. diamond Type diamond in the search box at the top of the Asset Browser dialog.

35. Locate the **Diamonds1** property set.

tile

Project Materials: All ▼

Search results for "tile"

Name
Tile, Mosaic, Gray

Tile, Porcelain, 4in

Vinyl Composition Tile - Diamond Pattern

Search Result ▼

Search Result

diamond

Appearance Library: Flooring: Vinyl

Search results for "diamond"

▼ Appearan...
  ▶ Ceramic
  ▼ Flooring
    Stone
    Tile
    Vinyl
  ▶ Metal
  ▶ Miscellaneous

Asset Name		Aspect	Type	Category
Diamond - Rosette 1		Appe...	Generic	Flooring: Vinyl
Diamonds		Appe...	Plastic	Flooring: Vinyl
Diamonds 1		Appe...	Generic	Flooring: Vinyl

At the far left, you should see a button that says replace the current asset in the editor with this asset.

36.

Identity | Graphics | Appearance | Physical | Thermal

Diamonds 1

▶ Information

▼ Generic

Color RGB 174 181 185

Image

Finishes.Flooring.VCT.Diamonds...

Image Fade 100

Glossiness 71

Highlights Non-Metallic

The Appearance tab will update with the new asset information.

37.  The Material Browser will update.

Close the Asset Browser.

Click **OK.**

38. Save as *ex4-7.rvt.*

Define a Glass Material

Drawing Name: glass.rvt
Estimated Time: 20 minutes

This exercise reinforces the following skills:

- Materials
- Use the Asset Browser
- Duplicate a Material

1.   Go to **Manage→Materials**.

Materials

2. | glass | Type glass in the search field.

3.

Locate the **Glass, Frosted** material in the Library pane.

Right click and select **Add to→ Document Materials** or use the Up arrow.

4.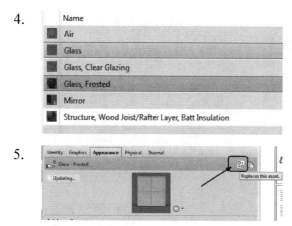

Highlight **Glass, Frosted** in the Documents pane.

5.

Activate the Appearance tab.

Select **Replace this asset**.

This will launch the Asset Browser.

6. The Asset Browser will launch.

Type **glass** in the search field.

Highlight **Glass** under Appearance Library and scroll through to see all the different glass colors.

Select **Clear - Amber**.

Select the Replace asset tool on the far right.

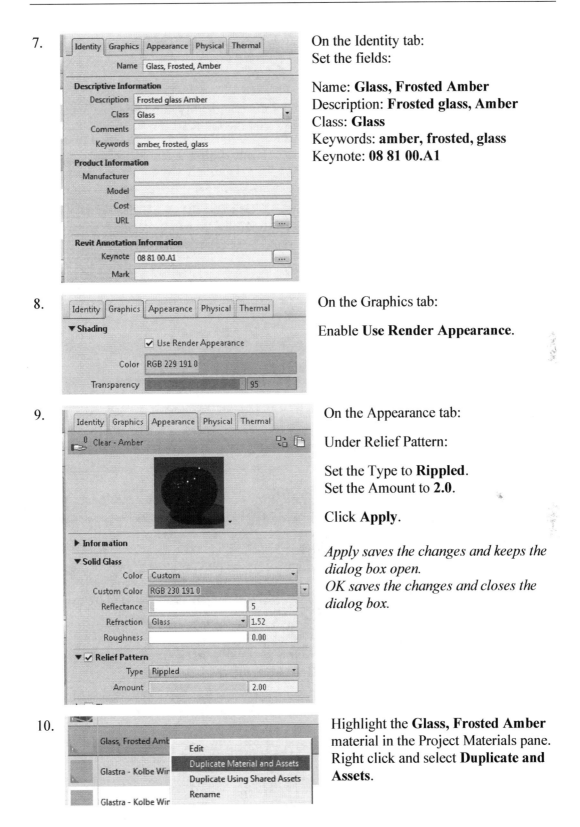

7.

On the Identity tab:
Set the fields:

Name: **Glass, Frosted Amber**
Description: **Frosted glass, Amber**
Class: **Glass**
Keywords: **amber, frosted, glass**
Keynote: **08 81 00.A1**

8.

On the Graphics tab:

Enable **Use Render Appearance**.

9.

On the Appearance tab:

Under Relief Pattern:

Set the Type to **Rippled**.
Set the Amount to **2.0**.

Click **Apply**.

Apply saves the changes and keeps the dialog box open.
OK saves the changes and closes the dialog box.

10.

Highlight the **Glass, Frosted Amber** material in the Project Materials pane.
Right click and select **Duplicate and Assets**.

11. 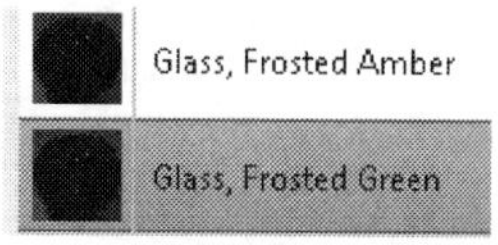 Change the name to **Glass, Frosted Green**.

12.

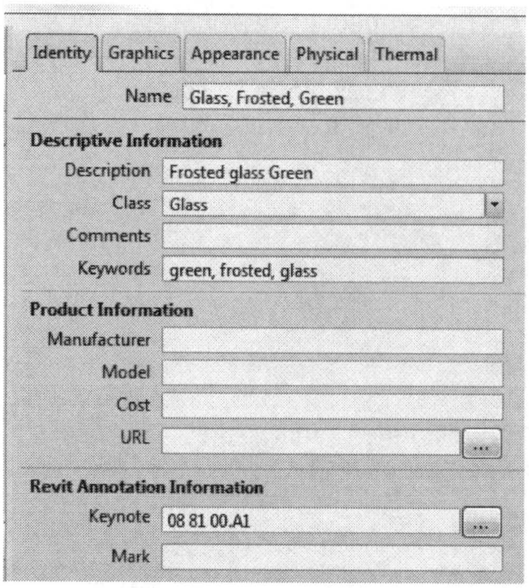

On the Identity tab:
Set the fields:

Name: **Glass, Frosted Green**
Description: **Frosted glass, Green**
Class: **Glass**
Keywords: **green, frosted, glass**
Keynote: **08 81 00.A1**

13. In the Asset Browser:
In the Appearance Library

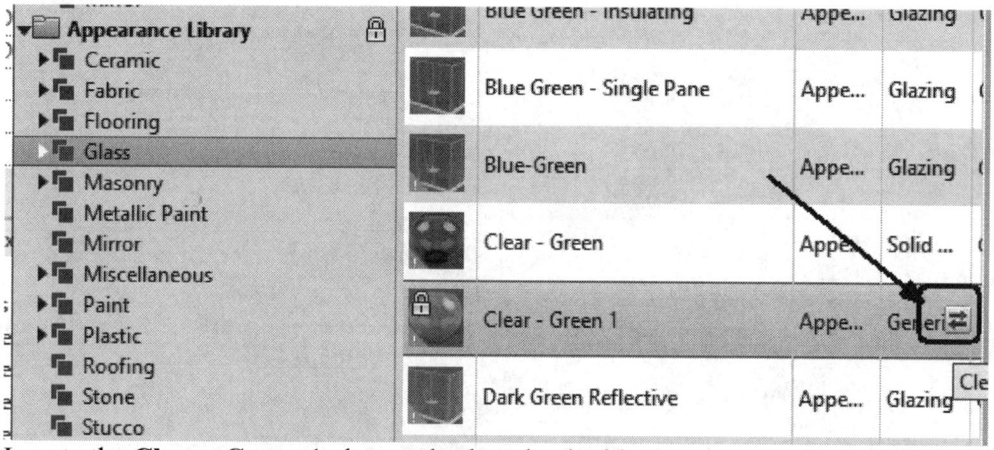

Locate the **Clear - Green 1** glass and select the double arrows to assign that material appearance to the Glass, Frosted Green material.

14. 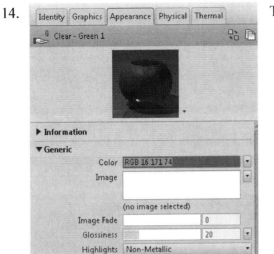 The Appearance properties will update.

15. Check the preview image and verify that the material looks good.

Adjust the appearance settings to your preferences.

16. Activate the Graphics tab.

Enable **Use Render Appearance**.

Click **Apply** to save the settings.

Click **OK** to save the new materials and close the dialogs.

17. Save as *ex4-8.rvt*.

Exercise 4-9
Defining Wood Materials

Drawing Name: wood.rvt
Estimated Time: 5 minutes

This exercise reinforces the following skills:

- Materials
- Appearance Settings
- Identity Settings
- Copying from Library to Project Materials

1.   ⬤   Go to **Manage→Materials**.

    Materials

2.   [oak]   Type **oak** in the search field.

3.   Locate the **Oak, Red** in the Project Materials.
    This material was added earlier.

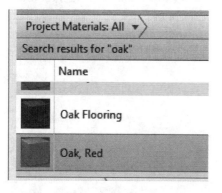

4.

In the Identity tab:

Name: **Oak, Red**
Description: **Oak, Red**
Class: **Wood**
Keywords: **red, oak**
Keynote: **09 64.00**

5.

On the Graphics tab:

Enable **Use Render Appearance**.

6.

Set the Foreground Surface Pattern to **Wood 1**.

Click **Apply** to save the settings.

7. Save as *ex4-9.rvt*.

Exercise 4-10
Defining Site Materials

Drawing Name: site materials.rvt
Estimated Time: 15 minutes

This exercise reinforces the following skills:

- ❑ Copying Materials from the Library to Document
- ❑ Editing Materials
- ❑ Using the Asset Browser
- ❑ Assigning a custom image to a material

1. Go to **Manage→Materials**.

2. Type **grass** in the search field at the top of the Material browser.

3. Grass is listed in the lower library pane.

Highlight **grass**.
Right click and select **Add to→ Document Materials**.

4.

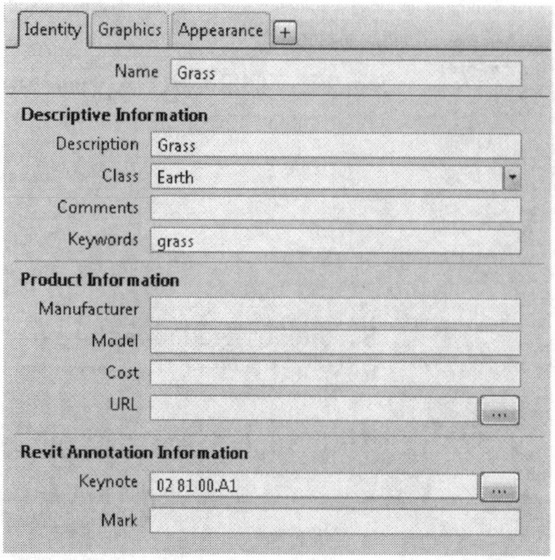

In the Identity tab:

Name: **Grass**
Description: **Grass**
Class: **Earth**
Keywords: **grass**
Keynote: **02 81 00.A1**

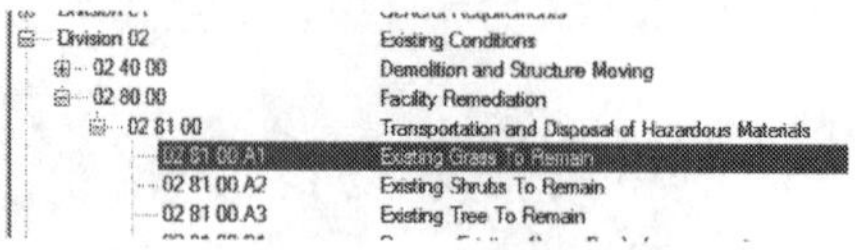

5.

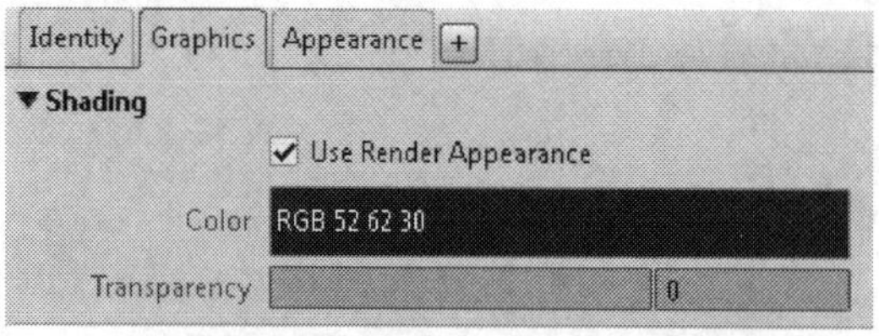

On the Graphics tab:

Enable **Use Render Appearance**.

6.

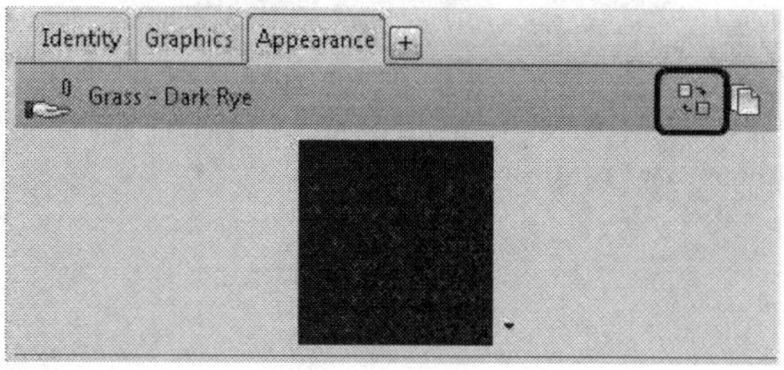

On the Appearance tab:

Select **Replaces this asset**.

7.  Type **grass** in the search field.

8.

You have a selection of grasses to select from.

9.

Select the **St Augustine** grass.

Right click and select **Replace in Editor**.

You can select your own choice as well.

Close the Asset Browser.

10.

The selected grass name should be listed under Assets.

Click **Apply**.

11. Do a search for **gravel** in the Material Browser.

12.

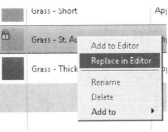

Locate the Gravel in the lower library panel.

Right click and **select Add to → Document Materials**.

13.

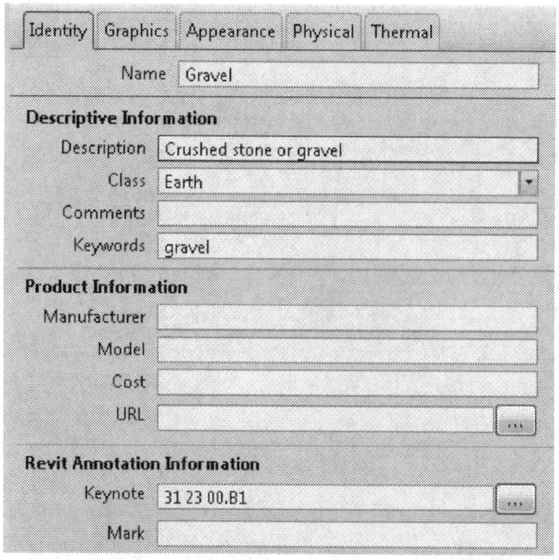

In the Identity tab:

Name: **Gravel**
Description: **Crushed stone or gravel**
Class: **Earth**
Keywords: **gravel**
Keynote: **31 23 00.B1**

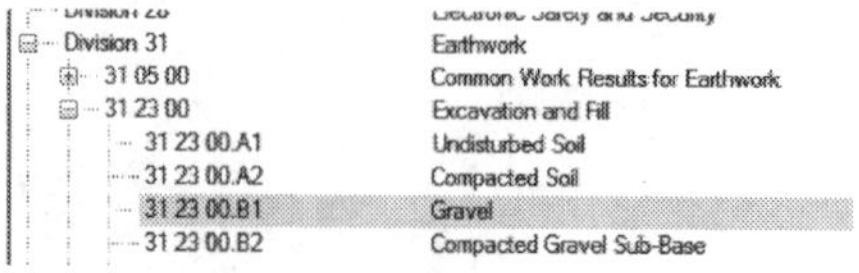

14.

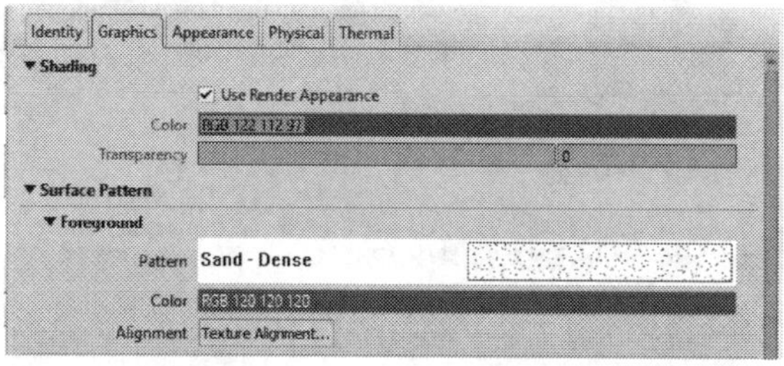

In the Graphics tab:

Enable **Use Render Appearance**.

Assign the **Sand Dense** hatch pattern to the Foreground Surface Pattern.

15.

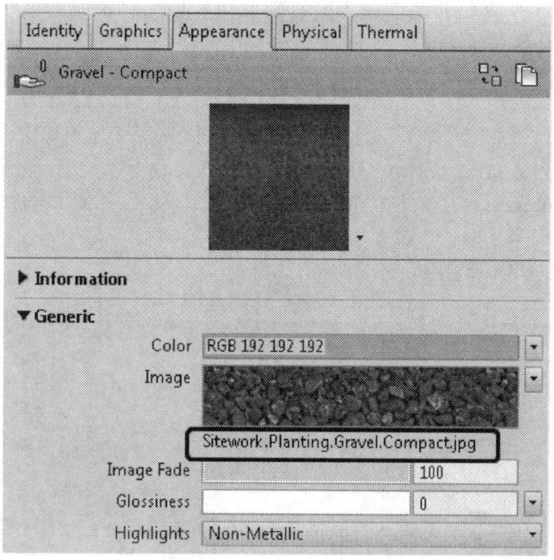

On the Appearance tab:

Change the color to **RGB 192 192 192**.

Left click on the image name.

16.

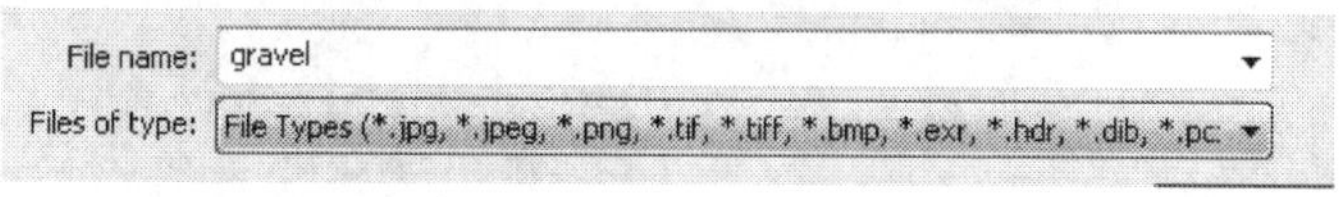

Locate the *gravel* image file in the exercise files.

Click **Open**.

17.

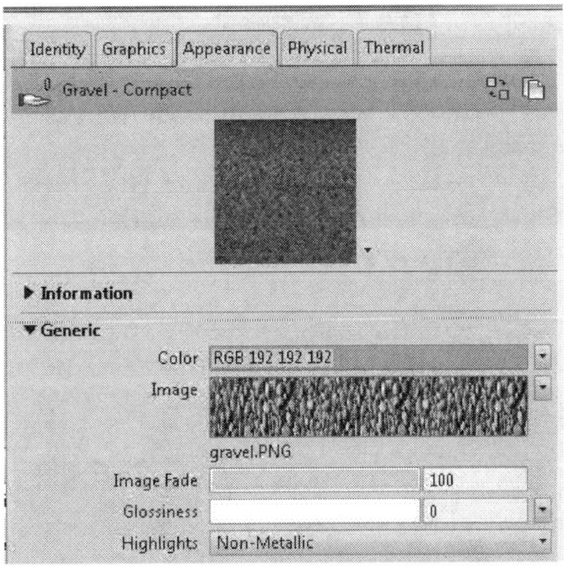

The Appearance will update.

Click **Apply** to save the changes.

18. Save as *ex4-10.rvt.*

Exercise 4-11
Defining Masonry Materials

Drawing Name: masonry.rvt
Estimated Time: 15 minutes

This exercise reinforces the following skills:

- ❑ Copying Materials from the Library to the Document Materials
- ❑ Using the Asset Browser

1.  Go to **Manage→Materials**.

    Materials

2.  Type **masonry** in the search field.

3. 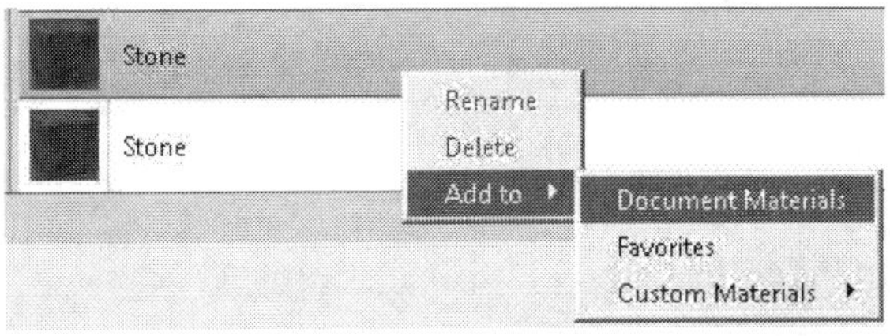 Highlight **Stone** in the Libraries pane.
    Right click and select **Add to →
    Document Materials**.

4.

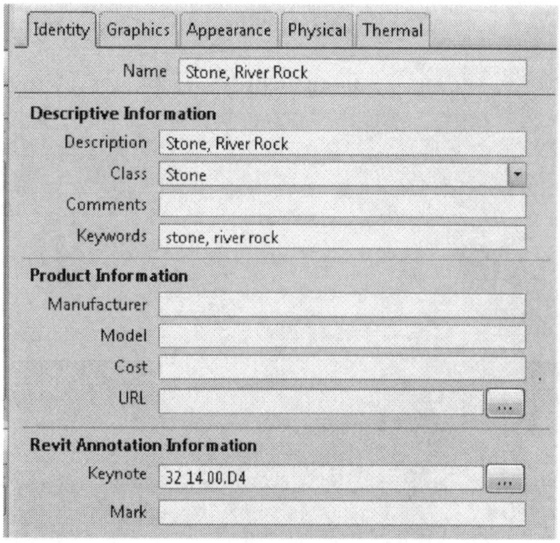

In the Identity tab:

Name: **Stone, River Rock**
Description: **Stone, River Rock**
Class: **Stone**
Keywords: **stone, river rock**
Keynote: **32 14 00.D4**

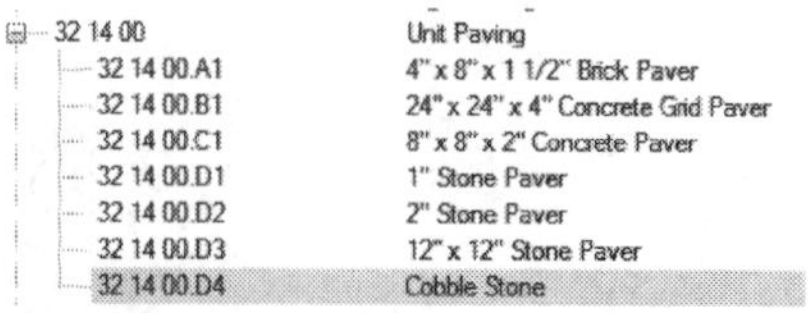

5.

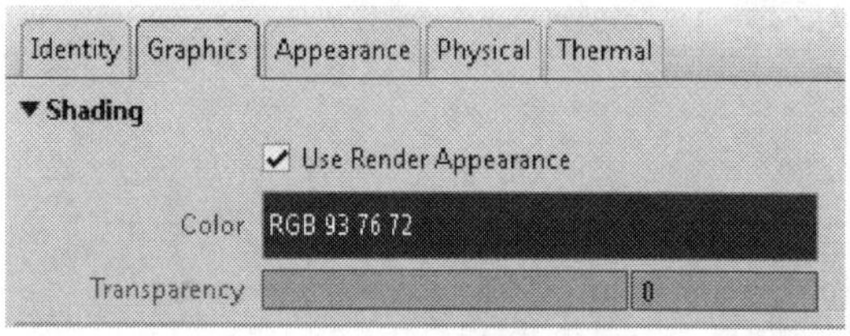

In the Graphics tab:

Enable **Use Render Appearance for Shading**.

6.

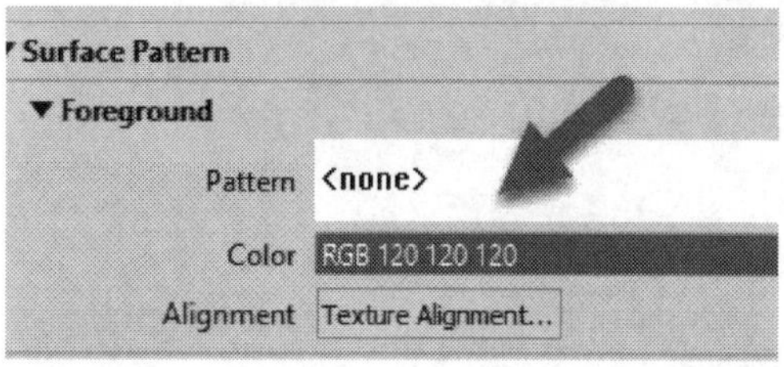

Select the Foreground Pattern button to assign a fill pattern.

7.

Select **New** to create a new fill pattern.

8.

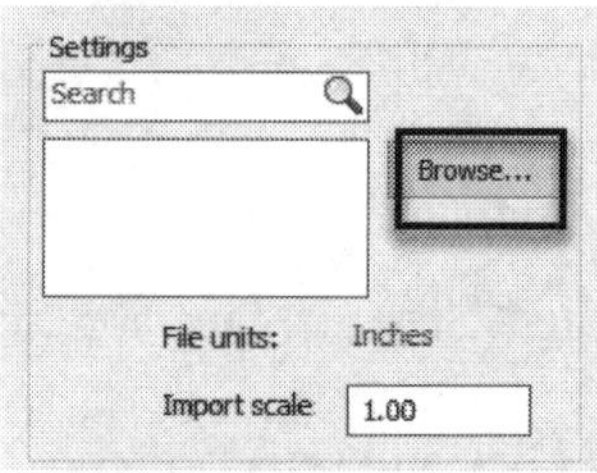

Type **Cobblestone** in the Name field.

Enable **Custom**.

Select **Browse**.

9.

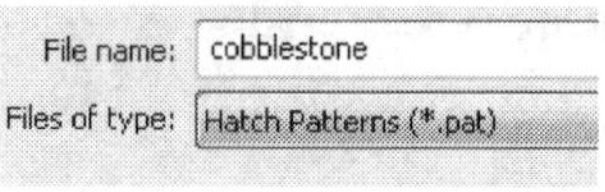

Locate the *cobblestone.pat* file in the exercise files downloaded from the publisher's website.

Click **Open**.

10.

You will see a preview of the pattern.

Verify that the name is set to **Cobblestone**.

Click **OK**.

Preview

Scale: 1" = 1"

Name: Cobblestone

Type: ○ Basic
● Custom

Settings

Search

Vespaio2   Browse...

File units: Inches

Import scale 1.00

Orientation in Host Layers:

Orient To View

11.

Highlight **Cobblestone**.

Click **OK**.

Revit allows you to use any AutoCAD pattern file as a fill pattern.

Pattern Type

● Drafting   ○ Model

Search

Name: Pattern:

No pattern   <none>

<Solid fill>

Aluminum

Cobblestone

Concrete

12.

On the Appearance tab:

Select **Replaces this asset**.

Identity | Graphics | Appearance | Physical | Thermal

Unfinished

13. Type **rock** in the search field.

Asset Browser ? ✕

rock ✕

Appearance Library ≡ ▾

Search results for "rock"

	Asset Name	▲ Aspect	Type	Ca
▾ 📁 Appearance Library 🔒				
🔳 Stone				
	Jagged Rock Wall	Appe...	Stone	St
🔒	Rubble - River Rock	Appe...	Maso...	⇄
	Rubble - River Rock			

Locate the **Rubble - River Rock**.

14. 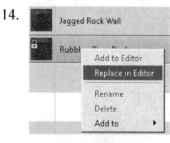 Highlight the **Rubble - River Rock**.

Right click and select **Replace in Editor**.

15. 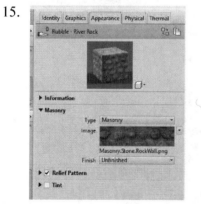 The Material Editor updates.

Click **OK** to save the changes and close the dialog.

16. Save as *ex4-11.rvt*.

Exercise 4-12
Assigning Materials to Stairs

Drawing Name: stairs2.rvt [m_stairs2.rvt]
Estimated Time: 20 minutes

This exercise reinforces the following skills:

- Floor
- Floor Properties

1.   Open *stairs2.rvt [m_stairs2.rvt]*.

2.   Activate **Level 1 – No Annotations**.

3.   Select the stairs.

4.   Select **Edit Type** on the Properties pane.

5.   Select **Duplicate**.

6.   Type **Stairs - Oak Tread with Painted Riser**.

Click **OK**.

7.

Under Construction:
Locate the Run Type and select the small …
button on the far right.

| Type: | Stairs - Oak Tread with Painted Riser | | Duplicate… |
| | | | Rename… |

Type Parameters

Parameter	Value	=	^
Calculation Rules			
Maximum Rise	190.0		
Minimum Trea	250.0		
Minimum Run	1000.0		
Calculation Rul	Edit…		
Construction			
Run Type	50mm Tread 13mm Riser		
Landing Type	Non-Monolithic Landing		
Function	Interior		
Supports			

8.

Parameter	Value
Materials and Finishes	
Tread Material	<By Category>
Riser Material	<By Category>

Select the **Tread Material** column.

9.

Type **oak** in the search field.
Select the **Oak, Red** material.

Click **OK**.

10.

Type Parameters

Parameter	Value
Materials and Finishes	
Tread Material	Oak, Red
Riser Material	<By Category>

Select the **Riser Material** column.

11.

Type **white** in the search field.

Locate the **Paint -
SW6126NavajoWhite** material.

Click **OK**.

12. Click **OK** twice to close the dialog.

13. Switch to a **3D** view.

14.

Graphic Display Options...
Wireframe
Hidden Line
Shaded
Consistent Colors
Textures
Realistic

1 : 100

Change the display to **Realistic**.

Orbit around the model to inspect the stairs.

15.

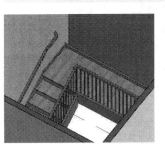

Observe how the assigned materials affect the stairs' appearance.

Some users may not see the wood pattern on their stairs. This is a graphics card issue. I use a high-end graphics card.

If you want the boards to be oriented horizontally instead of vertically on the treads:

16.

Image
▼ Position
Offset 0'-0"   X
        0'-0"   Y
Rotation   90.00°

Select the stairs.
Edit Type.

17.

Construction
Run Type   2" Tread 1" Nosing 1/4" Riser
Landing Type   Non-Monolithic Landing
Function   Interior

Left click in the Run Type column to open up the Run properties dialog.

18.

Materials and Finishes
Tread Material   Oak, Red

Select the Tread Material.

19.

Color
✓ Image
Edit Image...
0.03

Select the Appearance tab.

Next to the image picture, select Edit Image from the drop down list.

20.

▼ Transforms
        ☐ Link texture transforms
▼ Position
Offset 0'-0"   X
        0'-0"   Y
Rotation   90.00°

Expand the Transform area.
Set the Rotation to **90 degree**.
Click **Done**.

Click **OK** to close the dialogs.

21.   Save as *ex4-12.rvt*
       [m_ex4-12.rvt].

Exercise 4-13
Applying Paints and Wallpaper to Walls

Drawing Name: materials1.rvt [m_materials1.rvt]
Estimated Time: 30 minutes

This exercise reinforces the following skills:

- Views
- Section Box
- View Properties
- Materials
- Load From Library
- Render Appearance Library

1.   Open *materials1.rvt [m_materials1.rvt]*.

2.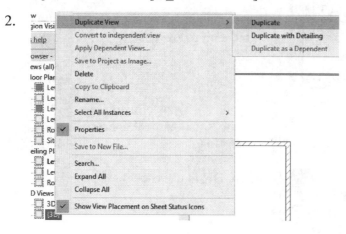

Highlight the {3D} view in
the Project Browser.

Right click and select
**Duplicate View →
Duplicate**.

3. Right click on the new view.

    Select **Rename**.

4. Type **3D-Lobby**.

5. Set Parts Visibility to **Show Original**.

6. Enable Section Box in the Properties panel.

7. 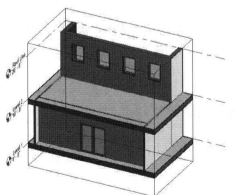 Using the grips on the section box, reduce the view to just the lobby area.

8.

Hide in View	>	Elements
Override Graphics in View	>	Category
		By Filter...
Create Similar		
Edit Family		
Select Previous		
Select All Instances	>	
Delete		
Reset Crop Boundary to Model		
Reset Crop Boundary to Screen		
Find in Project Browser		
Find Referring Views		
Zoom In Region		
Zoom Out (2x)		
Zoom To Fit		
Previous Pan/Zoom		
Next Pan/Zoom		
Browsers	>	
✓ Properties		

Level 1 - No Annotations

Select the floor on Level 2.

Right click and select **Hide in View→Elements**.

This will hide only the selected element.

9.

Hide in View	>	Elements
Override Graphics in View	>	Category
		By Filter...
Create Similar		

Select the ceiling.
Right click and select
**Hide in View →
Elements**.

This will hide only
the selected element.

10.

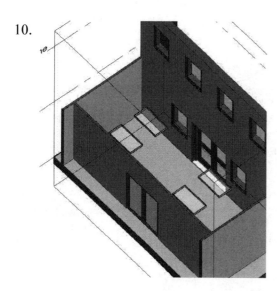

The lighting fixtures in the ceiling are still visible but can be ignored.

Or you can select the lighting fixtures, right click and select **Hide In View→Category**.

This will hide all lighting fixtures.

Select the section box.
Right click and select **Hide In View→Category**.

11. Activate the Modify ribbon.

Select the **Split Face** tool on the Geometry panel.

Select one of the side walls. You will see an orange outline to indicate the selection.

12.

Draw a line on the corner of the room to indicate where the wall should be split.

The end points of the vertical line should be coincident to the orange horizontal lines or you will get an error.

13. Select the **Green check**.

14. Select the Split Face tool on the Modify ribbon.

Draw a vertical line to divide the wall on the other side of the lobby.

15. Select the **Green check**.

16. Select the **Paint** tool on the Geometry panel.

17.

Set the Project Materials filter to **Paint**.

18.

Set the View Type to **List View**.

19.

Locate the **Finish – Paint - SW0068 Heron Blue** from the Material Browser list.

20.

Select the walls indicated.

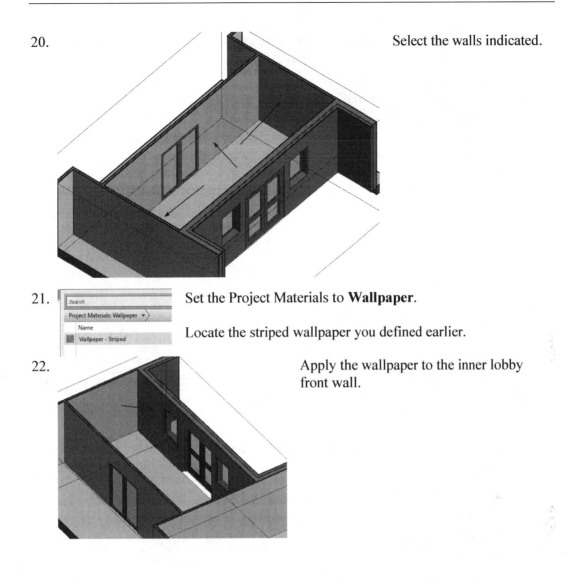

21.

Set the Project Materials to **Wallpaper**.

Locate the striped wallpaper you defined earlier.

22.

Apply the wallpaper to the inner lobby front wall.

23. **Material Browser - Wallpaper - Striped**

wallpap

Project Materials: All ▼

Search results for "wallpap"

Wallpaper -
Striped

Done

Select the **Done** button at the bottom of the Material Browser.

24.

Graphic Display Options...

Wireframe
Hidden Line
Shaded
Consistent Colors
Textures
Realistic

Change the display to **Realistic** so you can see the materials applied.

25. Save as *ex4-13.rvt [m_ex4-13.rvt]*.

Extra:

Select the other stairs and assign it to the new stairs type using the type selector.

Add all the custom materials to the Custom Materials library.

Define a custom wallpaper, a custom brick, and a custom wood material using assets from the Asset Browser.

Additional Projects

1) Create a material for the following Sherwin-Williams paints:

SW 6049 Gorgeous White
Interior/Exterior

Color Collection	Soft and Sheer
Color Family	Whites
Color Strip	8
RGB Value	R-232 | G-220 | B-212
Hexadecimal Value	# E8DCD4
LRV	73

SW 6050 Abalone Shell
Interior/Exterior

Color Family	Reds
Color Strip	8
RGB Value	R-219 | G-201 | B-190
Hexadecimal Value	# DBC9BE
LRV	60

SW 6052 Sandbank
Interior/Exterior

Color Family	Reds
Color Strip	8
RGB Value	R-196 | G-166 | B-152
Hexadecimal Value	# C4A698
LRV	41

2) Create a wallpaper material using appearances from the Asset Browser.

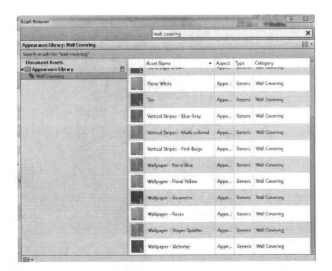

3) Create three different types of carpeting materials using appearances from the Asset Browser.

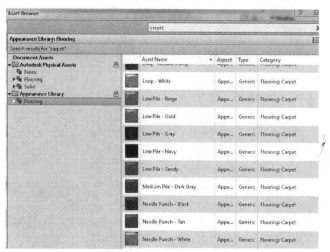

4) On Level 2, in the cafeteria area, apply paint to the walls and modify the floor to use a VCT material.

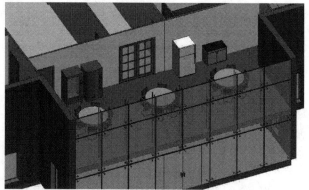

5) Apply paints to the walls in the conference rooms. Modify the floors to use a carpet material by creating parts.

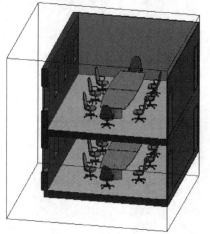

Lesson 4 Quiz

True or False

1. Materials in the default Material Libraries can be modified.
2. Only materials in the current project can be edited.
3. A material can have up to five assets, but only one of each type of asset.

Multiple Choice

4. Which of the following is NOT a Material asset?

   A. Identity
   B. Graphics
   C. Appearance
   D. Structural
   E. Thermal

5. Identify the numbers.

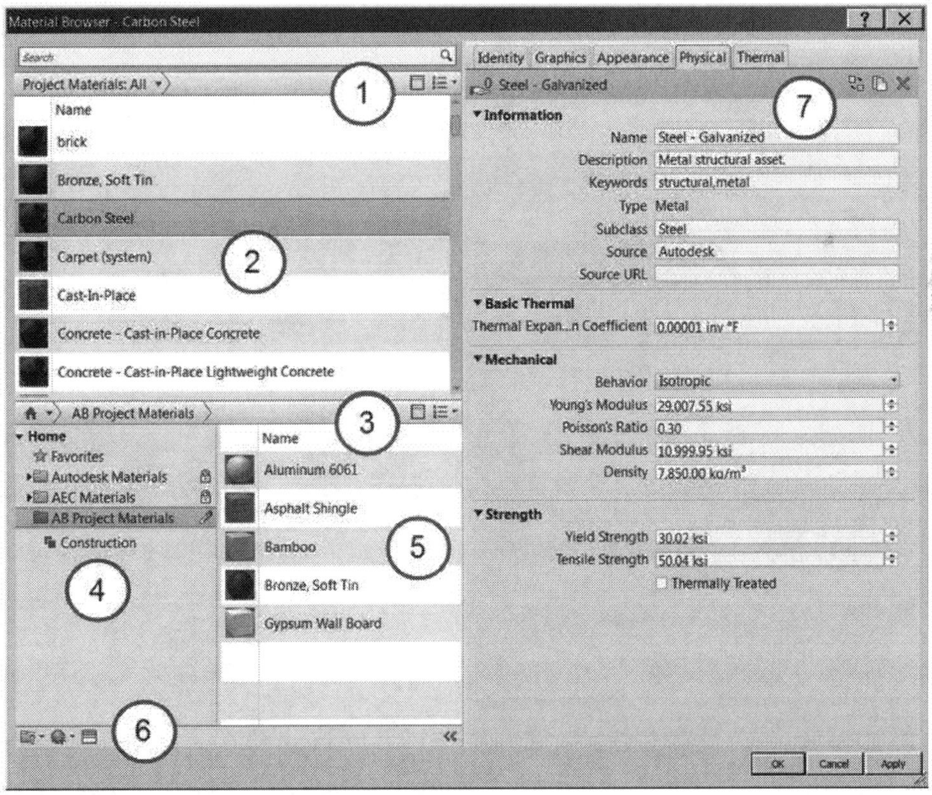

Material Browser

6. The lock icon next to the Material Library indicates the library is:

    A. Secure
    B. Closed
    C. "read-only"
    D. Unable to be moved

7. To access Materials, use this ribbon:

    A. Manage
    B. View
    C. Architecture
    D. Annotate
    E. Materials

8. To add a material to the current project:

    A. Double left click the material in the library list.
    B. Drag and drop the material from the library list into the project materials list.
    C. Right click the material and use the Add to Document materials short-cut menu.
    D. Select the material in the library list and click the Add button located on the far right.
    E. All of the above

ANSWERS:

1) F; 2) T; 3) T; 4) D; 5) 1- Show/Hide library panel button, 2- Property materials list, 3-Show/Hide library button, 4- Library list, 5- Library materials list, 6- Material Browser toolbar, 7- Asset panel; 6) C; 7) A; 8) E

Lesson 5
Floors and Ceilings

Ceiling Plans are used to let the contractor know how the ceiling is supposed to look.

When drafting a reflected ceiling plan, imagine that you are looking down on the floor, which is acting as a mirror showing a reflection of the ceiling. You want to locate lighting, vents, sprinkler heads, and soffits.

We start by placing the floors that are going to reflect the ceilings. (Reflected ceiling plans look down, not up. They are the mirror image of what you would see looking up, as if the ceiling were reflected in the floor.)

Exercise 5-1
Creating Floors

Drawing Name: floors.rvt [m_floors.rvt]
Estimated Time: 35 minutes

This exercise reinforces the following skills:

- ❏ Floors
- ❏ Floor Properties
- ❏ Materials

1. Open *floors.rvt [m_floors.rvt]*.

2. 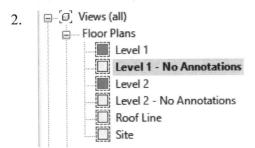   Activate **Level 1 – No Annotations.**

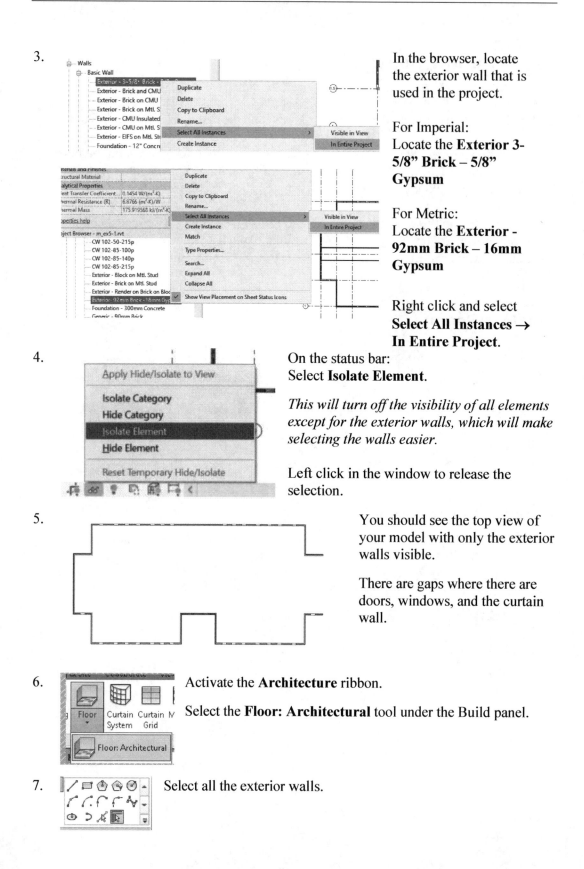

3.　In the browser, locate the exterior wall that is used in the project.

For Imperial:
Locate the **Exterior 3-5/8" Brick – 5/8" Gypsum**

For Metric:
Locate the **Exterior - 92mm Brick – 16mm Gypsum**

Right click and select **Select All Instances → In Entire Project**.

4.　On the status bar:
Select **Isolate Element**.

This will turn off the visibility of all elements except for the exterior walls, which will make selecting the walls easier.

Left click in the window to release the selection.

5.　You should see the top view of your model with only the exterior walls visible.

There are gaps where there are doors, windows, and the curtain wall.

6.　Activate the **Architecture** ribbon.

Select the **Floor: Architectural** tool under the Build panel.

7.　Select all the exterior walls.

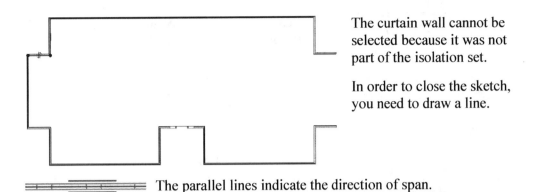

The curtain wall cannot be selected because it was not part of the isolation set.

In order to close the sketch, you need to draw a line.

The parallel lines indicate the direction of span.

By default, the first line placed/wall selected indicates the direction of span.
To change the span direction, select the Span Direction tool on the ribbon and select a different line.

8. Select the Line tool from the Draw panel.

Draw a line to close the floor boundary sketch.

9. Verify that your floor boundary has no gaps or intersecting lines.

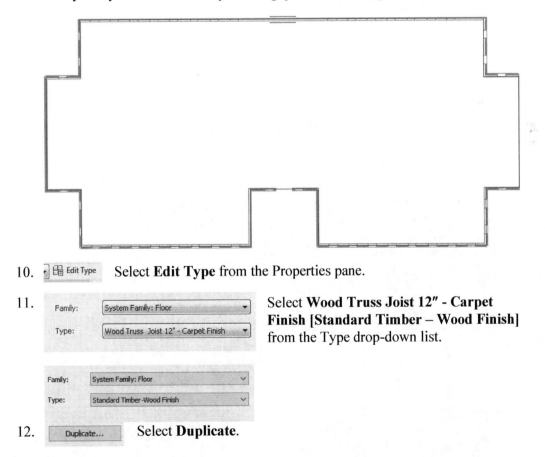

10. Edit Type   Select **Edit Type** from the Properties pane.

11.

Family: System Family: Floor
Type: Wood Truss Joist 12" - Carpet Finish

Select **Wood Truss Joist 12" - Carpet Finish [Standard Timber – Wood Finish]** from the Type drop-down list.

Family: System Family: Floor
Type: Standard Timber -Wood Finish

12. Duplicate...   Select **Duplicate**.

13.

Name: Wood Truss Joist 12" -VCT

Name: Standard Timber-VCT

OK   Cancel

Change the Name to **Wood Truss Joist 12" - VCT [Standard Timber – VCT]**.

Click **OK**.

14.

Type Parameters

Parameter	Value
Construction	
Structure	Edit...

Select **Edit** under Structure.

15.

	Function	Material
1	Finish 1 [4]	Carpet (1)
2	**Core Boundary**	**Layers Above**

Layers

	Function	Material	Thickness	V
1	Finish 2 [5]	Oak Flooring	9.0	
2	Core Boundar	Layers Above Wrap	0.0	
3	Structure [1]	Wood Sheathing, Chip	22.0	
4	Structure [1]	Structure, Timber Joist/	200.0	
5	Core Boundar	Layers Below Wrap	0.0	

For Layer 1: Finish 1 [4] [Finish 2[5]]: Select the *browse* button in the Material column.

16.

vct

Project Materials: All ▼ ▾

Search results for "vct"

Name

Vinyl Composition Tile - Diamond Pattern

Type **vct** in the search field.

vct

Project Materials: All ▾ ▶

Search results for "vct"

Name

Tile, Mosaic, Gray

Vinyl Composition Tile - Diamond Pattern

Highlight **the Vinyl Composition Tile – Diamond Pattern** to select.

Click **OK**.

17.

	Function	Material	Thickness
1	Finish 1 [4]	Vinyl Composition Tile - Diamond Pattern	0' 0 1/8"
2	Core Boundary	Layers Above Wrap	0' 0"
3	Structure [1]	Plywood, Sheathing	0' 0 3/4"

You should see the new material listed.

Click **OK**.

Layers

	Function	Material	Thickness	V
1	Finish 2 [5]	Vinyl Comp	19.0	
2	Core Boundar	Layers Above	0.0	
3	Structure [1]	Wood Sheat	22.0	
4	Structure [1]	Structure, Ti	200.0	
5	Core Boundar	Layers Below	0.0	

18.

Graphics	
Coarse Scale Fill Pattern	Diagonal crosshatch
Coarse Scale Fill Color	Black

Set the Coarse Scale Fill Pattern to **Diagonal crosshatch**.

Click **OK** twice to close the dialogs.

19. Select the **Green Check** under Mode to finish the floor.

If you get a dialog box indicating an error with the Floor Sketch, click the 'Show' button. Revit will zoom into the area where the lines are not meeting properly. Click 'Continue' and use Trim/Extend to correct the problem.

---

20. Switch to a 3D View.

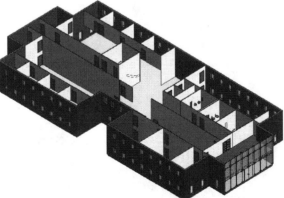

21. Activate **Level 1 – No Annotations**.

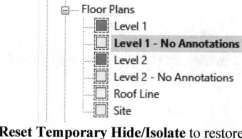

22. Select **Reset Temporary Hide/Isolate** to restore the elements in the Level 1 floor plan.

23. Save as *ex5-1.rvt [m_ex5-1.rvt]*.

Exercise 5-2
Copying Floors

Drawing Name: copy3.rvt [m_copy3.rvt]
Estimated Time: 5 minutes

This exercise reinforces the following skills:

- ❑ Copy
- ❑ Paste Aligned
- ❑ Opening
- ❑ Shaft Opening
- ❑ Opening Properties

1. Open *copy3.rvt* [m_copy3.rvt].

2. Switch to a 3D View.

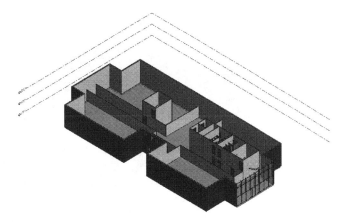

3.    Left click to select the floor.

Check the properties pane to verify the floor is selected.

4. 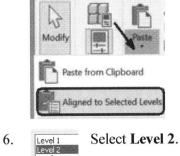   Select **Copy** from the Clipboard panel.

5. Select **Paste→Aligned to Selected Levels**.

6. Select **Level 2**.

Click **OK**.

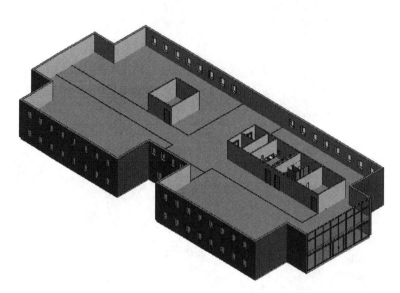

7.   Left click in the window to release the selection.

8.   Save as *ex5-2.rvt [m_ex5-2.rvt]*.

Floor Openings

There are two ways to create openings in floors.

Method #1 uses a Shaft. A shaft basically is a void (a mass that is negative space) that is in the vertical direction and is constrained to a top and bottom level. You can offset the shaft from a level, so it doesn't have to be aligned to a specific level. Shaft openings are a good choice for elevators or HVAC equipment or when the opening location is the same regardless of the level.

Method #2 modifies the sketch used to define the floor. This operation entails selecting the floor, using Edit Sketch on the ribbon, and then adding the outline for the desired opening to the sketch. This method is useful for stairs, loft areas, balconies, etc. where the opening is only desired for a single level. The floor sketch is tied to the assigned level and floor structure, so you cannot offset the opening. By default, the sketch opening goes through the entire floor structure and only the floor structure.

Exercise 5-3
Creating a Shaft Opening

Drawing Name: shaft opening.rvt [m_shaft opening.rvt]
Estimated Time: 15 minutes

This exercise reinforces the following skills:

- ❏ Shaft Opening
- ❏ Opening Properties
- ❏ Symbolic Lines

1. Open *shaft opening.rvt [m_shaft opening.rvt]*.

2. *We need to create openings in the Level 2 floor for the elevators.*
   Activate **Level 2 – No Annotations**.

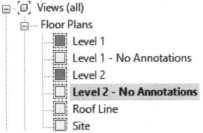

3. 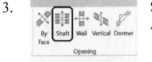 Select the **Shaft Opening** tool under the Opening panel on the Architecture ribbon.

4. Select the **Rectangle** tool from the Draw panel.

5.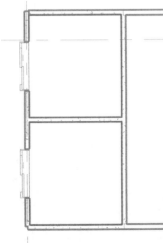

   Zoom into the elevator area.

   Draw two rectangles for the elevator shafts.

   Align the rectangles to the finish face of the inside walls and behind the elevator doors.

   The shaft will cut through any geometry, so be careful not to overlap any elements you want to keep.

6. 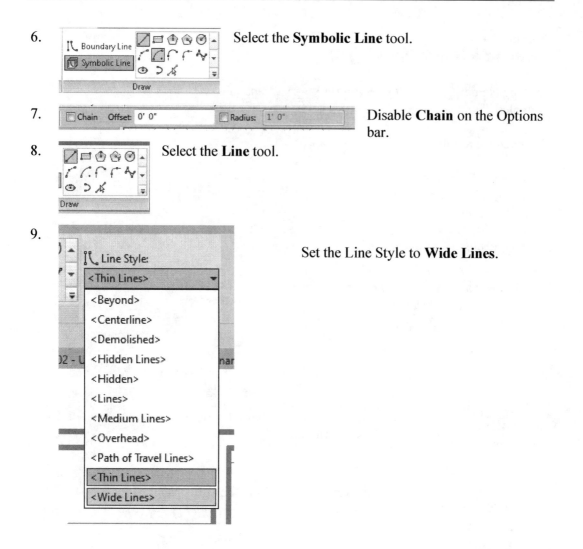 Select the **Symbolic Line** tool.

7. Disable **Chain** on the Options bar.

8. Select the **Line** tool.

9. Set the Line Style to **Wide Lines**.

10.

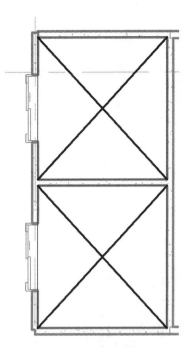

Place an X in each elevator shaft using the corners of the rectangle sketch.

The symbolic lines will appear in the floor plan to designate an opening.

11.

Shaft Openings	⌄	🔲 Edit Type
Constraints		⌃
Base Constraint	Level 1	
Base Offset	-4' 0"	
Top Constraint	Up to level: Level 2	
Unconnected Height	14' 0"	
Top Offset	0' 0"	
Phasing		⌃
Phase Created	New Construction	
Phase Demolished	None	

On the Properties pane:

Set the Base Offset at **-4' 0" [-600 mm]**. This starts the opening below the floor to create an elevator pit.

Set the Base Constraint at **Level 1**.
Set the Top Constraint at **Level 2**.

Constraints	
Base Constraint	Level 1
Base Offset	-600.0
Top Constraint	Up to level: Level 2
Unconnected Height	4600.0
Top Offset	0.0

12.

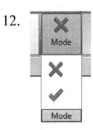

Select the **Green Check** under the Mode panel.

13. Switch to a 3D view.
Orbit around to inspect the model.

14.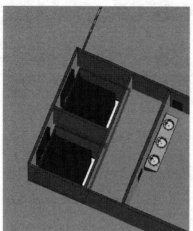

Inspect the elevator shafts.

Note that the shafts go through the bottom floor.

15. Activate **Level 1**.

The symbolic lines are visible.

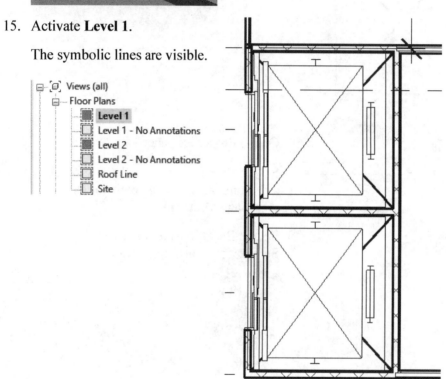

16. Save as *ex5-3.rvt [m_ ex5-3.rvt]*.

Exercise 5-4
Adding an Opening to a Floor

Drawing Name: floor opening.rvt [m_floor opening.rvt]

Estimated Time: 20 minutes

This exercise reinforces the following skills:

- ❑ Opening
- ❑ Modify Floor

1. Open *floor opening.rvt [m_floor opening.rvt]*.

2. We need to create openings in the Level 2 floor for the stairs.
   Activate **Level 2 – No Annotations**.

3.
   Select the floor.

   You can window around the model and use Filter to select the floor.

   If it is selected, you should see it listed in the Properties pane.

4. Select the **Edit Boundary** tool on the ribbon.

5.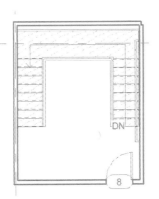
   Zoom into the first set of stairs.

6. Enable **Boundary Line**.

Select the **Rectangle** tool on the Draw panel.

7. 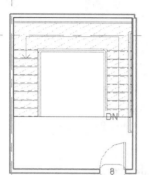 Place the rectangle so it is aligned to the inner walls and the stair risers.

8. Zoom and pan over to the second set of stairs.

9. 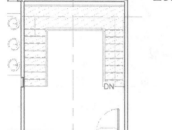 Place the rectangle so it is aligned to the inner walls and the stair risers.

10. Select the **Green Check** to finish modifying the floor.

11. Attaching to floor   ✕   Click **Don't attach**.

Would you like walls that go up to this floor's level to attach
to its bottom?

☐ Do not show me this message again   Attach   Don't attach

12. Revit   ✕   If you see this dialog, click **No**.

The floor/roof overlaps the highlighted wall(s).
Would you like to join geometry and cut the
overlapping volume out of the wall(s)?

Yes   No

13. Left click to release the selection.

Switch to a 3D view and inspect the openings in the
floor.

14. Save as *ex5-4.rvt [m_ex5-4.rvt]*.

Challenge Question:

When is it a good idea to use a Shaft Opening to cut a floor and when should you modify
the floor boundary?

Shaft Openings can be used to cut through more than one floor at a time. Modifying the
floor boundary creates an opening only in the floor selected.

Exercise 5-5
Creating Parts

Drawing Name: parts.rvt [m_parts.rvt]
Estimated Time: 30 minutes

This exercise reinforces the following skills:

- Parts
- Dividing an Element
- Assigning materials to parts
- Modifying materials

1.   Open *parts.rvt [m_parts.rvt]*.

   Switch to a 3D view.

2.   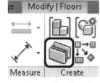   Select the floor on the second level.

3.   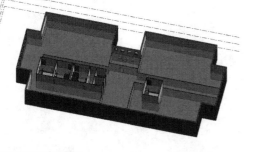   Select the **Create Parts** tool under the Create panel.

4.   Select **Divide Parts** from the Part panel.

Divide
Parts

5.   Activate **Level 2 – No Annotations**.

- Views (all)
  - Floor Plans
    - Level 1
    - Level 1 - No Annotations
    - Level 2
    - **Level 2 - No Annotations**
    - Roof Line
    - Site

6. 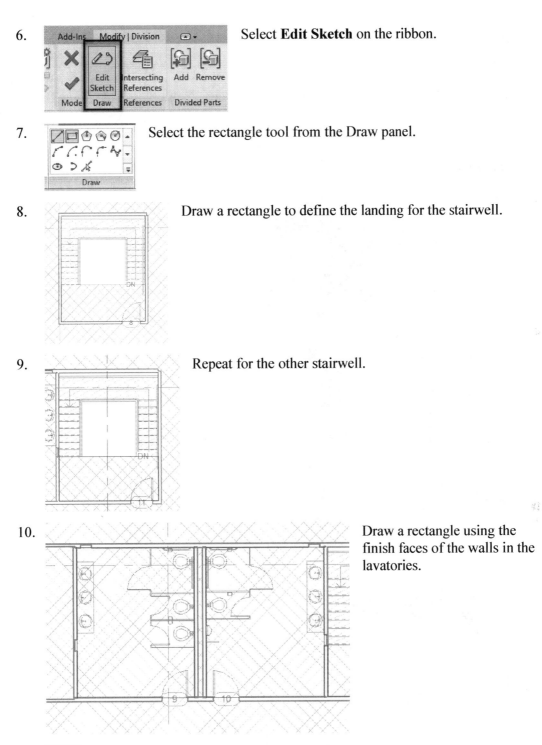 Select **Edit Sketch** on the ribbon.

7. Select the rectangle tool from the Draw panel.

8. Draw a rectangle to define the landing for the stairwell.

9. Repeat for the other stairwell.

10. Draw a rectangle using the finish faces of the walls in the lavatories.

11. Select the **Green Check** under Mode twice.

12.   Switch to a 3D View.

13.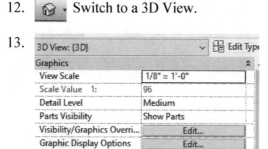

In the Properties pane for the view:

Under Parts Visibility: Enable **Show Parts**.

In order to see the parts in a view, you need to enable the visibility of the parts.

14.

Select the landing of the first stairwell.

You will see that it is identified as a part on the Properties pane.

15.

Original Type	Wood Truss Joist 12" - Vinyl
Material By Original	☐
Material	VCT - Vinyl Composition Tile Diamond Patte...
Construction	Finish

In the Properties pane:

Uncheck **Material by Original**.
Left click in the Material field.

16.

Project Materials: All ▼

Search results for "lamin"

Name
Laminate
Laminate - Ivory, Matte

Scroll up and select the **Laminate - Ivory, Matte** material.

Click **OK**.

17.

Original Type	Wood Truss Joist 12" -VCT
Material By Original	☐
Material	Laminate - Ivory, Matte
Construction	Finish

The new material is now listed in the Properties pane.
Left click in the window to release the selection.

18. Pan over to the lavatory area.

Hold down the Control key.

Select the lavatory floor in both lavatories.

Note that it is now identified as a part.

19.

Parts (2)		Edit Type
Original Family	Floor	
Original Type	Wood Truss Joist 12" -VCT	
Material By Original	☐	
Material	Vinyl Composition Tile - Diamond Pattern	
Construction	Finish	

Uncheck **Material by Original**. Left click inside the Material field.

20.

tile
Project Materials: All ▾
Search results for "tile"
Name
Iron, Ductile
Tile, Mosaic, Gray
Tile, Porcelain, 4in

Locate the **Tile, Mosaic, Gray** material.

Click **OK**.

21.

Parts (2)	
Original Family	Floor
Original Type	Wood Truss Joist 12" -VCT
Material By Original	☐
Material	Tile, Mosaic, Gray
Construction	Finish

The lavatory parts are now set to the Tile material.
Left click in the window to release the selection.

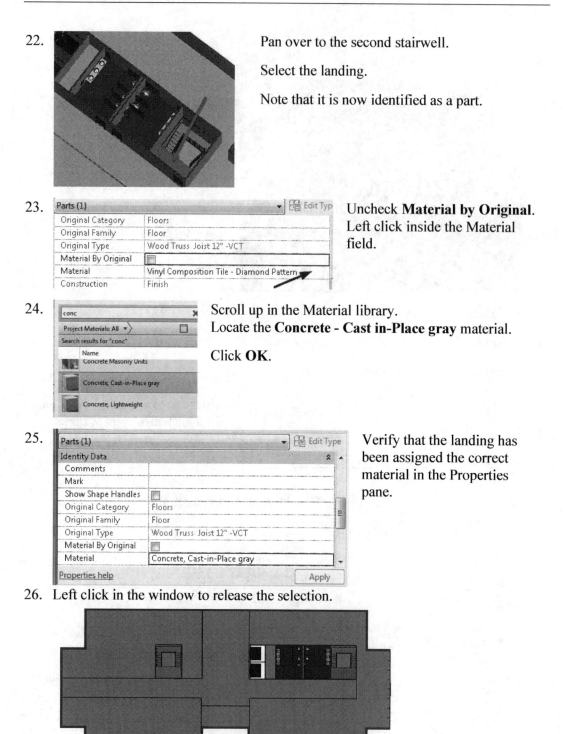

22. Pan over to the second stairwell.

Select the landing.

Note that it is now identified as a part.

23. Uncheck **Material by Original**. Left click inside the Material field.

24. Scroll up in the Material library.
Locate the **Concrete - Cast in-Place gray** material.

Click **OK**.

25. Verify that the landing has been assigned the correct material in the Properties pane.

26. Left click in the window to release the selection.

The different materials are displayed in the view.

27. Save as *ex5-5.rvt [m_ex5-5.rvt]*.

Extra: *Repeat for Level 1.*
In order to assign materials to the parts on the first level, you will need to select the parts on the second level and hide them.

Understanding View Range

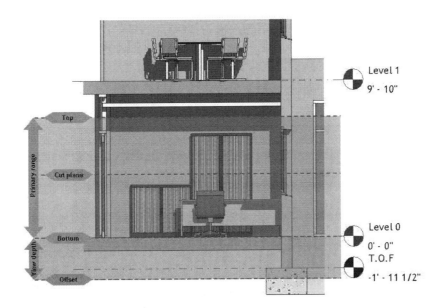

This figure shows an elevation view of a view range setting.

- Elements within the boundaries of the primary range that are not cut are drawn in the element's projection line weight.

- Elements that are cut are drawn in the element's cut line weight.

  Note: Not all elements can display as cut.

- Elements that are within the view depth are drawn in the beyond line style – that is a solid line.

  To modify line styles, go to the Manage ribbon→Additional Settings→Line Styles.

Line Styles

Category	Line Weight Projection	Line Color	Line Pattern
⊟ Lines	1	▓ RGB 000-166-000	Solid
<Area Boundary>	6	▓ RGB 128-000-255	Solid
<Beyond>	1	■ Black	Solid
<Centerline>	1	■ Black	Center 1/4"
<Demolished>	1	■ Black	Demolished 3/16"
<Fabric Envelope>	1	▓ RGB 127-127-127	Dash
<Fabric Sheets>	1	▓ RGB 064-064-064	Solid
<Hidden>	1	■ Black	Hidden 1/8"
<Overhead>	1	■ Black	Overhead 1/16"
<Room Separation>	1	■ Black	Solid

Object
Styles

To determine the line styles and weights set for different elements, go to the Manage ribbon and select Object Styles.

Filter list: Architecture ⌄

Category	Line Weight		Line Color	Line Pattern	Material
	Projection	Cut			
⊞ Detail Items	1	▨	■ Black	Solid	
⊟ Doors	2	2	■ Black	Solid	
Elevation Swing	1	1	■ Black	Dash 1/16"	
Frame/Mullion	1	3	■ Black	Solid	
Glass	1	4	■ Black	Solid	Glass
Handle	1	1	■ Black	Solid	
Hardware	1	1	■ Black	Solid	
Hidden Lines	2	2	▓ Blue	Dash	
Opening	1	3	■ Black	Solid	
Panel	1	3	■ Black	Solid	
Plan Swing	1	1	■ Black	Solid	
Threshold	1	1	■ Black	Solid	
Trim	1	1	■ Black	Solid	

Select the **Show** button on the View Range dialog.

The dialog will expand to show you how the fields correlate with how the view range is defined.

Exercise 5-6
Viewing Parts in a Floor Plan View

Drawing Name: view range.rvt [m_view range.rvt]
Estimated Time: 10 minutes

This exercise reinforces the following skills:

- ❑ View Range
- ❑ Parts

Parts can only be displayed in a view if they are within the view range.

1. Open *view range.rvt [m_view range.rvt]*.

2. Activate the **Level 2 –No Annotations** view.

3. The hatch patterns for the parts are not visible.

4. In the Properties pane:

Set the Parts Visibility to **Show Parts**.

The floor disappears.

5. Set the View Display to **Coarse** with **Hidden Lines**.

The view display does not change.

6. Scroll down the Properties pane.

Click **Edit** next to the View Range field.

7. Select the Show button located at the bottom of the dialog.

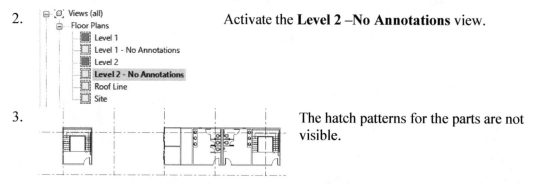

8.

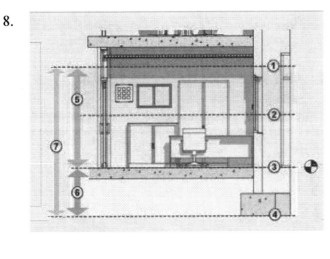

An image helps with understanding how to define the view range.

The numbers indicate the dialog inputs.

1: Top
2: Cut plane
3: Bottom
4. View Depth

9.

Set the Top Offset to **7' 6"** **[2300]**.
Set the Cut plane Offset to **4' 0"** **[1200]**.
Set the Bottom Offset to **0' 0"** **[0.0]**.
Set the View Depth Level to Level 1.
Set the View Depth Offset to **-1' 0"** **[-60.0]**.
Click **Apply**.

10. The parts now display the hatch patterns properly. Click **OK**.

If the materials are not defined to use a pattern fill, you will not see the pattern.

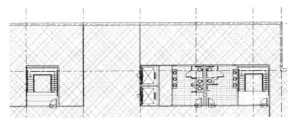

Save as *ex5-6.rvt [m_ ex5-6.rvt]*.

Exercise 5-7
Adding a Railing

Drawing Name: railing2.rvt [m_ railing2.rvt]
Estimated Time: 15 minutes

This exercise reinforces the following skills:

- ❑ Railing
- ❑ Railing Properties
- ❑ Re-Host

1. Open *railing2.rvt [m_ railing2.rvt]*.

2. 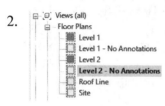 Activate the **Level 2 – No Annotations** view.

3.  Zoom into the concrete landing next to the lavatories.

4. Select the **Railing→Sketch Path** tool under the Circulation panel on the Architecture ribbon.

5. Draw the railing using the **Line** tool from the Draw panel.

    Use the end points of the existing rails to locate it.

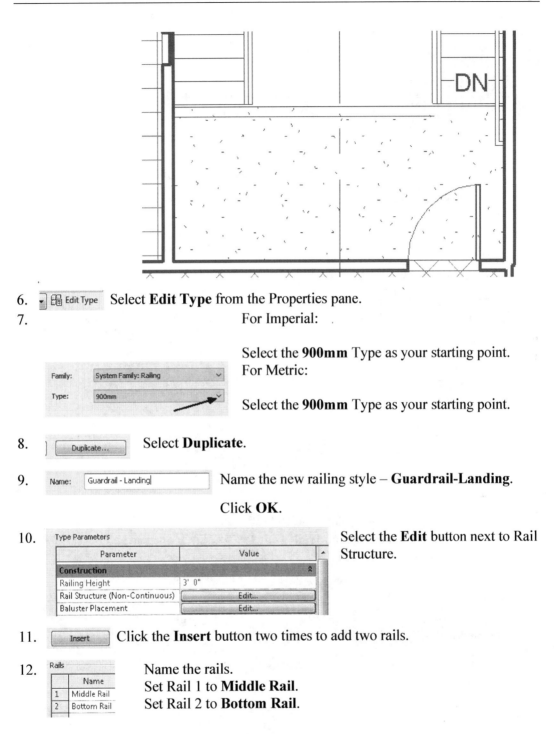

6. **Edit Type**   Select **Edit Type** from the Properties pane.

7.   For Imperial:

Select the **900mm** Type as your starting point.
For Metric:

| Family: | System Family: Railing |
| Type: | 900mm |

Select the **900mm** Type as your starting point.

8.   **Duplicate...**   Select **Duplicate**.

9.   Name: Guardrail - Landing   Name the new railing style – **Guardrail-Landing**.

Click **OK**.

10.

Type Parameters	
Parameter	Value
Construction	
Railing Height	3' 0"
Rail Structure (Non-Continuous)	Edit...
Baluster Placement	Edit...

Select the **Edit** button next to Rail Structure.

11.   **Insert**   Click the **Insert** button two times to add two rails.

12.

Rails	
	Name
1	Middle Rail
2	Bottom Rail

Name the rails.
Set Rail 1 to **Middle Rail**.
Set Rail 2 to **Bottom Rail**.

13.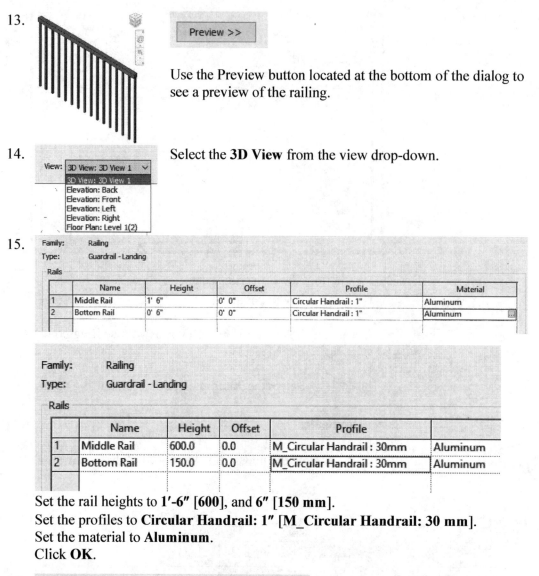

Preview >>

Use the Preview button located at the bottom of the dialog to see a preview of the railing.

14. Select the **3D View** from the view drop-down.

View: 3D View: 3D View 1

3D View: 3D View 1
Elevation: Back
Elevation: Front
Elevation: Left
Elevation: Right
Floor Plan: Level 1(2)

15.

Family: Railing
Type: Guardrail - Landing

Rails

	Name	Height	Offset	Profile	Material
1	Middle Rail	1' 6"	0' 0"	Circular Handrail : 1"	Aluminum
2	Bottom Rail	0' 6"	0' 0"	Circular Handrail : 1"	Aluminum

Family: Railing
Type: Guardrail - Landing

Rails

	Name	Height	Offset	Profile	
1	Middle Rail	600.0	0.0	M_Circular Handrail : 30mm	Aluminum
2	Bottom Rail	150.0	0.0	M_Circular Handrail : 30mm	Aluminum

Set the rail heights to **1'-6"** [**600**], and **6"** [**150 mm**].
Set the profiles to **Circular Handrail: 1"** [**M_Circular Handrail: 30 mm**].
Set the material to **Aluminum**.
Click **OK**.

16.

Type Parameters

Parameter	Value
Construction	⌃
Railing Height	3' 0"
Rail Structure (Non-Continuous)	Edit...
Baluster Placement	Edit...

Select **Edit** next to Baluster Placement.

17.

	Name	Baluster Family
1	Pattern start	N/A
2	Regular baluster	Baluster - Round : 1"
3	Pattern end	N/A

Main pattern

	Name	Baluster Family
1	Pattern start	N/A
2	Regular bal	M_Baluster - Round : 20mm
3	Pattern end	N/A

Under Main Pattern:

Set the Baluster Family to **Baluster - Round 1″ [M_Baluster – Round: 20mm]**.

18.

Posts

	Name	Baluster Family
1	Start Post	Baluster - Round : 1"
2	Corner Post	Baluster - Round : 1"
3	End Post	Baluster - Round : 1"

Under Posts:

Set the Baluster Family to **Baluster - Round 1″ [M_Baluster – Round: 20mm]**.

Posts

	Name	Baluster Family
1	Start Post	M_Baluster - Round : 20mm
2	Corner Post	M_Baluster - Round : 20mm
3	End Post	M_Baluster - Round : 20mm

19.

Dist. from previous
N/A
0' 4"
0' 0"

Dist. from previous	
N/A	N
150.0	0.
0.0	N

In the Dist. from previous column, set the distance to **4″ [150]**. Click **OK**.

20.

Rail Connections	Trim
Top Rail	
Height	3' 0"
Type	Circular - 1 1/2"
Handrail 1	

Set the Top Rail to **Circular -1 ½″ [Circular - 40mm]**.

Top Rail	
Use Top Rail	☑
Height	900.0
Type	Circular - 40mm

21.  Click on the **Apply** button to see how the changes you have made affect the preview.

Click **OK** to close the dialog.

22. Select the **Green Check** under Mode to finish the railing.

23. Switch to a 3D view to inspect the railing.

Repeat the exercise to place a railing at the other stair landing.

24. Save the file as *ex5-7.rvt [m_ex5-7.rvt]*.

Exercise 5-8
Creating Ceilings

Drawing Name: ceiling1.rvt [m_ceiling1.rvt]
Estimated Time: 10 minutes

This exercise reinforces the following skills:

- ❑ Ceilings
- ❑ Visibility of Annotation Elements

1.   Open *ceiling1.rvt [m_ceiling1.rvt]*.

2.   Activate **Level 1** under Ceiling Plans.

   Ceiling Plans
     Level 1
     Level 2
     Roof Line

3.   Type **VV**.

4.   Select the **Annotations Categories** tab.
   Disable visibility for Grids, Elevations and Sections.
   Click **OK**.

   ☑ Electrical Fixture Tags
   ☐ Elevations
   ☑ Floor Tags

   ☑ Scope Boxes
   ☑ Section Boxes
   ☐ Sections
   ☑ Site Tags

   ☑ Generic Annotations
   ☑ Generic Model Tags
   ☐ Grids
   ☑ Guide Grid

5.   The view display should update.

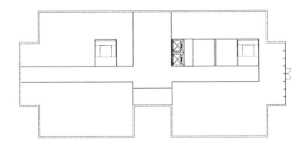

6.   Select the **Ceiling** tool from Build panel on the Architecture ribbon.

   Ceiling

7. 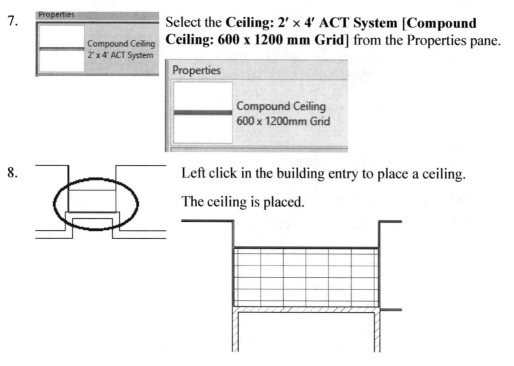 Select the **Ceiling: 2′ × 4′ ACT System [Compound Ceiling: 600 x 1200 mm Grid]** from the Properties pane.

8. Left click in the building entry to place a ceiling.

   The ceiling is placed.

9. Right click and select **Cancel** twice to exit the command.

10. Save as *ex5-8.rvt [m_ ex5-8.rvt]*.

Exercise 5-9
Adding Lighting Fixtures

Drawing Name: lighting1.rvt [m_lighting1.rvt]
Estimated Time: 10 minutes

This exercise reinforces the following skills:

❑   Add Component
❑   Load From Library

1.   Open *lighting1.rvt [m_lighting1.rvt]*.

2.   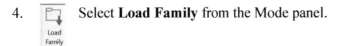   Activate **Level 1 Ceiling Plan**.

3.   Activate the **Architecture** ribbon.

     Select the **Component→Place a Component** tool from the Build panel.

4.   Select **Load Family** from the Mode panel.

5.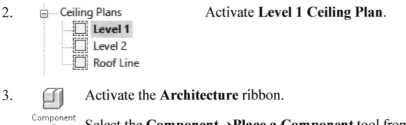

     Browse to the *Lighting/Architectural/Internal* folder.

6.   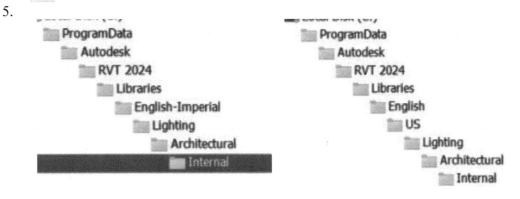   Locate the *Ceiling Light -Linear Box.rfa*
     [M_ Ceiling Light -Linear Box.rfa].
     Click **Open**.

7.

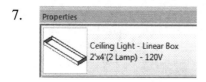

Select the **2′x 4′ (2 Lamp) -120V [0600x 1200mm (2 Lamp) – 277V]** for the element type.

8. Place fixtures on the grid.

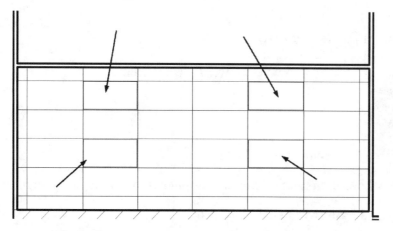

Use the ALIGN tool to position the fixtures in the ceiling grid.

9. Save the file as *ex5-9.rvt [m_ex5-9.rvt].*

Extra:

Add ceilings to Level 1 and Level 2.

Define new ceiling types using stucco material.

Additional Projects

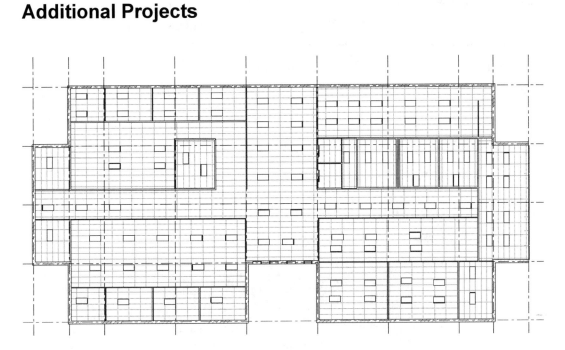

1) Add light fixtures and ceilings to the Second Level Ceiling Plan.
   Create a Ceiling Plan layout/sheet.

| File name: | Lighting Fixture Tag - Boxed.rfa |
| File of type: | Family Files [*.rfa] |

2) Load the Lighting Fixture Tag-Mark – Boxed (this family is in the Library) using File→Load Family.
   Use Tag All Not Tagged to tag all the lighting fixtures.
   Create a schedule of lighting fixtures.
   Add to your reflected ceiling plan sheet.

<Lighting Fixture Schedule>			
A	**B**	**C**	**D**
Type	Wattage	Level	Count
2'x4'(2 Lamp) - 120V	80 W	Level 2	93

Sort by Level and then by Type to get a schedule like this. Filter to only display Level 2 fixtures.

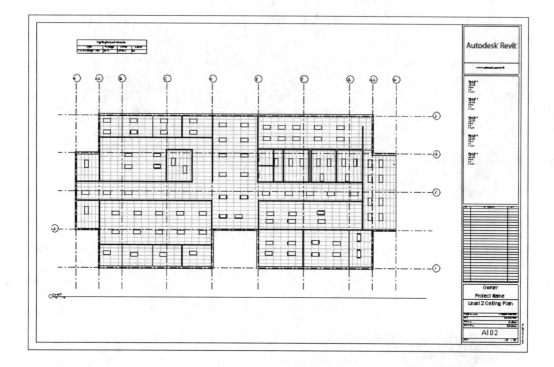

3) This is a department store in Japan. This exterior surface serves as both graphic pattern and structural system and is composed of 300mm-thick concrete and flush-mounted frameless glass. The resulting surface supports floor slabs spanning 10-15 meters without any internal columns.

How would you create a wall that looks like this in Revit?

You might be able to design a curtain wall, or you could modify a concrete wall and create parts of glass. Or you might combine both a curtain wall and a wall using parts.

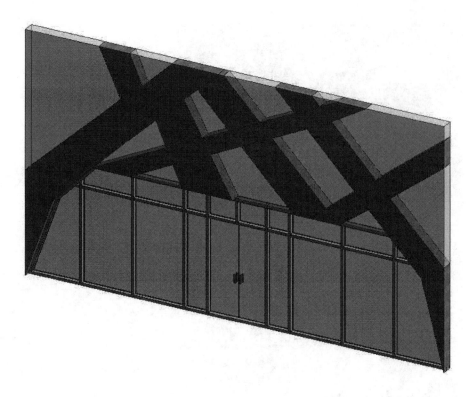

This wall was created by defining a wall using a 300 mm concrete structure, then dividing it into parts. The parts were then designated as either glass or concrete.

Here are the steps.

- Place a concrete wall going from Level 1 to Level 2.
- Edit the profile so that the sides are at an angle.
- Use the type selector to change the concrete wall to a curtain wall.
- Add the curtain wall door.
- Place a concrete wall going from Level 2 to Level 3.
- Edit the profile so that the wall fills in the angles missing from the curtain wall.
- Divide the concrete walls into parts. Sketch the angled sections that resemble tree branches. Assign a glass material to the parts.

See if you can do something similar.

Lesson 5 Quiz

True or False

1. Ceilings are not visible on the Floor Plan view.
2. Ceilings are visible on the Ceiling Plan view.
3. In order to place a ceiling-based light fixture, the model must have a ceiling.
4. To create an opening in a floor, use EDIT BOUNDARY.
5. Floors and ceilings are level-based.
6. Floors and ceilings can be offset from a level.
7. The boundaries for a floor or ceiling must be closed, non-intersecting loops.

Multiple Choice

8. When a floor is placed:

   A. The top of the floor is aligned to the level on which it is placed with the thickness projecting downward.
   B. The bottom of the floor is aligned to the level on which it is placed with the thickness projecting upwards.
   C. The direction of the thickness can be flipped going up or down.
   D. The floor offset can be above or below the level.

9. The structure of a floor is determined by its:

   A. Family type
   B. Placement
   C. Instance
   D. Geometry

10. Floors can be created by these two methods:
    Pick two answers.

    A. Sketching a closed polygon
    B. Picking Walls
    C. Place Component
    D. Drawing boundary lines

11. Ceilings are:
    Pick one answer.

    A. Level-based
    B. Floor-based
    C. Floating
    D. Non-hosted

12. Paint can be applied using the Paint tool which is located on this ribbon:

    A. Architecture
    B. Modify
    C. Rendering
    D. View

13. New materials are created using:

    A. Manage→Materials.
    B. View→Materials
    C. Architecture→Create→Materials
    D. Architecture→Render→Materials

14. This icon:

    A. Adjusts the brightness of a view.
    B. Temporarily isolates selected elements or categories.
    C. Turns off visibility of elements.
    D. Zooms into an element.

15. To place a Shaft Opening, activate the _____ ribbon.

    A. Modify
    B. View
    C. Manage
    D. Architecture

16. In order for parts to be visible in a view:

    A. Parts should be enabled on the Materials ribbon.
    B. Parts should be enabled in the Visibility/Graphics dialog.
    C. Show Parts should be enabled in the Properties pane for the view.
    D. The view display should be set to Shaded.

ANSWERS:

    1) T; 2) T; 3) T; 4) T; 5) T; 6) T; 7) T; 8) A; 9) A; 10) A & B; 11) A; 12) B; 13) A; 14) B; 15) D; 16) C

Lesson 6
Schedules

Revit proves its power in the way it manages schedules. Schedules are automatically updated whenever the model changes. In this lesson, users learn how to add custom parameters to elements to be used in schedules, how to create and apply keynotes, and how to place schedules on sheets.

Tags are annotations and they are **view-specific**. This means if you add tags in one view, they are not automatically visible in another view.

Tags are Revit families. Tags use parameter data. For example, a Room tag might display the room name and the room area. The room name and room area are both parameters.

A schedule is a tabular view of information, using the parameters of elements in a project. You can organize a schedule using levels, types of elements, etc. You can filter a schedule so it only displays the elements of a specific type or located on a specific level.

Because schedules are views, they are listed in the Project Browser in the view area. Schedules are placed on sheets, just like any other view; simply drag and drop from the Project Browser.

You cannot import schedules from Excel, but you can export schedules to Excel.

Exercise 6-1
Adding Door Tags

Drawing Name: door tags.rvt [m_door tags.rvt]
Estimated Time: 10 minutes

This exercise reinforces the following skills:

❑   Door Tags

1.   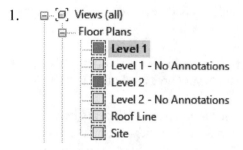   Activate **Level 1**.

2.   The floor plan has several doors.

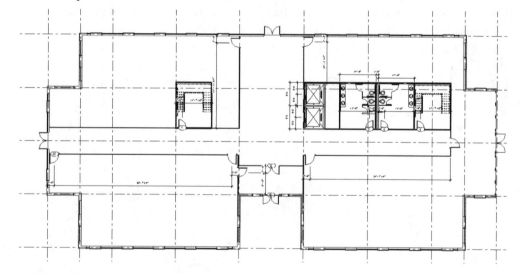

3.   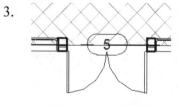

Select the exterior double door.

Your tag number may be different from mine.

4. Go to the **Properties** panel.

Locate the **Mark** parameter under Identity Data.

The Mark or Door Number can be modified in the Properties panel.

You cannot use the same door number twice. You will see an error message if you attempt to assign a door number already in use.

If you delete a door and add a new door, the door will be assigned the next mark number. It will not use the deleted door's mark number. Many users wait until the design is complete and then re-number all the doors on each level.

5. Zoom into the curtain wall door.

6. Go to the Annotate ribbon.

Select **Tag by Category**.

7. On the Options bar:
    Disable Leader.
    Verify the orientation of the tag is set to
    Horizontal.

8. Select the curtain wall door to add a door tag.

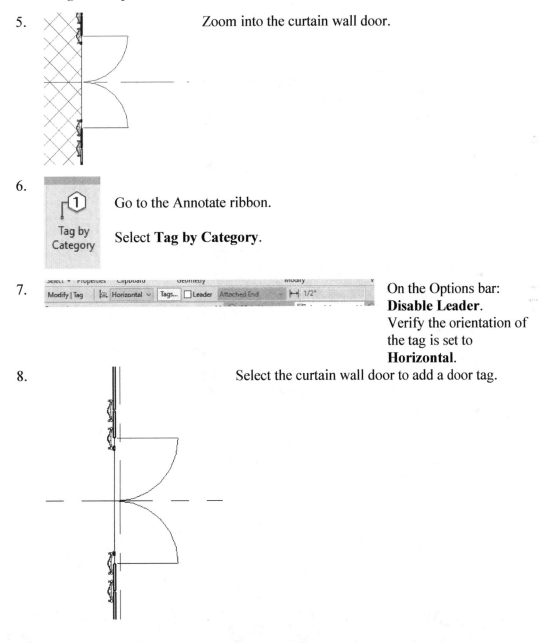

9.

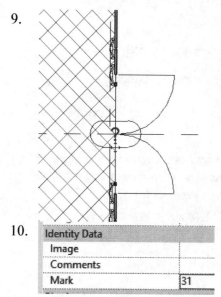

The tag has a question mark to indicate that there is no value in the Mark property for this door.

Escape out of the command.

Select the door.

10.

Identity Data	
Image	
Comments	
Mark	31

Type **31** in the Mark field.

The tag updates with the new information.

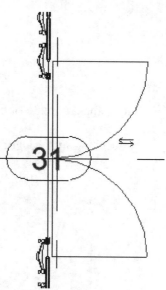

Not all the doors on Level 1 have door tags.

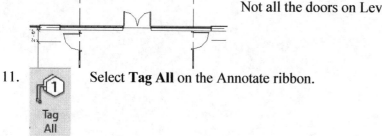

11.

Select **Tag All** on the Annotate ribbon.

12. 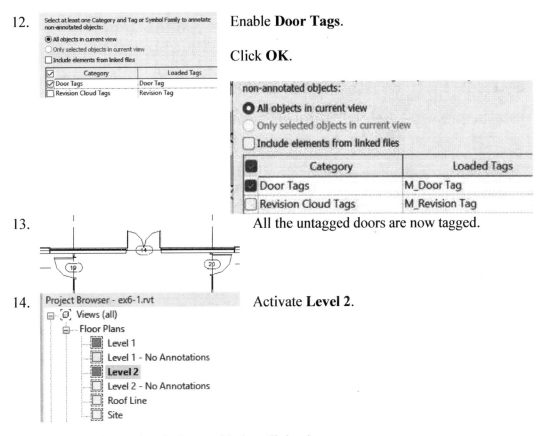 Enable **Door Tags**.

Click **OK**.

All the untagged doors are now tagged.

13.

14. Activate **Level 2**.

15. Door tags have already been added to all the doors.
    This is because Tag All was selected when we tagged the doors. All doors are tagged regardless of whether the view is active or not.

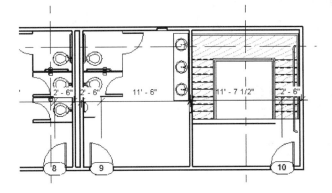

16. Save as *ex6-1.rvt [m_ex6-1.rvt]*.

Exercise 6-2
Creating a Door Schedule

Drawing Name: door schedule1.rvt [m_door schedule1.rvt]
Estimated Time: 30 minutes

This exercise reinforces the following skills:

- ❑ Door Tags
- ❑ Schedules

1.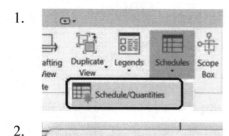

   Activate the View ribbon.

   Select **Schedules → Schedule/Quantities**.

2.

   Under Filter list:
   Uncheck all the disciplines except for Architecture.

3.

   Select **Doors**.

   Click **OK**.

The available fields are Parameters assigned to the door families.

You can add which parameters to use in the schedule by highlighting and selecting the Add button.

The order the fields appear in the right pane is the order the columns will appear in the schedule.

4.   Select the following fields in this order:

Mark	Height	Frame Material
Type	Thickness	Comments
Type Mark	Frame Type	Level
Width		

5.

Select the Sorting/ Grouping tab.

Sort by **Level**.
Then by **Mark**.

6.

The Formatting tab allows you to rename the Column Headers.

Select the Formatting tab.

Rename Mark to **No**.

7.

Highlight **Level**.

Enable **Hidden Field**.

This allows us to sort by the Level without showing it in the schedule.

8.

Select the **Appearance** tab.

Enable **Grid lines**.
Enable **Outline**.
Enable **Grid in headers/ footers/spacers**.

Set the Outline to **Wide Lines**.

Disable **Blank row before data**.

9.

Enable **Show Title**.
Enable **Show Headers**.
Set the Title Text to **1/4" Arial. [7mm Arial]**
Click **OK**.

10. A view opens with the schedule.

Tabs: 3D-Lobby | Level 1 | View 1 | Level 2 | **Door Schedule X**

<Door Schedule>

A	B	C	D	E	F	G	H	I
No.	Type	Type Mark	Width	Height	Thickness	Frame Type	Frame Material	Comments
1	36" x 84"	1	3' - 0"	7' - 0"	0' - 2"			
2	36" x 84"	1	3' - 0"	7' - 0"	0' - 2"			
3	36" x 84"	1	3' - 0"	7' - 0"	0' - 2"			
5	36" x 84"	1	3' - 0"	7' - 0"	0' - 2"			
6	72" x 82"	12	6' - 0"	6' - 10"	0' - 1 3/4"			
7	72" x 78"	8	6' - 0"	6' - 6"	0' - 2"			
21	Door-Curtain-Wall-	14	6' - 9 1/16"	8' - 0"				
8	36" x 84"	1	3' - 0"	7' - 0"	0' - 2"			
9	36" x 84"	1	3' - 0"	7' - 0"	0' - 2"			
10	36" x 84"	1	3' - 0"	7' - 0"	0' - 2"			
11	36" x 84"	1	3' - 0"	7' - 0"	0' - 2"			
12	12000 x 2290 Ope	13	2' - 11 7/16"	7' - 6 5/32"	0' - 1 3/16"			
13	12000 x 2290 Ope	13	2' - 11 7/16"	7' - 6 5/32"	0' - 1 3/16"			

Tabs: A101 - Landing Page | Level 1 | Level 2 | **Door Schedule X**

<Door Schedule>

A	B	C	D	E	F	G	H	I
No	Type	Type Mark	Width	Height	Thickness	Frame Type	Frame Material	Comments
1	0915 x 2134mm	20	915	2134	51			
2	0915 x 2134mm	20	915	2134	51			
3	0915 x 2134mm	20	915	2134	51			
4	0915 x 2134mm	20	915	2134	51			
5	1800 x 2100mm	42	1800	2100	45			
6	1800 x 2000mm	40	1800	2000	50			
14	1800 x 2000mm	40	1800	2000	50			
15	900 x 2400mm	45	900	2400	45			
16	900 x 2400mm	45	900	2400	45			
17	900 x 2400mm	45	900	2400	45			
18	900 x 2400mm	45	900	2400	45			
19	1800 x 2100mm	41	1800	2100	45			
20	900 x 2400mm	45	900	2400	45			
21	900 x 2400mm	45	900	2400	45			
22	900 x 2400mm	45	900	2400	45			
31	M_Door-Curtain-Wa	44	2054	2400				
7	0915 x 2134mm	20	915	2134	51			
8	0915 x 2134mm	20	915	2134	51			
9	0915 x 2134mm	20	915	2134	51			
10	0915 x 2134mm	20	915	2134	51			
11	12000 x 2290 Open	43	900	2290	30			
12	12000 x 2290 Open	43	900	2290	30			

Now, you can see where you may need to modify or tweak the schedule.

11.

Other	
Fields	Edit...
Filter	Edit...
Sorting/Grouping	Edit...

In the Properties pane:

Select **Edit** next to Sorting/Grouping.

12.

| Fields | Filter | Sorting/Grouping | Formatting | Appearance |

Sort by: [Mark ▼] ◉ Ascending   ○ Descending
  ☑ Header   ☑ Footer: [▼]   ☑ Blank line

Then by: [Level ▼]   ◉ Ascending   ○ Descending
  ☑ Header   ☑ Footer: [▼]   ☑ Blank line

Sort by **Mark** and then by **Level**.

Click **OK**.

13.

| 3D-Lobby | Level 1 | View 1 | Level 2 | Door Schedule ✕ |

You can use the tabs at the top of the display window to switch from one view to another.

Click on the Door Schedule to activate/open the view.

14.

Other	
Fields	Edit...
Filter	Edit...
Sorting/Gro...	Edit...

In the Properties Pane:

Select **Edit** next to Fields.

15.

Scheduled fields (in order):

Mark
Type
Type Mark
Width
Height
Thickness
Frame Type
Frame Material
Comments
Level

Highlight **Type** in the Scheduled fields list.

Select **Remove** (the red arrow) to remove from the list.

16.

| Fields | Filter | Sorting/Grouping | Formatting | Appearance |

Fields:
Mark
Type Mark
Width
Height
Thickness
Frame Type
Frame Material
Comments
Level

Heading:
Type

Heading orientation:
Horizontal

Alignment:
Center

Select the Formatting tab.

Highlight **Type Mark**.

Change the Heading to **Type**.

Set Alignment to **Center**.

17.

| Fields | Filter | Sorting/Grouping | Formatting | Appearance |

Fields:
Mark
Type Mark
Width
Height
Thickness
Frame Type
Frame Material
Comments
Level

Heading:
No.

Heading orientation:
Horizontal

Alignment:
Center

Highlight **Mark**.

Set the Alignment to **Center**.

Click **OK**.

18. The schedule updates.

<Door Schedule>

A	B	C	D	E	F	F
No.	Type	Width	Height	Thickness	Frame Type	Fi
1	1	3' - 0"	7' - 0"	0' - 2"		
2	1	3' - 0"	7' - 0"	0' - 2"		
3	1	3' - 0"	7' - 0"	0' - 2"		
4	1	3' - 0"	7' - 0"	0' - 2"		
5	8	6' - 0"	6' - 6"	0' - 2"		
6	8	6' - 0"	6' - 6"	0' - 2"		
7	1	3' - 0"	7' - 0"	0' - 2"		
8	1	3' - 0"	7' - 0"	0' - 2"		
9	1	3' - 0"	7' - 0"	0' - 2"		
10	1	3' - 0"	7' - 0"	0' - 2"		
11	13	2' - 11 3/8"	7' - 6 1/8"	0' - 1 1/8"		
12	13	2' - 11 3/8"	7' - 6 1/8"	0' - 1 1/8"		
21	14	6' - 9 1/8"	8' - 0"			

19.

B	C	D	E	F	F
Type	Width	Height	Thickness	Frame Type	Fr
1	3' - 0"	7' - 0"	0' - 2"		
1	3' - 0"	7' - 0"	0' - 2"		

Place the cursor over the **width** column.
Hold down the left mouse button.

Drag the mouse across the Width, Height, and Thickness columns.

20. Select **Group** on the Titles & Headers panel on the ribbon.

Group

21.

C	D	E
	Size	
Width	Height	Thickness
3' - 0"	7' - 0"	0' - 2"

Type **Size** as the header in the sub-header.

22. Place the cursor over the **Frame Type** column.
Hold down the left mouse button.
Drag the mouse across the Frame Type and Frame
Material columns.

23. Select **Group** on the Titles & Headers panel on the ribbon.

Group

24. Type **Frame** as the header in the sub-header.

25. Save as *ex6-2.rvt [m_ex6-2.rvt]*.

Exercise 6-3
Using Combined Parameters

Drawing Name: door schedule2.rvt [m_door schedule2.rvt]
Estimated Time: 10 minutes

This exercise reinforces the following skills:

- ❏ Combined Parameters
- ❏ Schedules

1. Open the **Door Schedule** view.

2. 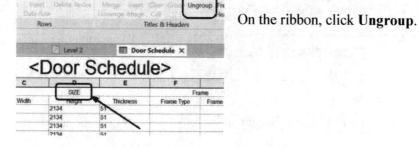 Highlight the **SIZE** header.

   On the ribbon, click **Ungroup**.

3. Click **Edit** next to Fields in the
Properties pane.

4.

Select **Combine parameters**.

5.

In the Combined Parameter Name field:

Type **Dimensions (W x H x THK).**

Select the Width, Height, and Thickness parameters from the left pane.

6. In the Suffix column: Type a space, then x, then a space for Width and Height.
Delete the / Separator from all three rows.
Look at the bottom of the dialog and check the preview to see if it looks correct.
Click **OK**.

7.

Scheduled fields (in order):
Mark
Type Mark
Dimensions (W x H x THK)
Width
Height
Thickness
Frame Type
Frame Material
Comments
Level

Use the Move up tool to locate the Dimensions combined parameter below the Type Mark parameter.

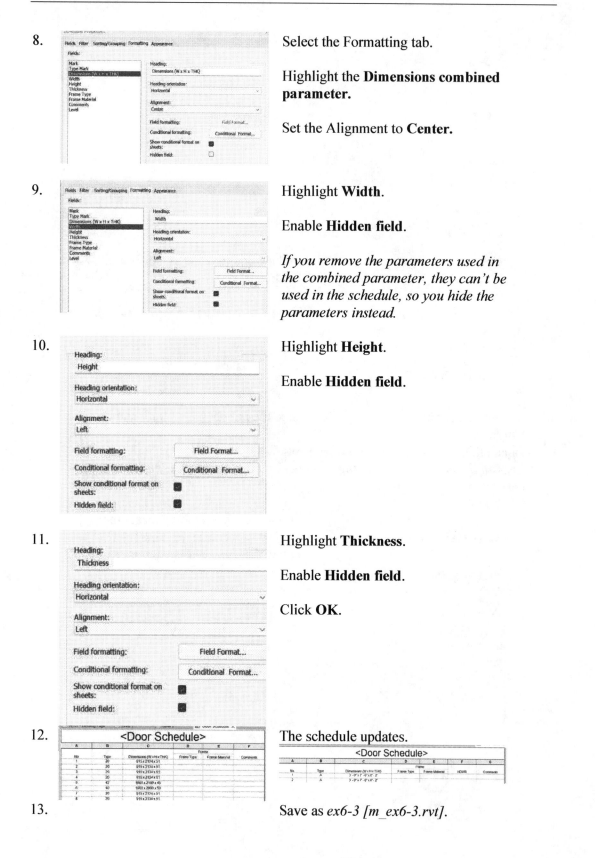

8. Select the Formatting tab.

    Highlight the **Dimensions combined parameter.**

    Set the Alignment to **Center.**

9. Highlight **Width.**

    Enable **Hidden field.**

    If you remove the parameters used in the combined parameter, they can't be used in the schedule, so you hide the parameters instead.

10. Highlight **Height.**

    Enable **Hidden field.**

11. Highlight **Thickness.**

    Enable **Hidden field.**

    Click **OK.**

12. The schedule updates.

13. Save as *ex6-3 [m_ex6-3.rvt].*

The door schedule is missing information. Many architects prefer to use letter designations for door types and not numbers. To modify the information and update the schedule, we need to update the parameters. The values displayed in schedules are the values stored in family parameters.

Revit uses two types of parameters: *instance* and *type*. Instance parameters can be modified in the Properties pane. They are unique to each **instance** of an element. Examples of instance parameters might be level, hardware, or sill height. Type parameters do not change regardless of where the element is placed. Examples of type parameters are size, material, and assembly code. To modify a type parameter, you must **edit type**.

Exercise 6-4
Modifying Family Parameters

Drawing Name: parameters.rvt [m_parameters.rvt]
Estimated Time: 10 minutes

This exercise reinforces the following skills:

- ❑ Family Parameters
- ❑ Schedules

1. 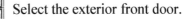 Select the Level 1 floor plan view.

2. 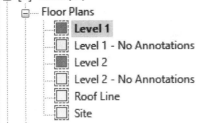 Select the exterior front door.

3.

Fill in the Frame Type as 1.

Fill in Frame Material as **WD**.
WD is the designator for wood.
Fill in Comments as **BY MFR**.

Click **Apply** to save the changes.

These are instance parameters, so they only affect each element selected.

4. Open the Door Schedule.

4		5 - 0	7 - 0	0 - 2				
5	8	6' - 0"	6' - 6"	0' - 2"	1		WD	BY MFR
6	8	6' - 0"	6' - 6"	0' - 2"				

The schedule has updated with the information, but it only affected the one door. That is because the parameters that were changed in the Properties were *instance* parameters.

5.

Select the Level 1 tab to activate the Level 1 floor plan view.

6.

Select the first exterior double door.

Note that this is the **Door-Exterior-Double-Two_Lite 2 [M_ Door-Exterior-Double-Two_Lite 2]**.

7.

Select the second exterior double door.

Note that this is the **Door-Double-Glass [M_ Door-Double-Glass]**.

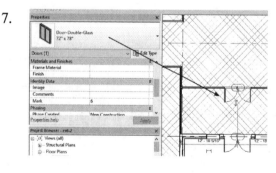

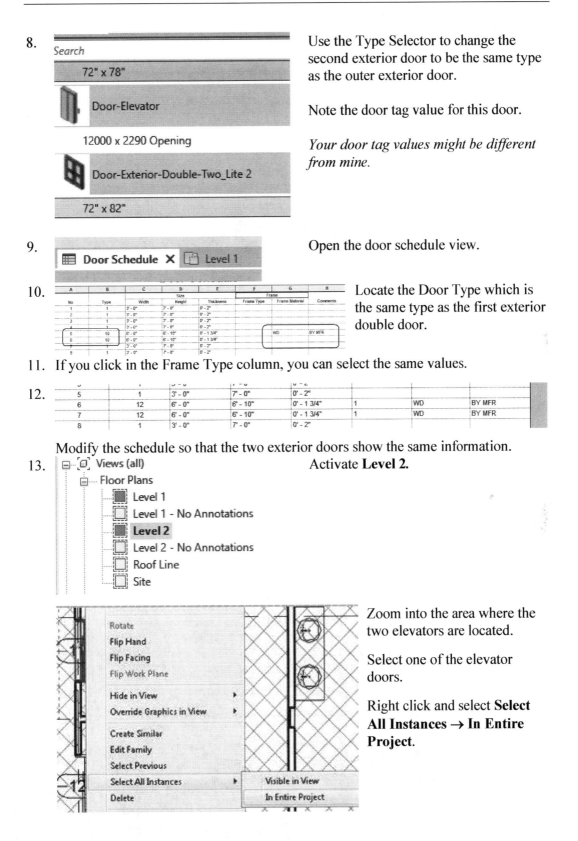

8. Use the Type Selector to change the second exterior door to be the same type as the outer exterior door.

    Note the door tag value for this door.

    Your door tag values might be different from mine.

9. Open the door schedule view.

10. Locate the Door Type which is the same type as the first exterior double door.

11. If you click in the Frame Type column, you can select the same values.

12.

No	Type	Width	Height	Thickness	Frame Type	Frame Material	Comments
5	1	3' - 0"	7' - 0"	0' - 2"			
6	12	6' - 0"	6' - 10"	0' - 1 3/4"	1	WD	BY MFR
7	12	6' - 0"	6' - 10"	0' - 1 3/4"	1	WD	BY MFR
8	1	3' - 0"	7' - 0"	0' - 2"			

    Modify the schedule so that the two exterior doors show the same information.

13. Activate **Level 2.**

    Zoom into the area where the two elevators are located.

    Select one of the elevator doors.

    Right click and select **Select All Instances → In Entire Project**.

14. 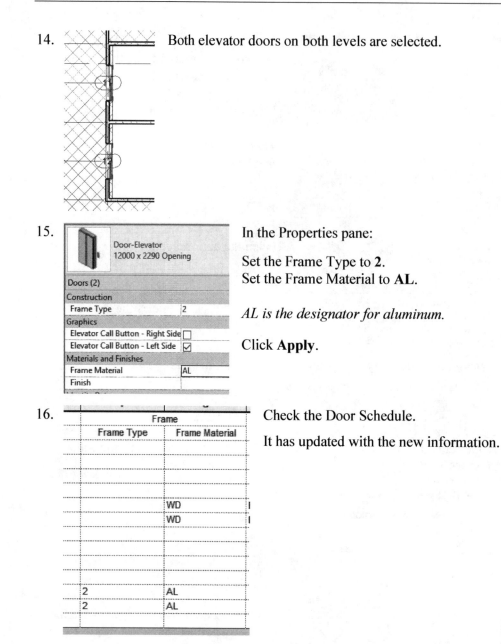 Both elevator doors on both levels are selected.

15.

	Door-Elevator 12000 x 2290 Opening
Doors (2)	
Construction	
Frame Type	2
Graphics	
Elevator Call Button - Right Side	☐
Elevator Call Button - Left Side	☑
Materials and Finishes	
Frame Material	AL
Finish	

In the Properties pane:

Set the Frame Type to **2**.
Set the Frame Material to **AL**.

AL is the designator for aluminum.

Click **Apply**.

16.

Frame	
Frame Type	Frame Material
	WD
	WD
2	AL
2	AL

Check the Door Schedule.

It has updated with the new information.

17.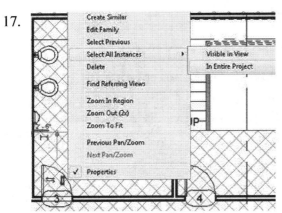

Activate **Level 1**.

Zoom into the area where the lavatories are located.

Select one of the interior single flush doors.

Right click and select **Select All Instances → In Entire Project**.

18.

Single-Flush 36" x 84"	
Doors (8)	
Construction	
Frame Type	3
Materials and Finishes	
Frame Material	WD
Finish	

In the Properties pane:

Set the Frame Type to **3**.
Set the Frame Material to **WD**.
WD is the designator for wood.
Click **Apply**.

19. Check the Door Schedule.

\<Door Schedule\>

A	B	C	D	E	F	G	H
		Size			Frame		
No.	Type	Width	Height	Thickness	Frame Type	Frame Material	Comments
1	1	3' - 0"	7' - 0"	0' - 2"			
2	1	3' - 0"	7' - 0"	0' - 2"			
3	1	3' - 0"	7' - 0"	0' - 2"			
4	1	3' - 0"	7' - 0"	0' - 2"			
5	10	6' - 0"	6' - 10"	0' - 1 3/4"	1	WD	BY MFR
6	10	6' - 0"	6' - 10"	0' - 1 3/4"	1	WD	BY MFR
7	1	3' - 0"	7' - 0"	0' - 2"			
8	1	3' - 0"	7' - 0"	0' - 2"			
9	1	3' - 0"	7' - 0"	0' - 2"			
10	1	3' - 0"	7' - 0"	0' - 2"			
11	11	2' - 11 1/2"	7' - 6 1/4"	0' - 1 1/4"	2	AL	
12	11	2' - 11 1/2"	7' - 6 1/4"	0' - 1 1/4"	2	AL	
13	9	6' - 0"	6' - 10"	0' - 1 3/4"			
14	9	6' - 0"	6' - 10"	0' - 1 3/4"			
15	13	3' - 6"	6' - 8"	0' - 1 3/4"	3	WD	
16	13	3' - 6"	6' - 8"	0' - 1 3/4"	3	WD	
18	13	3' - 6"	6' - 8"	0' - 1 3/4"	3	WD	
19	13	3' - 6"	6' - 8"	0' - 1 3/4"	3	WD	
20	13	3' - 6"	6' - 8"	0' - 1 3/4"	3	WD	
21	13	3' - 6"	6' - 8"	0' - 1 3/4"	3	WD	
22	13	3' - 6"	6' - 8"	0' - 1 3/4"	3	WD	
31	12	6' - 9"	8' - 0"				

It has updated with the new information.

20.

Properties	×
Single-Flush 36" x 84"	
Doors (49)	Edit Type

Activate **Level 1**.

Select one of the lavatory doors.

In the Properties Pane, select **Edit Type**.

Properties	×
M_Single-Flush 0915 x 2134mm	
Doors (1)	Edit Type

21.

Identity Data	
Assembly Code	C1020
Keynote	
Model	
Manufacturer	
Type Comments	
URL	
Description	
Assembly Description	Interior Doors
Type Mark	A
Fire Rating	
Cost	
OmniClass Number	23.30.10.00
OmniClass Title	Doors

Scroll down to locate the Type Mark.

Change the Type Mark to **A**.

This only changes the parameter value for the doors loaded in the project.

If you would like to change the parameter for the door family so that it will always show a specific value whenever it is loaded into a family, you have to edit the door family.

Click **OK**.

22. Release the door selection by left clicking anywhere in the display window.

23.

A	B	
No.	Type	
1	A	3' - 0"
2	A	3' - 0"
3	A	3' - 0"
4	A	3' - 0"
5	10	6' - 0"
6	10	6' - 0"
7	A	3' - 0"
8	A	3' - 0"
9	A	3' - 0"
10	A	3' - 0"
11	11	2' - 1"
12	11	2' - 1"
13	9	6' - 0"
14	9	6' - 0"
15	13	3' - 6"
16	13	3' - 6"
18	13	3' - 6"
19	13	3' - 6"

Check the Door Schedule.

It has updated with the new information.

Notice that all the single flush doors updated with the new Type Mark value.

*Because this was a **type** parameter, not an **instance** parameter, all doors that are the same type are affected.*

24.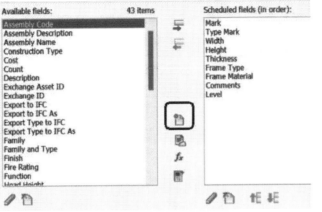

Select the **Edit** tab next to Fields.

We want to add a parameter for hardware.

If you look in the available fields, you will see that Hardware is not listed.

So, we will create that parameter.

Select the **Add Parameter** button.

25. Enable **Project parameter**.

This means the parameter is only available in this project. If you wanted to create a parameter which could be used in any project or family, you would use a shared parameter.

Under Parameter Data:
Type **Hardware** for the Name.
Select **Text** for the Type of Parameter.

By default, parameters are set to length, which is a dimension parameter. If you do not set the type of parameter to text, you will have to delete the parameter and re-create it.

Group the parameter under **Construction**.
Enable **Instance**.

26. Select **Edit Tooltip**.

27.

Delete the text in the box.
Type **Enter Hardware Assy Code for hardware to be installed**.

Click **OK**.

You have supplied a helpful tip to remind users what to enter for this parameter.

28. Click **OK**.

29.

Hardware is now listed in the Scheduled fields list.

Use the **Move Up** button to position the Hardware column before Comments.

30.

Select the Formatting tab.

Highlight **Hardware**.

Change the Heading to **HDWR**.

Change the Alignment to **Center**.

Click **OK**.

31. The Hardware column is added to the schedule.

<Door Schedule>

A	B	C	D	E	F	G	H	I
			Size		Frame			
No.	Type	Width	Height	Thickness	Frame Type	Frame Material	HDWR	Comments
1	A	3' - 0"	7' - 0"	0' - 2"				
2	A	3' - 0"	7' - 0"	0' - 2"				
3	A	3' - 0"	7' - 0"	0' - 2"				
4	A	3' - 0"	7' - 0"	0' - 2"				
5	10	6' - 0"	6' - 10"	0' - 1 3/4"	1	WD		BY MFR
6	10	6' - 0"	6' - 10"	0' - 1 3/4"	1	WD		BY MFR
7	A	3' - 0"	7' - 0"	0' - 2"				
8	A	3' - 0"	7' - 0"	0' - 2"				
9	A	3' - 0"	7' - 0"	0' - 2"				
10	A	3' - 0"	7' - 0"	0' - 2"				

32. Select a lavatory door on Level 1.

33. Note that Hardware is now available as a parameter in the Properties pane.

It is available in the Properties palette because it was designated as an **instance** *parameter.*

34. *Hover your mouse over the Hardware column to see your tool tip.*

35. Save as *ex6-3.rvt [m_ex6-3.rvt]*.

Many architectural firms have a specific format for schedules. The parameters for these schedules may not be included in the pre-defined parameters in Revit. If you are going to be working in multiple projects, you can define a single file to store parameters to be used in any schedule in any project.

Best practices has the CAD manager for the department create the shared parameters file and place it on the server in a location where all team members have access.

Exercise 6-5
Creating Shared Parameters

Drawing Name: shared parameters.rvt
Estimated Time: 30 minutes

This exercise reinforces the following skills:

- Shared Parameters
- Schedules
- Family Properties

1.   [Manage]   Activate the **Manage** ribbon.

2.   Select the **Shared Parameters** tool from the Settings panel.

3.   Click **Create** to create a file where you will store your parameters.

4.   Locate the folder where you want to store your file. Set the file name to *custom parameters.txt.* Click **Save**.

   Note that this is a txt file.

5.   Under Groups, select **New**.

6.   Enter **Door**. Click **OK**.

7.   Under Parameters, select **New**.

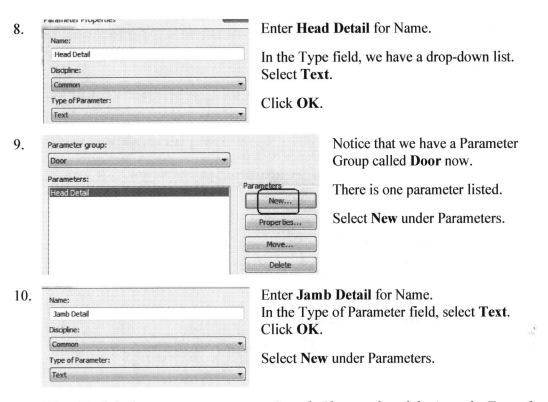

8. Enter **Head Detail** for Name.

In the Type field, we have a drop-down list. Select **Text**.

Click **OK**.

9. Notice that we have a Parameter Group called **Door** now.

There is one parameter listed.

Select **New** under Parameters.

10. Enter **Jamb Detail** for Name.
In the Type of Parameter field, select **Text**.
Click **OK**.

Select **New** under Parameters.

Hint: By default, parameters are set to Length. If you rush and don't set the Type of Parameter, you will have to delete it and re-do that parameter.

11. Enter **Threshold Detail** for Name.
In the Type field, select **Text**.
Click **OK**.

Select **New** under Parameters.

12. Enter **Hardware Group** for Name.
In the Type field, select **Text**.
Click **OK**.

13. Select **New** under Groups.

14. Name the new group **General**.
Click **OK**.

15. We now have two groups listed under Parameter group.

Select **Door**.

16. Highlight the **Hardware Group** field.

Select **Move**.

17. Select **General** from the drop-down list.
Click **OK**.

18. **Hardware Group** is no longer listed in the Door parameter group.

Select **General** from the Parameter group list.

19. We see **Hardware Group** listed.

Highlight **Hardware Group**.

Select **Properties**.

20. We see how **Hardware Group** is defined.
Click **OK**.

Notice that the properties cannot be modified.

Click **OK** to close the dialog.

21. Locate the *custom parameters.txt* file using Windows Explorer.

22. Right click and select **Open**.

This should open the file using Notepad and assumes that you have associated txt files with Notepad.

We see the format of the parameter file.

Note that we are advised not to edit manually.

However, currently this is the only place you can modify the parameter type from Text to Integer, etc.

23. Change the data type to **INTEGER** for Hardware Group.

Do not delete any of the spaces!

24. Save the text file and close.

25. Select the **Shared Parameters** tool on the Manage ribbon.

Shared
Parameters

If you get an error message when you attempt to open the file, it means that you made an error when you edited the file. Re-open the file and check it.

26.

Set the parameter group to **General**.

Highlight **Hardware Group**.
Select **Properties**.

27. Note that Hardware Group is now defined as an Integer.

Parameter Properties

Name:
Hardware Group

Discipline:
Common

Type of Parameter:
Integer

Click **OK** twice to exit the Shared Parameters dialog.

28. Save as *ex6-5.rvt [m_ex6-5.rvt]*.

Exercise 6-6
Adding Shared Parameters to a Schedule

Drawing Name: schedule1.rvt [m_schedule1.rvt]
Estimated Time: 30 minutes

This exercise reinforces the following skills:

❑ Shared Parameters
❑ Family Properties
❑ Schedules

1. Open *schedule1.rvt [m_schedule1.rvt]*.
   [View] Activate the **View** ribbon.

2. 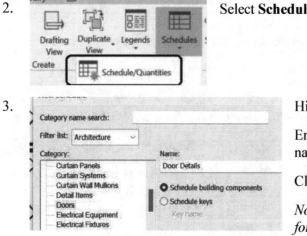 Select **Schedules→Schedule/ Quantities**.

3. Highlight **Doors**.

   Enter **Door Details** for the schedule name.

   Click **OK**.

   Note that you can create a schedule for each phase of construction.

4. Add **Mark, Type, Width, Height, Thickness, and Head Detail**.

The order of the fields is important. The first field is the first column, etc. You can use the Move Up and Move Down buttons to sort the columns.

5. Select the **Add Parameter** button.

6.

   Enable **Shared parameter**.

   Click **Select**.

If you don't see any shared parameters, you need to browse for the shared parameters .txt file to load it. Locate the file in your work folder or in the downloaded exercise files.

7.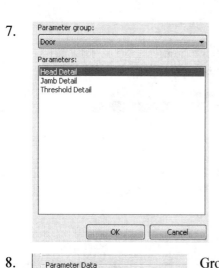

Select **Head Detail**.

Click **OK**.

8.

Group it under **Construction**.

9.

Place a check next to **Add to all elements in the category.**
This will add the parameter to the Type or Instance Properties for all the door elements in the project.

10.

Enable **Type**.

This means all the door elements of the same type will use this parameter.

Click **OK**.

11.

Head Detail is now listed as a column for the schedule.

Select **Add parameter**.

12.

Enable **Shared parameter**.

Click **Select**.

13.

Select **Jamb Detail**.

Click **OK**.

14.

Group it under **Construction**.

15.

Place a check next to **Add to all elements in the category.**

This will add the parameter to the Type or Instance Properties for all the door elements in the project.

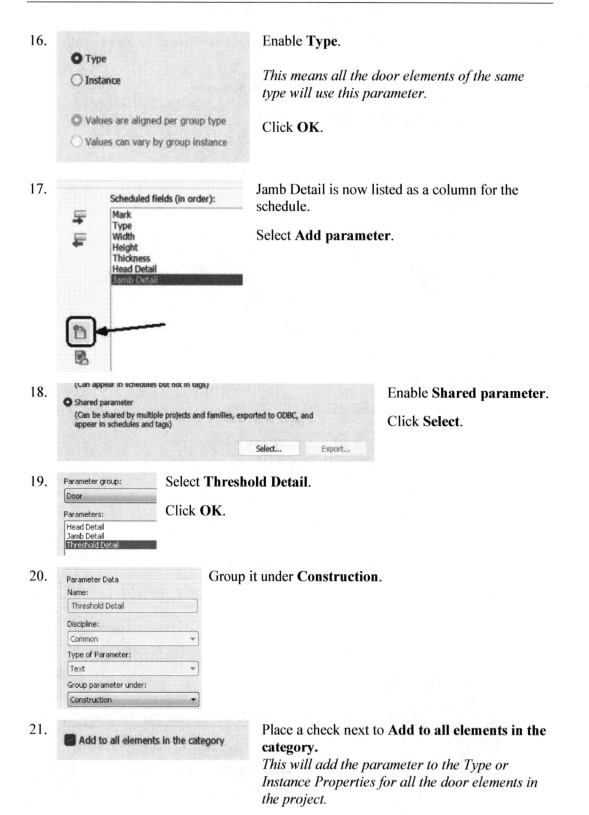

16. Enable **Type**.

This means all the door elements of the same type will use this parameter.

Click **OK**.

17. Jamb Detail is now listed as a column for the schedule.

Select **Add parameter**.

18. Enable **Shared parameter**.

Click **Select**.

19. Select **Threshold Detail**.

Click **OK**.

20. Group it under **Construction**.

21. Place a check next to **Add to all elements in the category**.

This will add the parameter to the Type or Instance Properties for all the door elements in the project.

22.

Type
Instance

Values are aligned per group type
Values can vary by group instance

Enable **Type**.
This means all the door elements of the same type will use this parameter.

Click **OK**.

Scheduled fields (in order):
Mark
Type
Width
Height
Thickness
Head Detail
Jamb Detail
Threshold Detail

The detail columns have now been added to the schedule.

23.

Heading:
Door No

Heading orientation:
Horizontal

Alignment:
Center

Select the **Formatting** tab.

Highlight **Mark**.
Change the Heading to **Door No**.
Change the Alignment to **Center**.

24.

Heading:
Door Type

Heading orientation:
Horizontal

Alignment:
Left

Change the Heading for Type to **Door Type**.

25.

Heading:
W

Heading orientation:
Horizontal

Alignment:
Center

Change the Heading for Width to **W**.

Change the Alignment to **Center**.

26.

Heading:
H

Heading orientation:
Horizontal

Alignment:
Center

Change the Heading for Height to **H**.

Change the Alignment to **Center**.

27.

Heading:
THK

Heading orientation:
Horizontal

Alignment:
Center

Change the Heading for Thickness to **THK**.

Change the Alignment to **Center**.

Click **OK**.

28. A window will appear with your new schedule.

\<Door Details\>

	A	B	C	D	E	F	G	H
	Door No.	Door Type	W	H	THK	Head Detail	Jamb Detail	Threshold Detail
1	36" x 84"	3' - 0"	7' - 0"	0' - 2"				
2	36" x 84"	3' - 0"	7' - 0"	0' - 2"				
3	36" x 84"	3' - 0"	7' - 0"	0' - 2"				
4	36" x 84"	3' - 0"	7' - 0"	0' - 2"				
5	72" x 82"	6' - 0"	6' - 10"	0' - 1 3/4"				
6	72" x 82"	6' - 0"	6' - 10"	0' - 1 3/4"				
7	36" x 84"	3' - 0"	7' - 0"	0' - 2"				
8	36" x 84"	3' - 0"	7' - 0"	0' - 2"				
9	36" x 84"	3' - 0"	7' - 0"	0' - 2"				
10	36" x 84"	3' - 0"	7' - 0"	0' - 2"				
11	12000 x 2290 Ope	2' - 11 1/2"	7' - 6 1/4"	0' - 1 1/4"				
12	12000 x 2290 Ope	2' - 11 1/2"	7' - 6 1/4"	0' - 1 1/4"				
21	Door-Curtain-Wall-	6' - 9"	8' - 0"					

Level 1　　　Level 1　　　Door Details　　**Door Details** ✕

\<Door Details\>

	A	B	C	D	E	F	G	H
	Door No	Door Type	W	H	THK	Head Detail	Jamb Detail	Threshold Detail
1	0915 x 2134mm	915	2134	51				
2	0915 x 2134mm	915	2134	51				
3	0915 x 2134mm	915	2134	51				
4	0915 x 2134mm	915	2134	51				
5	1800 x 2100mm	1800	2100	45				
6	1800 x 2100mm	1800	2100	45				
7	0915 x 2134mm	915	2134	51				
8	0915 x 2134mm	915	2134	51				
9	0915 x 2134mm	915	2134	51				
10	0915 x 2134mm	915	2134	51				
11	12000 x 2290 Ope	900	2290	30				
12	12000 x 2290 Ope	900	2290	30				
31	M_Door-Curtain-W	2054	2400					
14	1800 x 2000mm	1800	2000	50				
15	900 x 2400mm	900	2400	45				
16	900 x 2400mm	900	2400	45				
17	900 x 2400mm	900	2400	45				
18	900 x 2400mm	900	2400	45				
19	1800 x 2100mm	1800	2100	45				
20	900 x 2400mm	900	2400	45				
21	900 x 2400mm	900	2400	45				
22	900 x 2400mm	900	2400	45				

29. Select the W column, then drag your mouse to the right to highlight the H and THK columns.

30. Select **Titles & Headers→Group**.

Group

31. Type '**SIZE**' as the header for the three columns.

32. Select the Head Detail column, then drag your mouse to the right to highlight the Jamb Detail and Threshold Detail columns.

33. Select **Titles & Headers→Group**.

Group

34.

F	G	H
	DETAILS	
Head Detail	Jamb Detail	Threshold Detail

Type '**DETAILS**' as the header for the three columns.

35. Our schedule now appears in the desired format.

| 3D-Lobby | Level 1 | View 1 | Level 2 | Door Schedule | Door Details X |

<Door Details>

A	B	C	D	E	F	G	H
			SIZE			DETAILS	
Door No	Door Type	W	H	THK	Head Detail	Jamb Detail	Threshold Detail
1	36" x 84"	3' - 0"	7' - 0"	0' - 2"			
2	36" x 84"	3' - 0"	7' - 0"	0' - 2"			
3	36" x 84"	3' - 0"	7' - 0"	0' - 2"			
5	36" x 84"	3' - 0"	7' - 0"	0' - 2"			
6	72" x 82"	6' - 0"	6' - 10"	0' - 1 3/4"			
7	72" x 82"	6' - 0"	6' - 10"	0' - 1 3/4"			
8	36" x 84"	3' - 0"	7' - 0"	0' - 2"			
9	36" x 84"	3' - 0"	7' - 0"	0' - 2"			
10	36" x 84"	3' - 0"	7' - 0"	0' - 2"			
11	36" x 84"	3' - 0"	7' - 0"	0' - 2"			
12	12000 x 2290 Ope	2' - 11 7/16"	7' - 6 5/32"	0' - 1 3/16"			
13	12000 x 2290 Ope	2' - 11 7/16"	7' - 6 5/32"	0' - 1 3/16"			
21	Door-Curtain-Wall-	6' - 9 1/16"	8' - 0"				

<Door Details>

A	B	C	D	E	F	G	H
			SIZE			DETAILS	
Door No	Door Type	W	H	THK	Head Detail	Jamb Detail	Threshold Detail
1	0915 x 2134mm	915	2134	51			
2	0915 x 2134mm	915	2134	51			
3	0915 x 2134mm	915	2134	51			
4	0915 x 2134mm	915	2134	51			
5	1800 x 2100mm	1800	2100	45			
6	1800 x 2100mm	1800	2100	45			
7	0915 x 2134mm	915	2134	51			
8	0915 x 2134mm	915	2134	51			
9	0915 x 2134mm	915	2134	51			
10	0915 x 2134mm	915	2134	51			
11	12000 x 2290 Ope	900	2290	30			
12	12000 x 2290 Ope	900	2290	30			
31	M_Door-Curtain-W	2054	2400				
14	1800 x 2000mm	1800	2000	50			
15	900 x 2400mm	900	2400	45			
16	900 x 2400mm	900	2400	45			
17	900 x 2400mm	900	2400	45			
18	900 x 2400mm	900	2400	45			
19	1800 x 2100mm	1800	2100	45			
20	900 x 2400mm	900	2400	45			
21	900 x 2400mm	900	2400	45			
22	900 x 2400mm	900	2400	45			

36. Save as *ex6-6.rvt [m_ex6-6.rvt]*.

Exercise 6-7
Adding Shared Parameters to Families

Drawing Name: parameters2.rvt [m_parameters2.rvt]
Estimated Time: 25 minutes

In order to add data to the shared parameters, we have to add those shared parameters to the families used in the projects. We have several door families used in the project.

This exercise reinforces the following skills:

- ❑   Shared Parameters
- ❑   Families

1.   Open *parameters2.rvt [m_parameters2.rvt]*.

2.   
    In the Project Browser:

Locate the **Families** category.

Expand the **Doors** folder.

The door families used in the project are listed.

Not all of these families are in use.

> Doors
> - Door-Elevator
> - M_Door-Curtain-Wall-Double-Glass
> - M_Door-Double-Glass
> - M_Door-Exterior-Double-Two_Lite
> - M_Door-Exterior-Double-Two_Lite 2
> - M_Door-Passage-Single-One_Lite
> - M_Single-Flush

3.   Highlight the **Door-Curtain Wall Double Glass [M_Door-Curtain Wall Double Glass]** door.

Right click and select **Type Properties**.

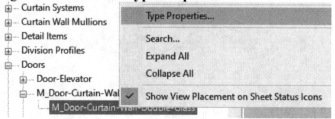

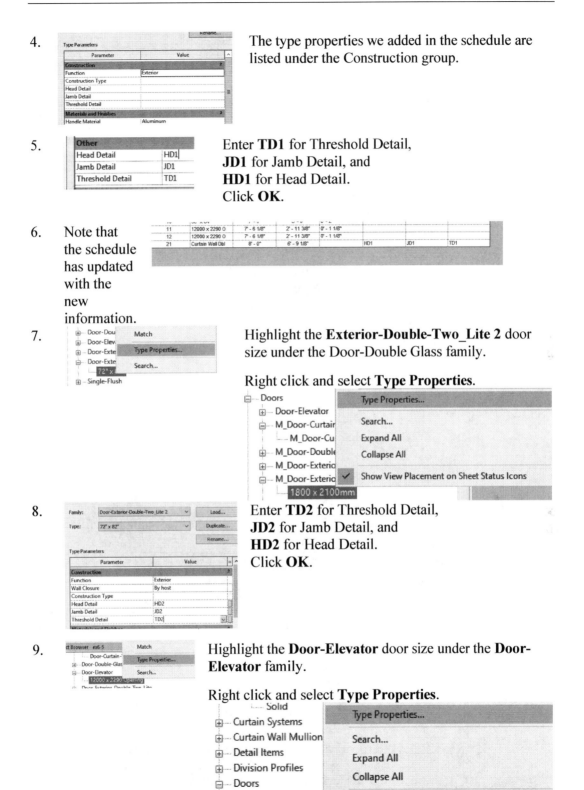

4. The type properties we added in the schedule are listed under the Construction group.

5. Enter **TD1** for Threshold Detail,
    JD1 for Jamb Detail, and
    HD1 for Head Detail.
    Click **OK**.

6. Note that the schedule has updated with the new information.

7. Highlight the **Exterior-Double-Two_Lite 2** door size under the Door-Double Glass family.

    Right click and select **Type Properties**.

8. Enter **TD2** for Threshold Detail,
    JD2 for Jamb Detail, and
    HD2 for Head Detail.
    Click **OK**.

9. Highlight the **Door-Elevator** door size under the **Door-Elevator** family.

    Right click and select **Type Properties**.

10.

Enter **TD3** for Threshold Detail,
JD3 for Jamb Detail, and
HD3 for Head Detail.
Click **OK**.

11.

In the first row of the schedule, type HD4 in the Head Detail column.

A dialog will come up saying that all single flush doors will now reference that head detail.

Click **OK**.

12.

On the first row, in the Jamb Detail column, type **JD4**.

13.

Click **OK**.

This change will be applied to all elements of type
M_Single-Flush: 0915 x 2134mm.

| OK | Cancel |

14.

Note that all the doors of the same type fill in with the information.

	G	
	DETAILS	
Detail	Jamb Detail	Threshol
	JD4	
	JD4	
	JD4	
	JD4	
	JD2	TD2
	JD4	
	JD4	
	JD4	
	JD4	
	JD3	TD3
	JD3	TD3
	JD1	TD1

F	G
DETAILS	
Head Detail	Jamb Detail
HD4	
HD4	
HD4	
HD4	
HD2	JD2
HD2	JD2
HD4	
HD4	
HD4	
HD4	
HD3	JD3
HD3	JD3
HD1	JD1

15.

On the first row, in the Threshold Detail column, type **TD4**.

H
Threshold Detail

> TD4 ⌄
> No matches
> TD2

16.

Revit ✕

This change will be applied to all elements of type
Single-Flush: 36" x 84".

OK   Cancel

Click **OK**.

This change will be applied to all elements of type
M_Single-Flush: 0915 x 2134mm.

OK   Cancel

17.

<Door Details>

			SIZE		DETAILS		
A	B	C	D	E	F	G	H
Door No	Door Type	W	H	THK	Head Detail	Jamb Detail	Threshold Detail
1	36" x 84"	3' - 0"	7' - 0"	0' - 2"	HD4	JD4	TD4
2	36" x 84"	3' - 0"	7' - 0"	0' - 2"	HD4	JD4	TD4
3	36" x 84"	3' - 0"	7' - 0"	0' - 2"	HD4	JD4	TD4
4	36" x 84"	3' - 0"	7' - 0"	0' - 2"	HD4	JD4	TD4
5	72" x 82"	6' - 0"	6' - 10"	0' - 1 3/4"	HD2	JD2	TD2
6	72" x 82"	6' - 0"	6' - 10"	0' - 1 3/4"	HD2	JD2	TD2
7	36" x 84"	3' - 0"	7' - 0"	0' - 2"	HD4	JD4	TD4
8	36" x 84"	3' - 0"	7' - 0"	0' - 2"	HD4	JD4	TD4
9	36" x 84"	3' - 0"	7' - 0"	0' - 2"	HD4	JD4	TD4
10	36" x 84"	3' - 0"	7' - 0"	0' - 2"	HD4	JD4	TD4
11	12000 x 2290 Ope	2' - 11 1/2"	7' - 6 1/4"	0' - 1 1/4"	HD3	JD3	TD3
12	12000 x 2290 Ope	2' - 11 1/2"	7' - 6 1/4"	0' - 1 1/4"	HD3	JD3	TD3
31	Door-Curtain-Wall	6' - 9"	8' - 8"		HD1	JD1	TD1
13	72" x 82"	6' - 0"	6' - 10"	0' - 1 3/4"			
14	72" x 82"	6' - 0"	6' - 10"	0' - 1 3/4"			
15	42" x 80"	3' - 6"	6' - 8"	0' - 1 3/4"			
16	42" x 80"	3' - 6"	6' - 8"	0' - 1 3/4"			
18	42" x 80"	3' - 6"	6' - 8"	0' - 1 3/4"			
19	42" x 80"	3' - 6"	6' - 8"	0' - 1 3/4"			
20	42" x 80"	3' - 6"	6' - 8"	0' - 1 3/4"			
21	42" x 80"	3' - 6"	6' - 8"	0' - 1 3/4"			
22	42" x 80"	3' - 6"	6' - 8"	0' - 1 3/4"			

Note each door type will have different details.

Can you fill in the blank areas quickly?

18.

<Door Details>

SIZE		DETAILS		
D	E	F	G	H
H	THK	Head Detail	Jamb Detail	Threshold Detail
7' - 0"	0' - 2"	HD4	JD4	TD4
7' - 0"	0' - 2"	HD4	JD4	TD4
7' - 0"	0' - 2"	HD4	JD4	TD4
7' - 0"	0' - 2"	HD4	JD4	TD4
6' - 10"	0' - 1 3/4"	HD2	JD2	TD2
6' - 10"	0' - 1 3/4"	HD2	JD2	TD2
7' - 0"	0' - 2"	HD4	JD4	TD4
7' - 0"	0' - 2"	HD4	JD4	TD4
7' - 0"	0' - 2"	HD4	JD4	TD4
7' - 0"	0' - 2"	HD4	JD4	TD4
7' - 6 1/4"	0' - 1 1/4"	HD3	JD3	TD3
7' - 6 1/4"	0' - 1 1/4"	HD3	JD3	TD3
8' - 8"		HD1	JD1	TD1
6' - 10"	0' - 1 3/4"	HD5	JD5	TD5
6' - 10"	0' - 1 3/4"	HD5	JD5	TD5
6' - 8"	0' - 1 3/4"	HD6	JD6	TD6
6' - 8"	0' - 1 3/4"	HD6	JD6	TD6
6' - 8"	0' - 1 3/4"	HD6	JD6	TD6
6' - 8"	0' - 1 3/4"	HD6	JD6	TD6
6' - 8"	0' - 1 3/4"	HD6	JD6	TD6
6' - 8"	0' - 1 3/4"	HD6	JD6	TD6
6' - 8"	0' - 1 3/4"	HD6	JD6	TD6

The detail information is listed for all the doors.

19.   Activate **Level 1** Floor Plan.

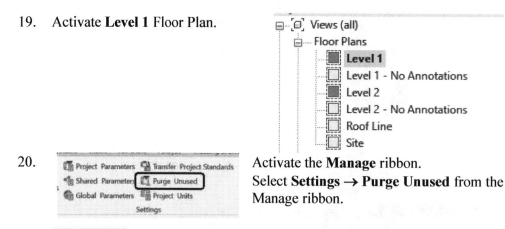

20.   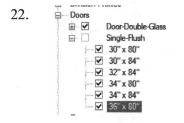   Activate the **Manage** ribbon.

Select **Settings** → **Purge Unused** from the Manage ribbon.

21.   [Check None]   Click **Check None**.

By checking none, you ensure that you don't purge symbols or families you may want to use later.

22.   Expand the Doors category.

Place a checkmark on the door families that are no longer used.

Note only families that are not in use will be listed. If you see families listed that should not be in use, go back and replace those families with the desired families.

Click **OK**.

23.   The browser now only lists the doors which are used.

24.   Activate the Door Details schedule.

<Door Details>

A	B	C	D	E	F	G	H
			SIZE			DETAILS	
Door No	Door Type	W	H	THK	Head Detail	Jamb Detail	Threshold Detail
1	36" x 84"	3' - 0"	7' - 0"	0' - 2"	HD4	JD4	TD4
2	36" x 84"	3' - 0"	7' - 0"	0' - 2"	HD4	JD4	TD4
3	36" x 84"	3' - 0"	7' - 0"	0' - 2"	HD4	JD4	TD4
4	36" x 84"	3' - 0"	7' - 0"	0' - 2"	HD4	JD4	TD4
5	72" x 82"	6' - 0"	6' - 10"	0' - 1 3/4"	HD2	JD2	TD2
6	72" x 82"	6' - 0"	6' - 10"	0' - 1 3/4"	HD2	JD2	TD2
7	36" x 84"	3' - 0"	7' - 0"	0' - 2"	HD4	JD4	TD4
8	36" x 84"	3' - 0"	7' - 0"	0' - 2"	HD4	JD4	TD4
9	36" x 84"	3' - 0"	7' - 0"	0' - 2"	HD4	JD4	TD4
10	36" x 84"	3' - 0"	7' - 0"	0' - 2"	HD4	JD4	TD4
11	12000 x 2290 Ope	2' - 11 1/2"	7' - 6 1/4"	0' - 1 1/4"	HD3	JD3	TD3
12	12000 x 2290 Ope	2' - 11 1/2"	7' - 6 1/4"	0' - 1 1/4"	HD3	JD3	TD3
31	Door-Curtain-Wall-	6' - 9"	8' - 0"		HD1	JD1	TD1
13	72" x 82"	6' - 0"	6' - 10"	0' - 1 3/4"	HD5	JD5	TD5
14	72" x 82"	6' - 0"	6' - 10"	0' - 1 3/4"	HD5	JD5	TD5
15	42" x 80"	3' - 6"	6' - 8"	0' - 1 3/4"	HD6	JD6	TD6
16	42" x 80"	3' - 6"	6' - 8"	0' - 1 3/4"	HD6	JD6	TD6
18	42" x 80"	3' - 6"	6' - 8"	0' - 1 3/4"	HD6	JD6	TD6
19	42" x 80"	3' - 6"	6' - 8"	0' - 1 3/4"	HD6	JD6	TD6
20	42" x 80"	3' - 6"	6' - 8"	0' - 1 3/4"	HD6	JD6	TD6
21	42" x 80"	3' - 6"	6' - 8"	0' - 1 3/4"	HD6	JD6	TD6
22	42" x 80"	3' - 6"	6' - 8"	0' - 1 3/4"	HD6	JD6	TD6

<Door Details>

A	B	C	D	E	F	G	H
			SIZE			DETAILS	
Door No	Door Type	W	H	THK	Head Detail	Jamb Detail	Threshold Detail
1	0915 x 2134mm	915	2134	51	HD4	JD4	TD4
2	0915 x 2134mm	915	2134	51	HD4	JD4	TD4
3	0915 x 2134mm	915	2134	51	HD4	JD4	TD4
4	0915 x 2134mm	915	2134	51	HD4	JD4	TD4
5	1800 x 2100mm	1800	2100	45	HD2	JD2	TD2
6	1800 x 2100mm	1800	2100	45	HD2	JD2	TD2
7	0915 x 2134mm	915	2134	51	HD4	JD4	TD4
8	0915 x 2134mm	915	2134	51	HD4	JD4	TD4
9	0915 x 2134mm	915	2134	51	HD4	JD4	TD4
10	0915 x 2134mm	915	2134	51	HD4	JD4	TD4
11	12000 x 2290 Ope	900	2290	30	HD3	JD3	TD3
12	12000 x 2290 Ope	900	2290	30	HD3	JD3	TD3
31	M_Door-Curtain-W	2054	2400		HD1	JD1	TD1
14	1800 x 2100mm	1800	2100	45	HD2	JD2	TD2
15	900 x 2400mm	900	2400	45	HD6	JD6	TD6
16	900 x 2400mm	900	2400	45	HD6	JD6	TD6
17	900 x 2400mm	900	2400	45	HD6	JD6	TD6
18	900 x 2400mm	900	2400	45	HD6	JD6	TD6
19	1800 x 2100mm	1800	2100	45	HD7	JD7	TD7
20	900 x 2400mm	900	2400	45	HD6	JD6	TD6
21	900 x 2400mm	900	2400	45	HD6	JD6	TD6
22	900 x 2400mm	900	2400	45	HD6	JD6	TD6

25.   Save as *ex6-7.rvt [m_ex6-7.rvt]*.

Exercise 6-8
Creating a Custom Window Schedule

Drawing Name: schedule2.rvt [m_schedule2.rvt]
Estimated Time: 25 minutes

This exercise reinforces the following skills:

❑ Shared Parameters
❑ Combined Parameters
❑ Schedule/Quantities

A	B	C	D	E	F	G	H	I
			SIZE			DETAILS		
WINDOW NO	FRAME TYPE	Type Image	W	H	SILL	HEAD	JAMB	REMARKS
1	36" x 48"		3' - 0"	4' - 0"				
36" x 48": 70								
18	36" x 48"		3' - 0"	4' - 0"	3' - 0"			
36" x 48": 2								
Qty for Each Type 72								

\<Glazing Schedule\>

We want our window schedule to appear as shown.

1. Open *schedule2.rvt [m_schedule2.rvt]*.

2. Activate the **View** ribbon.

   Select **Schedules** → **Schedule/Quantities**.

3. Type **WIND** in the Category name search.

   The category list now only shows Windows.

   Highlight **Windows**.

   Change the Schedule name to **Glazing Schedule**.

   Click **OK**.

4.   Add the following fields from the Available Fields.

Add:

Scheduled fields (in order):

| Type Mark |
| Type |
| Image |
| Width |
| Height |
| Sill Height |

Type Mark
Type
Image
Width
Height
Sill Height

5.   Select **Add Parameter**.

6.   ● Shared parameter
      (Can be shared by multiple projects and families, exported to ODBC, and
      appear in schedules and tags)

      Select...   Export...

      Enable **Shared Parameter**.
      Click **Select**.

7.   Parameter group:

      Door

      Parameters:

      Head Detail
      Jamb Detail
      Threshold Detail

      Select **Head Detail**.

      Click **OK**.

8.   Select...   Export...

      Parameter Data
      Name:
      Head Detail   ● Type
      Discipline: ○ Instance
      Common
      Data Type: ● Values are aligned per group type
      Text   ○ Values can vary by group instance
      Group parameter under:
      Construction

      Tooltip Description:
      <No tooltip description. Edit this parameter to write a custom tooltip. Custom tooltips have...

      ■ Add to all elements in the category

      Group it under **Construction**.
      Enable **Add to all elements in the category**.
      Enable **Type**.

      Click **OK**.

Head Detail is now listed as a column for the schedule.

Scheduled fields (in order):

| Type Mark |
| Type |
| Image |
| Width |
| Height |
| Sill Height |
| Head Detail |

9.

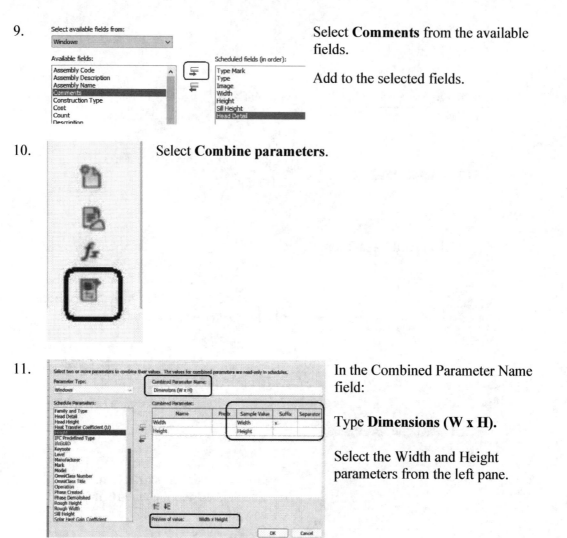

Select **Comments** from the available fields.

Add to the selected fields.

10.

Select **Combine parameters**.

11.

In the Combined Parameter Name field:

Type **Dimensions (W x H)**.

Select the Width and Height parameters from the left pane.

12. In the Suffix column: Type a space, then x, then a space for the Width row. Delete the / Separator from both rows.
Look at the bottom of the dialog and check the preview to see if it looks correct.
Click **OK**.

13.

Scheduled fields (in order):

Type Mark
Type
Image
Dimensions (W x H)
Width
Height
Sill Height
Head Detail
Comments

Use the Move up tool to locate the Dimensions combined parameter below the Type Mark parameter.

14.

Heading:

Dimensions (W x H)

Heading orientation:

Horizontal

Alignment:

Center

Select the Formatting tab.

Highlight the **Dimensions combined parameter.**

Set the Alignment to **Center.**

15.

Heading:

W

Heading orientation:

Horizontal

Alignment:

Center

Field formatting: Field Format...

Conditional formatting: Conditional Format...

Show conditional format on
sheets:

Hidden field:

Highlight **Width**.

Enable **Hidden field**.

If you remove the parameters used in the combined parameter, they can't be used in the schedule, so you hide the parameters instead.

16.

Highlight **Height**.

Enable **Hidden field**.

17.

Activate the Sorting/Grouping tab.

Sort by **Type**.
Enable **Footer**.
Select **Title, count and totals**.
Enable **Grand totals**.
Select **Title, count and totals**.
Type **Qty for Each Type** in the Custom grand total title.
Uncheck **Itemize every instance**.

18.

Select the Formatting tab.

Highlight Type Mark and change the Heading to **WINDOW NO**.

Change the Alignment to **Center**.

19.

Highlight Type and change the Heading to **FRAME TYPE**.

Change the Alignment to **Left.**

20.

Highlight Sill Height and change the Heading to **SILL**.

Change the Alignment to **Center**.

21.

Highlight Head Detail and change the Heading to **HEAD.**

Change the Alignment to **Center**.

22.

Change the heading for Comments to **REMARKS**.

Change the Alignment to **Left**.

23. Click **OK** to create the schedule.

We see that only one type of window is being used in the project and there are 70 windows of this type.

Your schedule may appear different depending on the number of windows placed and whether you placed different types.
Notice that SILL HEIGHT says <varies>.

When you have a schedule that is not itemized, if a value in a cell varies, then that cell has always just been blank in the past. Now, you have the option to choose what it says, based on the individual parameter per schedule or via a global override.

24.

> Other
> Fields   Edit...
> Filter   Edit...
> Sorting/Grouping   Edit...
> Formatting   [Edit...]
> Appearance   Edit...
> Properties help   Apply

Click **Edit** next to Formatting on the Properties palette.

25.

> Fields:
> Type Mark
> Type
> Image
> Width
> Height
> Sill Height
> Head Detail
> Comments

Highlight the Sill Height field.

26.

> Multiple values indication
> ○ Use project settings
> ○ Display as <varies>
> ● Display custom text:
>   REFER TO ELEVATION VIEW.

Enable **Display custom text**.

Type **REFER TO ELEVATION VIEW**.

Click **OK**.

> <Glazing Schedule>
>
> | A | B | C | D | E | F | G |
> |---|---|---|---|---|---|---|
> | WINDOW NO | FRAME TYPE | Image | Dimensions (W x H) | SILL | HEAD | REMARKS |
> | 15 | 0915 x 1220mm | | 915 x 1220 | REFER TO ELEVATION VIEW. | | |
> | 0915 x 1220mm 75 | | | | | | |

The schedule updates.

27. Save as *ex6-8.rvt [m_ex6-8.rvt]*.

Exercise 6-9
Creating an Image from a Family

Drawing Name: image1.rvt [m_image1.rvt]
Estimated Time: 10 minutes

This exercise reinforces the following skills:

- ❑ Images
- ❑ Families
- ❑ Project Browser search

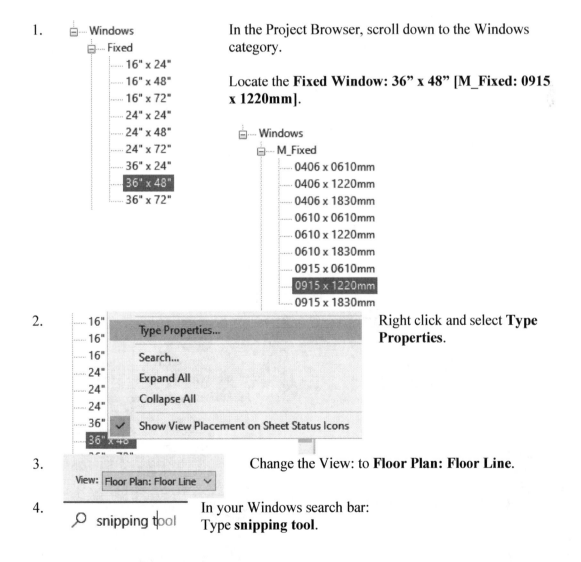

1.   In the Project Browser, scroll down to the Windows category.

Locate the **Fixed Window: 36" x 48" [M_Fixed: 0915 x 1220mm]**.

2.   Right click and select **Type Properties**.

3.   Change the View: to **Floor Plan: Floor Line**.

4.   In your Windows search bar:
Type **snipping tool**.

5.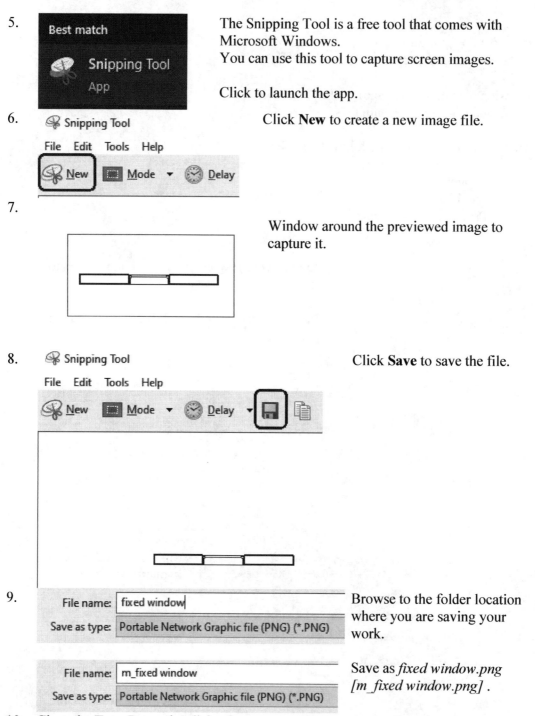
The Snipping Tool is a free tool that comes with Microsoft Windows.
You can use this tool to capture screen images.

Click to launch the app.

6.
Click **New** to create a new image file.

7.
Window around the previewed image to capture it.

8.
Click **Save** to save the file.

9.
Browse to the folder location where you are saving your work.

Save as *fixed window.png* *[m_fixed window.png]* .

10. Close the Type Properties dialog box.
11. Close without saving the Revit project file.

Exercise 6-10
Adding an Image to a Schedule

Drawing Name: schedule3.rvt [m_schedule3.rvt]
Estimated Time: 10 minutes

This exercise reinforces the following skills:

- ❑ Images
- ❑ Schedule/Quantities

Glazing Schedule							
Window No	Frame Type	Image	SIZE		Sill	Head	Remarks
			W	H			
1	36" x 48"		3'- 0"	4'- 0"	<varies>		
36" x 48": 10							
19	Corner Unit Double Pane Nosing		0'- 0"	0'- 0"	3'- 0"		
Corner Unit Double Pane Nosing: 20							
18	MC3046		4'- 6"	3'- 0"	3'- 0"		
MC3046: 62							
Qty for Each Type: 92							

We want our window schedule to appear as shown. We will be adding the images to the schedule.

1. Open *schedule3.rvt [m_schedule3.rvt]*.

2.

	A	B	C
	WINDOW NO	FRAME TYPE	Image
	1	36" x 48"	[...]
	36" x 48": 70		
	18	36" x 48"	
	36" x 48": 2		
	Qty for Each Type: 72		

Open the **Glazing Schedule** view.

Left click in the Image cell for Window No 1.

You will "wake up" the ... button.

Select the ... button.

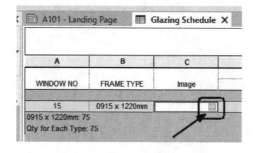

3. Select the **Add** button at the bottom of the dialog.

4. Select the fixed window image file located in the exercise files.

File name: fixed window.PNG

Files of type: All Image Files (*.bmp, *.jpg, *.jpeg, *.png, *.tif)

Click **Open**.

File name: m_fixed window.PNG

Files of type: All Image Files (*.bmp, *.jpg, *.jpeg, *.pdf, *.png, *.tif, *.tiff)

5. A preview of the image is shown.

Raster Image   Name

Click **OK**.

fixed window.PNG

Manage Images

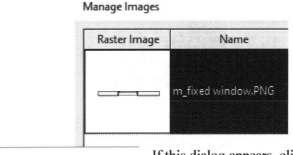

Raster Image	Name
	m_fixed window.PNG

Autodesk Revit 2019

Error - cannot be ignored

Changes to groups are allowed only in group edit mode. Use the Edit Group command to change to all instances of a group type. You may use the "Ungroup" option to proceed with this change by ungrouping the changed group instances.

Show More Info Expand >>

Ungroup OK Cancel

If this dialog appears, click **Ungroup**.

This error appears if the windows are part of a grouped array.

<Glazing Schedule>

A	B	C	D	E
			SIZE	
Window No	Frame Type	Image	W	H
1	36" x 48"	fixed window PNG	3' - 0"	4' -
36" x 48": 10				
19	Corner Unit Double		0' - 0"	0' -
Corner Unit Double Pane Nosing: 20				
18	MC3046		4' - 6"	3' -
MC3046: 62				

The image file is now listed in the schedule.

If you placed additional window types, you can add images for them as well if you have created the image files.

6.

36" x 48": 10			
19	Corner Unit Double		
Corner Unit Double Pane Nosing: 20			
18	MC3046		

Left click in the Image cell for Window No 19.

You will "wake up" the … button.

Select the … button.

7. Add... Delete

Select the **Add** button at the bottom of the dialog.

8.

File name: fixed corner.PNG

Files of type: All Image Files (*.bmp, *.jpg,

Select the fixed corner image file located in the exercise files.

Click **Open**.

9.

A preview of the image is shown.

Click **OK**.

10.

The schedule updates.

11.

Left click in the Image cell for Window No 18.

You will "wake up" the ... button.

Select the ... button.

12.

Select the **Add** button at the bottom of the dialog.

13.

Select the casement window image file located in the exercise files.

Click **Open**.

14.

A preview of the image is shown.

Click **OK**.

If this dialog appears, click **Ungroup**.

This error appears if the windows are part of a grouped array.

15.
```
⊟ 📄 Sheets (all)
   ⊞ A101 - First Level Floor Plan
   ⊟ A102 - Second Level Floor Plan
        📄 Floor Plan: Level 2
```

Open the **A101 – First Level Floor Plan [A102 First Level Floor Plan]** sheet.

```
⊟ 📄 Sheets (all)
     A101 - Landing Page
   ⊞ A102 - First Level Floor Plan
   ⊞ A103 - Second Level Floor Plan
```

16.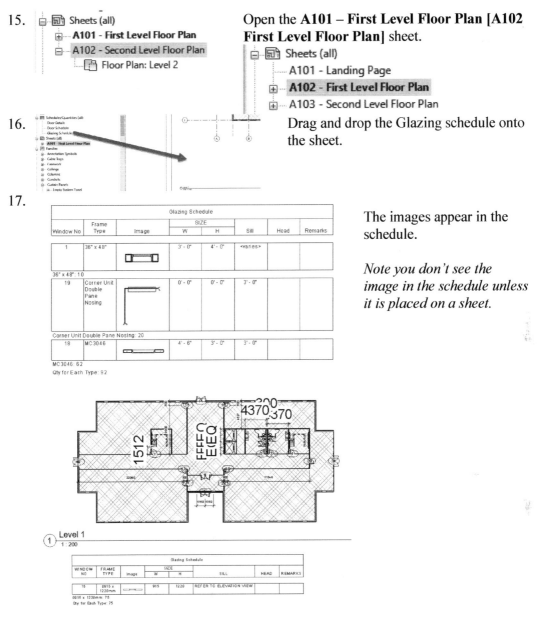

Drag and drop the Glazing schedule onto the sheet.

17.

The images appear in the schedule.

Note you don't see the image in the schedule unless it is placed on a sheet.

18. Save as *ex6-10.rvt [m_ex6-10.rvt]*.

Assembly Code	B2020.2040	
Type Image		
Keynote		
Cost		

Revit families have a Type Image parameter that can be used to attach an image file. This parameter can then be used in schedules and legends. The downside is that you need to modify the family for it to work across projects, but once done you will have the image available in any project in which the family is used.

Exercise 6-11
Adding an Image to a Family

Drawing Name: parameters3.rvt [m_parameters3.rvt]
Estimated Time: 10 minutes

This exercise reinforces the following skills:

❑ Images
❑ Parameters

1. Open *parameters3.rvt [m_parameters3.rvt]*.

2. Open **Level 1** floor plan.

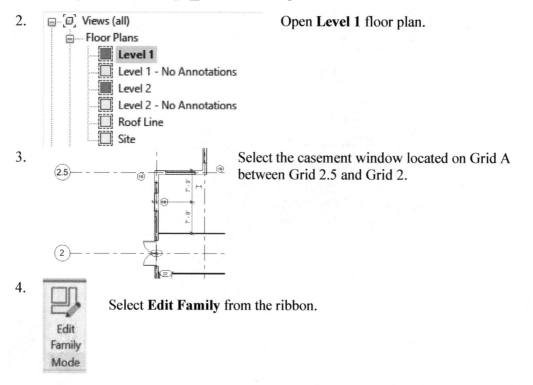

3. Select the casement window located on Grid A between Grid 2.5 and Grid 2.

4. Select **Edit Family** from the ribbon.

5. Select **Family Types** from the Properties panel on the ribbon.

6. Scroll down to locate the Type Image parameter.

Click in the field.

7. Select the **Add** button at the bottom of the dialog.

8. Select the casement window image file located in the exercise files.

Click **Open**.

9. A preview of the image is shown.

Click **OK**.

10. *The image file is listed in the Parameters panel.*

Click **OK**.

11. Save the file as *Window-Casement-2.rfa*.

12. Select **Load into Project and Close**.

The file will close and you will return to the project file.

Click **ESC** to exit the command.
Select the casement window.

13. Right click and **Select All Instances→In Entire Project**.

14.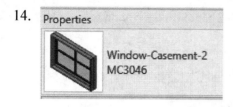

Use the Type Selector to assign all the selected components to the new Window-Casement-2.

Click **ESC** to release the selection.

15.

Save as *ex6-11.rvt*.

Extra: *Edit the remaining window families used in the project. Add the correct image created to the family.* **Verify that the size selected is the size used in the project. If the family has more than one size or type, you need to add the desired image to each type.** *Save the family as a new version. Replace all the window families used in the project with the new version. Create a schedule using the Type Image property as one of the columns. What is the advantage of simply modifying the schedule instead of modifying the families used in the project? What is the advantage of modifying the families?*

Exercise 6-12
Creating an Image of a Family

Drawing Name: image1.rvt [m_image1.rvt]
Estimated Time: 10 minutes

This exercise reinforces the following skills:

- ❑ Images
- ❑ Parameters

1. Open *image1.rvt [m_image1.rvt]*.

2. 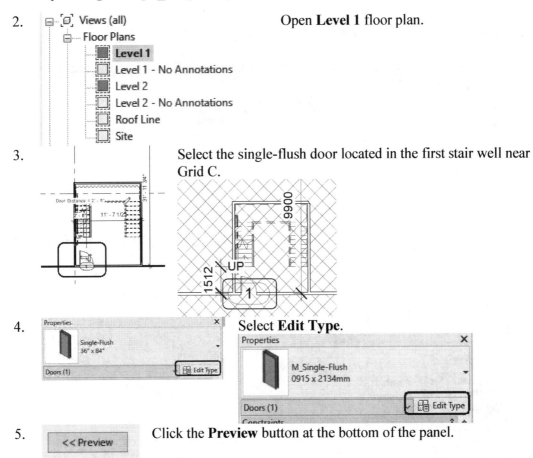   Open **Level 1** floor plan.

3. Select the single-flush door located in the first stair well near Grid C.

4. Select **Edit Type**.

5. ` << Preview `   Click the **Preview** button at the bottom of the panel.

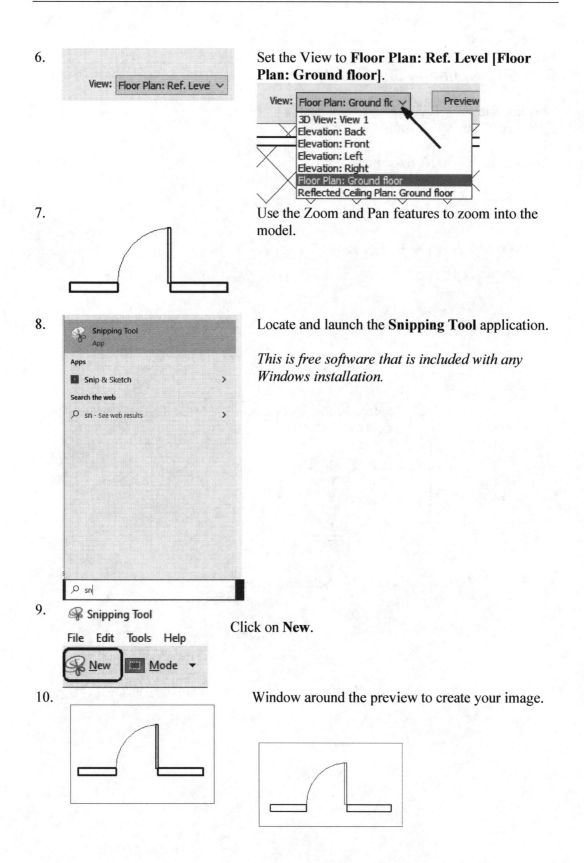

6. Set the View to **Floor Plan: Ref. Level [Floor Plan: Ground floor]**.

7. Use the Zoom and Pan features to zoom into the model.

8. Locate and launch the **Snipping Tool** application.

   This is free software that is included with any Windows installation.

9. Click on **New**.

10. Window around the preview to create your image.

11.

File   Edit   Tools   Help

New   Mode   ▾   Delay   ▾   🖫

Click on **Save**.

12.

File name: single flush door

Save as type: Portable Network Graphic file (PNG) (*.PNG)

Name the image **single flush door**
[m_single flush door].
Click **Save**.

13.

File name: m_single flush door

Save as type: Portable Network Graphic file (PNG) (*.PNG)

Click **OK** to close the Properties pane.

Extra: *Edit the Single Flush Door family and add the image created to the family. Save the family as a new version. Replace all the single flush doors used in the project with the new version. Repeat for all the doors used in the project. Create a schedule using the Type Image property as one of the columns.*

Exercise 6-13
Creating a Window Schedule Using Type Image

Drawing Name: schedule4.rvt [m_schedule4.rvt]
Estimated Time: 10 minutes

This exercise reinforces the following skills:

❑   Images
❑   Parameters

1.   Open *schedule4.rvt [m_schedule4.rvt]*.

2.

⊟ ▦ Schedules/Quantities (all)
└── Door Details
└── Door Schedule
└── **Glazing Schedule**

Open the **Glazing Schedule**.

3.

Other	
Fields	Edit...
Filter	Edit...
Sorting/Grouping	Edit...
Formatting	Edit...
Appearance	Edit...

Select **Edit** next to Fields.

4.

Scheduled fields (in order):

Type Mark
Type Image
Type
Width
Height
Sill Height
Head Detail
Comments

Add the **Type Image** field to the schedule.

Move the field up so it is below the Type Mark field.

5.

| Fields | Filter | Sorting/Grouping | Formatting | Appearance |

Fields:

Type Mark
Type Image
Type
Width
Height
Sill Height
Head Detail
Comments

Heading:

Symbol

Heading orientation:

Horizontal

Alignment:

Center

Click on the Formatting tab.

Highlight **Type Image**.
Change the Heading to **Symbol.**
Change the Alignment to **Center.**

Click **OK.**

6.

	A	B	C
	Window No	Symbol	Frame Type
	31	fixed window.PNG	36" x 48"
36" x 48": 10			
	21	fixed corner.PNG	Corner Unit Double
Corner Unit Double Pane Nosing: 20			
	20	casement window	MC3046
MC3046: 62			

You should see a file name in the rows for each window type.

A	B
Symbol	WINDOW NO
m_fixed window.P	32
0915 x 1220mm: 75	
Qty for Each Type: 75	

7.

Glazing Schedule

Sheets (all)
    A101 - First Level Floor Plan
    A102 - Second Level Floor Plan
Families

Activate the **A101 – First Level Floor Plan [A102 – First Level Floor Plan]** sheet.

Sheets (all)
    A101 - Landing Page
    A102 - First Level Floor Plan
    A103 - Second Level Floor Plan

8.

Glazing Schedule

Window No	Symbol	Frame Type	SIZE W	H	Sill	H
31		36" x 48"	3' - 0"	4' - 0"	<varies>	
36" x 48": 10						
21		Corner Unit Double Pane Nosing	0' - 0"	0' - 0"	3' - 0"	
Corner Unit Double Pane Nosing: 20						
20		MC 3046	4' - 6"	3' - 0"	3' - 0"	
MC3046: 62						
Qty for Each Type: 92						

Place the Glazing Schedule on the sheet.

Level 1
1 : 200

Glazing Schedule

Symbol	WINDOW NO	FRAME TYPE	SIZE W	H	SILL	HEAD	REMARKS
	32	0915 x 1220mm	915	1220	REFER TO ELEVATION VIEW		
0915 x 1220mm: 76							
Qty for Each Type: 75							

9.

Save as *ex6-13.rvt [m_ex6-13.rvt].*

Exercise 6-14
Using Keynotes

Drawing Name: keynotes.rvt [m_ keynotes.rvt]
Estimated Time: 30 minutes

This exercise reinforces the following skills:

- ❏ Keynotes
- ❏ Duplicate Views
- ❏ Cropping a View

1. Open *keynotes.rvt [m_ keynotes.rvt]*.

2. Activate the **View** ribbon.

3.  Select the **New Sheet** tool from the Sheet Composition panel.

4. **Select titleblocks:**

   D 22 x 34 Horizontal
   E1 30 x 42 Horizontal : E1 30x42 Horizontal
   None

   Highlight the **D 22 x 34 Horizontal [A3 metric]** titleblock.

   Click **OK**.

   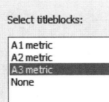

   Select titleblocks:

   A1 metric
   A2 metric
   A3 metric
   None

5.
Approved By	M Instructor
Designed By	J Student
Checked By	M Instructor
Drawn By	J. Student
Sheet Number	A103
Sheet Name	Lobby Keynotes
Sheet Issue Date	01/24/21
Appears In Sheet List	☑

   In the Properties pane:

   Change the Sheet Name to **LOBBY KEYNOTES**.
   Change the Drawn By to your name.
   Change the Checked By to your instructor's name.

6. ⊟ ⎯ 🗗 Views (all)
      ⊟ ⎯ Floor Plans
         ■ **Level 1**
         ☐ Level 1 - No Annotations
         ■ Level 2
         ☐ Level 2 - No Annotations
         ☐ Roof Line
         ☐ Site

   Activate the **Level 1** floor plan.

7.

Duplicate View	>	Duplicate
Convert to independent view		Duplicate with Detailing
Apply Dependent Views...		Duplicate as a Dependent
Save to Project as Image		

Highlight **Level 1**.
Right click and select
**Duplicate View →
Duplicate**.

*This creates a duplicate
view without any
annotations, such as
dimensions and tags.*

8.

Views (all)
  Floor Plans
    Level 1
    Level 1 - Loby Detail
    Level 1 - No Annotations
    Level 2
    Level 2 - No Annotations
    Roof Line
    Site

Rename the duplicate view **Level 1 - Lobby Detail**.
Click **OK**.

9.

Extents	
Crop View	☑
Crop Region Visible	☑
Annotation Crop	☐

In the Properties palette:

Enable **Crop View**.
Enable **Crop Region Visible**.

10.

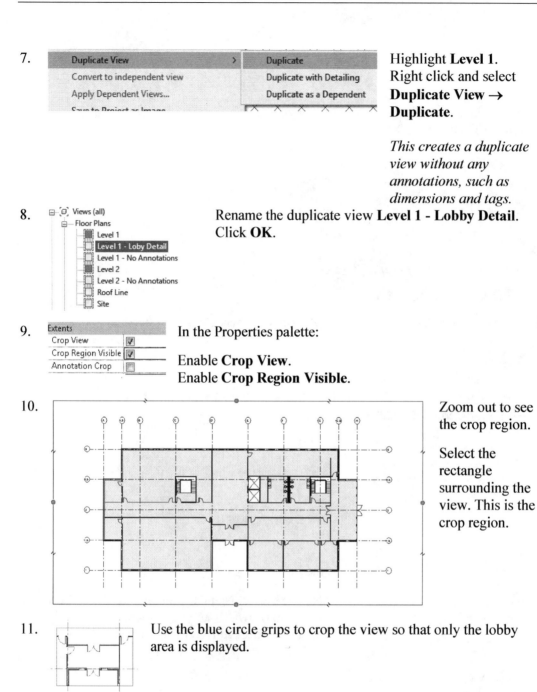

Zoom out to see
the crop region.

Select the
rectangle
surrounding the
view. This is the
crop region.

11.

Use the blue circle grips to crop the view so that only the lobby
area is displayed.

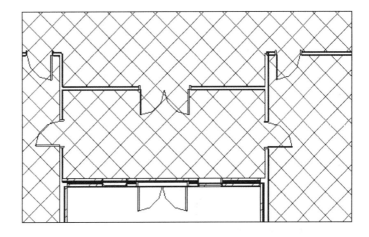

12. 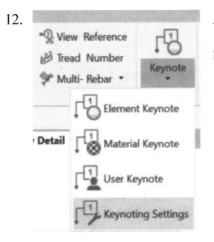 Activate the **Annotate** ribbon.

Select **Keynote → Keynoting Settings**.

13. *This shows the path for the keynote database.*

Keynote Table
File Location
C:\ProgramData\Autodesk\RVT 2024\Libraries\English-Imperial\RevitKeynotes_Imperial_

File Path (for local files)
○ Absolute   ● Relative   ○ At library locations

Numbering Method
● By keynote   ○ By sheet

This is a txt file that can be edited using Notepad. You can modify the txt file or select a different txt file for keynote use.

The file can be placed on a server in team environments.

Click **OK**.

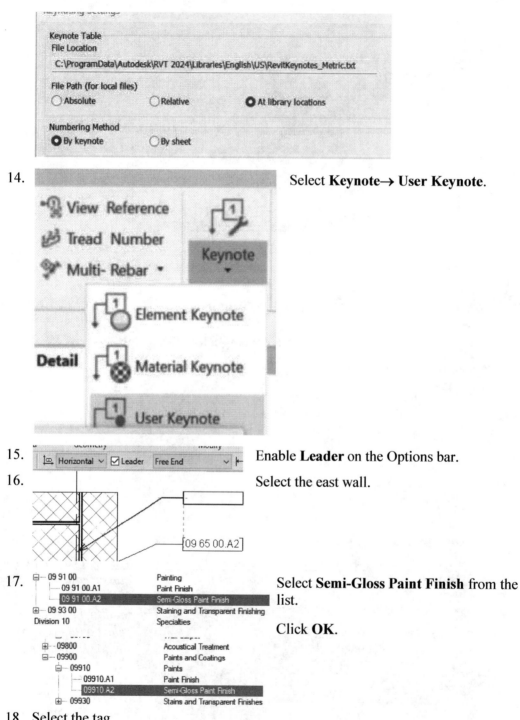

14. Select **Keynote→ User Keynote**.

15. Enable **Leader** on the Options bar.

16. Select the east wall.

17. Select **Semi-Gloss Paint Finish** from the list.

    Click **OK**.

18. Select the tag.

    In the Properties palette:
    Verify that the material displayed is **Semi-Gloss Paint Finish**.

19. Select **Keynote → User Keynote**.

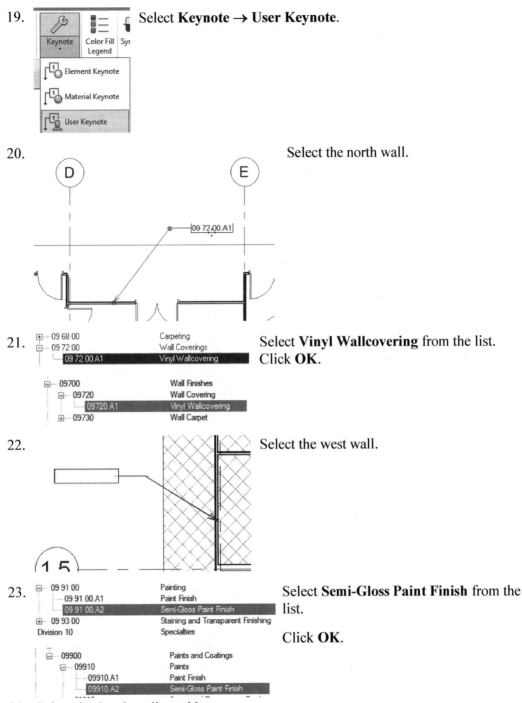

20. Select the north wall.

21. Select **Vinyl Wallcovering** from the list. Click **OK**.

22. Select the west wall.

23. Select **Semi-Gloss Paint Finish** from the list.

Click **OK**.

24. Select the South wall to add a tag.

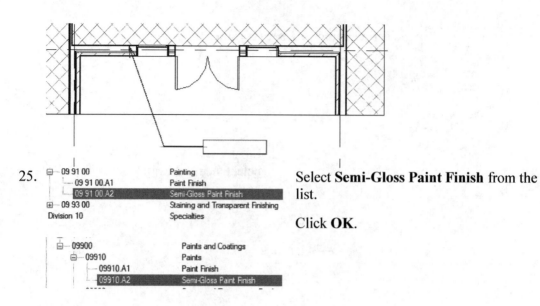

25. Select **Semi-Gloss Paint Finish** from the list.

Click **OK**.

26. The view should appear as shown.

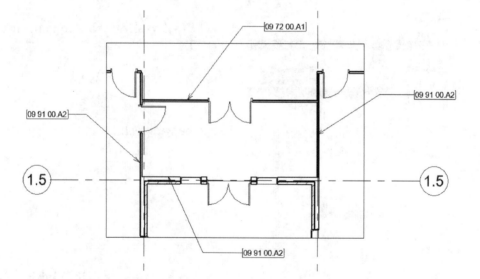

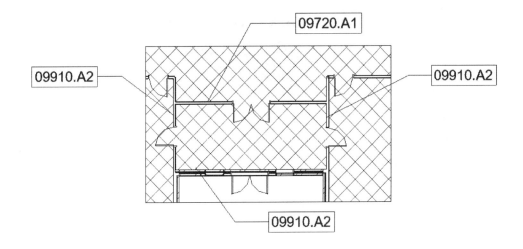

Save as *ex6-14.rvt [m_ex6-14.rvt]*.

Exercise 6-15
Create a Building Elevation with Keynotes

Drawing Name: keynotes2.rvt [m_keynotes2.rvt]
Estimated Time: 30 minutes

This exercise reinforces the following skills:

- ❑ Elevation View
- ❑ Cropping a View
- ❑ Keynote Tags
- ❑ Sheets
- ❑ Views

1. Open *keynotes2.rvt. [m_keynotes2.rvt]*.

2.    Activate the **Level 1 - Lobby Detail** view.

3.   Type **VV** to bring up the Visibility/Graphics dialog.

Activate the **Annotation Categories** tab.

Enable the visibility of Elevations.

Click **OK**.

4.   Activate the View ribbon.

Select **Elevation→ Elevation**. ⬆ Elevation ▾

5.   Check in the Properties palette.

You should see that you are placing a Building Elevation.

6.   Place an elevation marker as shown in front of the exterior front door.
Right click and select **Cancel** to exit the command.

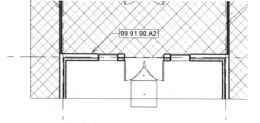

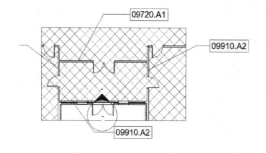

7.   Click on the triangle part of the elevation marker to adjust the depth of the elevation view.

Adjust the boundaries of the elevation view to contain the lobby room.

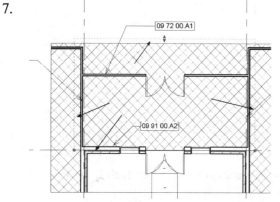

8. 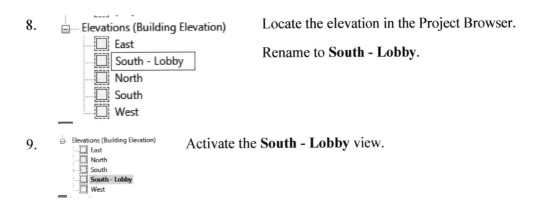 Locate the elevation in the Project Browser.

Rename to **South - Lobby**.

9. Activate the **South - Lobby** view.

10. Adjust the crop region to show the entire lobby including floor and ceiling.

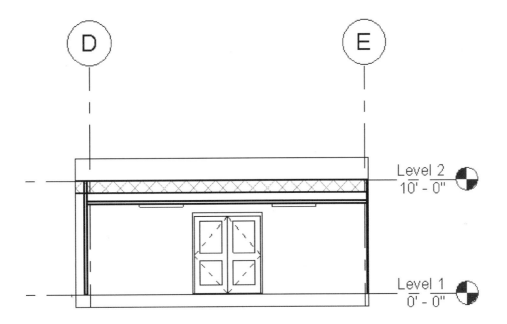

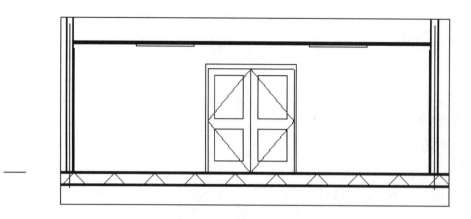

You may need to return to the Lobby Detail floor plan to adjust the location of the elevation view until it appears like the view shown.

This view is set to **Coarse Level Detail**, **Wireframe**.

11.   Activate the Annotate ribbon.

       Select **Keynote→Element Keynote**.

12. | Keynote Tag | Keynote Number | From the Properties palette:

       Set the Keynote Tag to **Keynote Number**.

13. 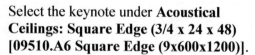   Select the ceiling.

14. 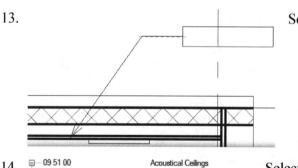   Select the keynote under **Acoustical Ceilings: Square Edge (3/4 x 24 x 48) [09510.A6 Square Edge (9x600x1200)]**.

Click **OK**.

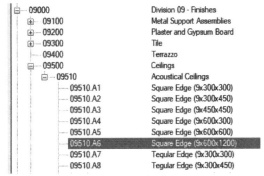

15. ?   Select the Lighting Fixture.

_____ *A ? indicates that Revit needs you to assign the keynote value.*

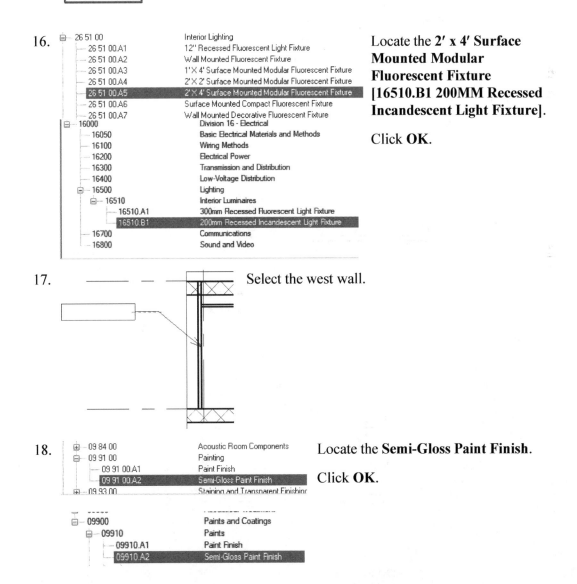

16.   Locate the **2' x 4' Surface Mounted Modular Fluorescent Fixture [16510.B1 200MM Recessed Incandescent Light Fixture]**.

Click **OK**.

17.   Select the west wall.

18.   Locate the **Semi-Gloss Paint Finish**.

Click **OK**.

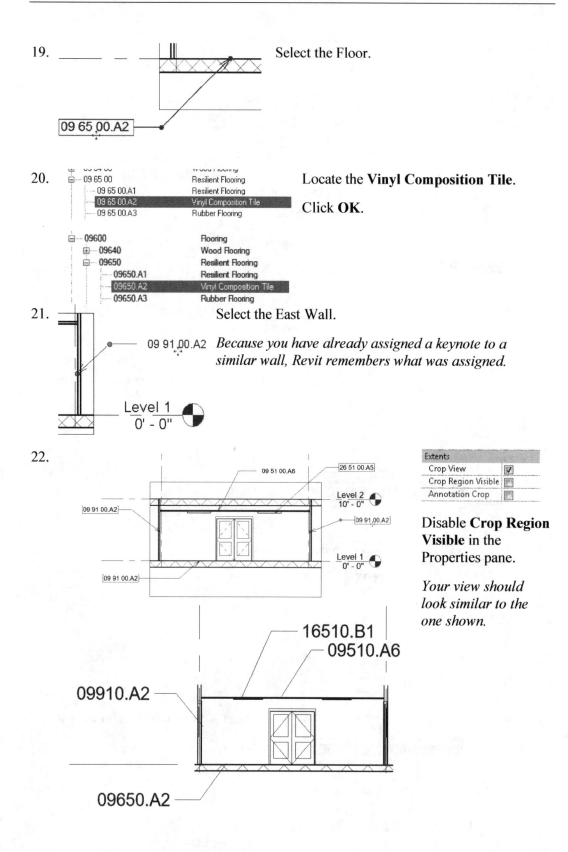

19. Select the Floor.

09 65 00.A2

20. Locate the **Vinyl Composition Tile**.

Click **OK**.

21. Select the East Wall.

09 91 00.A2 *Because you have already assigned a keynote to a similar wall, Revit remembers what was assigned.*

Level 1
0' - 0"

22.

Disable **Crop Region Visible** in the Properties pane.

Your view should look similar to the one shown.

23.

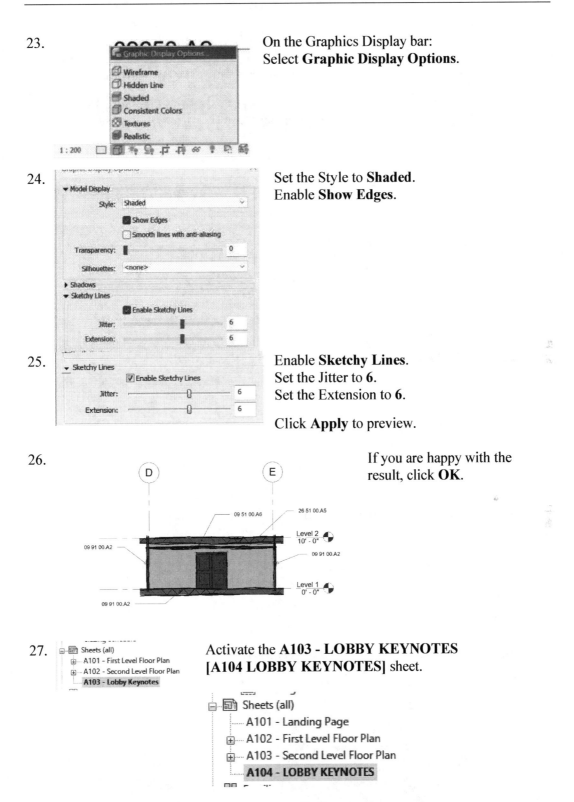

On the Graphics Display bar:
Select **Graphic Display Options**.

24.

Set the Style to **Shaded**.
Enable **Show Edges**.

25.

Enable **Sketchy Lines**.
Set the Jitter to **6**.
Set the Extension to **6**.

Click **Apply** to preview.

26.

If you are happy with the result, click **OK**.

27.

Activate the **A103 - LOBBY KEYNOTES**
[A104 LOBBY KEYNOTES] sheet.

28.

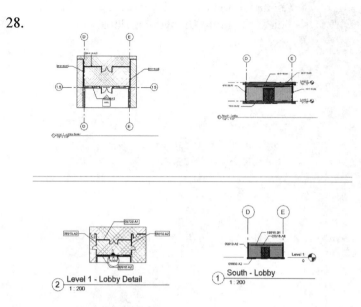

Add the **Level 1 - Lobby Detail** view to the sheet.

Add the **South Lobby Elevation** view to the sheet.

29. Save as *ex6-15.rvt [m_ex6-15.rvt]*.

Exercise 6-16
Create a Keynote Legend

Drawing Name: keynote legend.rvt [m_ keynote legend.rvt]
Estimated Time: 30 minutes

This exercise reinforces the following skills:

- Elevation View
- Cropping a View
- Keynote Tags
- Sheets
- Views

1.   Open *keynote legend.rvt [m_ keynote legend.rvt]*.

2.   Activate the **Level 1 - Lobby Detail** view.

3.   Activate the View ribbon.

Select **Legends→Keynote Legend**.

4.   Type **FINISH SCHEDULE KEYS**.

Click **OK**.

5.   Click **OK** to accept the default legend created.

6.   The legend appears in the Project Browser.

Legends
    FINISH SCHEDULE KEYS

7.   Open the **Lobby Keynotes** sheet.

Sheets (all)
    A101 - First Level Floor Plan
    A102 - Second Level Floor Plan
    A103 - Lobby Keynotes

Sheets (all)
    A101 - Landing Page
    A102 - First Level Floor Plan
    A103 - Second Level Floor Plan
    A104 - LOBBY KEYNOTES

8.   Drag and drop the FINISH SCHEDULE KEYS onto the sheet.

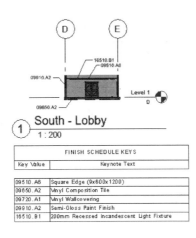

South - Lobby
1 : 200

FINISH SCHEDULE KEYS	
Key Value	Keynote Text
09510.A6	Square Edge (9x600x1200)
09650.A2	Vinyl Composition Tile
09720.A1	Vinyl Wallcovering
09910.A2	Semi-Gloss Paint Finish
16510.B1	200mm Recessed Incandescent Light Fixture

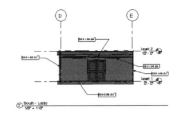

FINISH SCHEDULE KEYS	
Key Value	Keynote Text
09 51 00.A6	Square Edge (3/4 x 24 x 48)
09 65 00.A2	Vinyl Composition Tile
09 72 00.A1	Vinyl Wallcovering
09 91 00.A2	Semi-Gloss Paint Finish
26 51 00.A6	2' X 4' Surface Mounted Modular Fluorescent Fixture

9.

Activate the **Level 1 – Lobby Detail** floor plan. Zoom in to the view and you will see the elevation marker now indicates the Sheet Number and View Number.

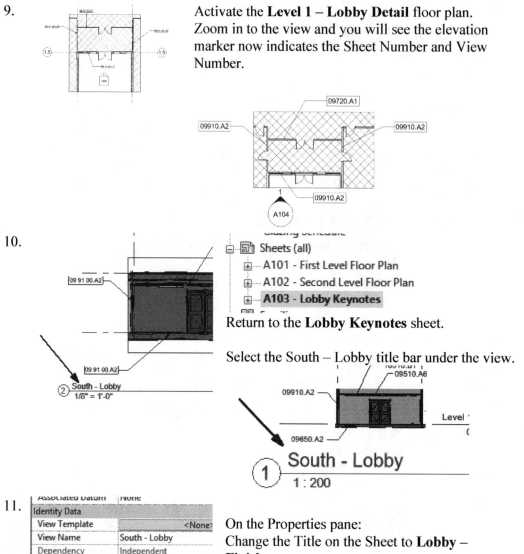

10.

Return to the **Lobby Keynotes** sheet.

Select the South – Lobby title bar under the view.

11.

On the Properties pane:
Change the Title on the Sheet to **Lobby – Finishes**.

Identity Data	
View Template	<None>
View Name	South - Lobby
Dependency	Independent
Title on Sheet	Lobby - Finishes
Sheet Number	A603
Sheet Name	Lobby Keynotes
Referencing Sheet	A603

Sheets (all)
A101 - First Level Floor Plan
A102 - Second Level Floor Plan
A103 - Lobby Keynotes

The view title
updates.

Notice that the view
name remains the
same.

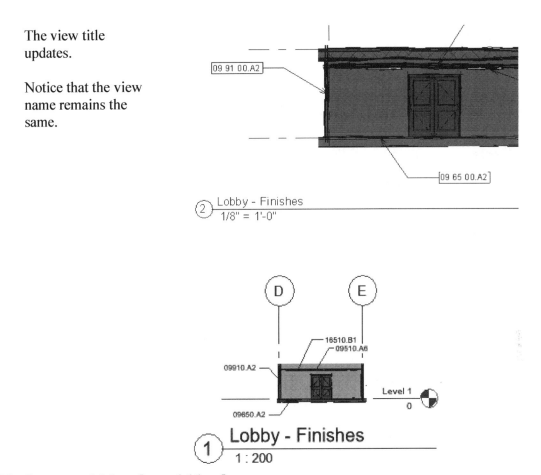

12. Save as *ex6-16.rvt [m_ex6-16.rvt]*.

Exercise 6-17
Create a Finish Schedule

Drawing Name: schedule5.rvt [m_schedule5.rvt]
Estimated Time: 15 minutes

This exercise reinforces the following skills:

- ❏ Schedule/Quantities
- ❏ Key Schedules
- ❏ Materials
- ❏ Images

1. Open *schedule5.rvt [m_schedule5.rvt]*.

2.

    Activate the View ribbon.

    Select **Schedules→Schedule/Quantities**.

3.

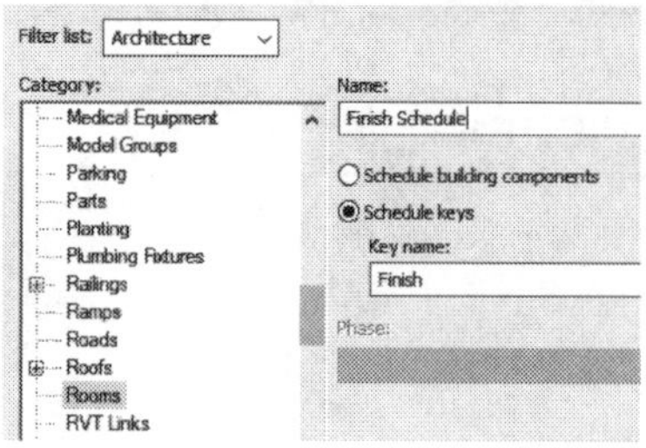

    Highlight **Rooms** in the Category pane.

    Type **Finish Schedule** for the Name.

    Enable **Schedule keys**.

    Under Key name: type **Finish**.

    Click **OK**.

4.

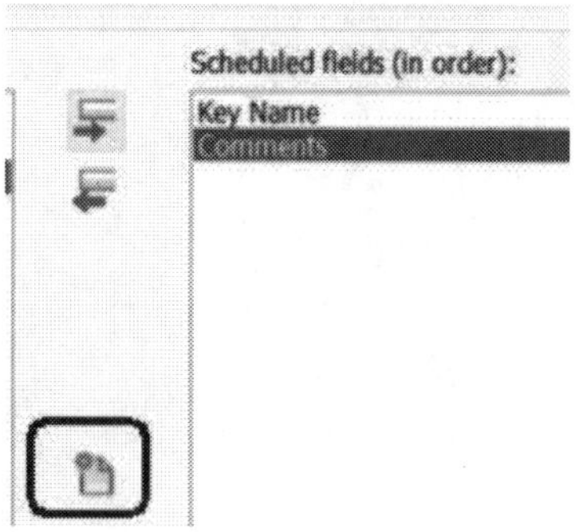

    Add the **Comments** fields.

    Click on **New Parameter**.

5.   Type **Image** for the Name.
Set the Type of Parameter to **Image**.
Group under Graphics.

Click **OK**.

6.   Position the Image parameter below Key Name.

Click on **New Parameter**.

7.   Type **Key Value** for the Name.
Set the Type of Parameter to **Text**.
Group under Identity Data.

Click **OK**.

8.   Position the Key Value parameter below Key Name.

Click on **New Parameter**.

9.   Type **Material** for the Name.
Set the Type of Parameter to **Material**.
Group under **Materials and Finishes**.

Click **OK**.

10.

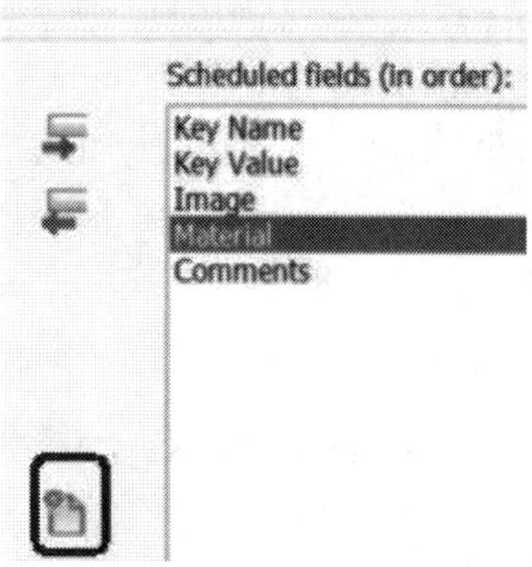

Position the Material parameter below Image.

Click on **New Parameter**.

Scheduled fields (in order):
- Key Name
- Key Value
- Image
- Material
- Comments

11.

Parameter Data
Name:
Location

Discipline:
Common

Data Type:
Text

Group parameter under:
Identity Data

○ Type
◉ Instance

◉ Values are aligned per group type
○ Values can vary by group instance

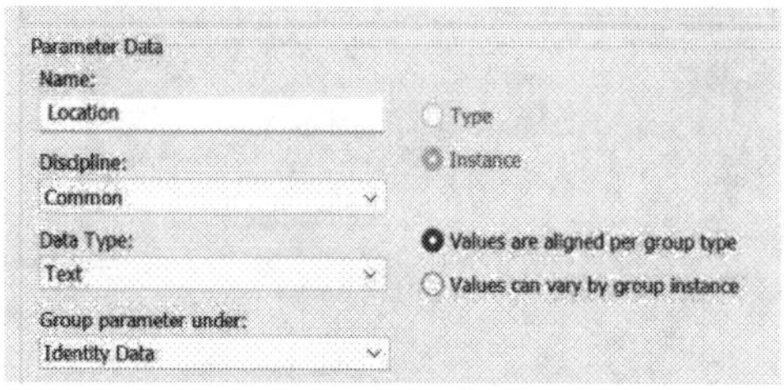

Type **Location** for the Name.
Set the Type of Parameter to **Text**.
Group under **Identity Data**.

Click **OK**.

12.

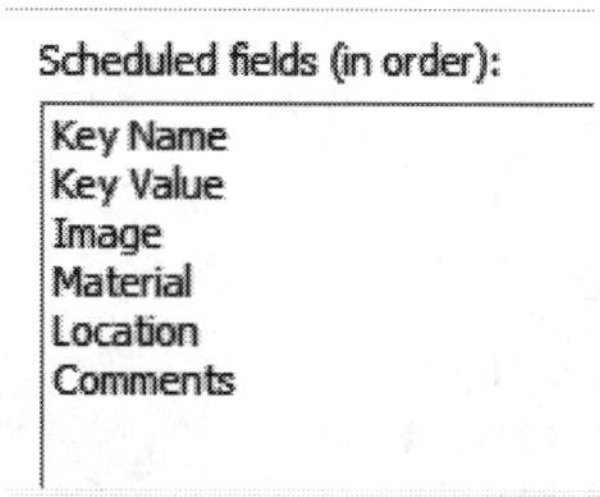

Scheduled fields (in order):
- Key Name
- Key Value
- Image
- Material
- Location
- Comments

The fields should appear as shown.

13.

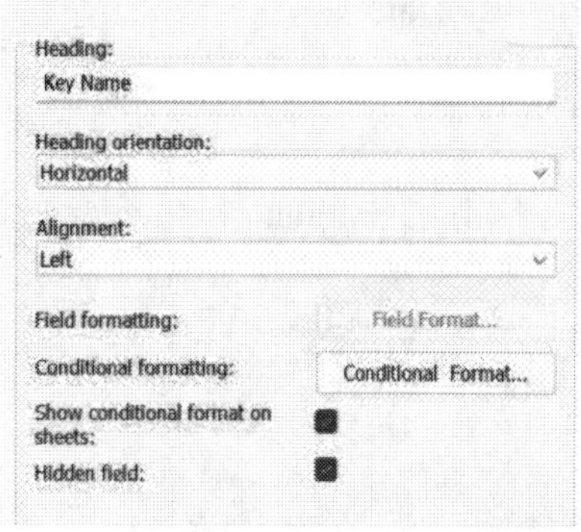

Heading:
Key Name

Heading orientation:
Horizontal

Alignment:
Left

Field formatting: Field Format...

Conditional formatting: Conditional Format...

Show conditional format on sheets:

Hidden field:

Select the Formatting tab.

Highlight Key Name.

Enable **Hidden field**.

Click **OK**.

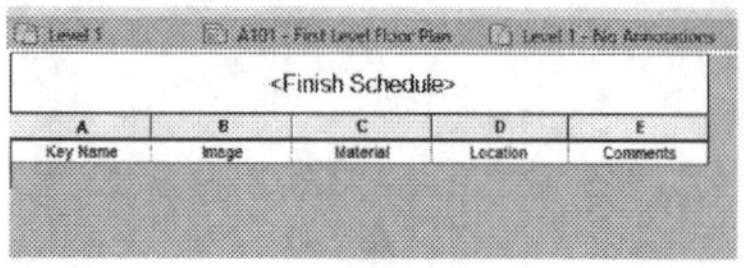

<Finish Schedule>

A	B	C	D	E
Key Name	Image	Material	Location	Comments

The schedule has no data.

14.

Insert Data Row

Click **Insert Data Row** on the ribbon.

15.

<Finish Schedule>				
A	B	C	D	E
Key Name	Image	Material	Location	Comments
1				
2				
3				
4				

Repeat to add four rows.

<Finish Schedule>				
A	B	C	D	E
Key Value	Image	Material	Location	Comments
09 51 00 A6		Acoustic Ceiling Tile 24 x 48	Lobby	Clg
09 65 00 A2		Vinyl Composition Tile – Diamond Pattern	Lobby	Floor
09 72 00 A1		Wallpaper-Striped	Lobby	S Wall
09 91 00 A2		Paint - Interior-SW0068 Heron Blue	Lobby	W/S/E Walls

<Finish Schedule>				
A	B	C	D	E
Key Value	Image	Material	Location	Comments
09510.A6		Ceiling Tile 600 x 1200	Lobby	Clg
09650.A2		Vinyl Composition Tile - Diamond Pa	Lobby	Floor
09720.A1		Wallpaper - Striped	Lobby	N Wall
09910.A2		Paint - Interior-SW0068 Heron Blue	Lobby	W/S/E Walls

16. Type in the keynote values in the Key Value cells.

Click in the Material cells and select the corresponding materials for each keynote.

Type Lobby for the Location.
Use Comments to indicate where the material/finish is applied.

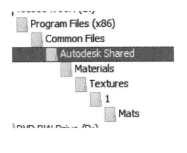

The images for the materials are located in the Materials folder.

Color	RGB 174 181 185
Image	Finishes.Flooring.VCT.Diamonds.jpg

If you aren't sure about the file name, you can locate it in the Materials Browser on the Appearance tab.

17.

You should have four images added.

The Sherwin Williams Heron Blue paint color was a custom material, so you will have to use the image provided with the exercise files. I also have included the image files for the other materials to make it easy for you to locate.

18.

The schedule should show images added for each key value.

A	B	
Key Value	Image	
09 51 00 A6	Finishes.Ceilings.Acoustical Tile.Exposed Grid.2x4.Pebble.White.jpg	Acou
09 65 00 A2	Finishes.Flooring.VCT.Diamonds.jpg	Vinyl
09 72 00 A1	Finishes.Wall Covering.Stripes.Vertical.Blue-Grey.jpg	Wallp
09 91 00 A2	SW0068 Heron Blue.PNG	Paint

<Finish Schedule>

A	B	
Key Value	Image	
09510.A6	Finishes.Ceilings.Acoustical Tile.Exposed Grid.2x4.Pebble.White.png	Ceil
09650.A2	Finishes.Flooring.VCT.Diamonds.jpg	Viny
09720.A1	Finishes.Wall Covering.Stripes.Vertical.Blue-Grey.jpg	Wal
09910.A2	SW0068 Heron Blue.PNG	Pair

19.

Open the **Lobby Keynotes** sheet.

- Sheets (all)
  - A101 - First Level Floor Plan
  - A102 - Second Level Floor Plan
  - **A103 - Lobby Keynotes**

- Sheets (all)
  - A101 - Landing Page
  - A102 - First Level Floor Plan
  - A103 - Second Level Floor Plan
  - **A104 - LOBBY KEYNOTES**

20.

Drag and drop the Finish Schedule onto the sheet.

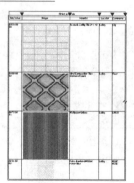

21. Select the Finish Schedule so it is highlighted.

On the Properties panel:
Select **All** in the Resize Rows drop-down list.

22. Type **1" [16]** as the desired row height.

Click **OK**.

23. Adjust the column widths and position on the sheet.

24. Save as *ex6-17.rvt [m_ex6-17.rvt]*.

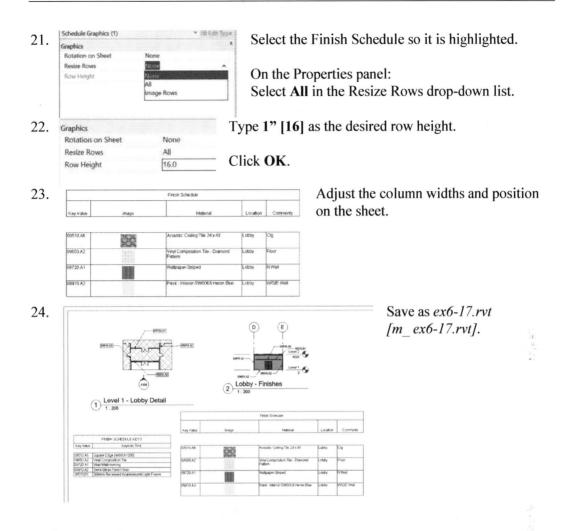

Exercise 6-18
Find and Replace Families

Drawing Name: schedule6.rvt [m_schedule6.rvt]
Estimated Time: 5 minutes

This exercise reinforces the following skills:

- ❑ Families
- ❑ Project Browser

1.   Open *schedule6.rvt [m_schedule6.rvt]*.

2.   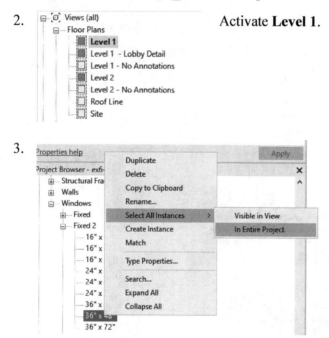   Activate **Level 1**.

3.   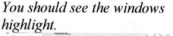

In the Project Browser, locate the *Windows* category under Families.

Highlight **36″ x 48″ Fixed 2 [0915 x 1220mm]**.

Right click and select **Select All Instances → In Entire Project**.

You should see the windows highlight.

4.      Select **Fixed 2: 36″ x 24″ [M_Fixed 2: 0915 x 0610mm]** from the Properties pane Type selector.

Left click in the display window to release the selection.

All the selected windows have been replaced with the new type.

5. 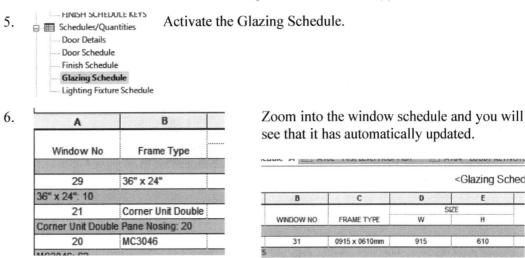 Activate the Glazing Schedule.

6. Zoom into the window schedule and you will see that it has automatically updated.

7. Save as *ex6-18.rvt [m_ex6-18.rvt]*.

Exercise 6-19
Modifying Family Types in a Schedule

Drawing Name: schedule7.rvt [m_ schedule7.rvt]
Estimated Time: 10 minutes

This exercise reinforces the following skills:

- ❑ Families
- ❑ Schedules

1. Open *schedule7.rvt [m_ schedule7.rvt]*.

2. 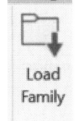 Activate the Insert ribbon.

   Select **Load Family**.

3. Locate the *Door-Interior-Single-Flush_Panel-Wood [M_Door-Interior-Single-Flush_Panel-Wood]* family in the Residential folder.

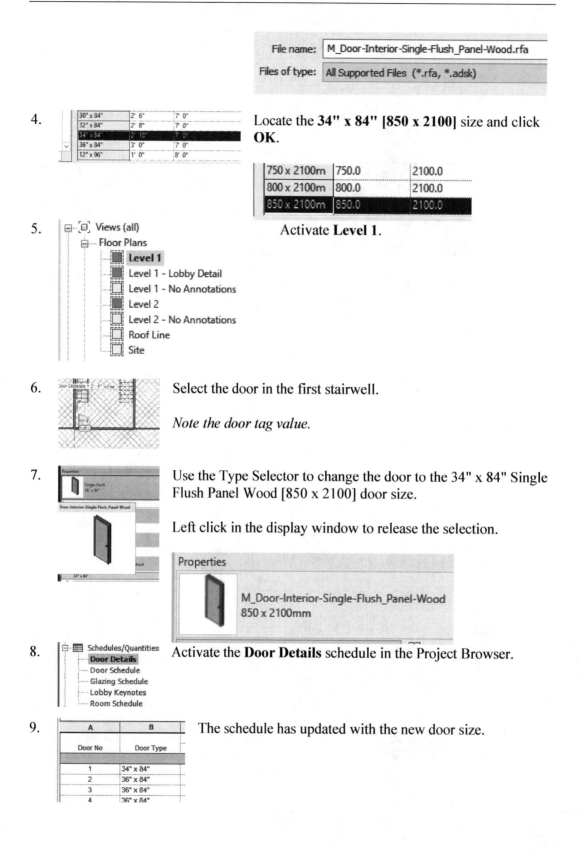

File name: M_Door-Interior-Single-Flush_Panel-Wood.rfa

Files of type: All Supported Files (*.rfa, *.adsk)

4. Locate the **34" x 84" [850 x 2100]** size and click **OK**.

750 x 2100m	750.0	2100.0
800 x 2100m	800.0	2100.0
850 x 2100m	850.0	2100.0

5. Activate **Level 1**.

6. Select the door in the first stairwell.

   Note the door tag value.

7. Use the Type Selector to change the door to the 34" x 84" Single Flush Panel Wood [850 x 2100] door size.

   Left click in the display window to release the selection.

   Properties

   M_Door-Interior-Single-Flush_Panel-Wood
   850 x 2100mm

8. Activate the **Door Details** schedule in the Project Browser.

9. The schedule has updated with the new door size.

A	B
Door No	Door Type
1	34" x 84"
2	36" x 84"
3	36" x 84"
4	36" x 84"

A	B	C	D	
			SIZE	
Door No	Door Type	W	H	
1	850 x 2100mm	850	2100	

10.

Activate the **View** ribbon.

Select **Close Inactive** on the Windows panel.

This closes any open views/files other than the active view.

Note that only the active tab remains available.

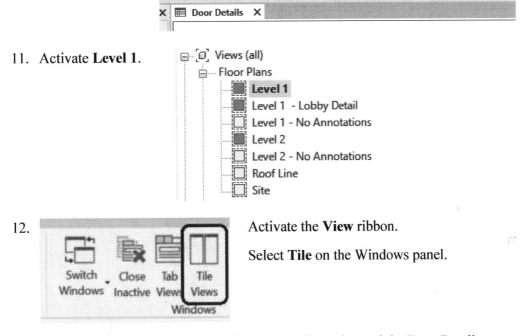

11. Activate **Level 1**.

12. Activate the **View** ribbon.

Select **Tile** on the Windows panel.

13. There should be two windows: The Level 1 floor plan and the Door Details schedule.

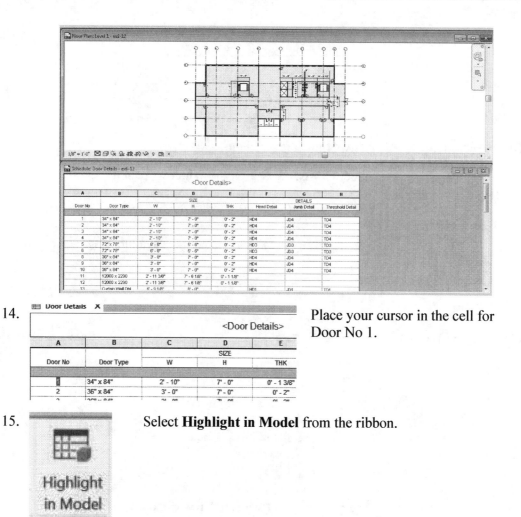

14.

Place your cursor in the cell for Door No 1.

15.

Select **Highlight in Model** from the ribbon.

16.

The floor plan view will zoom into the door. The door is automatically selected.

Click **Close**.

17.

In the Properties palette:

Note that the door is 34″ x 84″ [850 x 2100mm].

Properties

M_Door-Interior-Single-Flush_Panel-Wood
850 x 2100mm

18.

Properties

Single-Flush
36" x 84"

Use the Type Selector to change the door back to **36" x 84" Single Flush [0915 x 2134mm M_Single-Flush]**.

Left click in the window to release the selection.

900 x 2400mm

M_Single-Flush

0915 x 2134mm

19.

A	B
Door No.	Door Type
1	36" x 84"
2	36" x 84"
3	36" x 84"
4	36" x 84"
5	72" x 82"
6	72" x 82"
7	36" x 84"
8	36" x 84"
9	36" x 84"
10	36" x 84"
11	12000 x 2290 Opening
12	12000 x 2290 Opening
21	Door-Curtain-Wall-Double-Glass

Note that the Door Details schedule has updated.

A	B	
Door No	Door Type	
1	0915 x 2134mm	
2	0915 x 2134mm	
3	0915 x 2134mm	
4	0915 x 2134mm	
5	1800 x 2100mm	
6	1800 x 2100mm	

20.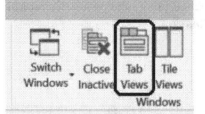

Click **Tab Views** on the View tab to return to a single window with tabs.

21. Save as *ex6-19.rvt [m_ex6-19.rvt]*.

Exercise 6-20
Create a Key Schedule to Calculate Occupancy

Drawing Name: art gallery.rvt [m_art gallery.rvt]
Estimated Time: 10 minutes

This exercise reinforces the following skills:

- ❑ Schedules
- ❑ Schedule Keys

1.   Open art gallery.*rvt [m_art gallery.rvt]*.

2.   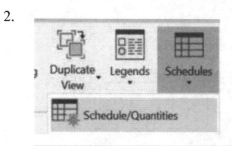   Go to the **View** ribbon.

Select **Schedules→Schedules/Quantities**.

3.   Type **room** in the Category name search field.

Highlight **Rooms** in the Category Panel on the left.

Enable **Schedule Keys**.

In the Name field, type **Space Occupancy Classification**.

In the Key Name field, type **Classification**.

Click **OK**.

4.   Add the following parameters to the Key Schedule:

- • Classification Description
- • Area per Occupant
- • Occupancy Gross or Net

These are custom project parameters that were added to the exercise file.

Click **OK**.

5.   Click **Insert Data Row** on the ribbon.

Insert
Data Row

<Space Occupancy Classification>			
A	**B**	**C**	**D**
Key Name	Classification Description	Area Per Occupant	Occupancy Gross or Net
A-3	Art Gallery	30 SF	Net

A	**B**	**C**	**D**
Key Name	Classification Description	Area Per Occupant	Occupancy Gross or Net
A-3	Art Gallery	3 m²	Net

6.   Enter the data:

Key Name: A-3
Classification Description: Art Gallery
Area Per Occupant: 30 [3]
Occupancy Gross or Net: Net

7.   Click **Insert Data Row** on the ribbon.

Insert
Data Row

A	**B**	**C**	**D**
Key Name	Classification Description	Area Per Occupant	Occupancy Gross or Net
A-3	Art Gallery	30 SF	Net
B-100	Office	150 SF	Net

<Space Occupancy Classification>			
A	**B**	**C**	**D**
Key Name	Classification Description	Area Per Occupant	Occupancy Gross or Net
A-3	Art Gallery	3 m²	Net
B-100	Office	14 m²	Net

8.   Enter the data:

Key Name: B-100

Classification Description: Office
Area Per Occupant: 150 [14]
Occupancy Gross or Net: Net

9.

Insert
Data Row

Click **Insert Data Row** on the ribbon.

A	B	C	D
Key Name	Classification Description	Area Per Occupant	Occupancy Gross or Net
A-1	Lobby	15 SF	Gross
A-3	Art Gallery	30 SF	Net
B-100	Office	150 SF	Net

<Space Occupancy Classification>			
A	B	C	D
Key Name	Classification Description	Area Per Occupant	Occupancy Gross or Net
A-1	Lobby	1 m²	Gross
A-3	Art Gallery	3 m²	Net
B-100	Office	14 m²	Net

10. Enter the data:

Key Name: A-1
Classification Description: Lobby
Area Per Occupant: 15 [1]
Occupancy Gross or Net: Gross

11.   Save as *ex6-20.rvt [m_ex6-20.rvt]*.

Exercise 6-21
Create a Schedule using Formulas

Drawing Name: occupancy_formula.rvt [m_occupancy_formula.rvt]
Estimated Time: 20 minutes

This exercise reinforces the following skills:

- Schedules
- Formulas

1.   Open *occupancy_formula.rvt [m_occupancy_formula.rvt]*.

2.   Open **Level 1** floor plan.

     Floor Plans

           Level 1

           Level 2

           Site

3.   Select the Room labeled Office 1.

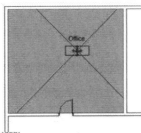

4.   Set the Classification to **B-100**.

Classification	B-100
Classification Description	Office
Occupancy Gross or Net	Net
Occupant	

Note that the other parameters fill in because they are using the schedule keys.

Dimensions	
Area	54.510 m²
Perimeter	29600.0
Unbounded Height	4000.0
Volume	Not Computed
Computation Height	0.0
Area Per Occupant	14.000 m²

5.   Select the Room labeled Office 2.

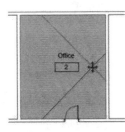

6.

Classification	B-100
Classification Description	Office
Occupancy Gross or Net	Net
Occupant	

Set the Classification to **B-100**.

7.

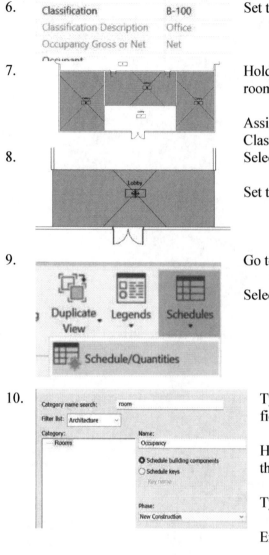

Hold down the CTL key and select the three rooms labeled Gallery.

Assign the rooms labeled Gallery to Classification **A-3 Art Gallery**.
Select the Lobby.

8.

Set the Classification to **A-1**.

9.

Go to the **View** ribbon.

Select **Schedules→Schedules/Quantities**.

10.

Type **room** in the Category name search field.

Highlight **Rooms** in the Category Panel on the left.

Type **Occupancy** in the Name field.

Enable Schedule Building Components

Click **OK**.

11.

Add the following fields:

- Name
- Number
- Area
- Area Per Occupant
- Classification

12.

Scheduled fields (in order):

Name
Number
Area
Area Per Occupant
Classification

Select **Add Calculated Parameter**.

13.

Name: Calculated Occupancy Load

○ Formula   ○ Percentage

Discipline: Common

Type: Integer

Formula: (Area/Area Per Occupant) + 0.499

Type **Calculated Occupancy Load** in the Name field.
Select Integer
For the formula, use:
(Area/Area Per Occupant) + 0.499
The 0.499 ensures the value is rounded up to the nearest whole number/integer.

Hint: *You can use the ... button to select the Area and Area Per Occupant parameters used in the formula.*

Click **OK**.

14.

Schedule Properties

Fields Filter Sorting/Grouping Formatting Appearance Embedded Schedule

Filter by: Area Per Occupant   has a value

And: (none)

Select the Filter tab.

Filter by: Area Per Occupant
Has a value.

The restrooms and hallway have no area per occupant assigned as they are considered transient areas. This will omit those rooms from the schedule.

15.

Schedule Properties

Fields Filter Sorting/Grouping Formatting Appearance Embedded Schedule

Sort by:	Number	⌄	⦿ Ascending	○ Descending
☐ Header	☐ Footer:	⌄		☐ Blank line
Then by:	(none)	⌄	⦿ Ascending	○ Descending
☐ Header	☐ Footer:	⌄		☐ Blank line
Then by:	(none)	⌄	⦿ Ascending	○ Descending
☐ Header	☐ Footer:	⌄		☐ Blank line
Then by:	(none)	⌄	⦿ Ascending	○ Descending
☐ Header	☐ Footer:	⌄		☐ Blank line

■ Grand totals: Title and totals   ⌄

Custom grand total title:

Grand total

■ Itemize every instance

Select the Sorting/Grouping tab.

Sort by **Number**.
Enable **Grand totals**.
Select **Title and totals**.
Enable **Itemize every instance.**

16.

Heading:

Area

Heading orientation:

Horizontal   ⌄

Alignment:

Left   ⌄

Field formatting: [Field Format...]

Conditional formatting: [Conditional Format...]

Show conditional format on sheets: ■

Hidden field: ☐

Calculate totals   ⌄

Select the Formatting tab.

Highlight **Area**.
Select **Calculate totals** from the drop down-list.

17.

Heading:

Calculated Occupancy Load

Heading orientation:

Horizontal

Alignment:

Left

Field formatting: Field Format...

Conditional formatting: Conditional Format...

Show conditional format on ▣
sheets:

Hidden field: ☐

Calculate totals

Highlight **Calculated Occupancy Load**. Select **Calculate totals** from the drop down-list.

Click **OK**.

18.

A	**B**	**C**	**D**	**E**	**F**
Name	Number	Area	Area Per Occupant	Classification	Calculated Occupan
Office	1	662 SF	150 SF	B-100	5
Office	2	444 SF	150 SF	B-100	3
Gallery	6	584 SF	30 SF	A-3	20
Gallery	7	679 SF	30 SF	A-3	23
Lobby	8	535 SF	15 SF	A-1	36
Gallery	9	667 SF	30 SF	A-3	23
Grand total		3572 SF			110

<Occupancy>

The schedule opens.

<Occupancy>

A	**B**	**C**	**D**	**E**	**F**
Name	Number	Area	Area Per Occupant	Classification	Calculated Occupancy Load
Office	1	55 m²	14 m²	B-100	4
Office	2	40 m²	14 m²	B-100	3
Gallery	6	76 m²	3 m²	A-3	28
Lobby	7	58 m²	1 m²	A-1	42
Gallery	8	62 m²	3 m²	A-3	30
Gallery	9	71 m²	3 m²	A-3	28
Grand total		382 m²			133

19.

Save as *ex6-21.rvt [m_ex6-21.rvt]*.

Exercise 6-22
Export a Schedule

Drawing Name: schedule_export.rvt [m_ schedule_export.rvt]
Estimated Time: 5 minutes

This exercise reinforces the following skills:

   ❑   Schedules

1.   Open *schedule_export.rvt [m_ schedule_export.rvt].*

2.   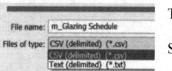   Activate the **Glazing Schedule**.

3.   Go to the Applications Menu.

Select **File → Export → Reports → Schedule**.

You will need to scroll down to see the Reports option.

4.   The schedule can be saved as a csv or txt file.

Select *.txt.

Browse to your exercise folder.
Click **Save**.

5.   Click **OK**.

6.   Launch **Excel**.

7.   📂 Open   Select **Open**.

8. | Text Files | Set the file types to **Text Files**.

9.

File name:	Glazing Schedule		Text Files	
		Tools ▼	Open ▼	Cancel

Browse to where you saved the file and select it. Click **Open**.

10.

The Text Wizard has determined that your data is Delimited.

If this is correct, choose Next, or choose the data type that best describes your data.

Original data type

Choose the file type that best describes your data:

⦿ Delimited   - Characters such as commas or tabs separate each field.

⦾ Fixed width   - Fields are aligned in columns with spaces between each field.

Start import at row: [1]   File origin: [Windows (ANSI)]

Preview of file E:\Schroff\Revit 2011 Basics\exercise files\Glazing Schedule.txt.

```
47 "8""36"" x 24"""3' – 0"""2' – 0"""""""3' – 0"""""""
48 "8""36"" x 24"""3' – 0"""2' – 0"""""""3' – 0"""""""
49 "8""36"" x 24"""3' – 0"""2' – 0"""""""3' – 0"""""""
50 "8""36"" x 24"""3' – 0"""2' – 0"""""""3' – 0"""""""
51 "8""36"" x 24"""3' – 0"""2' – 0"""""""3' – 0"""""""
```

Cancel   < Back   Next >   Finish

Click **Next**.

11.

This screen lets you set the delimiters your data contains. You can see how your text is affected in the preview below.

Delimiters

☑ Tab

☐ Semicolon   ☐ Treat consecutive delimiters as one

☐ Comma

☐ Space   Text qualifier: ["]

☐ Other: [ ]

Data preview

Glazing Schedule WINDOW NO	FRAME TYPE	SIZE W	H	DETAILS HEAD	JAMB	SILL	MULLIC
8	36" x 24"	3' – 0"	2' – 0"	this is a line			

Click **Next**.

12.

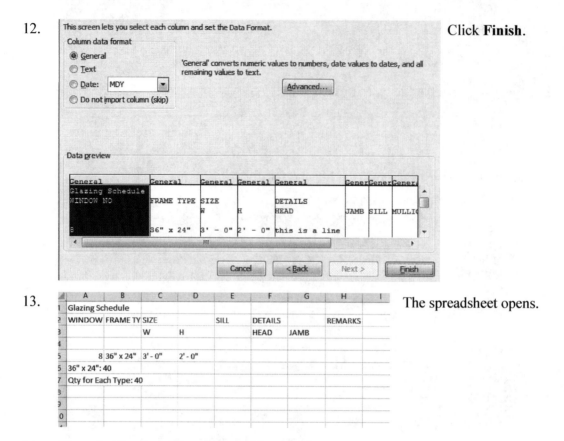

Click **Finish**.

13.

The spreadsheet opens.

14. Close the Revit and Excel files without saving.

Exercise 6-23
Assigning Fonts to a Schedule

Drawing Name: schedule_fonts.rvt [m_schedule fonts.rvt]
Estimated Time: 5 minutes

This exercise reinforces the following skills:

❑   Schedules
❑   Text Styles

1.   Open *schedule_fonts.rvt [m_schedule fonts.rvt]*.

2.   Activate the Annotate ribbon.

3.      Select the small arrow located in the bottom corner of the Text panel.

4. | Family: | System Family: Text | Load... |
     | Type: | 1/4" Arial | Duplicate... |
     | | | Rename... |

     Select **Duplicate**.

     | Family: | System Family: Text | Load... |
     | Type: | 5mm Arial | Duplicate... |
     | | | Rename... |

5. | Name: | Schedule Title |

     OK   Cancel

     Enter **Schedule Title**.
     Click **OK**.

6. | Text | |
     |---|---|
     | Text Font | Tahoma |
     | Text Size | 1/4" |
     | Tab Size | 1/2" |
     | Bold | ☐ |
     | Italic | ☑ |
     | Underline | ☐ |
     | Width Factor | 1.000000 |

     Set the Text Font to **Tahoma**.
     Set the Text Size to ¼"[5mm].
     Enable **Italic**.

     Any Windows font can be used to create a Revit text style.

     Click **Apply**.

Text	
Text Font	Tahoma
Text Size	5.0000 mm
Tab Size	12.7000 mm
Bold	☐
Italic	☑
Underline	☐
Width Fac	1.000000

7.   Duplicate...   Select **Duplicate**.

8.

| Name: | Schedule Header| |

Enter **Schedule Header**.
Click **OK**.

9.

Leader Arrowhead	Arrow 30 Degree
Text	
Text Font	Tahoma
Text Size	51/256"
Tab Size	1/2"
Bold	☐
Italic	☐
Underline	☐
Width Factor	1.000000

Set the Text Font to **Tahoma**.
Set the Text Size to **0.2″[4mm]**.
Disable **Italic**.

Click **Apply**.

Text	
Text Font	Tahoma
Text Size	4.0000 mm
Tab Size	12.7000 mm
Bold	☐
Italic	☐
Underline	☐
Width Fac	1.000000

10. [Duplicate...]   Select **Duplicate**.

11.

| Name: | Schedule Cell| |

Enter **Schedule Cell**.
Click **OK**.

12.

Text	
Text Font	Tahoma
Text Size	1/8"
Tab Size	1/2"
Bold	☐
Italic	☐
Underline	☐
Width Factor	1.000000

Set the Text Font to **Tahoma**.
Set the Text Size to **1/8″[2.5]**.

Click **OK**.

Text	
Text Font	Tahoma
Text Size	2.5 mm
Tab Size	12.7000 mm
Bold	☐
Italic	☐
Underline	☐
Width Fac	1.000000

13. Schedules/Quantities (all)
    Door Details
    Door Schedule
    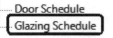 Glazing Schedule

Activate the **Glazing Schedule**.

14.

Other	
Fields	Edit...
Filter	Edit...
Sorting/Grouping	Edit...
Formatting	Edit...
Appearance	Edit...

Select **Edit** next to Appearance.

15.

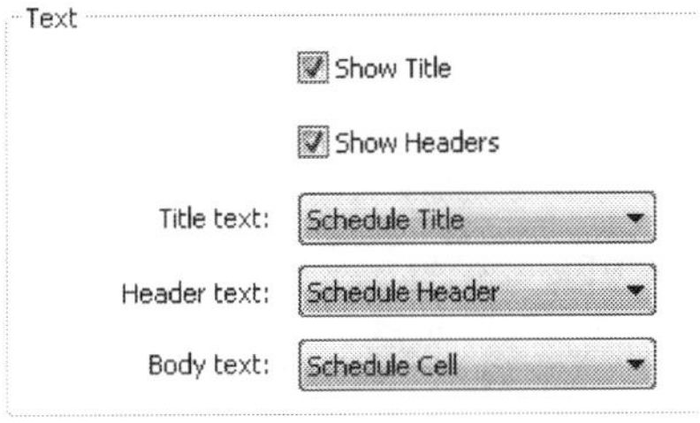

Set the Title text to **Schedule Title**.
Set the Header text to **Schedule Header**.
Set the Body text to **Schedule Cell**.

Click **OK.**

16.

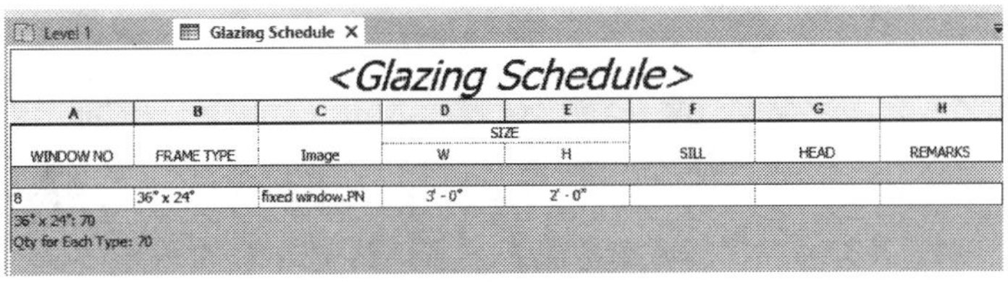

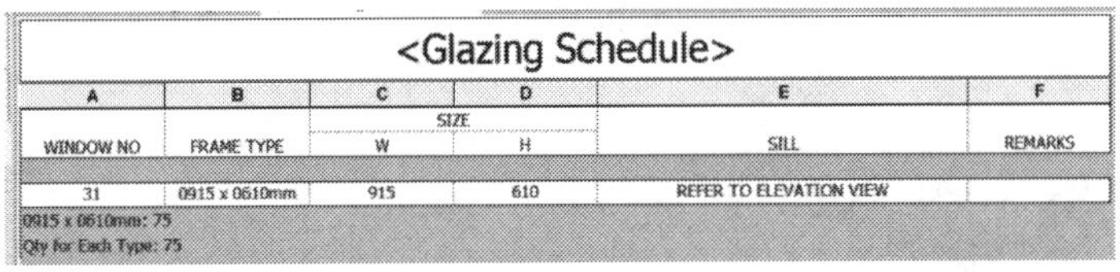

The schedule updates
with the new fonts
assigned.

17.

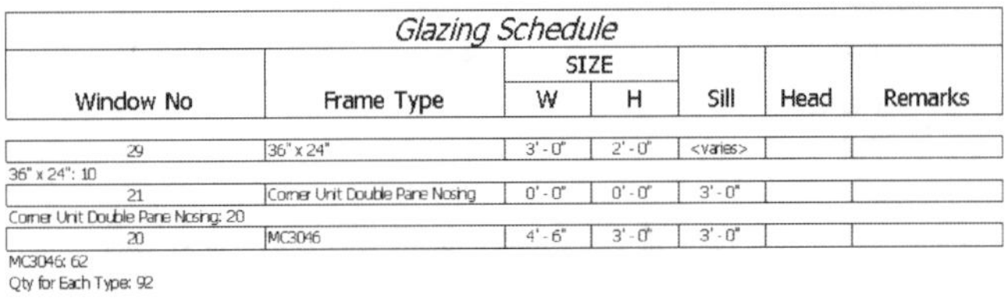

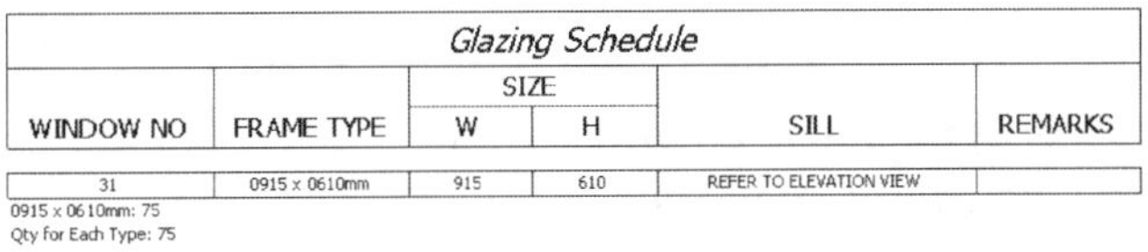

Place the schedule on
the First Level Floor
Plan sheet so you can
see the fonts clearly.

18.   Save as *ex6-23.rvt [m_ ex6-23.rvt].*

Exercise 6-24
Using a View Template for a Schedule

Drawing Name: view template1.rvt [m_ view template1.rvt]
Estimated Time: 5 minutes

This exercise reinforces the following skills:

- Schedules
- Text Styles

1.   Open *view template1.rvt [m_ view template1.rvt]*.

2.   Activate the **Glazing Schedule**.

3.   Activate the **View** ribbon.

    Select **Create Template from Current View**.

4.   Type **Schedule using Tahoma Fonts**.

    Click **OK**.

5.   Note that you can use a View Template to control not only the appearance of a schedule but for selecting fields, formatting, etc.

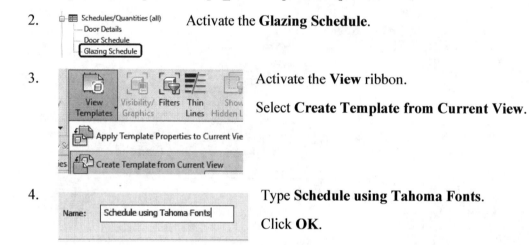

    Click **OK**.

6.   Activate the **Door Details** schedule.

7.

On the Properties palette:

Select the **\<None\>** button next to View Template.

Properties	×
▦ Schedule	▾
Schedule: Door Details ▾	⊞ Edit Type
Identity Data	⊗
View Template	\<None\>
View Name	Door Details
Dependency	Independent

8.

Names:

\<None\>
Schedule using Tahoma Fonts

Highlight the **Schedule using Tahoma Fonts** template.

Click **OK**.

9. The schedule updates with the new appearance.

▢ Level 1	▦ Door Details ✕	▦ Glazing Schedule	⬡ (3D)	▢ A101 - First Level Floor Plan

\<Door Details\>

A	B	C	D	E	F	G	H
			SIZE			DETAILS	
Door No	Door Type	W	H	THK	Head Detail	Jamb Detail	Threshold Detail
1	36" x 84"	3' - 0"	7' - 0"	0' - 2"	HD4	JD4	TD4
2	36" x 84"	3' - 0"	7' - 0"	0' - 2"	HD4	JD4	TD4
3	36" x 84"	3' - 0"	7' - 0"	0' - 2"	HD4	JD4	TD4
4	36" x 84"	3' - 0"	7' - 0"	0' - 2"	HD4	JD4	TD4
5	72" x 82"	6' - 0"	6' - 10"	0' - 1 3/4"	HD2	JD2	TD2
6	72" x 82"	6' - 0"	6' - 10"	0' - 1 3/4"	HD2	JD2	TD2

10. Save as *ex6-24.rvt [m_ex6-24.rvt]*.

Exercise 6-25
Exporting a View Template to another Project

Drawing Name: view template source.rvt, Office_2.rvt
Estimated Time: 30 minutes

This exercise reinforces the following skills:

- ❑ Schedules
- ❑ View Templates
- ❑ Transfer Project Standards

1.   Open *view template source.rvt*.

2.    Activate the **Glazing Schedule**.

3.    Activate the **View** ribbon.

          Select **Manage View Templates**.

4.

In the Discipline filter: Select **Architectural** from the drop-down list.

In the View type filter: Select **Schedules** from the drop-down list.

Highlight the **Schedule using Tahoma Fonts** template.

Select the **Duplicate** icon at the bottom of the dialog.

5.

Type **Schedule with Shared Parameters**.

Click **OK**.

6.

Verify that the new schedule template is highlighted.

7.

Select the **Edit** button next to Fields.

8.

Note the fields listed include the shared parameters.

Click **OK**.

9.

Parameter	Value
Phase Filter	Show All
Fields	Edit...
Filter	Edit...
Sorting/Grouping	Edit...
Formatting	Edit...
Appearance	Edit...

Select the **Edit** button next to Sorting/Grouping.

10.

☑ Grand totals: Title, count, and totals ⌄
Custom grand total title:
Qty for Each Type
☑ Itemize every instance

Place a check on **Itemize every instance**.

Click **OK**.

11.

Parameter	Value
Phase Filter	Show All
Fields	Edit...
Filter	Edit...
Sorting/Grouping	Edit...
Formatting	Edit...
Appearance	Edit...

Select the **Edit** button next to Appearance.

12.

Grid lines: ☑ <Thin Lines> ⌄   ☑ Grid in headers/footers/spacers
Outline: ☑ <Wide Lines> ⌄
☐ Blank row before data
Stripe Rows: ☑ First Row Stripe Color ⌄ ☐
☑ Show Stripe Rows on Sheets

Place a check next to **Grid in headers/footers/spacers**.
Place a check next to **Outline** and set the line style to **Wide Lines**.
Uncheck **Blank row before data**.
Enable **Stripe Rows**.
Set the color to **Yellow**.
Enable **Show Stripe Rows on Sheets**.

Click **OK**.

13.

Parameter	Value	Include
Phase Filter	Show All	☑
Fields	Edit...	☑
Filter	Edit...	☑
Sorting/Grouping	Edit...	☑
Formatting	Edit...	☑
Appearance	Edit...	☑

Note that there is a check to Include all the properties listed.

Click **OK** to close the dialog.

14. Open *Office_2.rvt*.

15.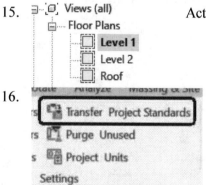

Activate **Level 1**.

16.

Activate the **Manage** ribbon.

Select **Transfer Project Standards**.

17.

Verify that the correct file is listed in the Copy from: list.

Select **Check None** to uncheck all the items in the list.

Scroll down and select **View Templates**.

Click **OK**.

18.

Select **New Only** to copy over only the new view templates.

19.

Activate the **Window Schedule** in the Project Browser.

20.

Note the fields included and the appearance of the schedule.

21.

In the Properties pane, select the **<None>** button next to View Template.

22.

Select the **Schedule with Shared Parameters** view template from the list.

Note that the two schedule view templates created were copied into the project.

Click **OK**.

23.

Note how the schedule updates.

Save as *Office_2_A.rvt*.

Note the list of items which can be copied from one project to another using Transfer Project Standards.

Additional Projects

FINISH SCHEDULE KEYS   Level 1   Level 1 - Lavatories   3D- Lavatory   Level 1   Lavatory   AP3-1 Office Layout   Level 1-Rooms   Room

			<Lavatory Finish Schedule>				
A	B	C	D	E	F	G	H
Room Name	E Wall	Name	Floor Finish	Ceiling Finish	N Wall	S Wall	W Wall
Restroom	Paint - SW6126NavajoWhite	Women's	VCT - Diamond Pattern	Acoustical Tile	Paint - SW6126NavajoWhite	Paint - SW6126NavajoWhite	Paint - SW6126NavajoWhite
Restroom	Paint - SW6126NavajoWhite	Men's	VCT - Diamond Pattern	Acoustical Tile	Paint - SW6126NavajoWhite	Paint - SW6126NavajoWhite	Paint - SW6126NavajoWhite

1) Using a key schedule, create a Finish Schedule for the lavatories.

Door Schedule - Ver 2			
Mark	Type	Count	Remarks
1	36" x 84"	1	Single-Flush
2	36" x 84"	1	Single-Flush
3	36" x 84"	1	Single-Flush
4	36" x 84"	1	Single-Flush
5	72" x 78"	1	Double-Exterior
6	72" x 78"	1	Double-Exterior
7	36" x 84"	1	Single-Flush
8	36" x 84"	1	Single-Flush
9	36" x 84"	1	Single-Flush
10	36" x 84"	1	Single-Flush
	Curtain Wall Dbl Glass	1	Double Glazed

2) Create a door schedule.

			Door Schedule 2				
			Rough Opening				
Mark	Type	Level	Rough Height	Rough Width	Material	Lock Jamb	Swing
1	36" x 84"	Level 1			ULTEX/WOOD	CENTER	RIGHT
2	36" x 84"	Level 1			ULTEX/WOOD	THROW	RIGHT
3	36" x 84"	Level 1			ULTEX/WOOD	THROW	LEFT
4	36" x 84"	Level 1			ULTEX/WOOD	THROW	LEFT
5	72" x 78"	Level 1			ULTEX/WOOD	THROW	CENTER
6	72" x 78"	Level 1			ULTEX/WOOD	THROW	CENTER
	Curtain Wall Dbl	Level 1			ULTEX/WOOD	THROW	CENTER
7	36" x 84"	Level 2			ULTEX/WOOD	THROW	LEFT
8	36" x 84"	Level 2			ULTEX/WOOD	THROW	RIGHT
9	36" x 84"	Level 2			ULTEX/WOOD	THROW	RIGHT
10	36" x 84"	Level 2			ULTEX/WOOD	THROW	RIGHT

3) Create this door schedule.

4) Create three new fonts for Schedule Title, Schedule Header, and Cell text. Apply them to a schedule.

Lavatory Finish Schedule						
Name	Floor Finish	Ceiling Finish	N Wall	E Wall	S Wall	W Wall
Women's	VCT - Diamond Pattern	Acoustical Tile	Paint - SW6126NavajoWhite	Paint - SW6126NavajoWhite	Paint - SW6126NavajoWhite	Paint - SW6126NavajoWhite
Men's	VCT - Diamond Pattern	Acoustical Tile	Paint - SW6126NavajoWhite	Paint - SW6126NavajoWhite	Paint - SW6126NavajoWhite	Paint - SW6126NavajoWhite

5) Create a view template using the schedule with your custom fonts.

6) Create three project parameters:

- Area Per Occupant
- Classification Description
- Occupancy Gross or Net

Refer to the images below on the parameter definitions.

Create a Key Schedule to use for Occupancy.

<Occupancy Keys>			
A	**B**	**C**	**D**
Key Name	Classification Description	Area Per Occupant	Occupancy Gross or Net
A-1	Lobby	15 SF	Gross
A-2	Cafeteria	15 SF	Net
B-100	Office	150 SF	Net
B-200	Laboratory	100 SF	Net
B-300	Conference Room	15 SF	Net

Create a Room schedule to show the occupancy load for the rooms on Level 1.

	A	B	C	D	E	F
				<Occupancy - Level 1>		
	Name	Number	Classification	Classification Description	Area	Occupancy Load
	Cafeteria	6	A-2	Cafeteria	709 SF	48
	Conference Room	9	B-300	Conference Room	233 SF	16
	Conference Room	10	B-300	Conference Room	479 SF	32
	Conference Room	11	B-300	Conference Room	470 SF	32
	Laboratory	4	B-200	Laboratory	427 SF	5
	Laboratory	5	B-200	Laboratory	429 SF	5
	Lobby	1	A-1	Lobby	252 SF	17
	Office	12	B-100	Office	158 SF	2
	Office	13	B-100	Office	271 SF	2
	Office	14	B-100	Office	267 SF	2
	Office	15	B-100	Office	265 SF	2
	Office	16	B-100	Office	235 SF	2
	Office	17	B-100	Office	372 SF	3
	Office	18	B-100	Office	364 SF	3
	Office	19	B-100	Office	176 SF	2
	Office	20	B-100	Office	176 SF	2
	Office	21	B-100	Office	248 SF	2
	Office	22	B-100	Office	237 SF	2
	Office	23	B-100	Office	340 SF	3
	Office	24	B-100	Office	281 SF	2
	Office	25	B-100	Office	277 SF	2

Lesson 6 Quiz

True or False

1. A schedule displays information about elements in a building project in a tabular form.
2. Each property of an element is represented as a field in a schedule.
3. If you replace a model element, the schedule for that element will automatically update.
4. Shared parameters are saved in an Excel spreadsheet.
5. A component schedule is a live view of a model.

Multiple Choice

6. Select the three types of schedules you can create in Revit:

   A. Component
   B. Multi-Category
   C. Key
   D. Symbol

7. Keynotes can be attached to the following:
   Select THREE answers.

   A. Model elements
   B. Detail Components
   C. Materials
   D. Datum

8. Select the tab that is not available on the Schedule Properties dialog:

   A. Fields
   B. Filter
   C. Sorting/Grouping
   D. Formatting
   E. Parameters

9. Schedules are exported in this file format:

   A. Excel
   B. Comma-delimited text
   C. ASCII
   D. DXF

10. To export a schedule:

    A. Right click on a schedule and select Export.
    B. Place a schedule on a sheet, select the sheet in the Project Browser, right click and select Export.
    C. Go to the Manage ribbon and select the Export Schedule tool.
    D. On the Application Menu, go to File→Export→Reports→Schedule.

11. To create a custom grand total title in a schedule:

    A. Enter the title in the custom grand total title text box on the Sorting/Grouping tab.
    B. Click in the schedule cell for the grand total title and modify the text.
    C. Select the schedule and type the desired text in the custom grand total title field on the Properties pane.
    D. All of the above.

12. To create a graphic display that looks like a view is sketched:

    A. Select Sketchy from the Graphics Display list.
    B. Enable Sketchy Lines in the Graphic Display Options dialog.
    C. Set the Graphic Display to Sketch.
    D. Apply a sketch view template.

ANSWERS:

    1) T; 2) T; 3) T; 4) F; 5) T; 6) A, B, and C; 7) A, B, and C; 8) E; 9) B; 10) D; 11) A; 12) B

Lesson 7
Roofs

Revit provides three methods for creating roofs: extrusion, footprint or by face. The extrusion method requires you to sketch an open outline. The footprint method uses exterior walls or a sketch. The face method requires a mass.

A roof cannot cut through windows or doors.

Roofs are system families.

Roof by footprint

- 2D closed-loop sketch of the roof perimeter

- Created when you select walls or draw lines in plan view

- Created at level of view in which it was sketched

- Height is controlled by Base Height Offset property

- Openings are defined by additional closed loops

- Slopes are defined when you apply a slope parameter to sketch lines

Roof by extrusion

- Open-loop sketch of the roof profile

- Created when you use lines and arcs to sketch the profile in an elevation view

- Height is controlled by the location of the sketch in elevation view

- Depth is calculated by Revit based on size of sketch, unless you specify start and end points.

Roofs are defined by material layers, similar to walls.

Exercise 7-1

Creating a Roof Using Footprint

Drawing Name: roof_1.rvt [m_roof_1.rvt]
Estimated Time: 15 minutes

This exercise reinforces the following skills:

- ❑ Level
- ❑ Roof
- ❑ Roof Properties
- ❑ Roof Options
- ❑ Isolate Element
- ❑ Select All Instances
- ❑ 3D View

1. Open *roof_1.rvt [m_roof_1.rvt]*.

2.
⊟ ⟨ O ⟩ Views (all)	
⊟ Floor Plans	
☐ Level 1	
☐ Level 1 - Lobby Detail	
☐ Level 1 - No Annotations	
☐ Level 2	
☐ Level 2 - No Annotations	
☐ **Roof Line**	
☐ Site	

Activate the **Roof Line** view under Floor Plans by double clicking on it.

3.
Underlay	
Range: Base Level	Level 1
Range: Top Level	Roof Line
Underlay Orientation	Look down

In the Properties pane:

Scroll down to the Underlay heading.
Set the Base Level Range to **Level 1**.
Set the Top Level Range to **Roof Line**.

4.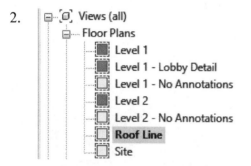

In the browser, locate the exterior wall that is used in the project.

Right click and select **Select All Instances → In Entire Project**.

5.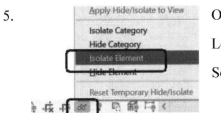

On the Display bar:

Left click on the icon that looks like sunglasses.

Select **Isolate element**.

This will turn off the visibility of all elements except for the exterior walls.
This will make selecting the walls easier.

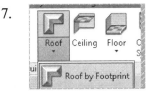

You should see the top view of your model with only the exterior walls visible.

You won't see the curtain wall because it is a different wall style.

6. Left click in the display window to release the selection set.

7.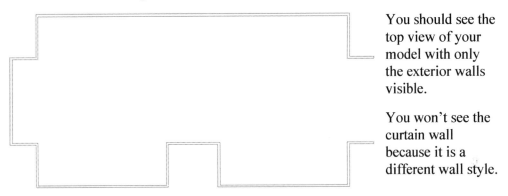

Select **Build → Roof → Roof by Footprint** from the Architecture ribbon.

Notice that **Pick Walls** is the active mode.

8.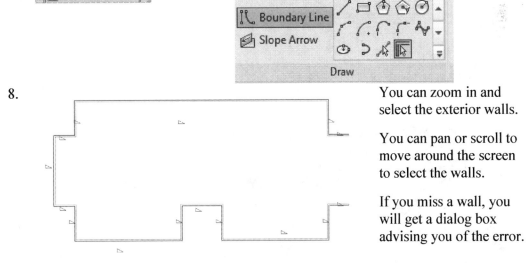

You can zoom in and select the exterior walls.

You can pan or scroll to move around the screen to select the walls.

If you miss a wall, you will get a dialog box advising you of the error.

You can use the ALIGN tool to align the roof edge with the exterior side of the wall.

9.

☑ Defines slope	Overhang: 0' 6"	☑ Extend to wall core

☑ Defines slope	Overhang: 60.0	☑ Extend to wall core

Hold down the CTL key and select all the roof lines.
On the options bar, enable **Defines slope**.
Set the Overhang to **6″ [60]**.
Enable **Extend to wall core**.

It is very important to enable the **Extend into wall (to core)** if you want your roof overhang to reference the structure of your wall.

For example, if you want a 2′-0″ overhang from the face of the stud, not face of the finish, you want this box checked. The overhang is located based on your pick point. If your pick point is the inside face, the overhang will be calculated from the interior finish face.

The angle symbols indicate the roof slope.

10. Select the **Line** tool from the Draw panel.

11. Draw a line to close the sketch.

Check the sketch to make sure there are no gaps or intersecting lines.

12. On the Properties palette:

Select the **Wood Rafter 8″ – Asphalt Shingle – Insulated [Warm Roof - Timber]** from the Type Selector.

13. 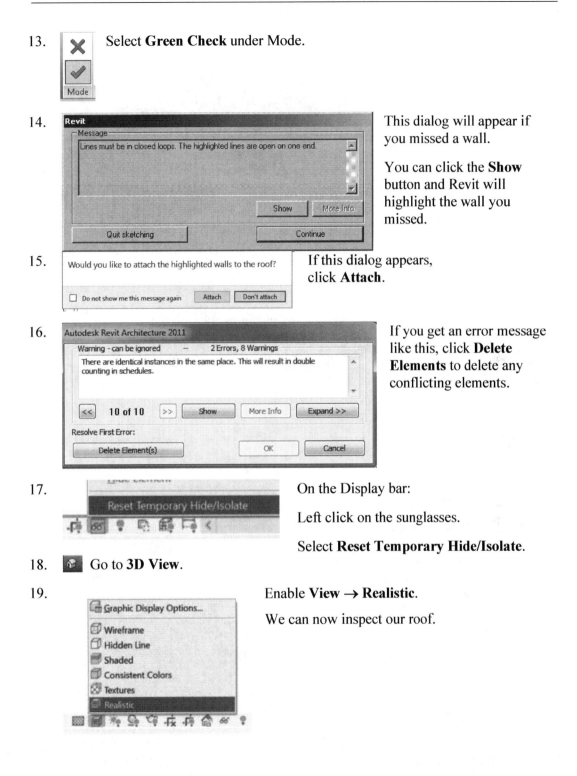 Select **Green Check** under Mode.

14. This dialog will appear if you missed a wall.

You can click the **Show** button and Revit will highlight the wall you missed.

15. If this dialog appears, click **Attach**.

16. If you get an error message like this, click **Delete Elements** to delete any conflicting elements.

17. On the Display bar:

Left click on the sunglasses.

Select **Reset Temporary Hide/Isolate**.

18. Go to **3D View**.

19. Enable **View → Realistic**.

We can now inspect our roof.

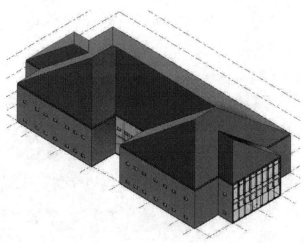

20.   Save as *ex7-1.rvt [m_ ex7-1.rvt]*.

The Cutoff Level Property defines the distance above or below the level at which the roof is cut off.

Exercise 7-2
Modifying a Roof

Drawing Name: roof_2.rvt [m_ roof_2.rvt]
Estimated Time: 10 minutes

This exercise reinforces the following skills:

- ☐   Modifying Roofs
- ☐   Edit Sketch
- ☐   Align Eaves
- ☐   Roof
- ☐   Work Plane

1.   Open *roof_2.rvt [m_ roof_2.rvt]*.

2.   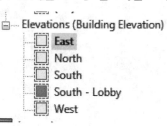   Activate the East elevation.

3. Select the **Level** tool from the Datum panel on the Architecture ribbon.

4. Select the **Pick Line** tool from the Draw panel.

5. Set the Offset to **10' 6" [2700mm]**.

6. Select the Roof Line level.

Cancel out of the command.

7. Rename the Level **Roof Cutoff**.

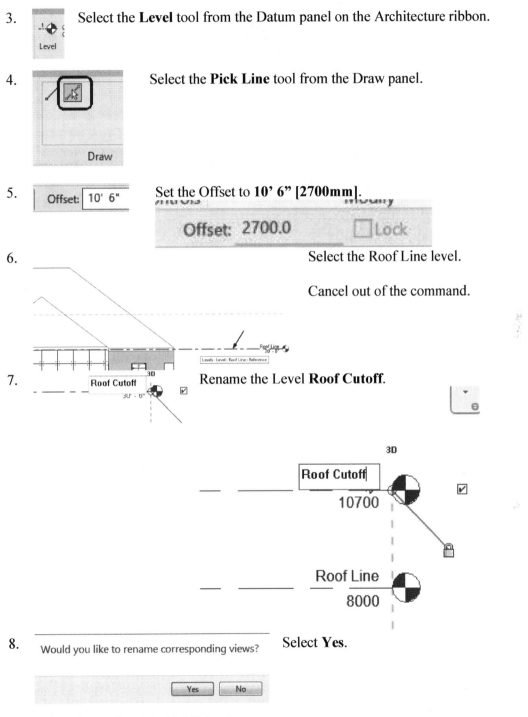

8. Would you like to rename corresponding views? Select **Yes**.

9. Select the roof so it is highlighted.

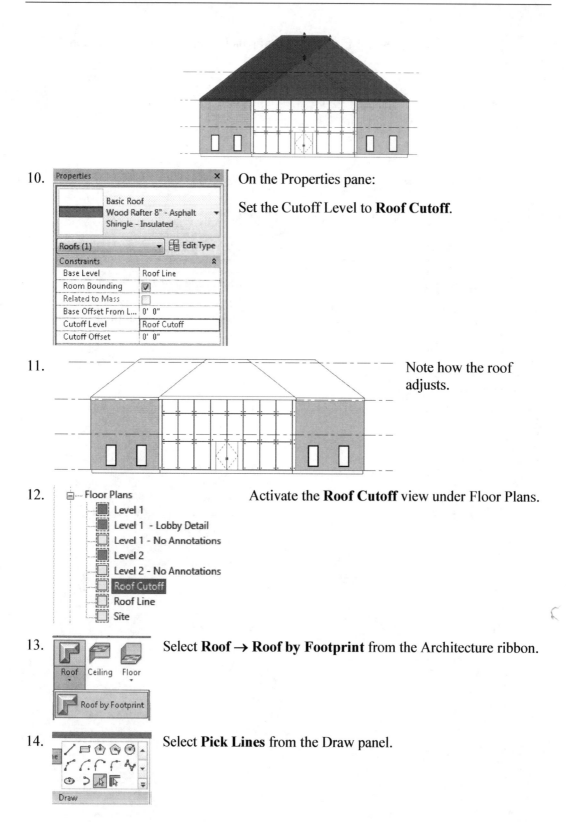

10. On the Properties pane:

Set the Cutoff Level to **Roof Cutoff**.

11. Note how the roof adjusts.

12. Activate the **Roof Cutoff** view under Floor Plans.

13. Select **Roof → Roof by Footprint** from the Architecture ribbon.

14. Select **Pick Lines** from the Draw panel.

15. 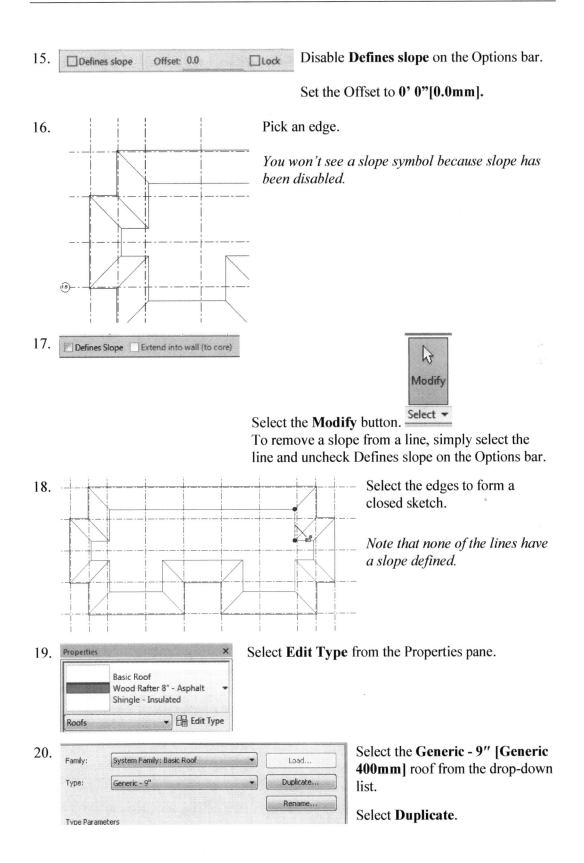 Disable **Defines slope** on the Options bar.

Set the Offset to **0' 0"[0.0mm]**.

16. Pick an edge.

You won't see a slope symbol because slope has been disabled.

17. Select the **Modify** button.
To remove a slope from a line, simply select the line and uncheck Defines slope on the Options bar.

18. Select the edges to form a closed sketch.

Note that none of the lines have a slope defined.

19. Select **Edit Type** from the Properties pane.

Basic Roof
Wood Rafter 8" - Asphalt
Shingle - Insulated

Roofs   Edit Type

20. Select the **Generic - 9" [Generic 400mm]** roof from the drop-down list.

Family: System Family: Basic Roof   Load...

Type: Generic - 9"   Duplicate...

                                           Rename...

Type Parameters

Select **Duplicate**.

Family:	System Family: Basic Roof	▼	Load...
Type:	Generic - 400mm	▼	Duplicate...
			Rename...

21. Name: Tar and Gravel

Enter **Tar and Gravel**.

Click **OK**.

OK

22. Type Parameters

Parameter	Value
Construction	
Structure	Edit...
Default Thickness	0' 9"

Select **Edit** next to Structure.

23. Insert three layers.

Layers

	Function	Material	Thickness
1	Finish 1 [4]	<By Catego	0' 0"
2	**Core Boundary**	**Layers Above**	**0' 0"**
3	Membrane Layer	<By Catego	0' 0"
4	Substrate [2]	<By Catego	0' 0"
5	Structure [1]	<By Catego	0' 9"
6	**Core Boundary**	**Layers Below**	**0' 0"**

Arrange the layers so:
Layer 1 is set to Finish 1 [4].
Layer 2 is Core Boundary.
Layer 3 is Membrane Layer.
Layer 4 is Substrate [2]
Layer 5 is Structure [1]
Layer 6 is Core Boundary.

Layers

	Function	Material	Thickness
1	Finish 1 [4]	<By Category>	0.0
2	**Core Boundary**	**Layers Above Wrap**	**0.0**
3	Membrane Layer	<By Category>	0.0
4	Substrate [2]	<By Category>	0.0
5	Structure [1]	<By Category>	400.0
6	**Core Boundary**	**Layers Below Wrap**	**0.0**

24. Select the Material column for **Finish 1 [4]**.

25. gravel

Project Materials: All ▼

Search results for "gravel"

	Name
	Gravel

Locate the **gravel** material in the Document materials.

Click **OK**.

26. Layers

	Function	Material	Thickness	\
1	Finish 1 [4]	Gravel	0' 1/4"	

Set the thickness to **1/4" [6mm]**.

	Material	Thickness
1	Gravel	6.0
2	**Layers Above Wrap**	**0.0**
3	Roofing, EPDM Membrane	0.0
4	Plywood, Sheathing	16.0
5	Structure, Wood Joist/Rafter L	200.0
6	**Layers Below Wrap**	**0.0**

27. Select the Material column for **Membrane Layer**.

    Do a search for **membrane**.

28. 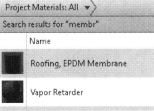 Set the Material to **Roofing - EPDM Membrane**. Click **OK**.

    Set the Membrane thickness to **0″[0.0mm]**.

	Function	Material	Thickness
1	Finish 1 [4]	Gravel	0' 0 1/4"
2	**Core Boundary**	**Layers Above Wrap**	**0' 0"**
3	Membrane Layer	Roofing - EPDM Membrane	0' 0"
4	Substrate [2]	<By Category>	0' 0"
5	Structure [1]	<By Category>	0' 9"
6	**Core Boundary**	**Layers Below Wrap**	**0' 0"**

	Material	Thickness
1	Gravel	6.0
2	**Layers Above Wrap**	**0.0**
3	Roofing, EPDM Membrane	0.0
4	Plywood, Sheathing	16.0
5	Structure, Wood Joist/Rafter L	200.0
6	**Layers Below Wrap**	**0.0**

29. 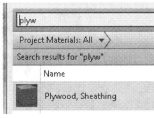 Set Layer 4 Substrate [2] to **Plywood, Sheathing** with a thickness of **5/8″ [16mm]**.

	Material	Thickness
1	Gravel	6.0
2	**Layers Above Wrap**	**0.0**
3	Roofing, EPDM Membrane	0.0
4	Plywood, Sheathing	16.0
5	Structure, Wood Joist/Rafter L	200.0
6	**Layers Below Wrap**	**0.0**

30.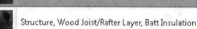

Set Layer 5 Structure [1] to **Structure-Wood Joist/ Rafter Layer [Structure, Timber Joist/Rafter Layer]** with a thickness of **7 ½″ [200mm]**.

Project Materials: All ▼

Search results for "wood joi"

Name
Structure, Wood Joist/Rafter Layer
Structure, Wood Joist/Rafter Layer, Batt Insulation

Search results for "wood joi"

Name
Structure, Timber Joist/Rafter Layer

31. Click **OK**.

Layers

	Function	Material	Thickness	Wraps	Vari
1	Finish 1 [4]	Gravel	0' 0 1/4"	☐	☐
2	Core Boundary	Layers Above Wrap	0' 0"		
3	Membrane Lay	Roofing, EPDM Membr	0' 0"	☐	☐
4	Substrate [2]	Plywood, Sheathing	0' 0 5/8"	☐	☐
5	Structure [1]	Structure, Wood Joist/R	0' 7 1/2"	☐	☐
6	Core Boundary	Layers Below Wrap	0' 0"		

Layers

	Material	Thickness
1	Gravel	6.0
2	Layers Above Wrap	0.0
3	Roofing, EPDM Membrane	0.0
4	Plywood, Sheathing	16.0
5	Structure, Wood Joist/Rafter L	200.0
6	Layers Below Wrap	0.0

32.

Graphics ☆

Coarse Scale Fill Pattern	
Coarse Scale Fill Color	■ Black

Click the Browse button in the Coarse Scale Fill Pattern to assign a fill pattern.

33. Click **New** to define a new fill pattern.

34. Name: New pattern name

Type: ○ Basic
      ● Custom

Settings
Search 🔍

Browse...

Enable **Custom**.

Click **Browse**.

35. File name: gravel

Files of type: Hatch Patterns (*.pat)

Select the *gravel* hatch pattern in the exercise files. Click **Open**.

36.

Change the Name to **Gravel**.

Set the Import scale to **0.03**.

Note how the preview changes.

Click **OK**.

37. Highlight Gravel.

Click **OK**.

38. Assign the **Gravel** pattern to the Coarse Scale Fill Pattern.

Click **OK**.

Click **OK** to close the Type Properties dialog box.

39. Select the **Green Check** under the Model panel to **Finish Roof**.

40. Zoom in to see the hatch pattern.

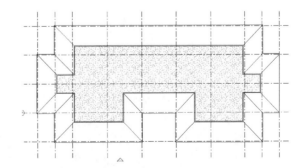

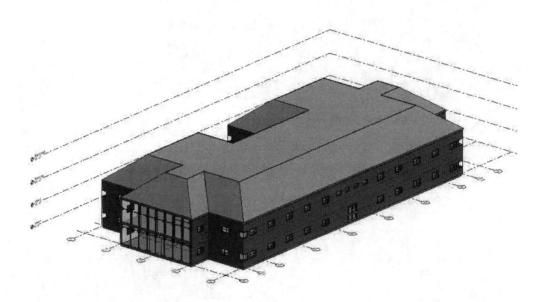

41. Save as *ex7-2.rvt [m_ ex7-2.rvt]*.

Exercise 7-3
Modifying a Roof Form

Drawing Name: roof3.rvt [m_ roof3.rvt]
Estimated Time: 10 minutes

This exercise reinforces the following skills:

☐ Shape Editing

1. Open *roof3.rvt [m_ roof3.rvt]*.

2. 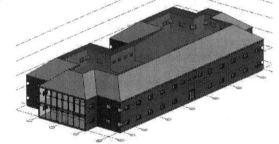 Activate a **3D** view.

3.  Select the tar and gravel roof.

    It will highlight in blue and appear to be transparent.

4.   Select the **Add Point** tool on the **Shape Editing** panel on the ribbon.

5. Place points as shown.

6. Select **Modify Sub Elements**.

7.

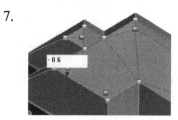

Select each point and set them to **-0' 6" [10800]**.

These will be drain holes.

Right click and select Cancel when you are done.

8. The roof surface will show the drainage areas.

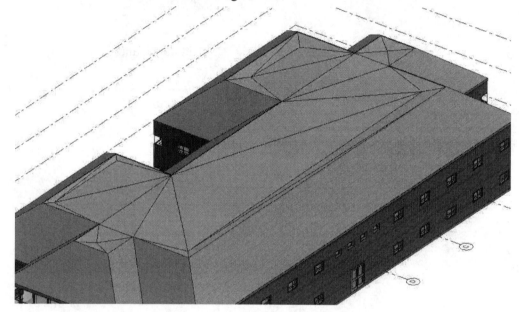

9. Save as *ex7-3.rvt [m_ ex7-3.rvt]*.

Exercise 7-4
Adding Roof Drains

Drawing Name: roof drain.rvt [m_ roof drain.rvt]
Estimated Time: 10 minutes

This exercise reinforces the following skills:

- ❑ Component
- ❑ Load Family

1.   Open *roof drain.rvt [m_ roof drain.rvt]*.

2.   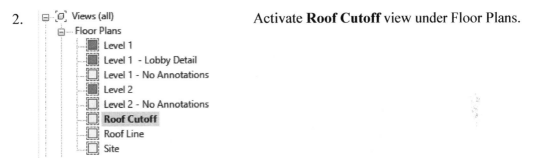   Activate **Roof Cutoff** view under Floor Plans.

3.   1/8" = 1'-0"   Change the display settings to **Medium, Wireframe**.

4.   Select **Place a Component** from the Build panel on the Architecture ribbon.

   Component

5.   Select **Load Family** under the Mode panel.

   Load Family

6.   File name: Drain-Roof

   Files of type: All Supported Files (*.rfa, *.adsk)

   Locate the *Drain-Roof* from the Class Files downloaded from the publisher's website.

   Click **Open**.

7.   There is no tag loaded for Plumbing Fixtures. Do you want to load one now?

   Yes   No

   If a dialog appears asking if you want to load a tag for Plumbing Fixtures, click **No**.

8. Place roof drains coincident to the points placed earlier.

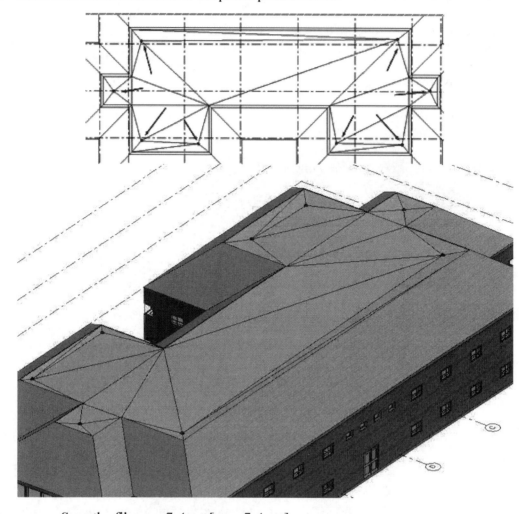

9. Save the file as *ex7-4.rvt [m_ex7-4.rvt]*.

Exercise 7-5
Adding a Gable by Modifying a Roof Sketch

Drawing Name: add_gable.rvt [m_add gable.rvt]
Estimated Time: 10 minutes

This exercise reinforces the following skills:

    ❑   Modify a Roof Sketch

    A video for this exercise is available on my MossDesigns station on Youtube.com.

1.   Open *add_gable.rvt [m_add gable.rvt]*.

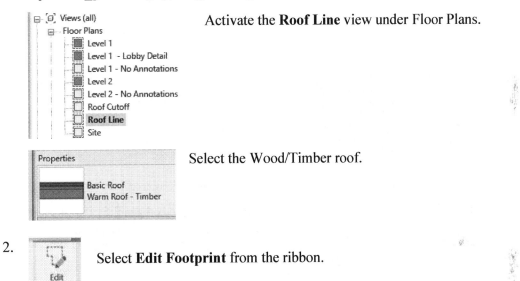

Activate the **Roof Line** view under Floor Plans.

Select the Wood/Timber roof.

2.   Select **Edit Footprint** from the ribbon.

      This returns you to sketch mode.

3.   Select the **Split** tool on the Modify panel.

4.   Split the roof on the south side at the Grids B & C.

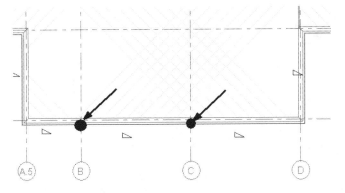

5. Select the new line segment created by the split.

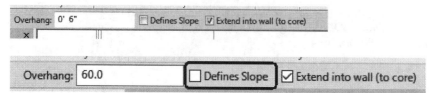

Uncheck **Defines Slope** on the option bar.

6.   Slope Arrow   Select the **Slope Arrow** tool on the ribbon.

7. 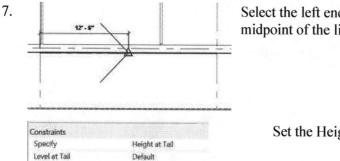 Select the left end point of the line and the midpoint of the line to place the slope.

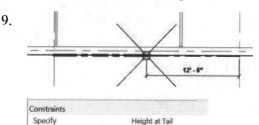 Set the Height Offset at Head to **2700**.

8.   Slope Arrow   Select the **Slope Arrow** tool on the ribbon.

9. Select the right end point of the line and the midpoint of the line to place the slope.

Set the Height Offset at Head to **2700**.

10. The sketch should appear as shown.

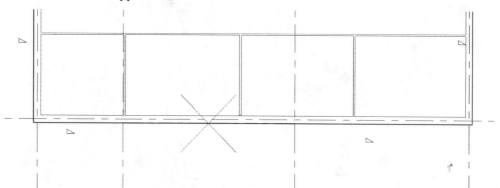

11. 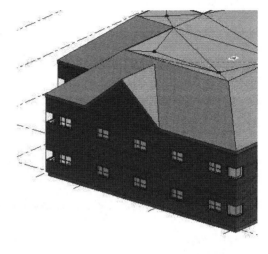 Select the green check to exit sketch mode.

12. Switch to a 3D view and orbit to inspect the new gable.

Note that the wall automatically adjusted.

Save as *ex7-5.rvt [m_ex7-5.rvt]*.

Exercise 7-6
Create a Shed Roof

Drawing Name: shed_roof.rvt
Estimated Time: 10 minutes

This exercise reinforces the following skills:

- ❑ Create Roof
- ❑ Modify Slope

1. Open *shed_roof.rvt*.

2.  Activate the **Level 2** floor plan.

3. Switch to the Architecture ribbon.

Select **Roof by Footprint**.

4. 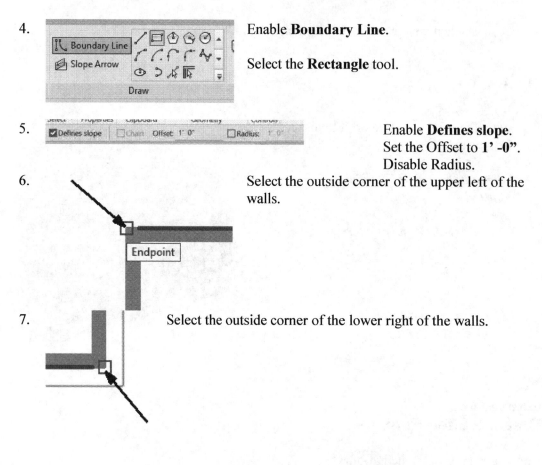 Enable **Boundary Line**.

    Select the **Rectangle** tool.

5. Enable **Defines slope**.
    Set the Offset to **1' -0"**.
    Disable Radius.

6. Select the outside corner of the upper left of the walls.

7. Select the outside corner of the lower right of the walls.

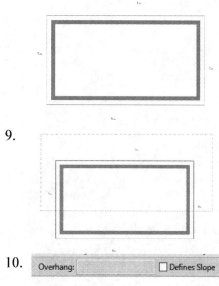

8. Cancel out of the command.

    You should see an offset rectangle with slope symbols on each side.

9. Use a crossing to select the two vertical lines and the top horizontal line.

10. Disable the **Defines Slope** option to remove the slope for those three sides.

    Click ESC to release the selection.

11.

<Sketch> (1)	
Constraints	
Defines Roof Slope	☑
Offset From Roof Base	0' 0"
Dimensions	
Slope	4 1/2" / 12"
Length	32' 7 61/64"

Click on the lower horizontal line.
This line still has a slope defined.
Change the value of the Slope to **4.5"/12"**.
This is approximately a 20% pitch.

12.

Select the **Green Check** to complete the roof.

13.

Attaching to roof ✕
Would you like to attach the highlighted walls to the roof?
☐ Do not show me this message again   [Attach] [Don't attach]

Click **Attach**.

14. 3D Views
{3D}

Switch to a 3D view.

15.

The roof is created.

Save as *ex7-6.rvt*.

Exercise 7-7
Create a Gable Roof

Drawing Name: gable_roof.rvt
Estimated Time: 10 minutes

This exercise reinforces the following skills:

- ❑ Create Roof
- ❑ Modify Slope

1. Open *gable_roof.rvt*.

2. 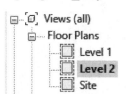   Activate the **Level 2** floor plan.

3. Switch to the Architecture ribbon.

    Select **Roof by Footprint**.

4. Enable **Boundary Line**.

    Select the **Rectangle** tool.

5. 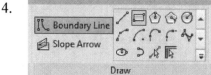   Enable **Defines slope**.
    Set the Offset to **1.0**.
    Disable Radius.

6. Select the outside corner of the upper left of the walls.

    Endpoint

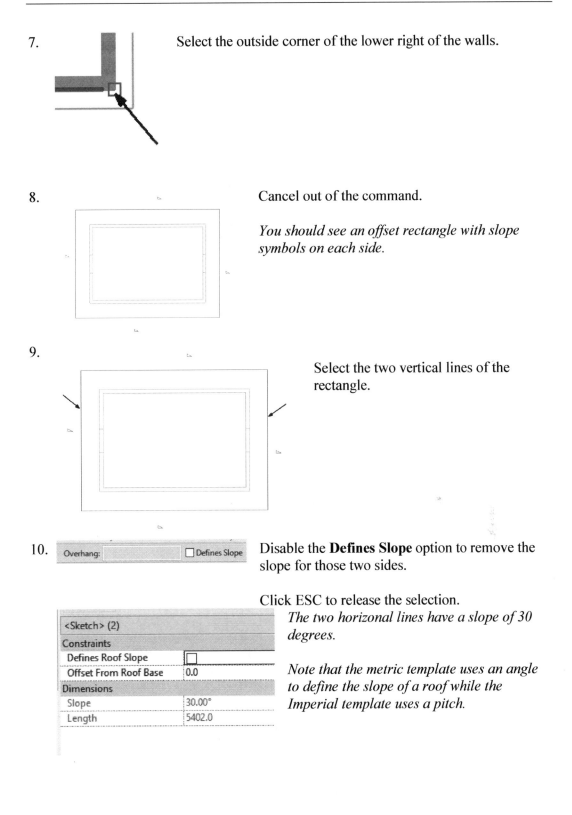

7. Select the outside corner of the lower right of the walls.

8. Cancel out of the command.

   You should see an offset rectangle with slope symbols on each side.

9. Select the two vertical lines of the rectangle.

10. Disable the **Defines Slope** option to remove the slope for those two sides.

    Click ESC to release the selection.
    The two horizonal lines have a slope of 30 degrees.

    Note that the metric template uses an angle to define the slope of a roof while the Imperial template uses a pitch.

11. 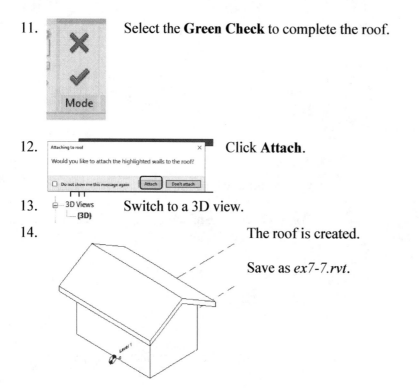 Select the **Green Check** to complete the roof.

12. Click **Attach**.

13. Switch to a 3D view.

14. The roof is created.

Save as *ex7-7.rvt*.

Exercise 7-8
Create a Gable Roof with Cat Slide

Drawing Name: gable_roof_with_cat_slide.rvt
Estimated Time: 10 minutes

This exercise reinforces the following skills:

- ❑ Create Roof
- ❑ Modify Slope

1.   Open *gable_roof_with_cat_slide.rvt*.

2.   Activate the **Level 2** floor plan.

3.   Switch to the Architecture ribbon.

     Select **Roof by Footprint**.

4.   Enable **Boundary Line**.

     Select the **Rectangle** too.

5.   Enable **Defines slope**.
     Set the Offset to **1.0**.
     Disable Radius.

6.   Place a rectangle by selecting the upper left and lower right corners of the building.

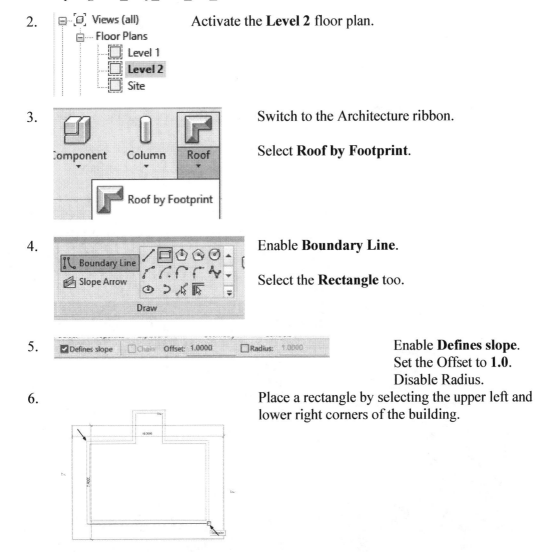

7.

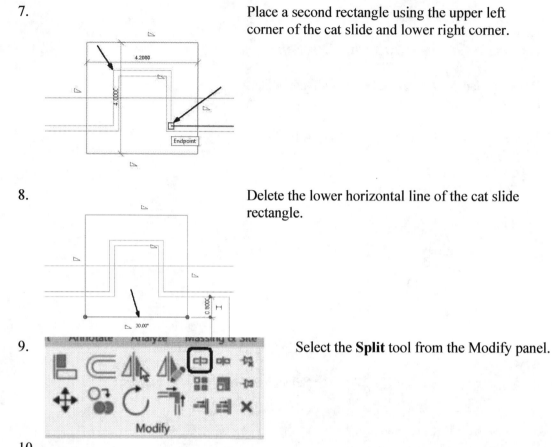

Place a second rectangle using the upper left corner of the cat slide and lower right corner.

8.

Delete the lower horizontal line of the cat slide rectangle.

9.

Select the **Split** tool from the Modify panel.

10.

Select a point on the lower line above the cat slide to split that line.

11.

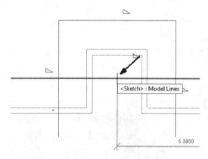

Select the **Trim** tool to create corners from the overlapping lines.

12.

Select the lines indicated to trim.

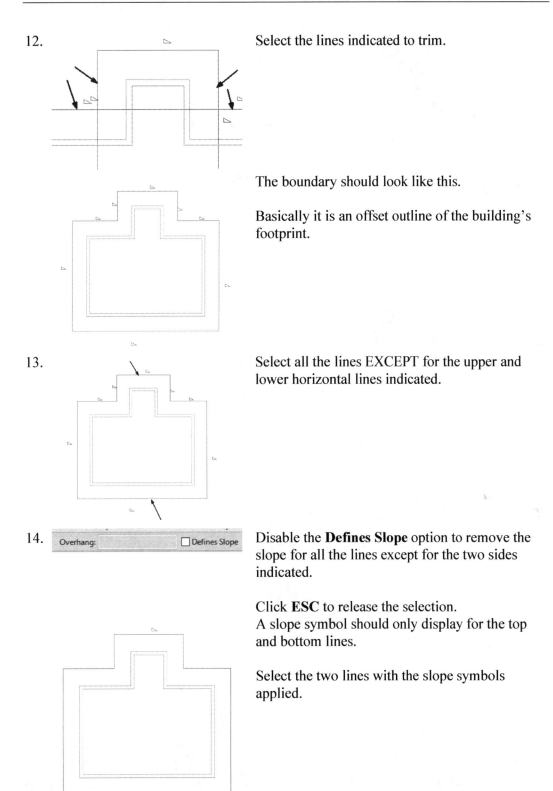

The boundary should look like this.

Basically it is an offset outline of the building's footprint.

13.

Select all the lines EXCEPT for the upper and lower horizontal lines indicated.

14.

Overhang: ☐ Defines Slope

Disable the **Defines Slope** option to remove the slope for all the lines except for the two sides indicated.

Click **ESC** to release the selection.
A slope symbol should only display for the top and bottom lines.

Select the two lines with the slope symbols applied.

15.

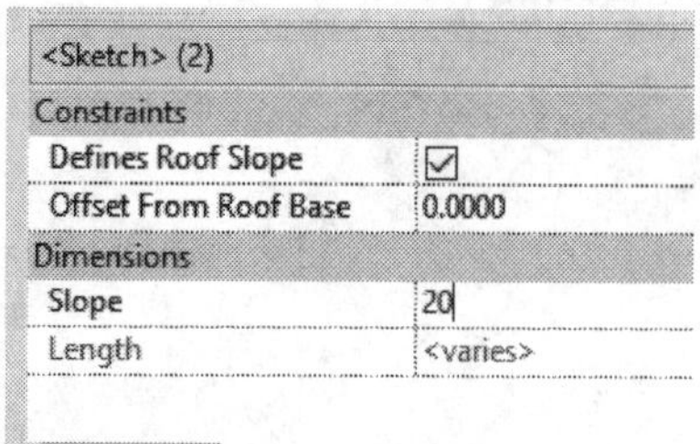

<Sketch> (2)	
Constraints	
Defines Roof Slope	☑
Offset From Roof Base	0.0000
Dimensions	
Slope	20
Length	<varies>

Change the slope to **20 degrees**.

Click **Apply**.

Click **ESC** to release the selection.

16.

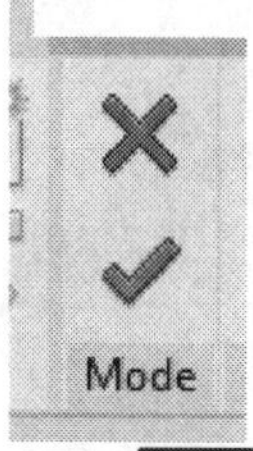

Select the **Green Check** to complete the roof.

17.

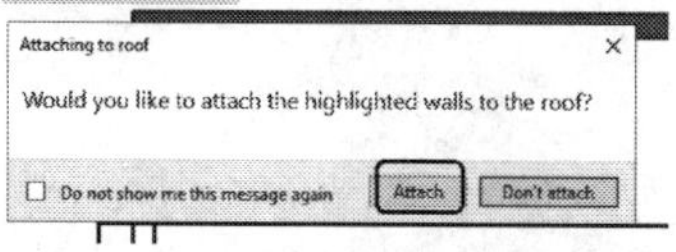

Attaching to roof ✕

Would you like to attach the highlighted walls to the roof?

☐ Do not show me this message again   Attach   Don't attach

Click **Attach**.

18.

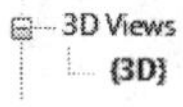

⊟ 3D Views
   {3D}

Switch to a 3D view.

19.

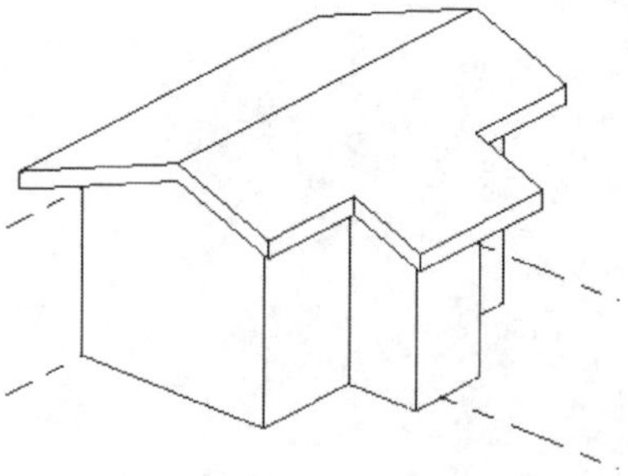

The roof is created.

Save as *ex7-8.rvt*.

Exercise 7-9
Create a Gambrel Roof

Drawing Name: gambrel_roof.rvt
Estimated Time: 20 minutes

This exercise reinforces the following skills:

- ❏ Create Roof by Extrusion
- ❏ Modify Slope
- ❏ Reference Plane
- ❏ Section View
- ❏ Mirror-Pick Axis
- ❏ Attach Walls

1. Open *gambrel_roof.rvt*.

2. 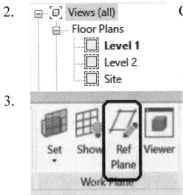 Open the **Level 1** Floor Plan.

3.   Switch to the Architecture tab.

    Select the **Ref Plane** tool.

4. 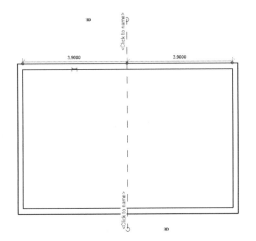 Draw a reference plane through the middle of the building model.

    Notice that there is a distance of 3.9m on each side of the reference plane.

5.

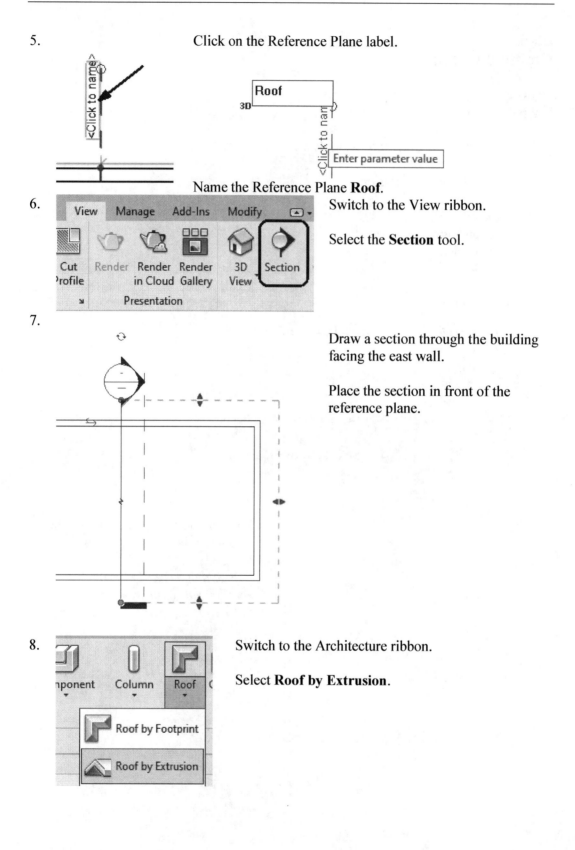

Click on the Reference Plane label.

Name the Reference Plane **Roof**.

6. Switch to the View ribbon.

Select the **Section** tool.

7. Draw a section through the building facing the east wall.

Place the section in front of the reference plane.

8. Switch to the Architecture ribbon.

Select **Roof by Extrusion**.

9. Specify a new Work Plane

- ⦿ Name   Reference Plane : Roof
- ○ Pick a plane
- ○ Pick a line and use the work plane it was sketched in

Enable **Name**.
Select the Reference Plane that was placed.

This only works if you added a name to the reference plane. If you didn't add a name, you can enable Pick a plane and select the plane.

Click **OK**.

10. Elevation: East
Elevation: West
Section: Section 1

Open View

Highlight **Section 1**.

This is the section you created.

Click **Open View**.

11. Level: Level 2

Offset: 0.0000

OK

Click **OK**.

12. Draw

Select the **Pick Line** tool on the Draw panel.

13. Level 2

4

Walls : Basic Wall : Generic - 200mm : Shape handle

Select the top of the east wall.

14. 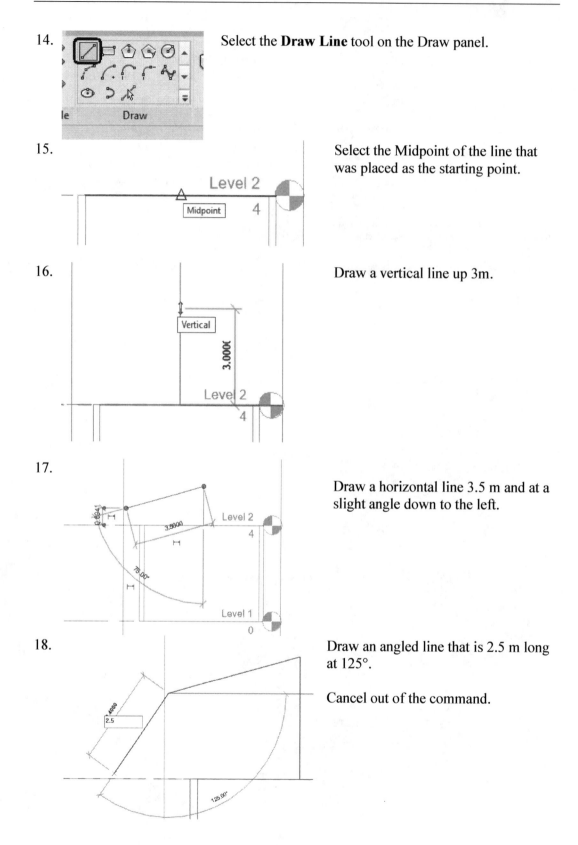 Select the **Draw Line** tool on the Draw panel.

15. Select the Midpoint of the line that was placed as the starting point.

16. Draw a vertical line up 3m.

17. Draw a horizontal line 3.5 m and at a slight angle down to the left.

18. Draw an angled line that is 2.5 m long at 125°.

Cancel out of the command.

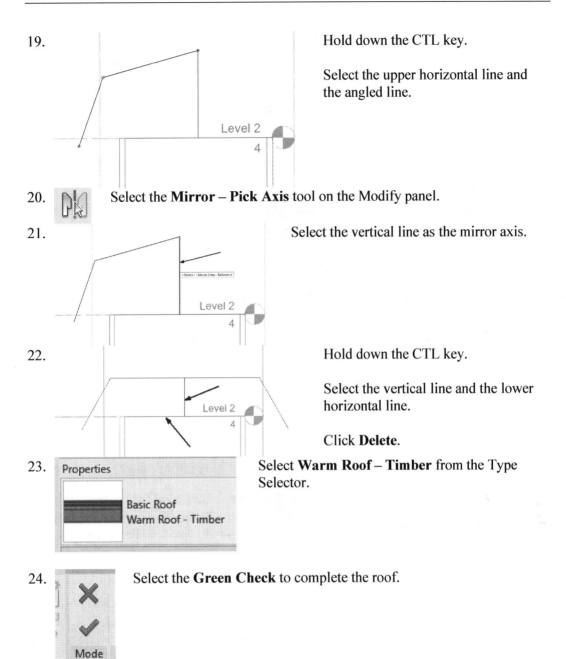

19. Hold down the CTL key.

   Select the upper horizontal line and the angled line.

20. Select the **Mirror – Pick Axis** tool on the Modify panel.

21. Select the vertical line as the mirror axis.

22. Hold down the CTL key.

   Select the vertical line and the lower horizontal line.

   Click **Delete**.

23. Select **Warm Roof – Timber** from the Type Selector.

24. Select the **Green Check** to complete the roof.

25. Switch to a 3D view.

26.

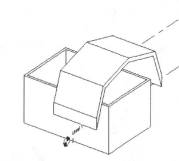

The roof is created.

Select the roof.

27.

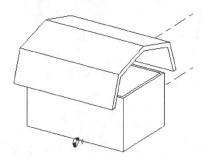

Change the Extrusion Start to **-4.2**.
Change the Extrusion End to **4.2**.

Change the Rafter Cut to **Two Cut – Plumb.**

Click **Apply** to see the updates applied.

28.

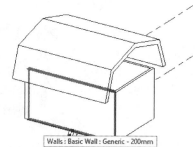

Click ESC to release the selection.

29.

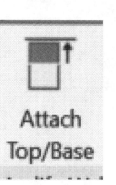

Hover the mouse over one of the walls.

Click the TAB key to select all the walls.

30.

Select **Attach Top/Base** on the ribbon.

Select the roof as the boundary.

31.

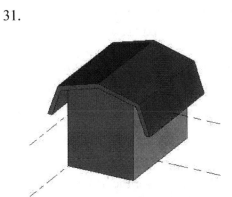

The walls will adjust.

Orbit the model to inspect the roof.

Save as *ex7-9.rvt*.

Exercise 7-10
Create a Clerestory Roof

Drawing Name: clerestory_roof.rvt
Estimated Time: 45 minutes

This exercise reinforces the following skills:

- ☐ Create Roof by Footprint
- ☐ Modify Slope
- ☐ Section View
- ☐ Add Level
- ☐ Attach Walls

1. Open *celerestory_roof.rvt*.

2. Activate the **Level 2** floor plan.

   Views (all)
   Floor Plans
   Level 1
   Level 2
   Site

3. Switch to the Architecture ribbon.

   Select **Roof by Footprint**.

4.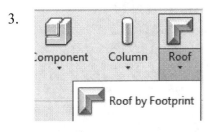

   Enable **Boundary Line**.

   Select the **Rectangle** too.

5.

Enable **Defines slope**.
Set the Offset to **1.5**.
Disable Radius.

6.

Select the upper left corner as the start point for the rectangle.

7.

Hover over the inside right corner and click SPACE on the keyboard.

This will offset the rectangle inside the building model.

Select the corner.

Click ESC to exit the command.

8.

Select the **Align** tool on the Modify panel.

9.

Disable **Multiple Alignment.**
Prefer **Wall faces**.

10.

Align the left side of the rectangle to the outside left/west wall.

11.

Align the right side of the rectangle to the outside right/east wall.

12.

Hold down the CTL key.
Select the two vertical lines.

13.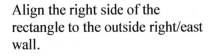

Disable the **Defines Slope** option to remove the slope for the selected lines.

Click **ESC** to release the selection.

The two horizontal lines should have a slope of 30°.

<Sketch> (1)	
Constraints	
Defines Roof Slope	☑
Offset From Roof Base	0.0000
Dimensions	
Slope	30.00°
Length	8.2000

14.

Select the **Green Check** to complete the roof.

Mode

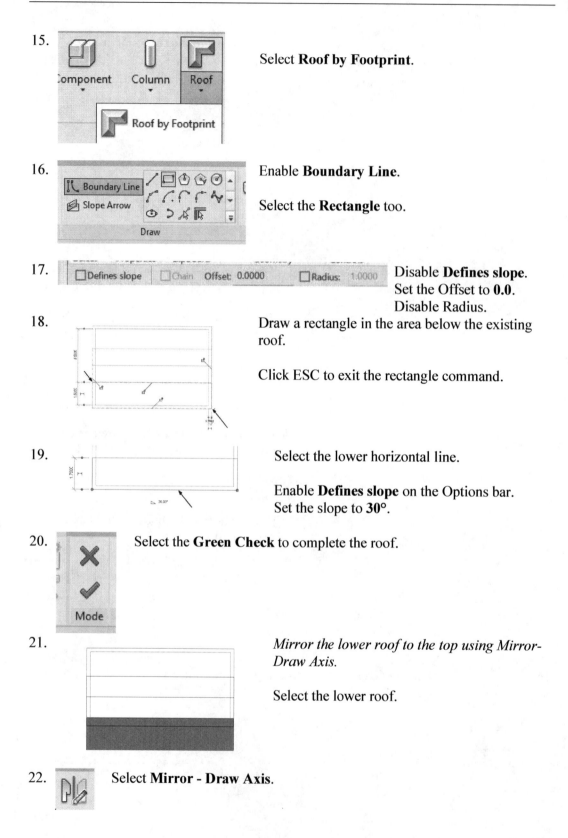

15. Select **Roof by Footprint**.

16. Enable **Boundary Line**.

    Select the **Rectangle** too.

17. Disable **Defines slope**.
    Set the Offset to **0.0**.
    Disable Radius.

18. Draw a rectangle in the area below the existing roof.

    Click ESC to exit the rectangle command.

19. Select the lower horizontal line.

    Enable **Defines slope** on the Options bar.
    Set the slope to **30°**.

20. Select the **Green Check** to complete the roof.

21. *Mirror the lower roof to the top using Mirror-Draw Axis.*

    Select the lower roof.

22. Select **Mirror - Draw Axis**.

23.

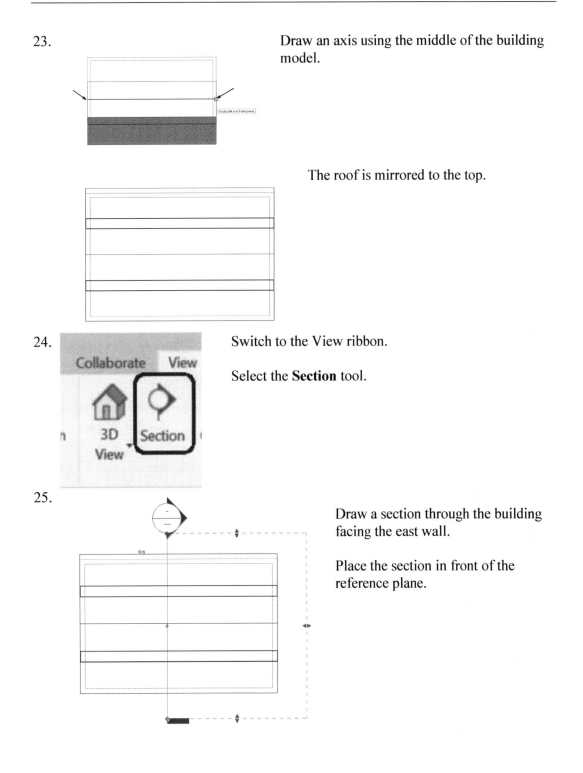

Draw an axis using the middle of the building model.

The roof is mirrored to the top.

24.

Switch to the View ribbon.

Select the **Section** tool.

25.

Draw a section through the building facing the east wall.

Place the section in front of the reference plane.

26.

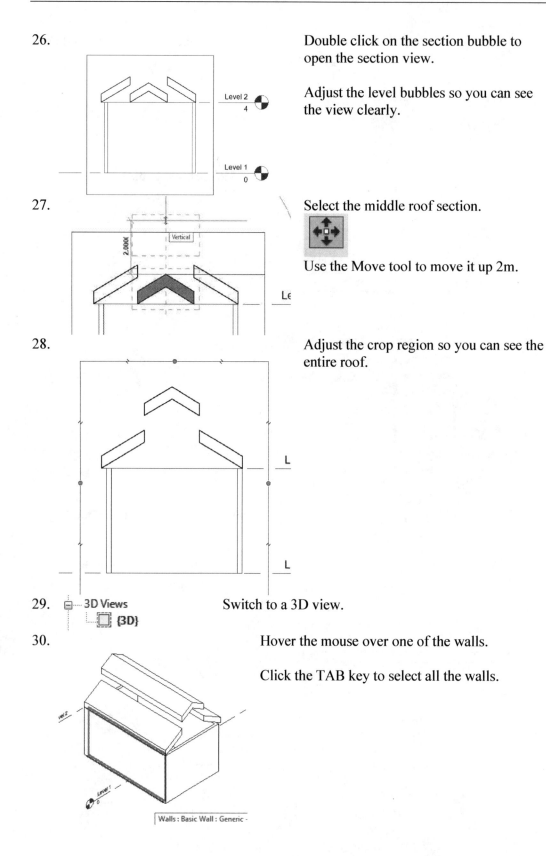

Double click on the section bubble to open the section view.

Adjust the level bubbles so you can see the view clearly.

27.

Select the middle roof section.

Use the Move tool to move it up 2m.

28.

Adjust the crop region so you can see the entire roof.

29. 3D Views
      {3D}

Switch to a 3D view.

30.

Hover the mouse over one of the walls.

Click the TAB key to select all the walls.

Walls : Basic Wall : Generic -

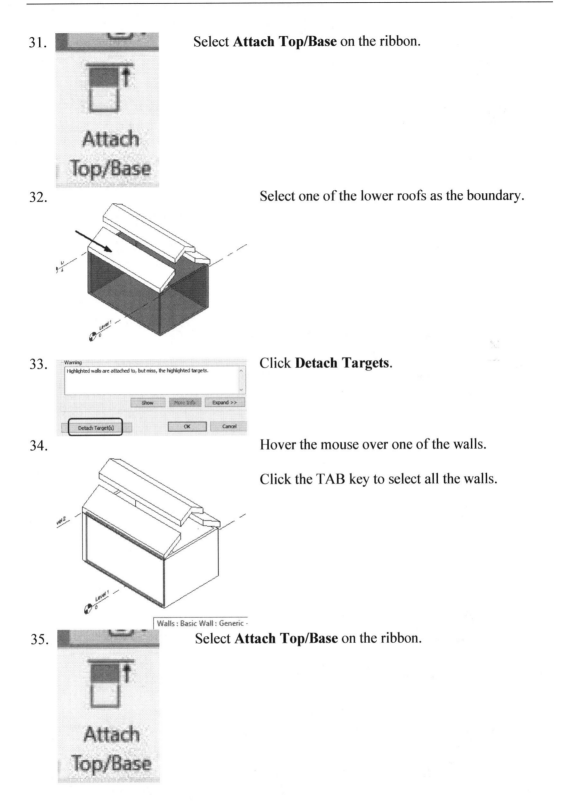

31.   Select **Attach Top/Base** on the ribbon.

32.   Select one of the lower roofs as the boundary.

33.   Click **Detach Targets**.

34.   Hover the mouse over one of the walls.

       Click the TAB key to select all the walls.

35.   Select **Attach Top/Base** on the ribbon.

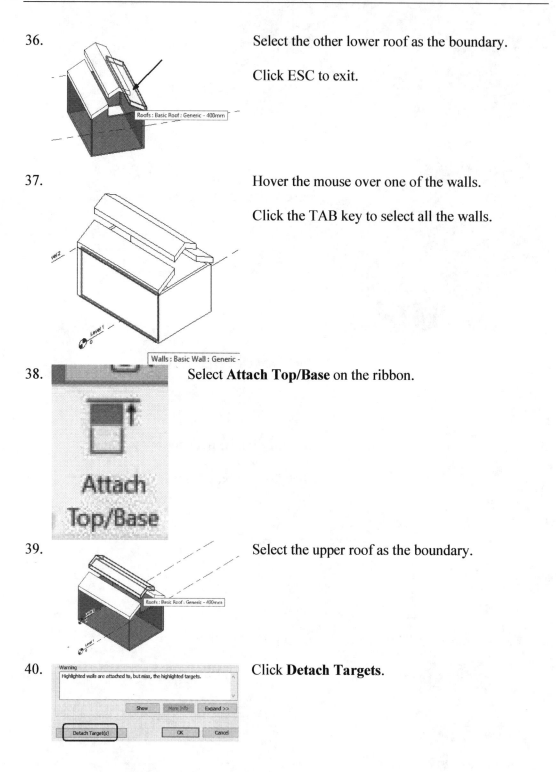

36.   Select the other lower roof as the boundary.

Click ESC to exit.

37.   Hover the mouse over one of the walls.

Click the TAB key to select all the walls.

38.   Select **Attach Top/Base** on the ribbon.

39.   Select the upper roof as the boundary.

40.   Click **Detach Targets**.

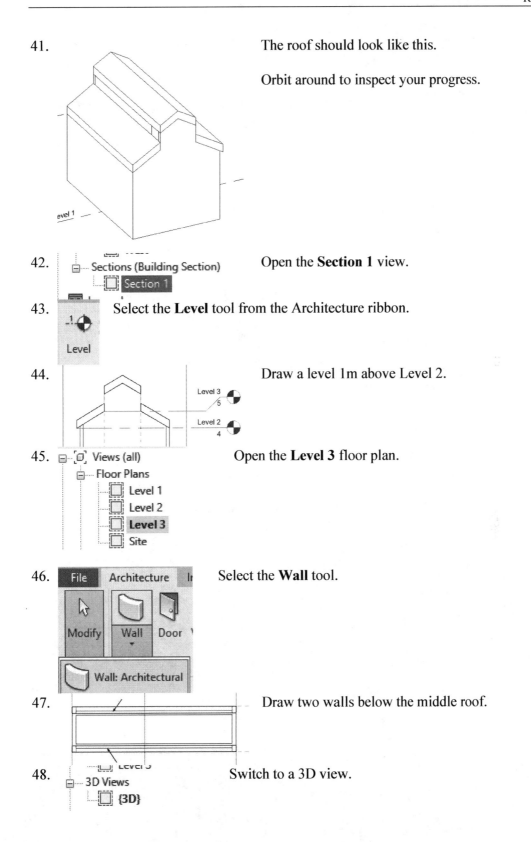

41. The roof should look like this.

Orbit around to inspect your progress.

42. Sections (Building Section) — Section 1

Open the **Section 1** view.

43. Select the **Level** tool from the Architecture ribbon.

Level

44. Draw a level 1m above Level 2.

Level 3
5

Level 2
4

45. Views (all)
Floor Plans
Level 1
Level 2
Level 3
Site

Open the **Level 3** floor plan.

46. File Architecture I

Modify Wall Door

Wall: Architectural

Select the **Wall** tool.

47. Draw two walls below the middle roof.

48. Level 3
3D Views
{3D}

Switch to a 3D view.

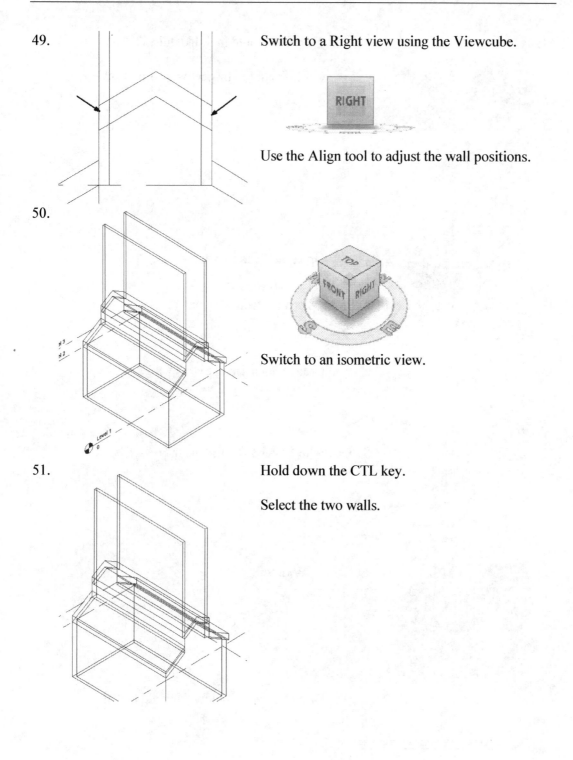

49.

Switch to a Right view using the Viewcube.

Use the Align tool to adjust the wall positions.

50.

Switch to an isometric view.

51.

Hold down the CTL key.

Select the two walls.

52. 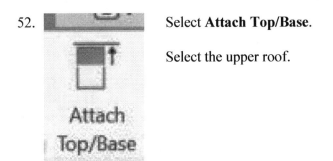 Select **Attach Top/Base**.

Select the upper roof.

53. 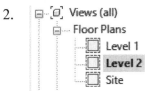 Save as *ex7-10.rvt*.

Exercise 7-11
Create a Half-Hip Roof

Drawing Name: half-hip_roof.rvt
Estimated Time: 20 minutes

This exercise reinforces the following skills:

- ❑ Create Roof
- ❑ Modify Slope
- ❑ Reference Planes

1. Open *half-hip_roof.rvt*.

2.   Views (all)
       Floor Plans
         Level 1
         Level 2
         Site

Activate the **Level 2** floor plan.

3.

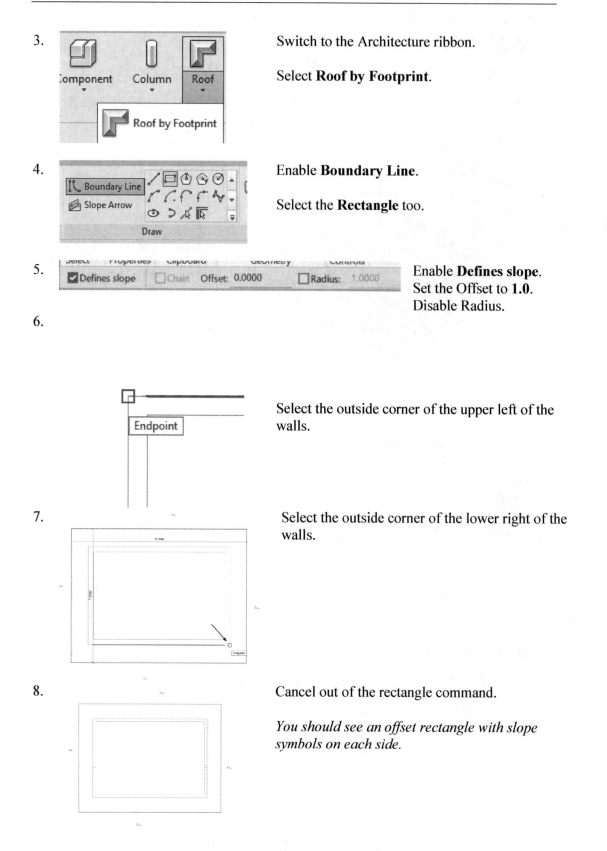

Switch to the Architecture ribbon.

Select **Roof by Footprint**.

4.

Enable **Boundary Line**.

Select the **Rectangle** too.

5.

Enable **Defines slope**.
Set the Offset to **1.0**.
Disable Radius.

6.

Select the outside corner of the upper left of the walls.

7.

Select the outside corner of the lower right of the walls.

8.

Cancel out of the rectangle command.

You should see an offset rectangle with slope symbols on each side.

9. Select the Ref. Plane tool from the ribbon.

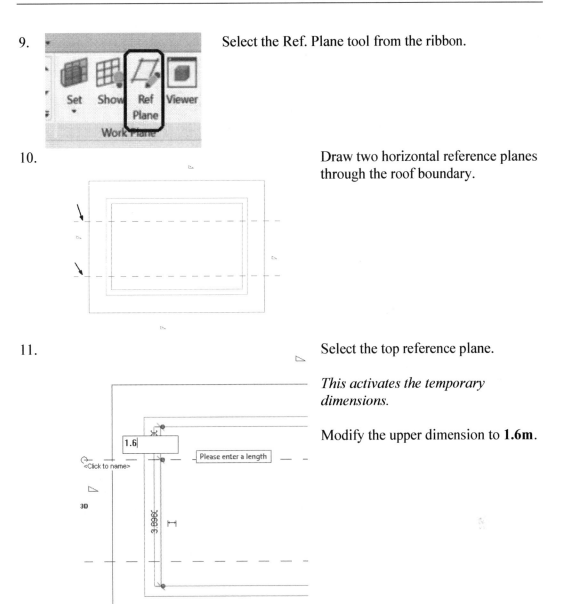

10. Draw two horizontal reference planes through the roof boundary.

11. Select the top reference plane.

This activates the temporary dimensions.

Modify the upper dimension to **1.6m**.

12.

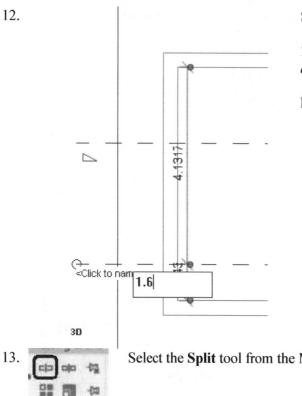

3D

Select the lower reference plane.

This activates the temporary dimensions.

Modify the lower dimension to **1.6m**.

13.

Select the **Split** tool from the Modify panel.

14.

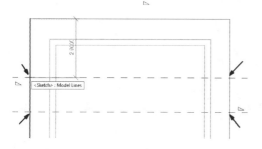

Add splits to the boundary at the intersection points of the reference planes and the roof boundary.

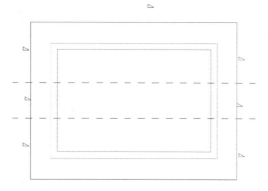

Notice that slope symbols were added to the new segments that were created by the split.

15.

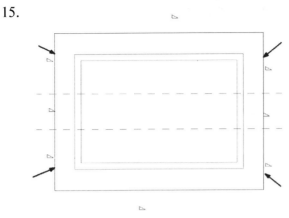

Select the upper and lower vertical segments of the roof boundary.

You can select by using a crossing or by holding down the CTL key and selecting each segment.

☐ Defines Slope

Disable **Defines slope** on the Options bar.

16.

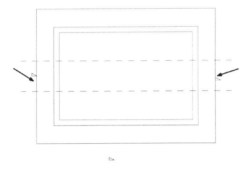

Select the two middle vertical segments.

Constraints	
Defines Roof Slope	☑
Offset From Roof Base	1.0000
Dimensions	
Slope	30.00°
Length	2.0000

Modify the Offset from Roof Base to **1m**.

This raises the middle section of the roof 1m above the level the roof is placed on.

17.

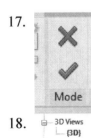

Select the **Green Check** to complete the roof.

18. 3D Views
{3D}

Switch to a 3D view.

The roof is created.

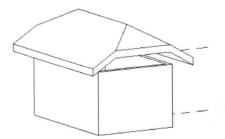

19.

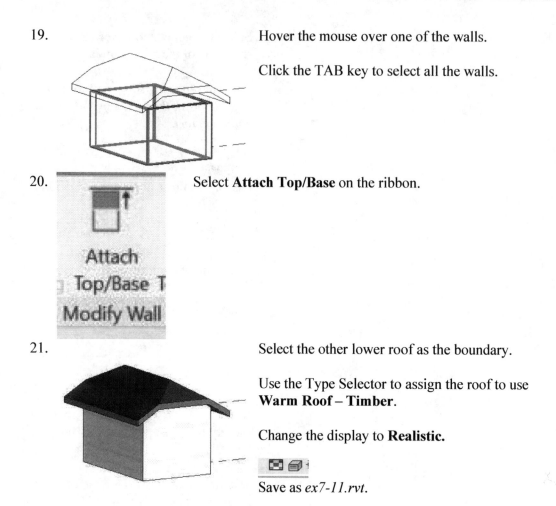

Hover the mouse over one of the walls.

Click the TAB key to select all the walls.

20.

Select **Attach Top/Base** on the ribbon.

21.

Select the other lower roof as the boundary.

Use the Type Selector to assign the roof to use **Warm Roof – Timber**.

Change the display to **Realistic.**

Save as *ex7-11.rvt*.

Exercise 7-12
Create a Dutch Gable Roof

Drawing Name: dutch_gable_roof.rvt
Estimated Time: 45 minutes

This exercise reinforces the following skills:

- ❑ Create Roof
- ❑ Modify Slope
- ❑ Section View
- ❑ Walls

1. Open *dutch_gable_roof.rvt*.

2. 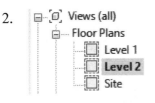 Activate the **Level 2** floor plan.

3. Switch to the Architecture ribbon.

   Select **Roof by Footprint**.

4. Set the roof type to **Warm Roof – Timber** using the Type Selector on the Properties pane.

5. Enable **Boundary Line**.

   Select the **Rectangle** too.

6. Enable **Defines slope**.
   Set the Offset to **1.0**.
   Disable Radius.

7.

Select the outside corner of the upper left of the walls.

8.

Select the outside corner of the lower right of the walls.

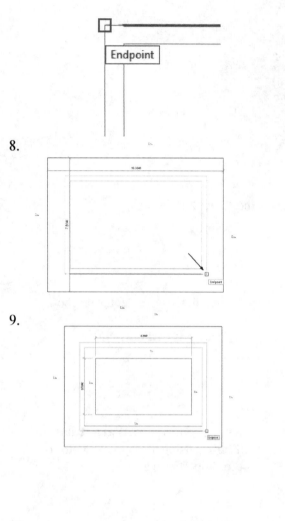

9.

Place a second rectangle inside the building with the 1m offset.

Select the upper left corner as before.

Click the SPACE key to change the direction of the offset.

Select the lower right corner as before.

10.

Cancel out of the rectangle command.

You should see two offset rectangles with slope symbols on each side.

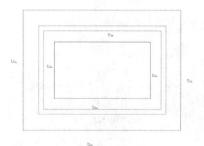

11.

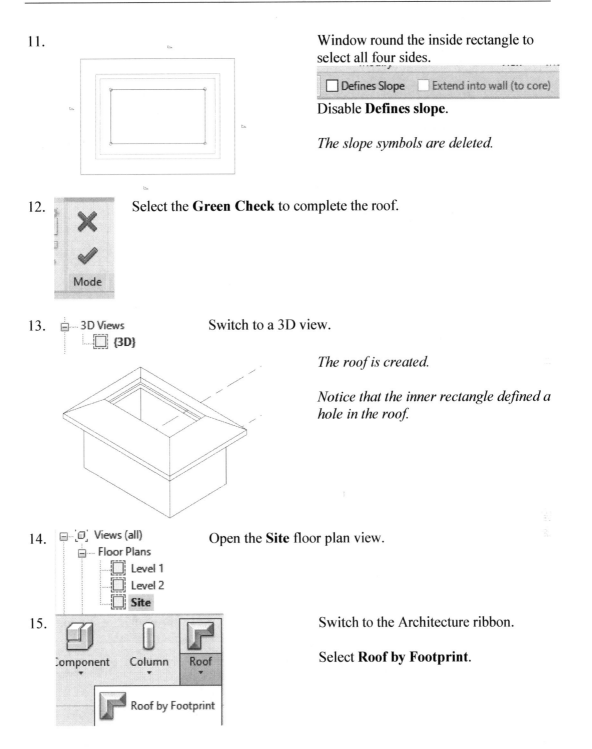

Window round the inside rectangle to select all four sides.

☐ Defines Slope   ☐ Extend into wall (to core)

Disable **Defines slope**.

The slope symbols are deleted.

12. Select the **Green Check** to complete the roof.

Mode

13. 3D Views
{3D}

Switch to a 3D view.

The roof is created.

Notice that the inner rectangle defined a hole in the roof.

14. Views (all)
Floor Plans
Level 1
Level 2
Site

Open the **Site** floor plan view.

15. Component   Column   Roof

Roof by Footprint

Switch to the Architecture ribbon.

Select **Roof by Footprint**.

16.

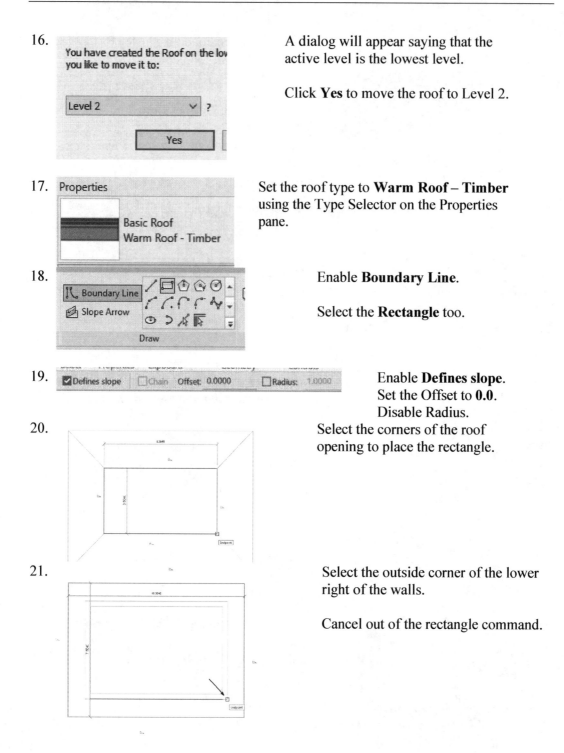

A dialog will appear saying that the active level is the lowest level.

Click **Yes** to move the roof to Level 2.

17.

Set the roof type to **Warm Roof – Timber** using the Type Selector on the Properties pane.

18.

Enable **Boundary Line**.

Select the **Rectangle** too.

19.

Enable **Defines slope**.
Set the Offset to **0.0**.
Disable Radius.

20.

Select the corners of the roof opening to place the rectangle.

21.

Select the outside corner of the lower right of the walls.

Cancel out of the rectangle command.

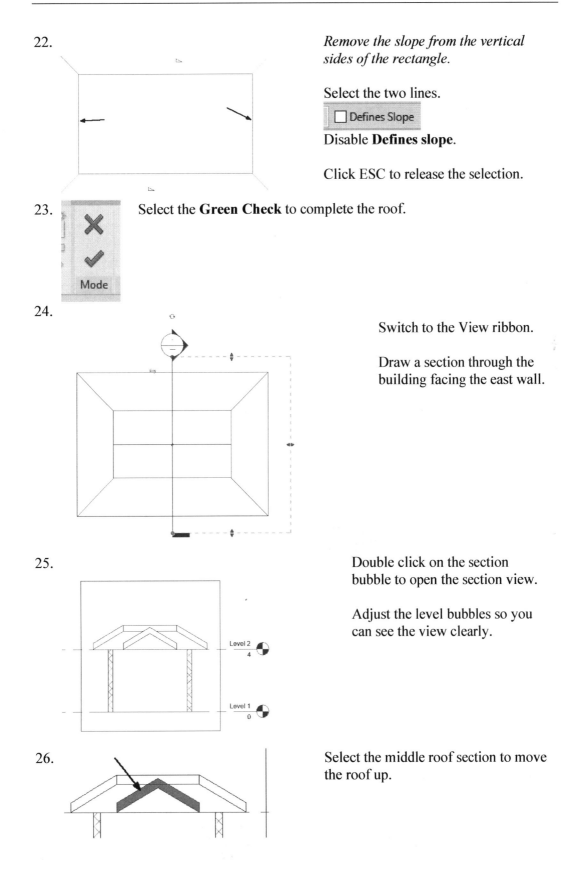

22.

Remove the slope from the vertical sides of the rectangle.

Select the two lines.

☐ Defines Slope

Disable **Defines slope**.

Click ESC to release the selection.

23. Select the **Green Check** to complete the roof.

Mode

24. Switch to the View ribbon.

Draw a section through the building facing the east wall.

25. Double click on the section bubble to open the section view.

Adjust the level bubbles so you can see the view clearly.

Level 2
4

Level 1
0

26. Select the middle roof section to move the roof up.

27. Select the **Move** tool from the ribbon.

28. ☑ Constrain ☐ Disjoin ☐ Multiple   Enable **Constrain** on the Options bar.

29.

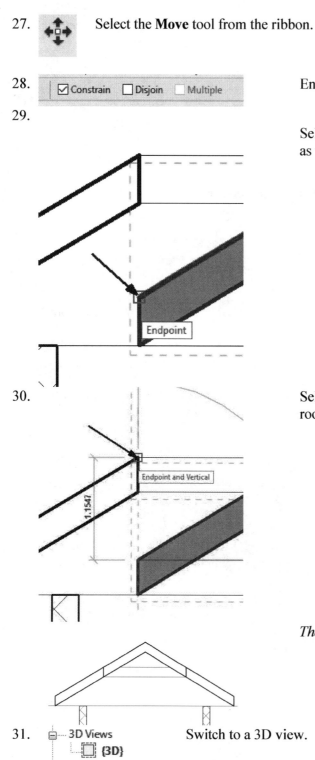

Select the upper left vertex of the roof as the base point.

Endpoint

30. Select the left upper part of the outer roof as the destination point.

Endpoint and Vertical

1.1547

The roof should appear as shown.

31. 3D Views {3D}   Switch to a 3D view.

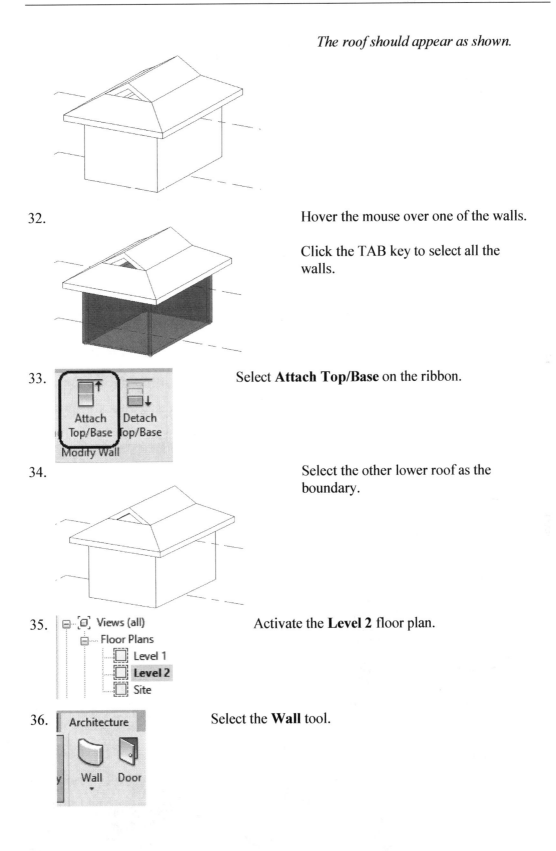

The roof should appear as shown.

32. Hover the mouse over one of the walls.

Click the TAB key to select all the walls.

33. Select **Attach Top/Base** on the ribbon.

34. Select the other lower roof as the boundary.

35. Activate the **Level 2** floor plan.

36. Select the **Wall** tool.

37.

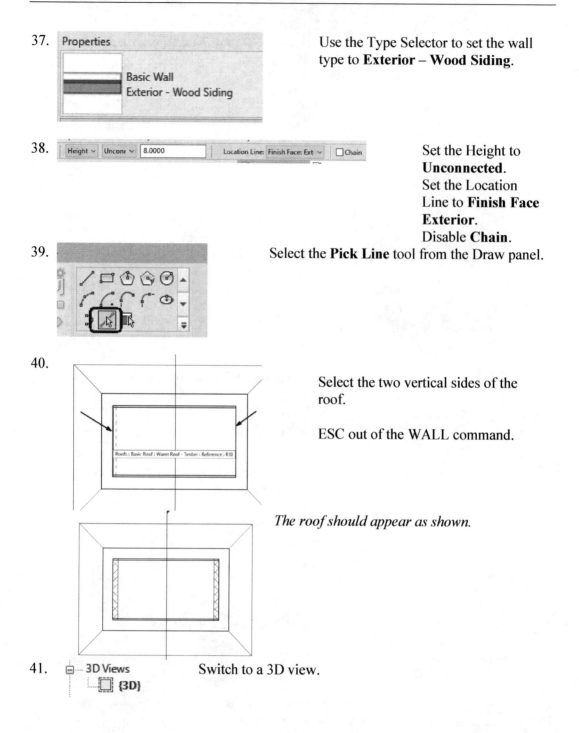

Use the Type Selector to set the wall type to **Exterior – Wood Siding**.

38.

Set the Height to **Unconnected**.
Set the Location Line to **Finish Face Exterior**.
Disable **Chain**.

39.

Select the **Pick Line** tool from the Draw panel.

40.

Select the two vertical sides of the roof.

ESC out of the WALL command.

The roof should appear as shown.

41. 3D Views
     {3D}

Switch to a 3D view.

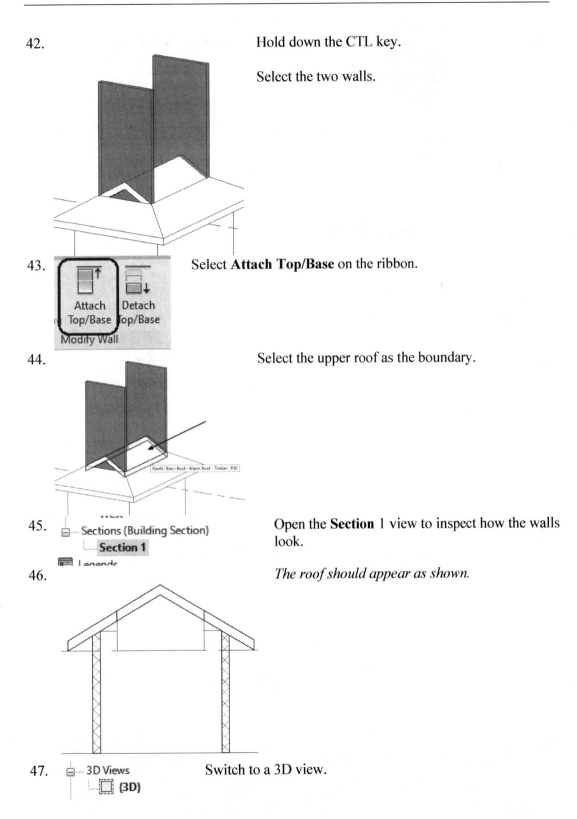

42.	Hold down the CTL key.

	Select the two walls.

43.	Select **Attach Top/Base** on the ribbon.

44.	Select the upper roof as the boundary.

	Roofs: Basic Roof : Warm Roof – Timber : R10

45.	Sections (Building Section)	Open the **Section** 1 view to inspect how the walls
	  Section 1	look.
	  Legends

46.		*The roof should appear as shown.*

47.	3D Views	Switch to a 3D view.
	  {3D}

48.

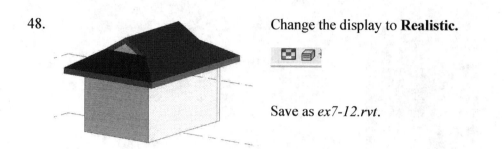

Change the display to **Realistic.**

Save as *ex7-12.rvt.*

Exercise 7-13
Create a Cross Gabled Roof

Drawing Name: cross_gabled_roof.rvt
Estimated Time: 15 minutes

This exercise reinforces the following skills:

❑   Create Roof

1.   Open *cross_gabled_roof.rvt.*

2.   Activate the **Level 2** floor plan.

3.   Switch to the Architecture ribbon.

Select **Roof by Footprint.**

4.   Enable **Boundary Line.**

Select the **Rectangle** too.

5.   Enable **Defines slope.**
Set the Offset to **1.0.**
Disable Radius.

6.

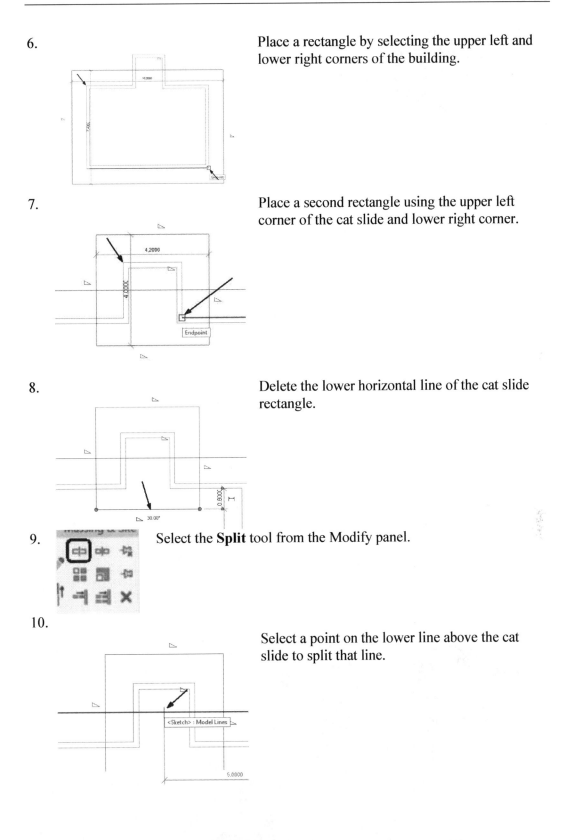

Place a rectangle by selecting the upper left and lower right corners of the building.

7.

Place a second rectangle using the upper left corner of the cat slide and lower right corner.

8.

Delete the lower horizontal line of the cat slide rectangle.

9.

Select the **Split** tool from the Modify panel.

10.

Select a point on the lower line above the cat slide to split that line.

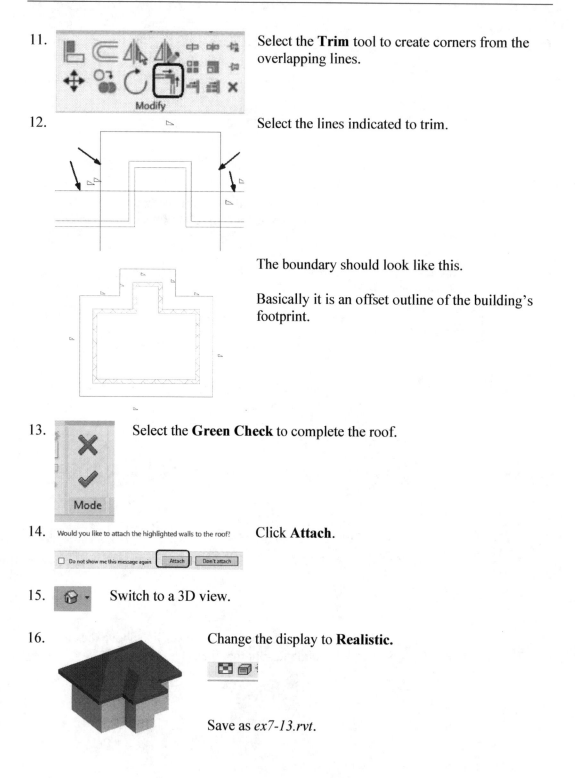

11. Select the **Trim** tool to create corners from the overlapping lines.

12. Select the lines indicated to trim.

The boundary should look like this.

Basically it is an offset outline of the building's footprint.

13. Select the **Green Check** to complete the roof.

14. Would you like to attach the highlighted walls to the roof?   Click **Attach**.

☐ Do not show me this message again   Attach   Don't attach

15. Switch to a 3D view.

16. Change the display to **Realistic.**

Save as *ex7-13.rvt*.

Exercise 7-14
Create a Butterfly Roof

Drawing Name: butterfly_roof.rvt
Estimated Time: 20 minutes

This exercise reinforces the following skills:

- ❑ Create Roof by Extrusion
- ❑ Modify Slope
- ❑ Section View
- ❑ Reference Plane
- ❑ Equal dimensions

1.   Open *butterfly_roof.rvt*.

2.   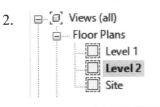   Activate the **Level 2** floor plan.

3.   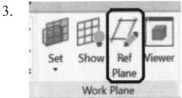   Switch to the Architecture tab.

Select the **Ref Plane** tool.

4.   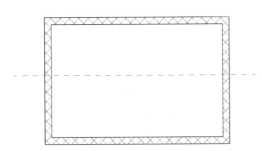   Draw a horizonal reference plane through the middle of the building model.

5.

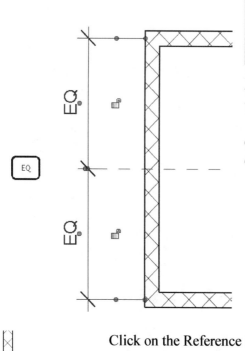

Place a continuous aligned dimension to center the reference plane on the building model.

Click on the EQ toggle to set the dimensions equal.

6.

Click on the Reference Plane label.

Name the Reference Plane **Roof**.

7.

Switch to the View ribbon.

Select the **Section** tool.

8.

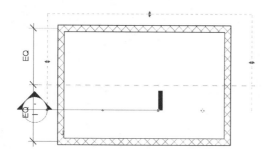

Draw a section through the building facing the north wall.

Place the section below the reference plane.

9.

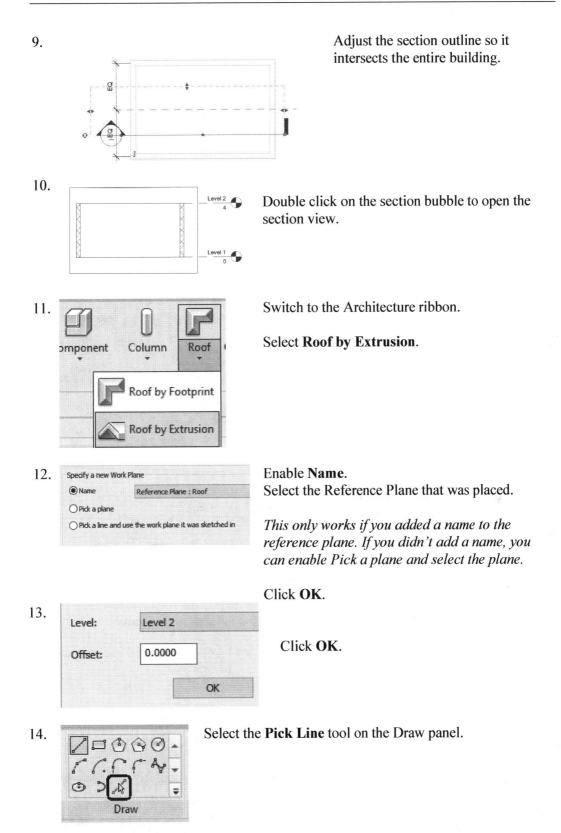

Adjust the section outline so it intersects the entire building.

10.

Double click on the section bubble to open the section view.

11.

Switch to the Architecture ribbon.

Select **Roof by Extrusion**.

12.

Enable **Name**.
Select the Reference Plane that was placed.

This only works if you added a name to the reference plane. If you didn't add a name, you can enable Pick a plane and select the plane.

Click **OK**.

13.

Click **OK**.

14.

Select the **Pick Line** tool on the Draw panel.

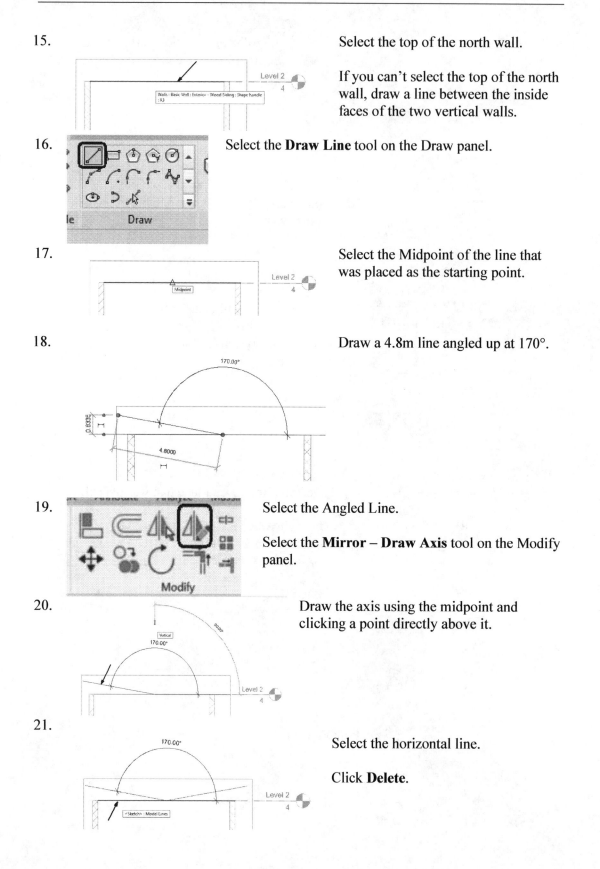

15. Select the top of the north wall.

If you can't select the top of the north wall, draw a line between the inside faces of the two vertical walls.

16. Select the **Draw Line** tool on the Draw panel.

17. Select the Midpoint of the line that was placed as the starting point.

18. Draw a 4.8m line angled up at 170°.

19. Select the Angled Line.

Select the **Mirror – Draw Axis** tool on the Modify panel.

20. Draw the axis using the midpoint and clicking a point directly above it.

21. Select the horizontal line.

Click **Delete**.

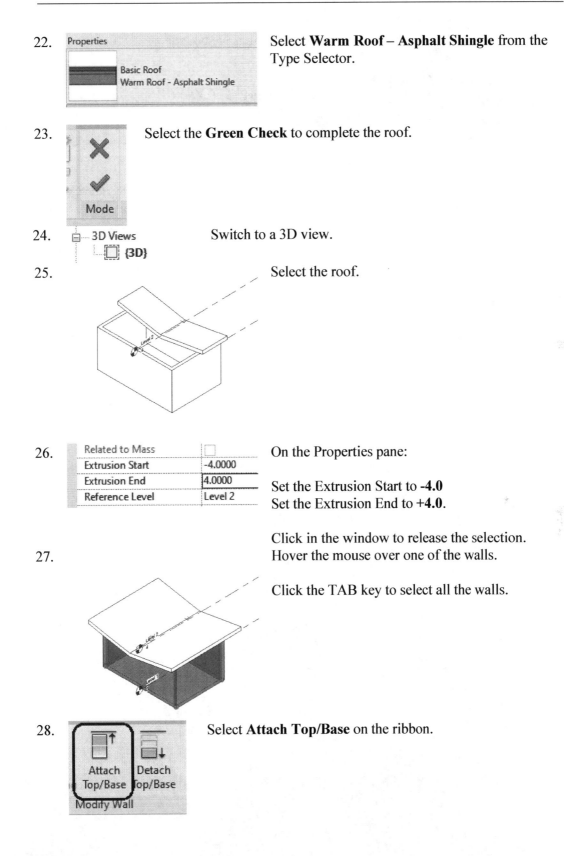

22. Select **Warm Roof – Asphalt Shingle** from the Type Selector.

23. Select the **Green Check** to complete the roof.

24. Switch to a 3D view.

25. Select the roof.

26. On the Properties pane:

Set the Extrusion Start to **-4.0**
Set the Extrusion End to **+4.0**.

27. Click in the window to release the selection. Hover the mouse over one of the walls.

Click the TAB key to select all the walls.

28. Select **Attach Top/Base** on the ribbon.

29.

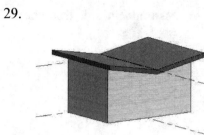

Select the roof as the boundary.

Change the display to **Realistic.**

Save as *ex7-14.rvt*.

Additional Projects

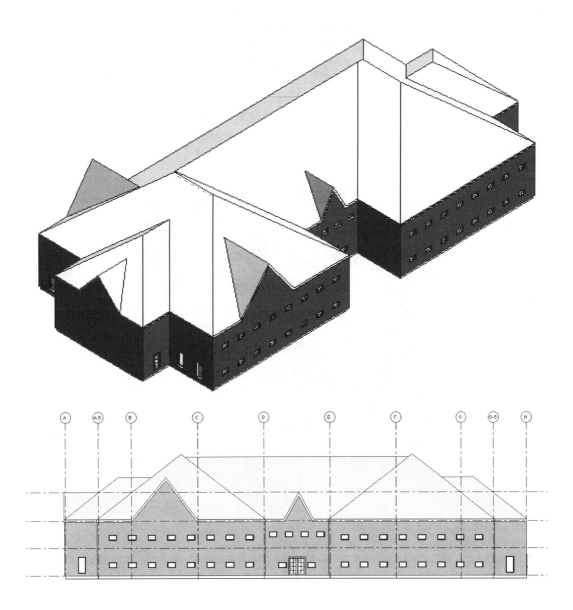

1) Create a roof with several dormers.

Family: Basic Roof
Type: Spanish Tile
Total thickness: 0' 8 3/8" (Default)
Resistance (R): 43.9645 (h·ft²·°F)/BTU
Thermal Mass: 1.9943 BTU/°F

Layers

	Function	Material	Thickness	Wraps	Variabl
1	Finish 1 [4]	Roofing, Tile	0' 0 1/4"	☐	☐
2	**Core Boundar**	**Layers Above Wrap**	**0' 0"**		
3	Membrane	Roofing, EPDM Me	0' 0"	☐	☐
4	Substrate [2	Roofing, EPDM Me	0' 0 5/8"	☐	☐
5	Structure [1	Structure, Wood Jois	0' 7 1/2"	☐	☐
6	**Core Boundar**	**Layers Below Wrap**	**0' 0"**		

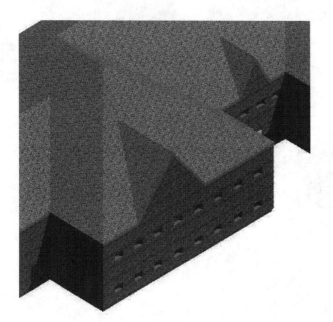

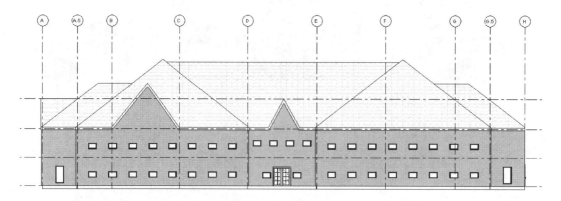

2) Create a roof family using Spanish tile material.

Lesson 7 Quiz

True or False

1. A roof footprint is a 2-D sketch.
2. When creating a roof by extrusion, the sketch must be closed.
3. When placing a roof by footprint, you must specify the work plane to place the sketch.
4. When you create a roof by extrusion, you can extend the extrusion toward or away from the view.
5. The Join/Unjoin Roof tool can be used to join two roofs together.

Multiple Choice

6. Roofs can be created using the following options:
   Select three:

   A. Roof by footprint
   B. Roof by extrusion
   C. Roof by face
   D. Roof by sketch

7. The Join/Unjoin roof tool is located:

   A. On the Modify panel of the Modify ribbon.
   B. On the Build panel of the Home ribbon.
   C. On the Geometry panel of the Modify ribbon.
   D. On the right shortcut menu when the roof is selected.

8. When creating a roof sketch you can use any of the following EXCEPT:

   A. Pick wall
   B. Lines
   C. Arcs
   D. Circles

9. In order to attach a wall to a roof that is above the wall, you should use this option:

   A. Base Attach
   B. Top Attach
   C. Edit Profile
   D. Join/Unjoin

10. The Properties pane for a roof displays the roof:

    A. Area
    B. Weight
    C. Elevation
    D. Function

ANSWERS:
    1) T; 2) F; 3) F; 4) T; 5) T; 6) A, B, & C; 7) C; 8) D; 9) B; 10) A

Lesson 8
Elevations, Details & Plans

An "elevation" is a drawing that shows the front or side of something. A floor plan, by contrast, shows a space from above – as if you are looking down on the room from the ceiling. Thus, you see the tops of everything, but you cannot view the front, side or back of an object. An elevation gives you the chance to see everything from the other viewpoints.

Elevations are essential in kitchen design, as well as other detailed renovations. Without elevation drawings, you cannot see the details of your new cabinetry, the size of each drawer or the location of each cabinet. A floor plan simply cannot communicate all of this information adequately.

While an elevation is not required for every renovation or redecorating project, they are very useful when designing items like a fireplace, bathroom vanities, bars, or any location with built-in cabinetry, such as an office or entertainment space. The information shown on an elevation drawing will give you a chance to make small changes to the design before anything is built or ordered – you don't want to be surprised during the installation!

While every detail isn't typically shown on an elevation (such as the exact cabinet door style you plan to use), the major elements will be there, including cabinet locations, the direction each cabinet door opens (hint: look at the "arrows" on the doors – the arrow points to the hinge, so you know which way the door opens!), appliance locations, height of cabinets and more. On elevations intended for use during the preliminary design stage of your project, you will find much less detail. Drawings intended for use as a guide for construction will include numerous notes and dimensions on the page.

Because Revit is a BIM software, when you update elements in the plan view, the elevation views will automatically update and vice-versa. This is called "bi-directional associativity."

Exercise 8-1
Creating Elevation Documents

Drawing Name: add_view_1.rvt [m_add_view_1.rvt]
Estimated Time: 10 minutes

This exercise reinforces the following skills:

- Sheet
- Add View
- Changing View Properties

1.   📂   Open *add_view_1.rvt [m_add_view_1.rvt]*.

2.   [View]   Activate the **View** ribbon.

3.   Select **Sheet** from the Sheet Composition panel to add a new sheet.

4.   Select **D 22 x 34 Horizontal [A3 metric]** title block. Click **OK**.

Select a titleblock
D 22 x 34 Horizontal
E1 30 x 42 Horizontal : E1 30x42 Horizontal

Select titleblocks:
A1 metric
A2 metric
A3 metric
None

In the Properties pane:

Approved By	M. Instructor
Designed By	J. Student
Checked By	M. Instructor
Drawn By	J. Student
Sheet Number	A201
Sheet Name	Exterior Elevations

Change Approved By to your instructor's name.
Change Designed By to your name.
Change Checked By to your instructor's name.
Change Drawn by to your name.
Change the Sheet Number to **A201**.
Change the Sheet Name to **Exterior Elevations**.

5.   Elevations (Building Elevation)
East
North
South
South - Lobby
West

Locate the North and South Elevations in the Project Browser.

6.

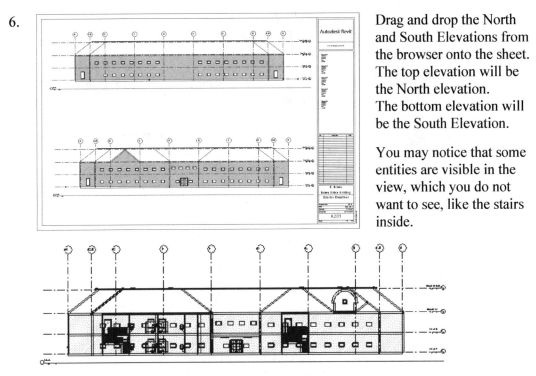

Drag and drop the North and South Elevations from the browser onto the sheet. The top elevation will be the North elevation. The bottom elevation will be the South Elevation.

You may notice that some entities are visible in the view, which you do not want to see, like the stairs inside.

Let's look at the **North** Elevation first.

7.

Viewports (1)	⌄
Graphics	
View Scale	1 : 200
Scale Value 1:	200

Select the **North** Elevation View on the sheet.

For metric:

Change the Scale to: **1:200**.

8.

For metric:

Adjust the title line by dragging the grip under the elevation view.

9.

Rotation on Sheet	None
Visibility/Graphics Overri...	Edit...
Graphic Display Options	Edit...
Hide at scales coarser than	1" = 400'-0"
Discipline	Architectural

In the Properties pane:
Select the **Edit** button next to Graphic Display Options.

10.

Model Display	
Style:	Hidden Line
	☑ Show Edges
Transparency:	0
Silhouettes:	<none>

Under Model Display:

Set Style to **Hidden Line**.

Click **OK**.

11.

Rotation on Sheet	None
Visibility/Graphics Overri...	Edit...
Graphic Display Options	Edit...
Hide at scales coarser than	1" = 400'-0"

Select **Edit** next to Visibility/Graphics Overrides.

12.

Under Annotation Categories, disable Grids and Sections.

Hint: You can perform a search for grids and sections to make it easier to locate those categories.

Click **OK**.

13.

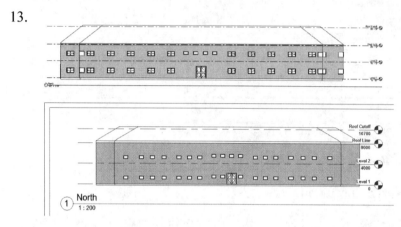

The North Elevation now looks more appropriate.

14.

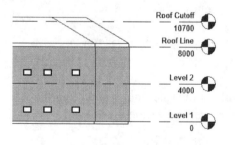

In the metric project:

To adjust the position of the level bubbles, right click on the view.
Select Activate View.
Move the level bubble for one of the levels. All the other level bubbles will remain aligned.
Right click and select Deactivate View.

15.

Select the **South** Elevation View on the sheet.
For metric:

Change the Scale to: **1:200**.

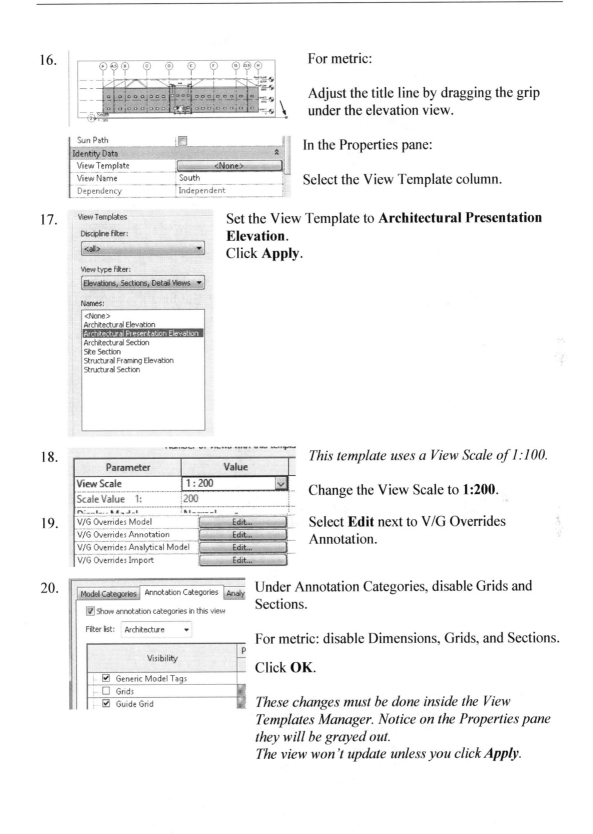

16.

For metric:

Adjust the title line by dragging the grip under the elevation view.

In the Properties pane:

Select the View Template column.

17.

Set the View Template to **Architectural Presentation Elevation**.
Click **Apply**.

18.

This template uses a View Scale of 1:100.

Change the View Scale to **1:200**.

19.

Select **Edit** next to V/G Overrides Annotation.

20.

Under Annotation Categories, disable Grids and Sections.

For metric: disable Dimensions, Grids, and Sections.

Click **OK**.

These changes must be done inside the View Templates Manager. Notice on the Properties pane they will be grayed out.
*The view won't update unless you click **Apply**.*

21.

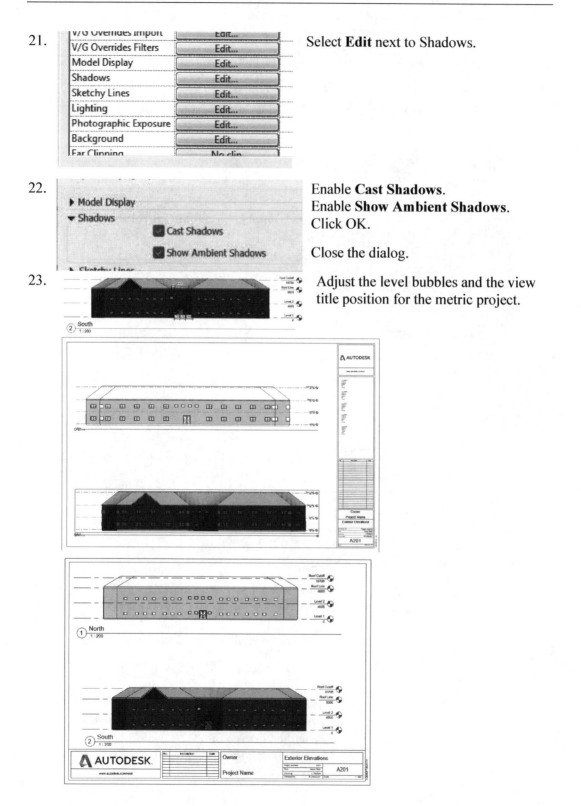

Select **Edit** next to Shadows.

22.

Enable **Cast Shadows**.
Enable **Show Ambient Shadows**.
Click OK.

Close the dialog.

23.

Adjust the level bubbles and the view title position for the metric project.

24. Save as *ex8-1.rvt [m_ ex8-1.rvt]*.

Exercise 8-2
Using Line Work

Drawing Name: linework.rvt [m_ linework.rvt]
Estimated Time: 15 minutes

This exercise reinforces the following skills:

- ❑ Activate View
- ❑ Linework tool
- ❑ Line Styles
- ❑ Line Weight

1.   📂   Open *linework.rvt [m_ linework.rvt]*.

2.   ⊟ 🗐 Sheets (all)
         ⊞ A101 - First Level Floor Plan
         ⊞ A102 - Second Level Floor Plan
         ⊞ A103 - Lobby Keynotes
         ⊞ **A201 - Exterior Elevations**

   Activate the **Exterior Elevations** sheet.

3.   Cancel
     Repeat [Delete]
     Recent Commands
     Activate View

   Select the top elevation view (**North**).
   Right click and select **Activate View**.

Activate View works similarly to Model Space/Paper Space mode in AutoCAD.

4.   Modify   Select the **Modify** ribbon.

5.      Select the **Linework** tool from the View panel.

6. 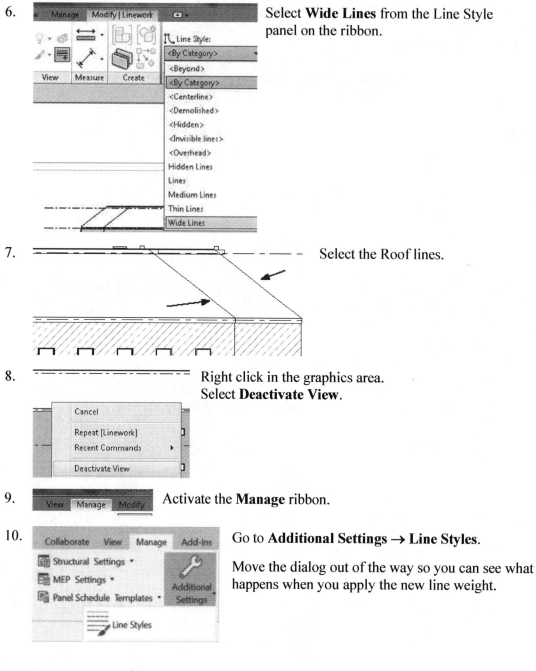 Select **Wide Lines** from the Line Style panel on the ribbon.

7. Select the Roof lines.

8. Right click in the graphics area.
   Select **Deactivate View**.

9. Activate the **Manage** ribbon.

10. Go to **Additional Settings → Line Styles**.

    Move the dialog out of the way so you can see what happens when you apply the new line weight.

11.

Thin Lines	1
Wide Lines	16

Set the Line Weight for Wide Lines to **16**.
Click **Apply**.

12. Our rooflines have dramatically changed.

13.

Medium Lines	3
Thin Lines	1
Wide Lines	12

Set the Wide Lines Line Weight to **12**.
Click **Apply**.
Click **OK**.

14. Note that both views updated with the new line style applied.

15.

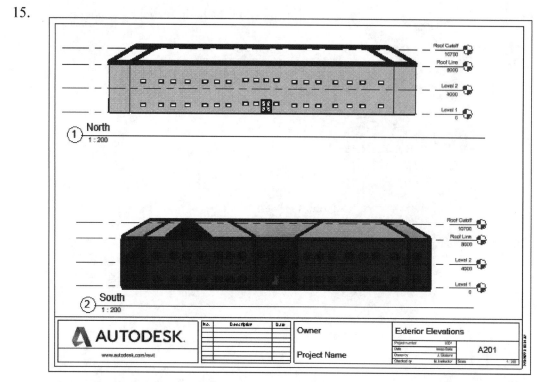

Save the file as *ex8-2.rvt [m_ ex8-2.rvt]*.

You can use invisible lines to hide any lines you don't want visible in your view. Linework is specific to each view. If you duplicate the view, any linework edits you applied will be lost.

Wall Section Views

The wall section view is used to show the various wall construction materials and act as a guide for the construction of the wall.

Exercise 8-3
Creating a Section View

Drawing Name: section_view.rvt [m_section_view.rvt]
Estimated Time: 15 minutes

This exercise reinforces the following skills:

- Activate View
- Add Section Symbol

1.   Open *section_view.rvt [m_section_view.rvt]*.

2. 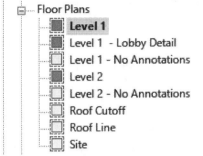   Activate **Level 1**.

3.   View   Activate the **View** ribbon.

4.   Select the **Section** tool from the Create panel.

5. 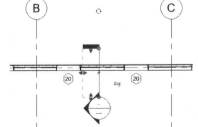 Place a section on the north exterior wall.

   Adjust the clip range to show just the wall.

   The first left click places the bubble.
   The second click places the tail.
   Use the flip orientation arrows to change the
   direction of the section view.

6. On the Properties panel:

   Select **Wall Section** from the Type Selector
   drop-down list.

7. Sections (Wall Section) The new section appears in your Project Browser.
     Section 1

8. Sections (Wall Section) Rename the section view **North Exterior Wall**.
   North Exterior Wall

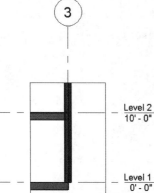

You can rename the view in the Properties panel or in the browser.

Activate the section view.

9. 1/8" = 1'-0" Set the Detail Level to **Fine**.
        Set the Visual Style to **Consistent Colors**.

10.

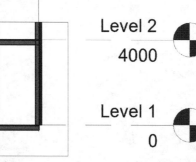

Adjust the crop region to display the wall as shown.
Select the crop region and use the bubble grips to re-size.

11. Switch to the Annotate ribbon.

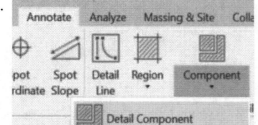

Select the **Detail Component** tool.

12. Select **Load Family**.

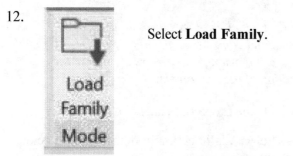

13.

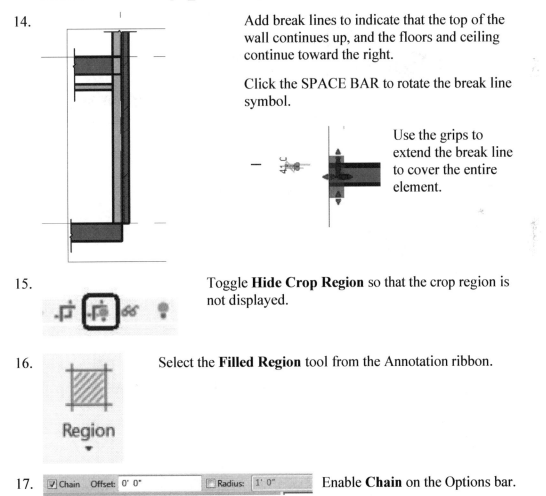

Browse to the *Detail Items\Div 01-General* folder.

Select the *Break Line [M_Break Line]*.

14.

Add break lines to indicate that the top of the wall continues up, and the floors and ceiling continue toward the right.

Click the SPACE BAR to rotate the break line symbol.

Use the grips to extend the break line to cover the entire element.

15.

Toggle **Hide Crop Region** so that the crop region is not displayed.

16.

Select the **Filled Region** tool from the Annotation ribbon.

17.

Enable **Chain** on the Options bar.

18.

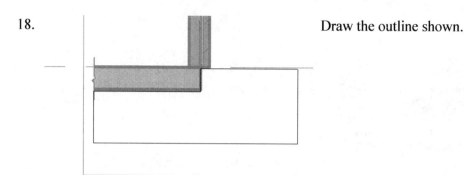

Draw the outline shown.

19. Click the Green Check to finish the filled region.

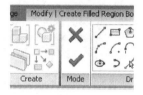

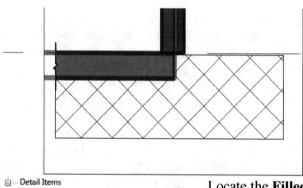

A filled region is placed.

20.

Detail Items
- Break Line
- Brick Standard
- Filled region
  - Diagonal Crosshatch
  - Diagonal Crosshatch - Transparent
  - Diagonal Down
  - Diagonal Down - Transparent
  - Diagonal Up
  - Horizontal Lines
  - Ortho Crosshatch
  - Solid Black
  - Vertical Lines
  - Wood 1
  - Wood 2

Locate the **Filled region** under *Families/Detail Items* in the Project Browser.

21.

Filled region
- Diagonal Crosshatch
- Diagonal Crossl   Duplicate
- Diagonal Down
- Diagonal Down   Delete

Highlight the **Diagonal Crosshatch**.
Right click and select **Duplicate**.

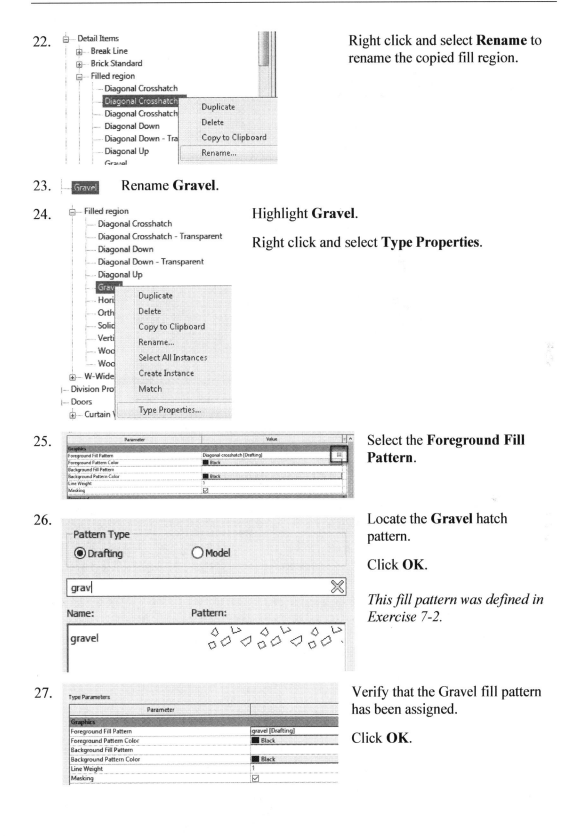

22. Right click and select **Rename** to rename the copied fill region.

23. Rename **Gravel**.

24. Highlight **Gravel**.

Right click and select **Type Properties**.

25. Select the **Foreground Fill Pattern**.

26. Locate the **Gravel** hatch pattern.

Click **OK**.

This fill pattern was defined in Exercise 7-2.

27. Verify that the Gravel fill pattern has been assigned.

Click **OK**.

28.

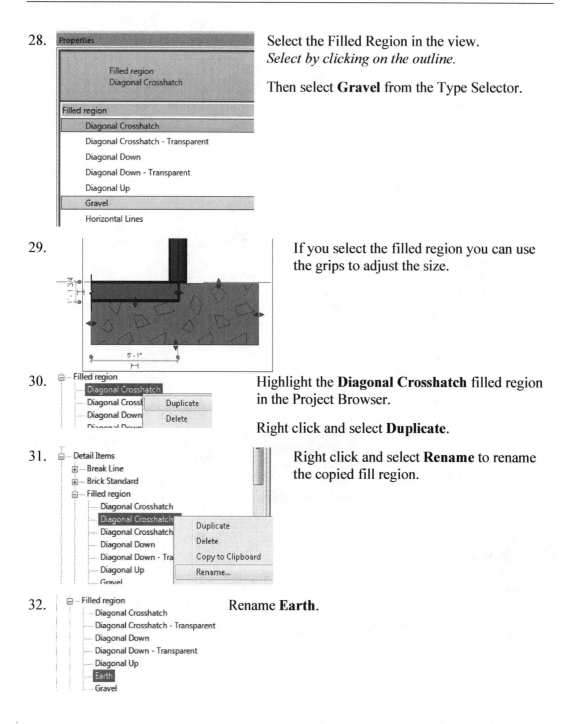

Select the Filled Region in the view.
Select by clicking on the outline.

Then select **Gravel** from the Type Selector.

29.

If you select the filled region you can use the grips to adjust the size.

30.

Highlight the **Diagonal Crosshatch** filled region in the Project Browser.

Right click and select **Duplicate**.

31.

Right click and select **Rename** to rename the copied fill region.

32.

Rename **Earth**.

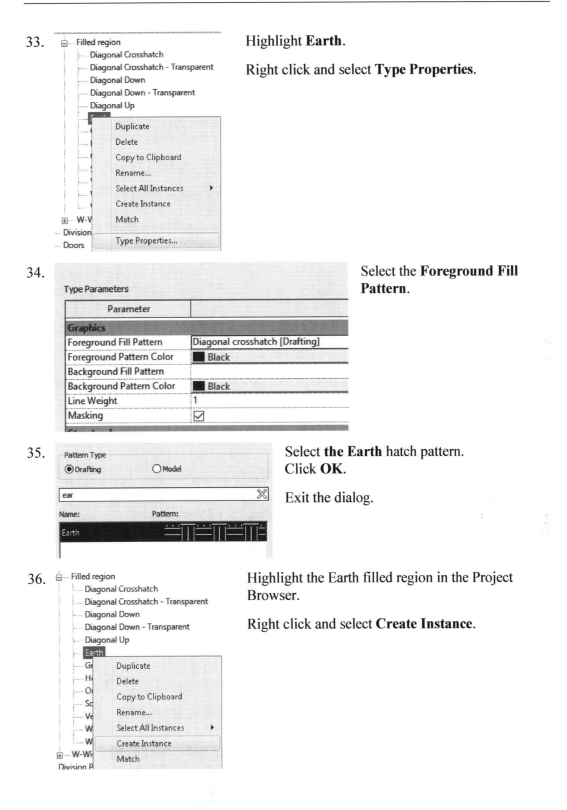

33.   Highlight **Earth**.

Right click and select **Type Properties**.

34.   Select the **Foreground Fill Pattern**.

35.   Select **the Earth** hatch pattern.
Click **OK**.

Exit the dialog.

36.   Highlight the Earth filled region in the Project Browser.

Right click and select **Create Instance**.

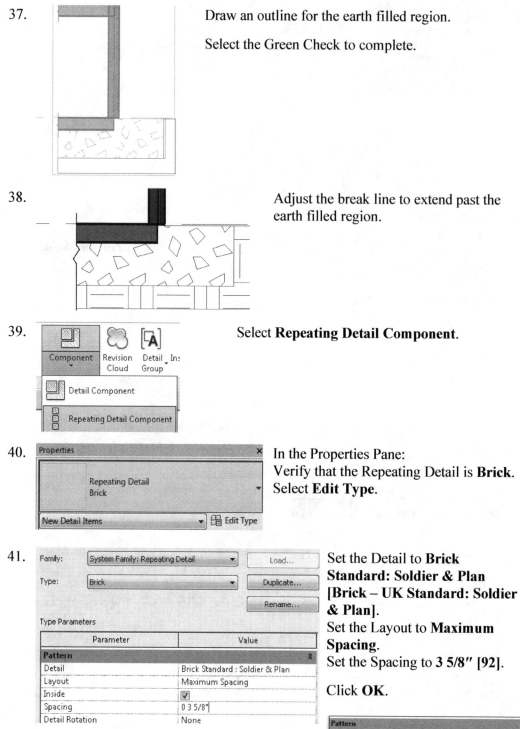

37. Draw an outline for the earth filled region.

Select the Green Check to complete.

38. Adjust the break line to extend past the earth filled region.

39. Select **Repeating Detail Component**.

40. In the Properties Pane:
Verify that the Repeating Detail is **Brick**.
Select **Edit Type**.

41. Set the Detail to **Brick Standard: Soldier & Plan [Brick – UK Standard: Soldier & Plan]**.
Set the Layout to **Maximum Spacing**.
Set the Spacing to **3 5/8″ [92]**.

Click **OK**.

Pattern	
Detail	Brick – UK Standard : Soldier & Plan
Layout	Maximum Spacing
Inside	☐
Spacing	92.0
Detail Rotation	None

42.

Select the interior bottom side for the start point of the brick detail.

Select the interior top side for the end point.

43.

Select the **Material Tag** tool from the Annotate ribbon.

44.

Place the material callouts.

The material displays the description of the material.

The materials are determined by the layers defined by the wall type used.

You should see a preview of the tag before you click to place.

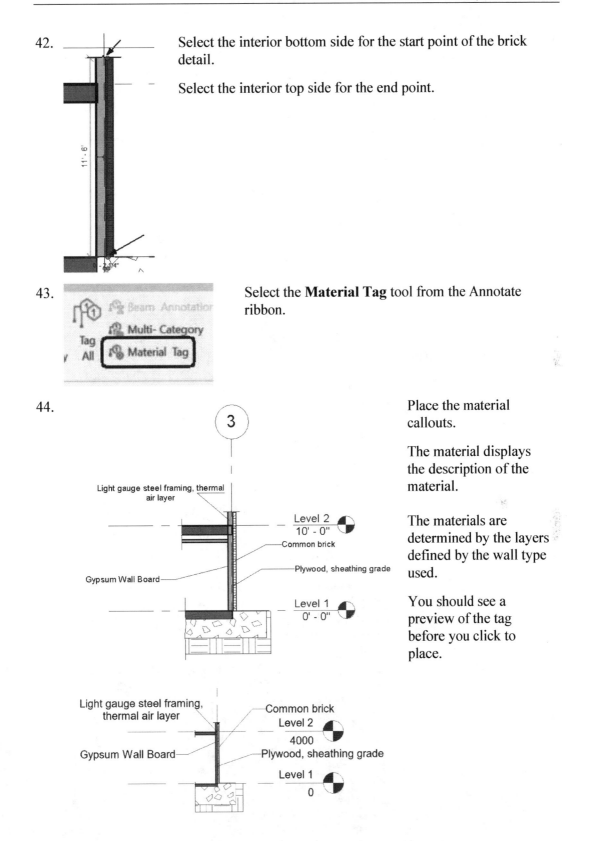

45. [View]   Activate the **View** ribbon.

46.    Select **New Sheet** from the Sheet Composition panel.

47. Select titleblocks:

D 22 x 34 Horizontal
E1 30 x 42 Horizontal : E1 30x42 Horizontal
None

Select **D 22 x 34 Horizontal [A1 Metric]** titleblock.
Click **OK**.

Select titleblocks:

A1 metric
A2 metric
A3 metric
None

48. In the Properties pane:

Approved By	M. Instructor
Designed By	J. Student
Checked By	M. Instructor
Drawn By	J. Student
Sheet Number	AD501
Sheet Name	Details

Change the Sheet Number to **AD501**.

Change the Sheet Name to **Details**.

Modify the Drawn By field with your name.

Modify the Approved By field with your instructor's name.

49. Drag and drop the North Exterior Wall section view onto the sheet.

50.

Viewports (1)	▾ Edit Type
Graphics	⟨
View Scale	1/4" = 1'-0"
Scale Value 1:	48

Select the view.
In the Properties pane:
Set the View Scale to **1/4″ = 1′-0″ [1:100]**.

Graphics	
View Scale	1 : 100
Scale Value 1:	100

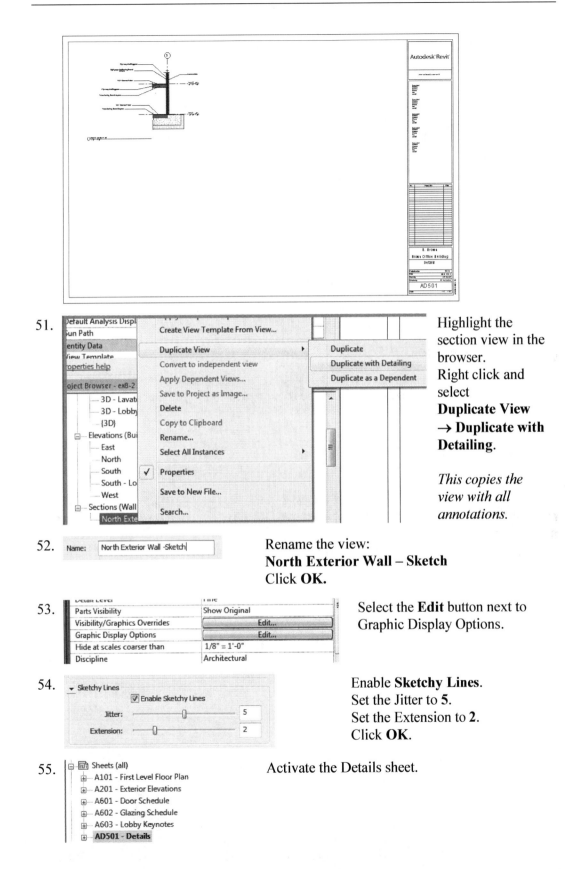

51. Highlight the section view in the browser.
Right click and select
Duplicate View → Duplicate with Detailing.

This copies the view with all annotations.

52. Name: North Exterior Wall -Sketch

Rename the view:
North Exterior Wall – Sketch
Click **OK.**

53.

Detail Level	Fine
Parts Visibility	Show Original
Visibility/Graphics Overrides	Edit...
Graphic Display Options	Edit...
Hide at scales coarser than	1/8" = 1'-0"
Discipline	Architectural

Select the **Edit** button next to Graphic Display Options.

54.
Sketchy Lines
☑ Enable Sketchy Lines
Jitter: 5
Extension: 2

Enable **Sketchy Lines**.
Set the Jitter to **5**.
Set the Extension to **2**.
Click **OK.**

55.
Sheets (all)
A101 - First Level Floor Plan
A201 - Exterior Elevations
A601 - Door Schedule
A602 - Glazing Schedule
A603 - Lobby Keynotes
AD501 - Details

Activate the Details sheet.

56.

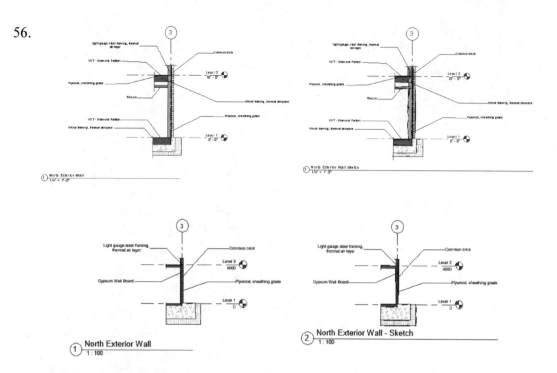

Drag and drop the sketch version of the section next to the first section.

Compare the views.

57. Save the file as *ex8-3.rvt [m_ ex8-3.rvt]*.

Any blue symbol will behave like a hyperlink on a web page and can be double clicked to switch to the referenced view. (Note: The blue will print black on your plots if you want it to.)

Exercise 8-4
Modifying Keynote Styles

Drawing Name: keynotes3.rvt [m_ keynotes3.rvt]
Estimated Time: 15 minutes

This exercise reinforces the following skills:

  ❏ Keynotes
  ❏ Materials
  ❏ Edit Family

1. Open *keynotes3.rvt [m_keynotes3.rvt]*.

2. 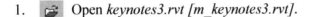 Activate the **Exterior Elevations** sheet.

3. Activate the Southern Elevation view.
    Select the view.
    Right click and select **Activate View**.

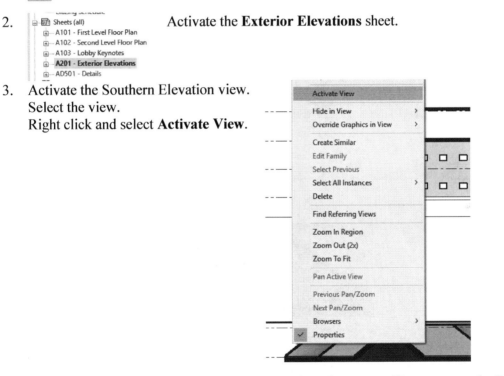

*If you do not add the notes in Model Space, then the notes will not automatically
scale and move with the view.*

4. Switch to the **Annotate** ribbon.

5. Select the **User Keynote** tool from the Tag panel.

Note: Revit only uses fonts available in the Windows fonts folder. If you are an AutoCAD user with legacy shx fonts, you need to locate ttf fonts that are equivalent to the fonts you want to use and load them into the Windows font folder. To convert shx fonts over to ttf, check with www.tcfonts.com.

6. Select **Keynote Text** from the Properties pane Type Selector.

7. Place a leader starting on the roof.

8. Locate the **Asphalt Shingles** material.

Click **OK**.

Cancel out of the command.

Architects are particular about the size of text, and often about the alignment. Notes should not interfere with elevations, dimensions, etc. You can customize the appearance of keynotes.

9.

Properties
×

Keynote Tag
Keynote Text

Keynote Tags ▼ ⊞ Edit Type

Select the keynote that was placed.

Select **Edit Type** from the Properties panel.

10. [Duplicate...]   Select **Duplicate**.

11.

Name

Name: Keynote Text - Custom

OK   Cancel

Enter **Keynote Text - Custom**.

Click **OK**.

12.

Type Parameters

Parameter	
Graphics	
Keynote Text	☑
Keynote Number	☐
Boxed	☐
Leader Arrowhead	Arrow Filled 30 Degree

Change the Leader Arrowhead to **Arrow Filled 30 Degree**.

Click **OK**.

13.

Asphalt Shingles

Apply the new type to the keynote.

Notice that the arrowhead changes.

14.

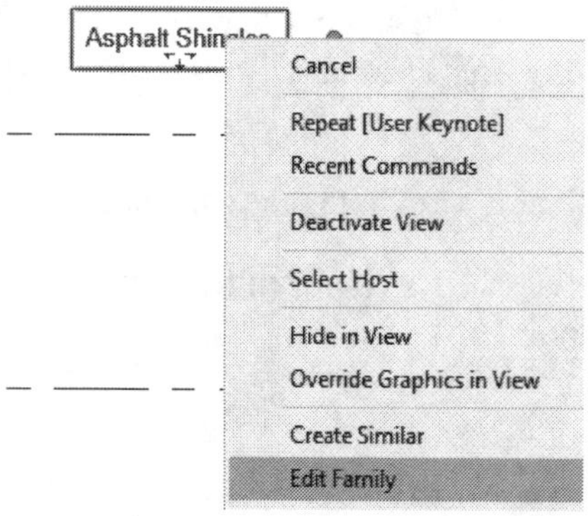

Asphalt Shin~~~

Cancel

Repeat [User Keynote]
Recent Commands

Deactivate View

Select Host

Hide in View
Override Graphics in View

Create Similar

Edit Family

Select the keynote.
Right click and select **Edit Family**.

15.

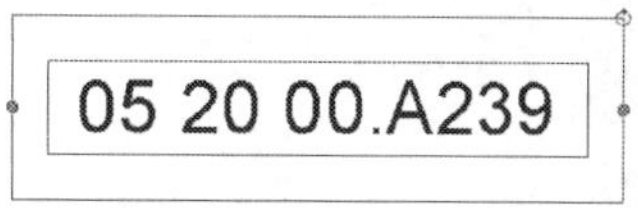

05 20 00.A239

Select the text.

1

16.

Label
3/32"

Other (1)   Edit Type

Select **Edit Type**.

17.  Duplicate...   Select **Duplicate**.

18.

Change the name to **3/32″ Tahoma
[2.5mm Tahoma]**.

Click **OK**.

Name: 2.5mm Tahoma

19.

Type Parameters	
Parameter	
Graphics	
Color	■ Blue
Line Weight	1
Background	Opaque
Show Border	☐
Leader/Border Offset	5/64"
Text	
Text Font	Tahoma
Text Size	3/32"
Tab Size	1/2"
Bold	☐
Italic	☐
Underline	☐
Width Factor	1.000000

Change the Color to **Blue**.

Change the Text Font to **Tahoma**.

Click **OK**.

Type Parameters	
Parameter	
Graphics	
Color	■ Blue
Line Weight	1
Background	Opaque
Show Border	☐
Leader/Border Offset	2.0320 mm
Text	
Text Font	Tahoma
Text Size	2.5000 mm
Tab Size	12.7000 mm

20. Change the label text to use the new label type using the Type Selector.

21.

05 20 00.A239

0

There are actually two labels overlapping each other. One label is the Keynote number and one is the Keynote text. Move the label above its current position to reveal the label below.

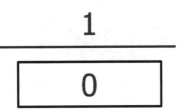

22.

Label
3/32"
3/32" Tahoma

Select the keynote text label.

Using the Type Selector, change the label text to use the **3/32" Tahoma [2.5mm Tahoma]**.

Properties
Tag Label
2.5mm Tahoma

23.

_05 20 00.A239_____

0

The two labels should both use the **3/32" Tahoma [2.5mm Tahoma]**font.

Move the keynote number label back to its original position overlapping the keynote text label.

24.

Load into Project and Close

File name: M_Keynote Tag Tahoma.rfa

Files of type: Family Files (*.rfa)

Save the file as *Keynote Tag Tahoma [M_ Keynote Tag Tahoma]*.

Select **Load into Project and Close** under the Family Editor panel.

25.

Load into Projects

Check the open Projects/Families you want to load the edited Family into

☑ ex8-3.rvt

If this dialog appears:

Place a check next to the project and click **OK**.

26.

Sheets (all)
- A101 – First Level Floor Plan
- A102 – Second Level Floor Plan
- A103 – Lobby Keynotes
- **A201 – Exterior Elevations**
- AD501 – Details

Return to the **Exterior Elevations** sheet.

Activate the South Elevation view.

27.

Asphalt Shingles

Select the keynote placed for the roof.

28.

Keynote Tag Tahoma
- Keynote Number
- Keynote Number - Boxed - Large
- Keynote Number - Boxed - Small
- Keynote Text

Using the Type Selector:

Locate the **Keynote Tag Tahoma [M_ Keynote Tag Tahoma]**.

Select the **Keynote Text** type for the Keynote Tag Tahoma family.

M_Keynote Tag Tahoma
- Keynote Number
- Keynote Number - Boxed
- Keynote Text

29.

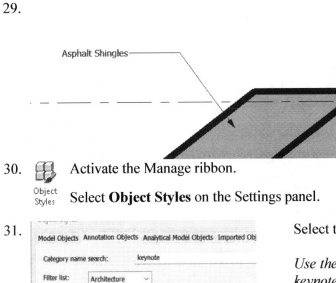

The keynote font updates.

You may need to use Edit Type to reset the arrowhead on the leader.

30. Activate the Manage ribbon.

Object Styles Select **Object Styles** on the Settings panel.

31.

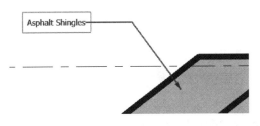

Select the Annotation Objects tab.

Use the name search field to locate the keynote tags.

Set the Line Color for Keynote Tags to **Blue**.

Click **OK**.

32.

The leader changes to blue.

33. Add additional keynotes.

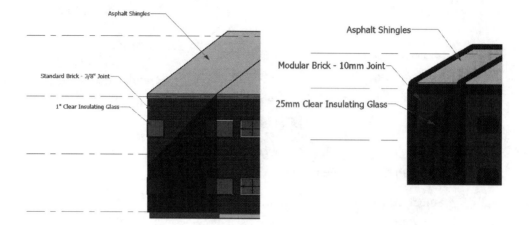

34. Deactivate the view.

35. Save the file *ex8-4.rvt [m_ ex8-4.rvt]*.

 Once you assign an elevation view to a sheet, the elevation marker will update with the sheet number.

Window Tags

By default, Window Tags are associated with window types, not instances. Window tags are annotations that generally identify particular types of windows in a drawing by displaying the value of the window's Type Mark property.

You can specify that window tags are attached automatically as you place windows or you can attach them later, either individually or all at once.

A window tag does not display if any part of the tagged window is outside the annotation crop region.

Exercise 8-5
Adding Window Tags

Drawing Name: window_tags.rvt [m_ window_tags.rvt]
Estimated Time: 5 minutes

This exercise reinforces the following skills:

- ❑ Tag All Not Tagged
- ❑ Window Schedules
- ❑ Schedule/Quantities
- ❑ Schedule Properties

1.  Open *window_tags.rvt [m_ window_tags.rvt]*.

2. 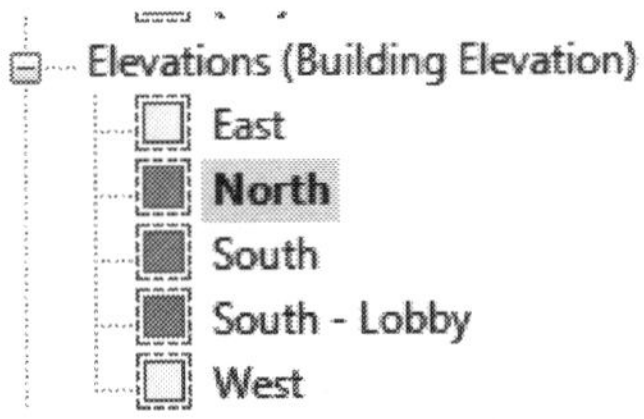 Activate the **North Building Elevation**.

    Elevations (Building Elevation)
    - East
    - **North**
    - South
    - South - Lobby
    - West

    Can you tell which building elevation views have been placed on sheets?

3. 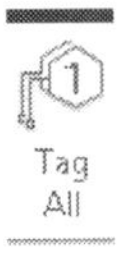 Activate the Annotate ribbon.

    Tag All   Select the **Tag All** tool from the Tag panel.

4. 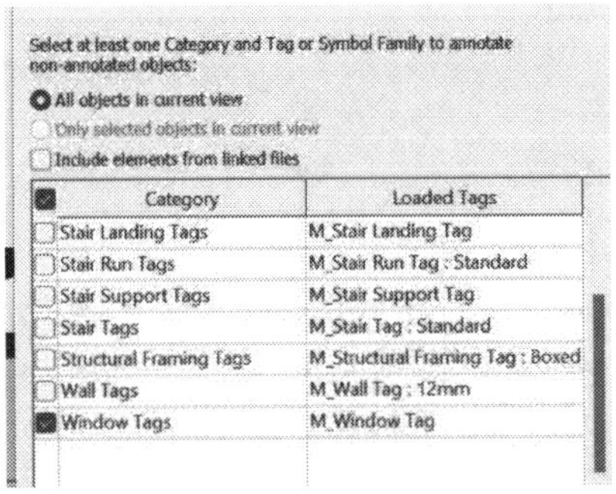 Enable the **Window Tags**.

    Click **OK**.

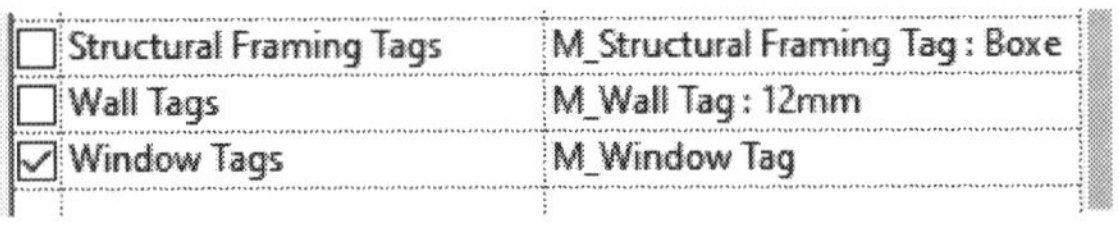

    Select at least one Category and Tag or Symbol Family to annotate non-annotated objects:

    ◉ All objects in current view
    ◯ Only selected objects in current view
    ☐ Include elements from linked files

Category	Loaded Tags
Stair Landing Tags	M_Stair Landing Tag
Stair Run Tags	M_Stair Run Tag : Standard
Stair Support Tags	M_Stair Support Tag
Stair Tags	M_Stair Tag : Standard
Structural Framing Tags	M_Structural Framing Tag : Boxed
Wall Tags	M_Wall Tag : 12mm
Window Tags	M_Window Tag

	Category	Loaded Tags
☐	Structural Framing Tags	M_Structural Framing Tag : Boxe
☐	Wall Tags	M_Wall Tag : 12mm
☑	Window Tags	M_Window Tag

5.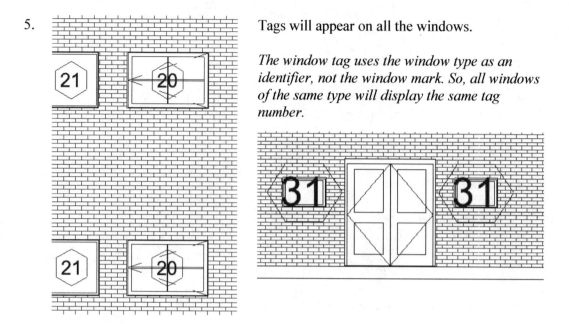

Tags will appear on all the windows.

The window tag uses the window type as an identifier, not the window mark. So, all windows of the same type will display the same tag number.

6. Save the file as *ex8-5.rvt [m_ ex8-5.rvt]*.

Exercise 8-6
Changing Window Tags from Type to Instance

Drawing Name: window_tag_family.rvt [m_ window_tag_family.rvt]
Estimated Time: 20 minutes

This exercise reinforces the following skills:

- ❑ Family Types
- ❑ Parameters
- ❑ Tags

In Exercise 8-5, we added a window tag that was linked to window type. So, all the tags displayed the same number for all windows of the same type. Some users want their tags to display by instance (individual placement). That way they can specify in their schedule the location of each window as well as the type. In this exercise, you learn how to modify Revit's window tag to link to an instance instead of a type.

1.   📂   Open *window_tag_family.rvt [m_ window_tag_family.rvt]*.

2.   ⊟ Elevations (Building Elevation)
        ☐ East
        ▣ **North**
        ▣ South
        ▣ South - Lobby
        ☐ West

Activate the **North Building Elevation**.

3. 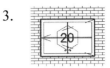 Select one of the window tags.

   Make sure you don't select a window.
   You can use the TAB key to cycle through the selections.

The Window Tag will display in the Properties pane if it is selected.

Properties

M_Window Tag

Properties ✕

Window Tag ▾

Window Tags (1) ▾ ▣ Edit Type
Graphics ☆
Leader Line ☐
Orientation Horizontal

4.  Right click and select **Edit Family**.

Edit Family
Select Previous
Select All Instances
Delete

Find Referring Views

Zoom In Region
Zoom Out (2x)
Zoom To Fit

Previous Pan/Zoom
Next Pan/Zoom

✓ Properties

5. Select the label/text located in the center of the tag.
   Select **Edit Label** on the ribbon bar.

6. Highlight the Type Mark listed as the Label Parameters.

   Label Parameters
   Parameter Name
   1 Type Mark

   This is linked to the type of window placed.

   Select the **Remove** button.

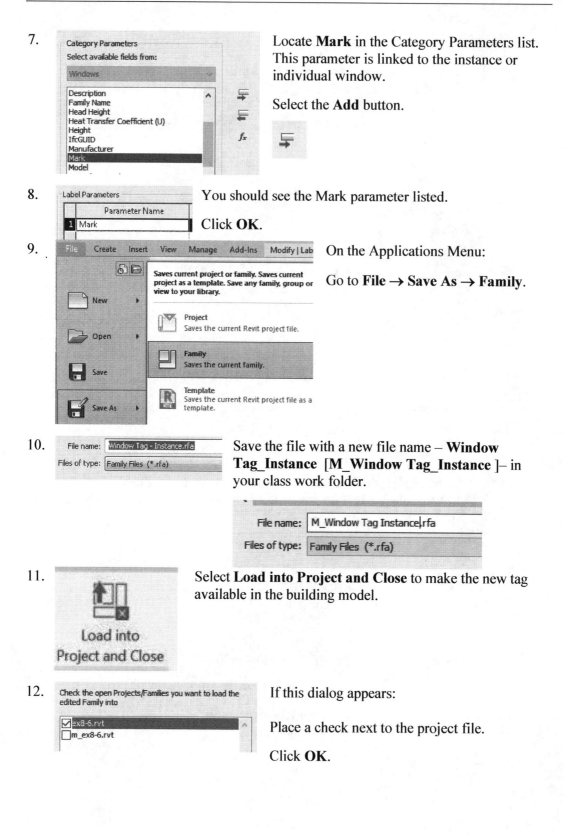

7. Locate **Mark** in the Category Parameters list. This parameter is linked to the instance or individual window.

Select the **Add** button.

8. You should see the Mark parameter listed.

Click **OK**.

9. On the Applications Menu:

Go to **File → Save As → Family**.

10. Save the file with a new file name – **Window Tag_Instance [M_Window Tag_Instance]**– in your class work folder.

11. Select **Load into Project and Close** to make the new tag available in the building model.

12. If this dialog appears:

Place a check next to the project file.

Click **OK**.

13. 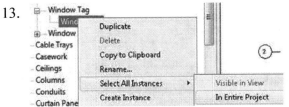 In the Project Browser:
Locate the **Window Tag** under the Annotation Symbols category.

Right click and select **Select All Instances → In Entire Project**.

14. The window tags will be selected in all views.

In the Properties Pane:

Use the Type Selector to select the **Window Tag – Instance [M_Window Tag – Instance]**.

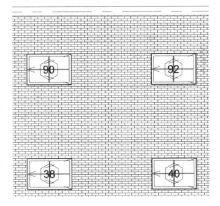

15. Activate the North Exterior elevations.

Zoom in to see that the windows are now renumbered using the Mark parameter.

16. Save as *ex8-6.rvt [m_ex8-6.rvt]*.

Exercise 8-7
Creating a Plan Region View

Drawing Name: plan region view.rvt (This can be located in the Class Files
                            downloaded for the text.)
Estimated Time: 20 minutes

Thanks to John Chan, one of my Revit students at SFSU, for this project!

This exercise reinforces the following skills:

- ❑ Plan Region View
- ❑ Split Level Views
- ❑ Linework

Some floor plans are split levels. In those cases, users need to create a Plan Region View in order to create a good floor plan.

1.   📂 Open *plan region view.rvt*.

    If you switch to a 3D view, you
    see that this is a split-level floor
    plan.

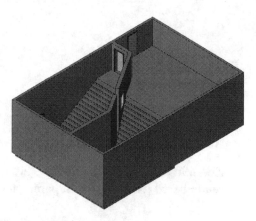

2.   Activate the **1/F Level 1** view under floor plans.

Floor Plans
- **1/F Level 1**
- 1/F Level 2
- 2/F
- Site

3.   You see the door and window tags but no doors or windows.

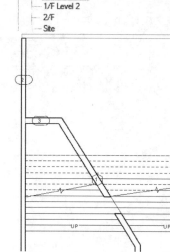

4. **View** Activate the **View** ribbon.

5. Select **Plan Views → Plan Region** tool on the Create panel.

6. Select the **Rectangle** tool from the ribbon.

7. Draw a rectangle around the region where you want the doors and windows to be visible.

Be sure to extend the rectangle across the entire plan or the stairs may not appear properly.

8. Select the **Edit** button next to View Range on the Properties pane.

9. Set the Offset for the Top plane to **10′ 0″**.
Set the Cut Plane Offset to **10′ 0″**.

Click **OK**.

Note: Your dimension values may be different if you are using your own model.

10. 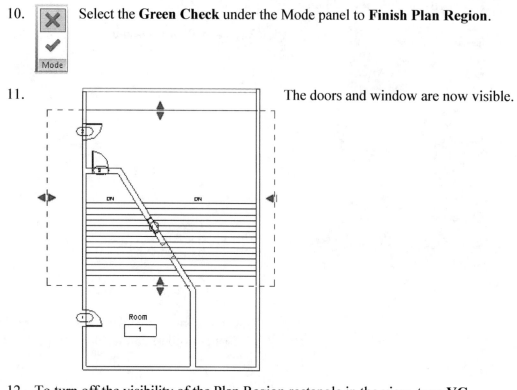 Select the **Green Check** under the Mode panel to **Finish Plan Region**.

11. The doors and window are now visible.

12. To turn off the visibility of the Plan Region rectangle in the view, type **VG**.

☑ Parking Tags
☑ Part Tags
☑ Path of Travel Tags
☐ Plan Region
☑ Planting Tags
☑ Plumbing Fixture Tags

Select the Annotations tab.
Disable **Plan Region**.
Click **OK**.

13. 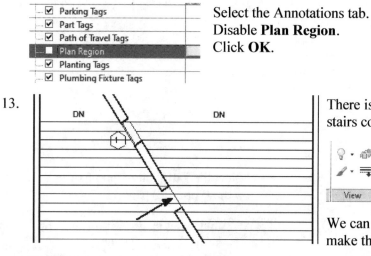 There is a line where the two stairs connect.

We can use the Linework tool to make this line invisible.

14. Modify | Activate the Modify ribbon.

15. Select the **Linework** tool under the View panel.

16. Select **Invisible Lines** from the Line Style drop-down list.

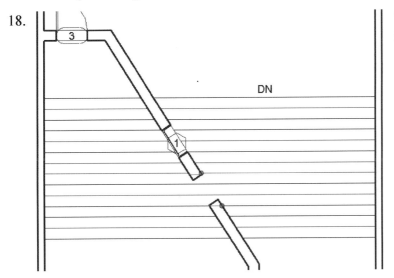

17. Select the lines you wish to make invisible.

You may need to pick the line more than once as there are overlapping lines.

18. The lines are now invisible.

19. Close without saving.

Drafting Views

Use a drafting view to create unassociated, view-specific details that are not part of the modeled design.

Rather than create a callout and add details to it, you may want to create detail conditions where the model is not needed (for example, a carpet-transition detail which shows where carpet switches to tile, or roof-drain details not based on a callout on the roof). For this purpose, create a drafting view.

In a drafting view, you create details at differing view scales (coarse, medium, or fine) and use 2D detailing tools: detail lines, detail regions, detail components, insulation, reference planes, dimensions, symbols, and text. These are the exact same tools used in creating a detail view. However, drafting views do not display any model elements. When you create a drafting view in a project, it is saved with the project.

Exercise 8-8
Creating a Drafting View

Drawing Name: drafting_view1.rvt [m_ drafting_view1.rvt]
Estimated Time: 40 minutes

This exercise reinforces the following skills:

- Detail Components
- Filled Regions
- Notes

1.   Open *drafting_view1.rvt [m_ drafting_view1.rvt]*.

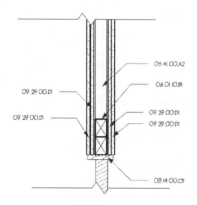

You will create this view of a Head Detail for an interior door using detail components, filled regions, and key notes.

Drafting views are 2D views which are not generated from the Revit model. They are re-usable views which can be used in any project to detail construction views. Once the view is imported or created, you link it to a call out on a model view.

2.   Activate the **View** ribbon.

    Select **Drafting View**.

3.   Type **Head Detail – Single Flush Door** for the name.

Click **OK**.

Name:	Head Detail – Single Flush Door
Scale:	1 : 10
Scale value 1:	10

A blank view is opened.

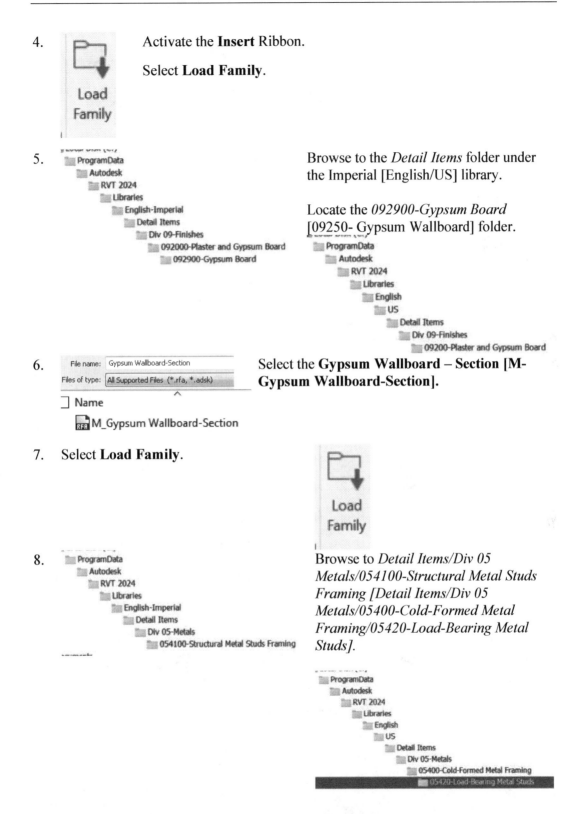

4. Activate the **Insert** Ribbon.

    Select **Load Family**.

5. Browse to the *Detail Items* folder under the Imperial [English/US] library.

    Locate the *092900-Gypsum Board* [09250- Gypsum Wallboard] folder.

6. Select the **Gypsum Wallboard – Section [M-Gypsum Wallboard-Section]**.

7. Select **Load Family**.

8. Browse to *Detail Items/Div 05 Metals/054100-Structural Metal Studs Framing [Detail Items/Div 05 Metals/05400-Cold-Formed Metal Framing/05420-Load-Bearing Metal Studs]*.

9.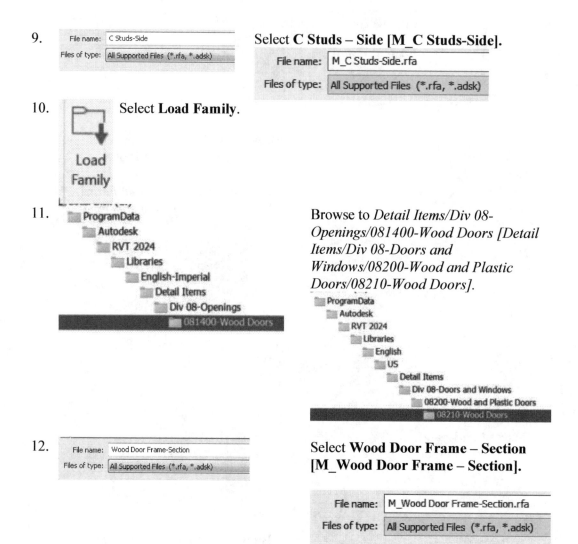

File name: C Studs-Side

Files of type: All Supported Files (*.rfa, *.adsk)

Select **C Studs – Side [M_C Studs-Side].**

File name: M_C Studs-Side.rfa

Files of type: All Supported Files (*.rfa, *.adsk)

10. Select **Load Family**.

Load Family

11. ProgramData
   Autodesk
      RVT 2024
         Libraries
            English-Imperial
               Detail Items
                  Div 08-Openings
                    081400-Wood Doors

Browse to *Detail Items/Div 08-Openings/081400-Wood Doors [Detail Items/Div 08-Doors and Windows/08200-Wood and Plastic Doors/08210-Wood Doors].*

ProgramData
   Autodesk
      RVT 2024
         Libraries
            English
               US
                  Detail Items
                    Div 08-Doors and Windows
                       08200-Wood and Plastic Doors
                         08210-Wood Doors

12. File name: Wood Door Frame-Section

Files of type: All Supported Files (*.rfa, *.adsk)

Select **Wood Door Frame – Section [M_Wood Door Frame – Section].**

File name: M_Wood Door Frame-Section.rfa

Files of type: All Supported Files (*.rfa, *.adsk)

When you place the detail components, be sure to use the Type Selector to select the correct size of component to be used.

The structure of the interior wall used in our project:

Layers		EXTERIOR SIDE	
	Function	Material	Thickness
1	Finish 2 [5]	Gypsum Wall Board	0' 0 5/8"
2	Finish 2 [5]	Gypsum Wall Board	0' 0 5/8"
3	**Core Boundary**	**Layers Above Wrap**	**0' 0"**
4	Structure [1]	Metal - Stud Layer	0' 2 1/2"
5	**Core Boundary**	**Layers Below Wrap**	**0' 0"**
6	Finish 2 [5]	Gypsum Wall Board	0' 0 5/8"
7	Finish 2 [5]	Gypsum Wall Board	0' 0 5/8"

Layers			EXTERIOR SIDE
	Function	Material	Thickness
1	Finish 2 [5]	Gypsum Wall Board	15.5
2	Core Boundary	Layers Above Wrap	0.0
3	Structure [1]	Metal Stud Layer	92.0
4	Core Boundary	Layers Below Wrap	0.0
5	Finish 2 [5]	Gypsum Wall Board	15.5
6	Finish 2 [5]	Gypsum Wall Board	15.5

13. Component Select the **Detail Component** tool from the Annotate ribbon.

14. Properties ✕

Gypsum Wallboard-Section 5/8"

Select **Gypsum Wallboard – Section 5/8″ [M-Gypsum Wallboard-Section 16mm]** on the Properties palette.

M_Gypsum Wallboard-Section 16mm

15. Place two instances of gypsum wall board by drawing two vertical lines **1' 0" [200mm]** tall.

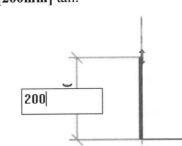

200

16. Component Select the **Detail Component** tool from the Annotate ribbon.

17. Properties ✕

C Studs-Side 2 1/2"

New Detail Items ▾ Edit Type

M_C Studs-Side 92mm

Select **C Studs – Side 2 1/2″ [M_C Studs- Side 92mm]** on the Properties palette.

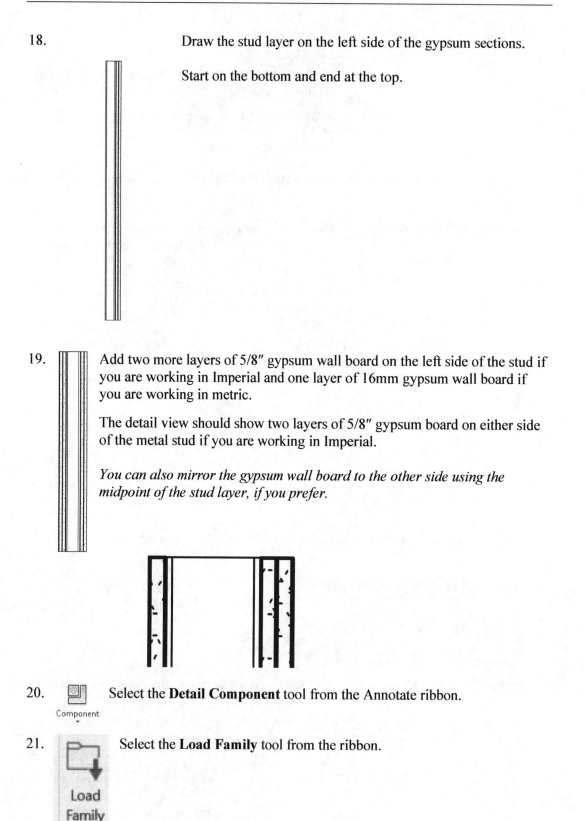

18.   Draw the stud layer on the left side of the gypsum sections.

Start on the bottom and end at the top.

19.   Add two more layers of 5/8″ gypsum wall board on the left side of the stud if you are working in Imperial and one layer of 16mm gypsum wall board if you are working in metric.

The detail view should show two layers of 5/8″ gypsum board on either side of the metal stud if you are working in Imperial.

You can also mirror the gypsum wall board to the other side using the midpoint of the stud layer, if you prefer.

20.   Select the **Detail Component** tool from the Annotate ribbon.

21.   Select the **Load Family** tool from the ribbon.

22.

Browse to the *061100 – Wood Framing folder* under Detail Items.

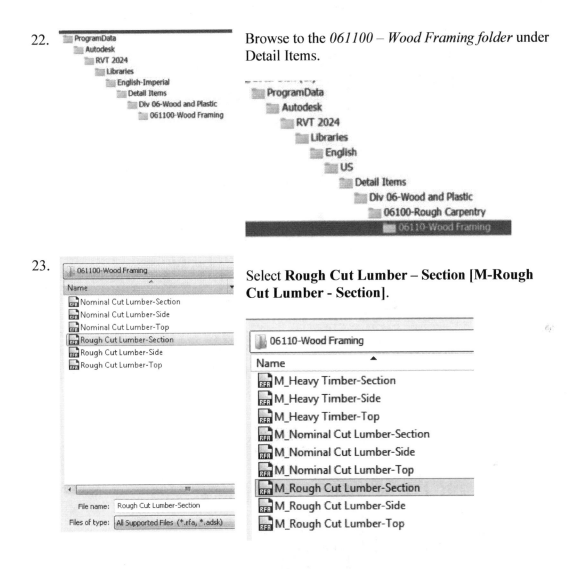

23.

Select **Rough Cut Lumber – Section [M-Rough Cut Lumber - Section]**.

24.

Locate the **2x3R [50 x 75mm]** size.

Click **OK**.

25. 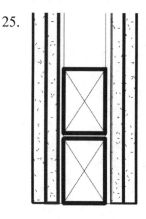 Place the headers between the gypsum board assemblies at the bottom.

For the metric detail, rotate the rough lumber.

26. 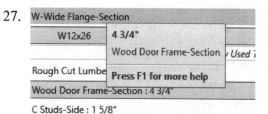 Select the **Detail Component** tool from the Annotate ribbon.

Component

27. Select **Wood Door Frame Section 4 3/4″ [M-Wood Door Frame - Section 110mm]** from the Properties palette.

W-Wide Flange-Section

W12x26 | 4 3/4″

Wood Door Frame-Section | Used i

Rough Cut Lumbe | Press F1 for more help

Wood Door Frame-Section : 4 3/4″

C Studs-Side : 1 5/8″

Gypsum Wallboard-Section : 3/4″

M_Wood Door Frame-Section 110mm

28. 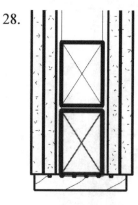 Place the door frame below the other detail components.

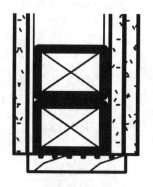

29. Select the **Filled Region** tool.

Annotate Analyze Massin

Detail Line | Region

Filled Region

30. 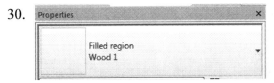 Select **Filled region: Wood 1** from the Properties palette.

31. Select **rectangle** from the Draw panel on the ribbon.

32. 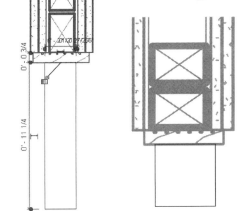 Draw the rectangle below the door frame.

33. Select **Green Check** on the ribbon to exit filled region mode.

34. Place the door using a filled region with a wood fill pattern.

35.

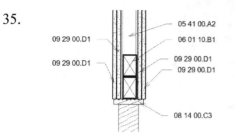

Add the keynotes.

The element keynote tool will be able to identify all the detail components placed.

36.

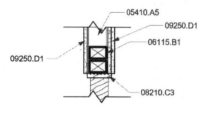

Add the break lines.

Component

Select the Detail Component tool.

Select **Break Line [M_Break Line]** using the Type Selector.

37. Save as *ex8-8.rvt [m_ ex8-8.rvt]*.

Callouts

You can create reference callouts, detail callouts, and view callouts.

View Callouts

Use a view callout when you want to provide more or different information about a part of the parent view.

For example, you can use a view callout to provide a more detailed layout of fixtures in a bathroom.

When you add a view callout to a view, Revit creates a view that has the same view type as the parent view.

Exercise 8-9
Adding a Callout

Drawing Name: callout1.rvt [m_ callout1.rvt]
Estimated Time: 10 minutes

This exercise reinforces the following skills:

- ❑ Detail Components
- ❑ Notes
- ❑ Callouts

Keep in mind that any changes to the model are not reflected in the drafting view, so if you create a drafting view of a wall construction and then change the wall materials in the model, the drafting view does not update.

1.   📂   Open *callout1.rvt [m_ callout1.rvt]*.

2.   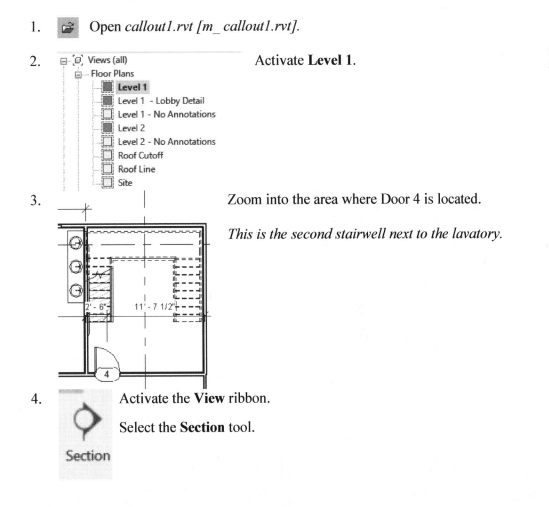   Activate **Level 1**.

3.   Zoom into the area where Door 4 is located.

     This is the second stairwell next to the lavatory.

4.   Activate the **View** ribbon.

     Select the **Section** tool.

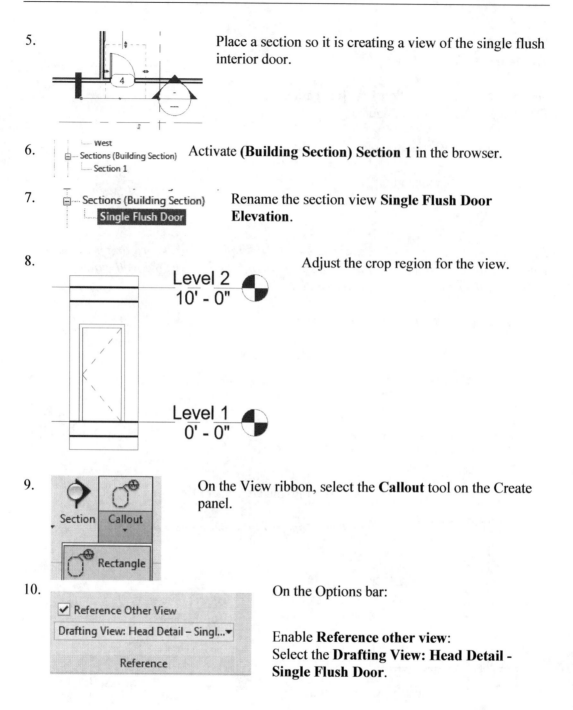

5.   Place a section so it is creating a view of the single flush interior door.

6.   Activate **(Building Section) Section 1** in the browser.

7.   Rename the section view **Single Flush Door Elevation**.

8.   Adjust the crop region for the view.

9.   On the View ribbon, select the **Callout** tool on the Create panel.

10.   On the Options bar:

Enable **Reference other view**:
Select the **Drafting View: Head Detail - Single Flush Door**.

11.

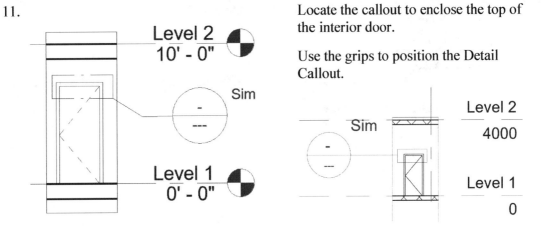

Locate the callout to enclose the top of the interior door.

Use the grips to position the Detail Callout.

The detail number and sheet number will be filled out when you place the view on a sheet automatically.

Green check to complete the view definition.

12. Double click on the callout bubble to open the detail view.

13. Save as *ex8-9.rvt [m_ex8-9.rvt]*.

Exercise 8-10
Adding a Detail to a Sheet

Drawing Name: detail_1.rvt [m_ detail_1.rvt]
Estimated Time: 15 minutes

This exercise reinforces the following skills:

- ❑ Views
- ❑ Sheets
- ❑ Schedules
- ❑ Keynote Legends

1.   📂   Open *detail_1.rvt [m_ detail_1.rvt]*.

2. Activate **AD501 - Details** sheet.

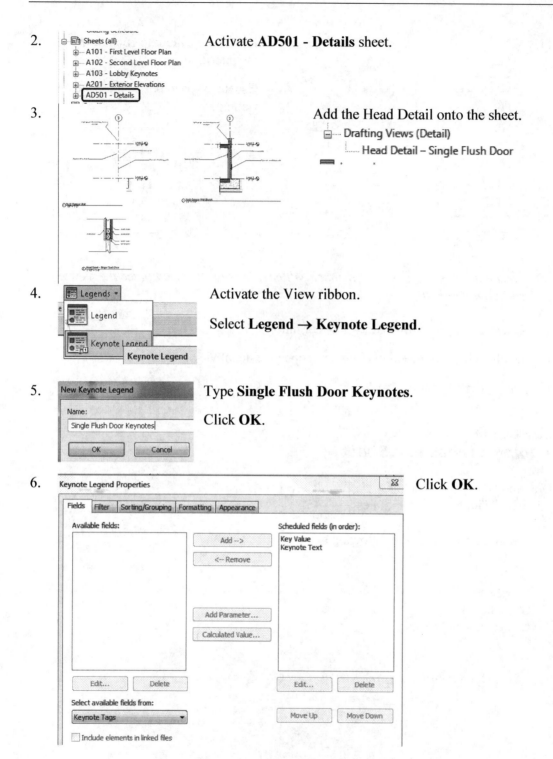

3. Add the Head Detail onto the sheet.

  Drafting Views (Detail)
   Head Detail – Single Flush Door

4. Activate the View ribbon.

Select **Legend → Keynote Legend**.

5. Type **Single Flush Door Keynotes**.

Click **OK**.

6. Click **OK**.

7.

<Single Flush Door Keynotes>	
A	**B**
Key Value	Keynote Text
04 21 00.A1	Standard Brick - 3/8" Joint
06 01 10.B1	2x3R
07 31 00.A1	Asphalt Shingles
08 14 00.C3	Wood Door Frame
08 81 00.F1	1" Clear Insulating Glass
09 29 00.D1	5/8" Gypsum Wallboard
09 51 00.A6	Square Edge (3/4 x 24 x 48)
09 65 00.A2	Vinyl Composition Tile
09 72 00.A1	Vinyl Wallcovering
09 91 00.A2	Semi-Gloss Paint Finish
26 51 00.A5	2' X 4' Surface Mounted Modular Fluorescent Fixture

The Keynote Legend appears in the view.

Note there are some materials listed we don't need to call out for the door. All keynotes which have been added in the project are listed.

8.

Other	
Fields	Edit...
Filter	Edit...
Sorting/Grouping	Edit...
Formatting	Edit...
Appearance	Edit...

Select the **Edit** button next to Filter.

9.

Fields Filter Sorting/Grouping Formatting Appearance

Filter by: (none)
And: (none)
And: (none)
And: (none)
And: (none)
And: (none)
And: (none)
And: (none)

☑ Filter by sheet

Enable **Filter by sheet** at the bottom of the dialog.

Click **OK**.

This ensures that when the legend is placed on the sheet only the keynotes used in any views on the sheet will be listed.

10.

Schedule: Single Flush Door Keynotes   ⌄   🔲 Edit Type

Identity Data	
View Template	Schedule using Tahoma Fonts
View Name	Single Flush Door Keynotes
Dependency	Independent
Other	

Apply the **Schedule using Tahoma Fonts** View Template.

11.

- Sheets (all)
  - ⊞ A101 - First Level Floor Plan
  - ⊞ A201 - Exterior Elevations
  - ⊞ A601 - Door Schedule
  - ⊞ A602 - Glazing Schedule
  - ⊞ A603 - LOBBY KEYNOTES
  - ⊞ **AD501 - Details**
- Families

Activate **AD501 - Details** sheet.

12.

- Legends
  - FINISH SCHEDULE KEYS
  - Single Flush Door Keynotes

Locate the Keynote Legend for the Single Flush Door.

13.

Drag and drop it onto the Details sheet.

Single Flush Door Keynotes	
Key Value	Keynote Text
05 41 00.A2	2-1/2" Metal Stud
06 01 10.B1	2x3R
08 14 00.C3	Wood Door Frame
09 29 00.D1	5/8" Gypsum Wallboard

Single Flush Door Keynotes	
Key Value	Keynote Text
06115.B1	50x75R
08210.C3	110mm Wood Door Frame
09250.D1	16mm Gypsum Wallboard

Notice that the legend automatically changes to reflect the keynotes on the view on the sheet.

Legend views can be placed on multiple sheets.

14.
(3) Head Detail - Single Flush Door
1 1/2" = 1'-0"

Zoom into the Head Detail.

The bubble is filled in with the number 3. This is the detail number.

(3) Head Detail – Single Flush Door
1 : 10

15.
Single Flush Door Keynotes
Schedules/Quantities (all)
Door Details
Door Schedule
Glazing Schedule

Activate the **Door Details** schedule.

16. Modify the detail number for the Head Detail for the interior single flush doors.

\<Door Details>

A	B	C	D	E	F	G	H
			SIZE			DETAILS	
Door No	Door Type	W	H	THK	Head Detail	Jamb Detail	Threshold Detail
1	36" x 84"	3' - 0"	7' - 0"	0' - 2"	HD3	JD4	TD4
2	36" x 84"	3' - 0"	7' - 0"	0' - 2"	HD3	JD4	TD4
3	36" x 84"	3' - 0"	7' - 0"	0' - 2"	HD3	JD4	TD4
4	36" x 84"	3' - 0"	7' - 0"	0' - 2"	HD3	JD4	TD4
5	72" x 82"	6' - 0"	6' - 10"	0' - 1 3/4"	HD2	JD2	TD2
6	72" x 82"	6' - 0"	6' - 10"	0' - 1 3/4"	HD2	JD2	TD2
7	36" x 84"	3' - 0"	7' - 0"	0' - 2"	HD3	JD4	TD4
8	36" x 84"	3' - 0"	7' - 0"	0' - 2"	HD3	JD4	TD4
9	36" x 84"	3' - 0"	7' - 0"	0' - 2"	HD3	JD4	TD4
10	36" x 84"	3' - 0"	7' - 0"	0' - 2"	HD3	JD4	TD4
21	Door-Curtain-Wall-	6' - 9 1/16"	8' - 0"		HD1	JD1	TD1

17. Click **OK**.

> Revit ✕
>
> This change will be applied to all elements of type
> Single-Flush: 36" x 84".
>
> OK   Cancel

18.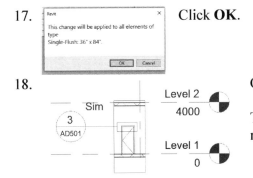

Open the **Single Flush Door Elevation** view.

The callout bubble has updated with the detail number and the sheet number.

19. Save as *ex8-10.rvt [m_ ex8-10.rvt]*.

Exercise 8-11
Importing a Detail View

Drawing Name: import_detail.rvt [m_ import_detail.rvt]
Estimated Time: 20 minutes

This exercise reinforces the following skills:

- ❑ Import CAD
- ❑ Sheets
- ❑ Detail Views

1.   Open *import_detail.rvt [m_ import_detail.rvt]*.
Activate Level 1.

2.   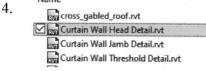   Activate **Level 1**.

3.   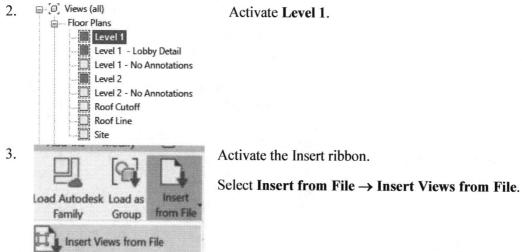   Activate the Insert ribbon.

                                  Select **Insert from File → Insert Views from File**.

If your active view is a schedule or legend, you will not be able to insert a view from a file.

4.   Locate the **Curtain Wall Head Detail** file
available in the downloaded Class Files.
Click **Open**.

5.   A preview of the detail will appear.

Click **OK**.

6.

Click **OK**.

The Drafting View is listed in the Project Browser.

7.

Activate the **AD501 - Details** sheet.

8.

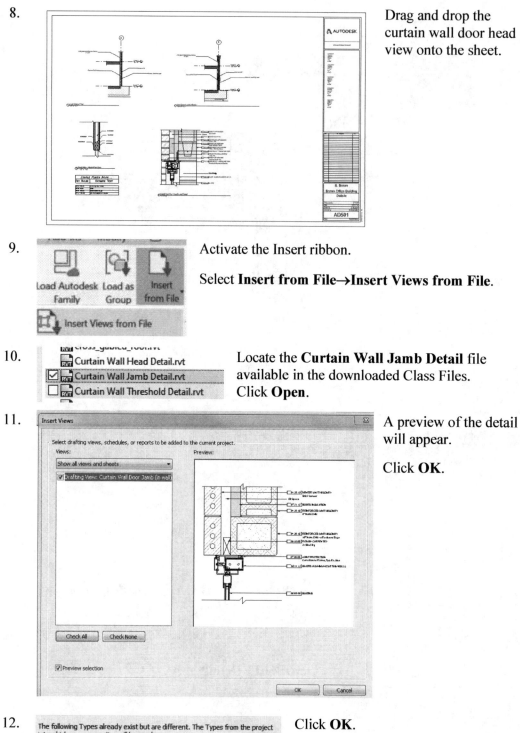

Drag and drop the curtain wall door head view onto the sheet.

9.

Activate the Insert ribbon.

Select **Insert from File→Insert Views from File**.

10.

Locate the **Curtain Wall Jamb Detail** file available in the downloaded Class Files. Click **Open**.

11.

A preview of the detail will appear.

Click **OK**.

12.

The following Types already exist but are different. The Types from the project into which you are pasting will be used.

Generic(2)
Materials : Poche

Click **OK**.

Drafting Views (ARCxl)
  Curtain Wall Door Head (in wall) Detail
  Curtain Wall Door Jamb (in wall) Detail
Drafting Views (Detail)
  Head Detail – Single Flush Door

The Drafting View is listed in the Project Browser.

13. Sheets (all)
  A101 - First Level Floor Plan
  A102 - Level 2 Ceiling Plan
  A201 - Exterior Elevations
  A601 - Door Schedule
  A602 - Glazing Schedule
  A603 - Lobby Keynotes
  AD501 - Details

Activate the **AD501 - Details** sheet.

14.

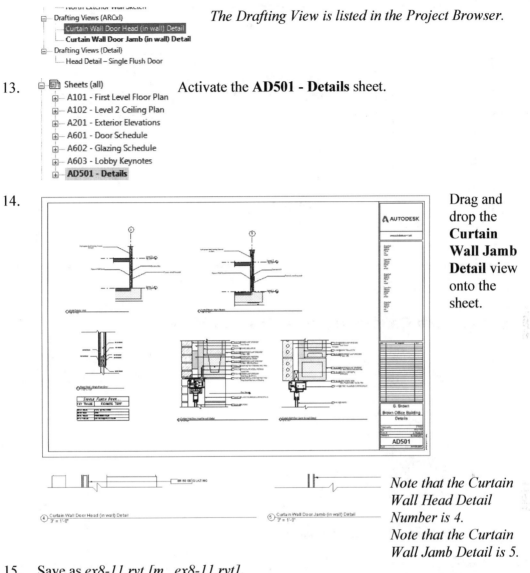

Drag and drop the **Curtain Wall Jamb Detail** view onto the sheet.

Note that the Curtain Wall Head Detail Number is 4.
Note that the Curtain Wall Jamb Detail is 5.

15. Save as *ex8-11.rvt [m_ ex8-11.rvt]*.

Extra: *Change the Door Details schedule to reflect the correct detail numbers.*
Create a section view of the curtain wall door.
Add callouts to the section view of the curtain wall door and link them to the correct drafting views.

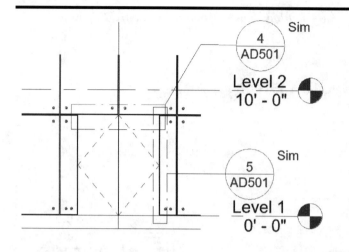

Exercise 8-12
Reassigning a Callout to a new Drafting View

Drawing Name: callout2.rvt [m_ callout_2.rvt]
Estimated Time: 15 minutes

This exercise reinforces the following skills:

-   Callouts
-   Drafting Views
-   Detail Items
-   Views

1.   Open *callout2.rvt [m_ callout_2.rvt]*.

2.   Activate **Head Detail – Single Flush Door** detail view.

3.   Right click and select **Duplicate View → Duplicate with Detailing**.

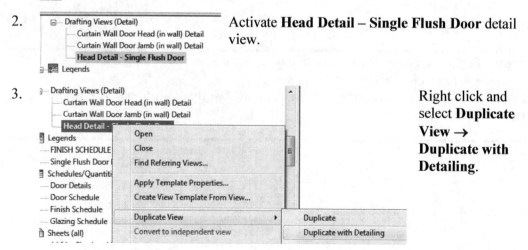

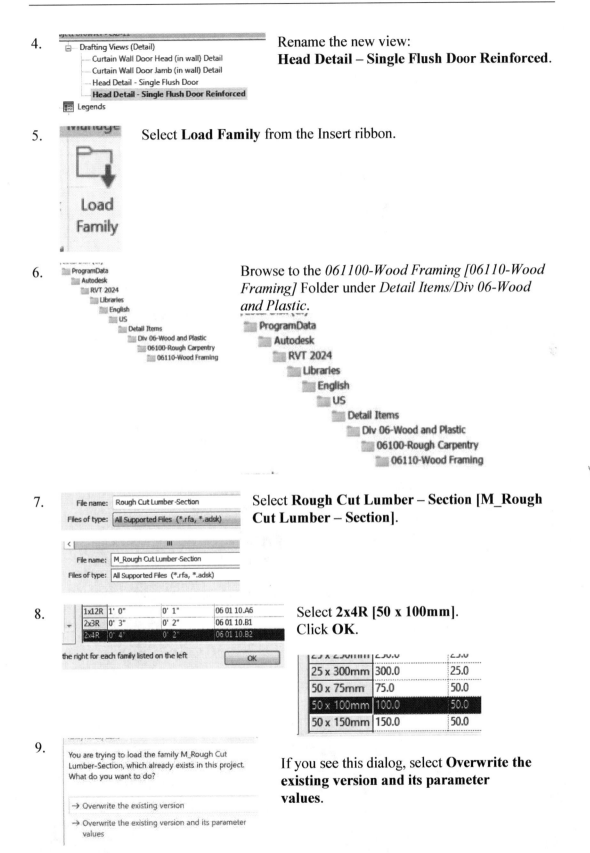

4.

Rename the new view:
Head Detail – Single Flush Door Reinforced.

5.

Select **Load Family** from the Insert ribbon.

6.

Browse to the *061100-Wood Framing [06110-Wood Framing]* Folder under *Detail Items/Div 06-Wood and Plastic*.

7.

Select **Rough Cut Lumber – Section [M_Rough Cut Lumber – Section]**.

8.

Select **2x4R [50 x 100mm]**.
Click **OK**.

9.

If you see this dialog, select **Overwrite the existing version and its parameter values**.

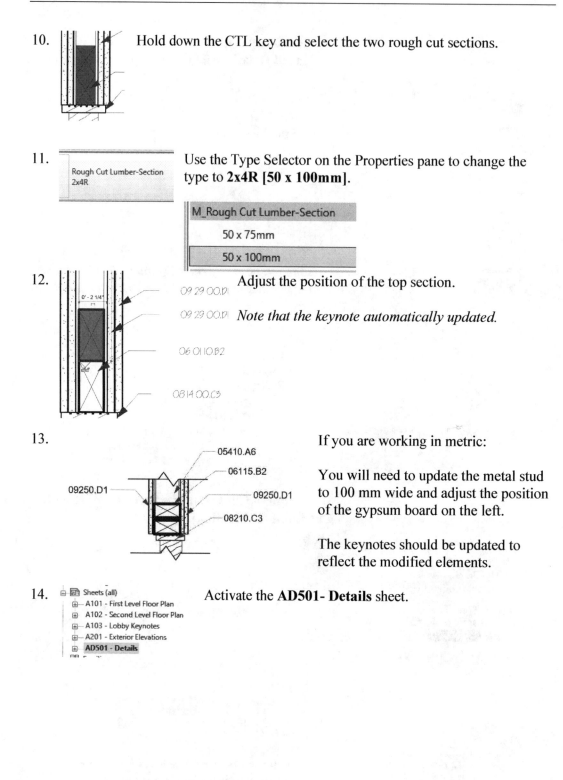

10. Hold down the CTL key and select the two rough cut sections.

11. Use the Type Selector on the Properties pane to change the type to **2x4R [50 x 100mm]**.

Rough Cut Lumber-Section
2x4R

M_Rough Cut Lumber-Section

50 x 75mm

50 x 100mm

12. Adjust the position of the top section.

Note that the keynote automatically updated.

13. If you are working in metric:

You will need to update the metal stud to 100 mm wide and adjust the position of the gypsum board on the left.

The keynotes should be updated to reflect the modified elements.

14. Activate the **AD501- Details** sheet.

Sheets (all)
- A101 - First Level Floor Plan
- A102 - Second Level Floor Plan
- A103 - Lobby Keynotes
- A201 - Exterior Elevations
- **AD501 - Details**

15.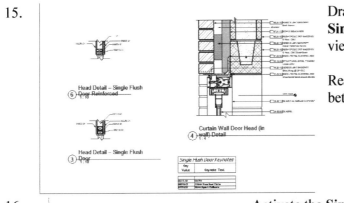

Drag and drop the **Head Detail –
Single Flush Door Reinforced**
view on the sheet.

Rearrange the views, so they look
better.

16.
Sections (Building Section)
Curtain Wall Door Elevation
Single Flush Door Elevation

Activate the **Single Flush Door Elevation**.

17. Select the callout in the view.
Hint: Click on the rectangle.
On the Options bar:
Select the Drafting view for the **Head Detail – Single Flush Door Reinforced**.

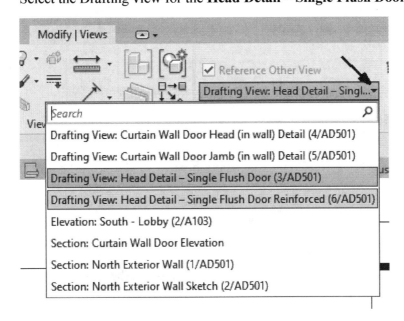

Note that the detail and sheet number are displayed for each view.

18.

6
AD501

The callout should update with the new view number.

19. Save as *ex8-12.rvt [m_ex8-12.rvt]*.

Exercise 8-13
Using a Matchline

Drawing Name: matchline.rvt [m_ matchline.rvt]
Estimated Time: 30 minutes

This exercise reinforces the following skills:

- ❑   Matchline
- ❑   Sheets
- ❑   Views
- ❑   View Reference Annotation

1.   Open *matchline.rvt [m_ matchline.rvt]*

2.   Activate **Level 1**.

- Views (all)
  - Floor Plans
    - **Level 1**
    - Level 1 - Lobby Detail
    - Level 1 - No Annotations
    - Level 2
    - Level 2 - No Annotations
    - Roof Cutoff
    - Roof Line
    - Site

3.   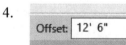 Select the **Grid** tool from the Architecture ribbon.

4.   On the Options ribbon, set the Offset to **12' 6"** **[3810mm]**.

Offset: 12' 6"

Offset: 3810

5.   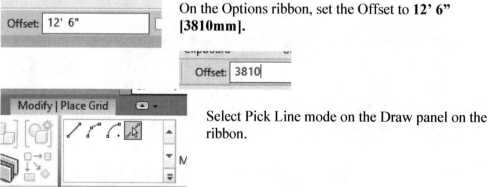 Select Pick Line mode on the Draw panel on the ribbon.

6.

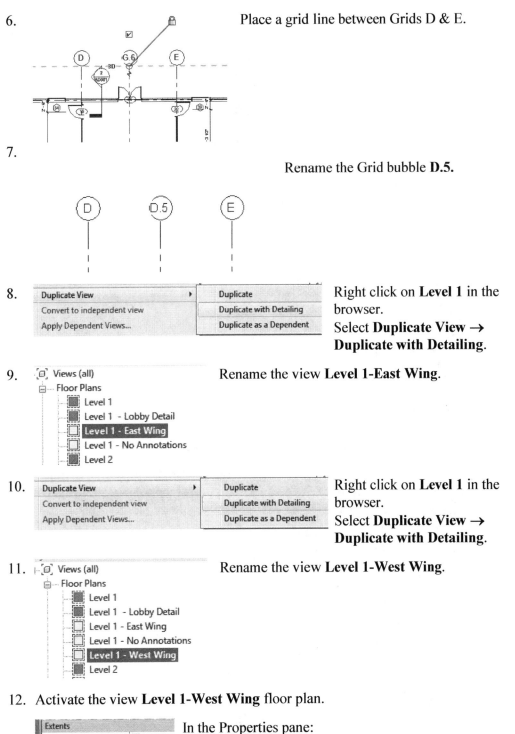

Place a grid line between Grids D & E.

7.

Rename the Grid bubble **D.5.**

8.

Duplicate View ▶	Duplicate
Convert to independent view	Duplicate with Detailing
Apply Dependent Views...	Duplicate as a Dependent

Right click on **Level 1** in the browser.

Select **Duplicate View →
Duplicate with Detailing**.

9.

Views (all)
Floor Plans
Level 1
Level 1 - Lobby Detail
Level 1 - East Wing
Level 1 - No Annotations
Level 2

Rename the view **Level 1-East Wing.**

10.

Duplicate View ▶	Duplicate
Convert to independent view	Duplicate with Detailing
Apply Dependent Views...	Duplicate as a Dependent

Right click on **Level 1** in the browser.

Select **Duplicate View →
Duplicate with Detailing**.

11.

Views (all)
Floor Plans
Level 1
Level 1 - Lobby Detail
Level 1 - East Wing
Level 1 - No Annotations
Level 1 - West Wing
Level 2

Rename the view **Level 1-West Wing.**

12. Activate the view **Level 1-West Wing** floor plan.

Extents	
Crop View	☑
Crop Region Visible	☑
Annotation Crop	☐

In the Properties pane:
Enable **Crop View**.
Enable **Crop Region Visible**.

13.

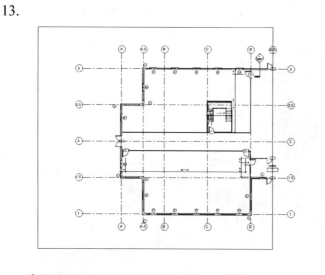

Use the grips on the crop region to only show the west side of the floor plan.

You may need to zoom out to see the crop region.

14.

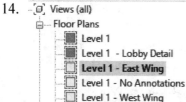

Activate **Level 1 - East Wing**.

15.

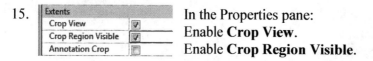

In the Properties pane:
Enable **Crop View**.
Enable **Crop Region Visible**.

16.

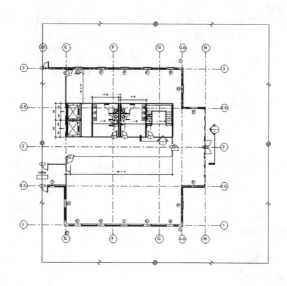

Use the grips on the crop region to only show the east side of the floor plan.

17. Activate the **View** ribbon.

18.  Under the Sheet Composition panel:

Select the **Matchline** tool.

19.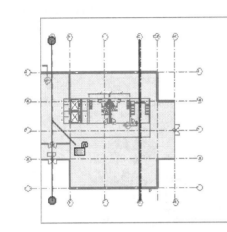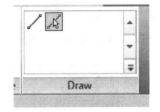

Activate **Pick Line** mode.

Select **Grid D.5**.

20. 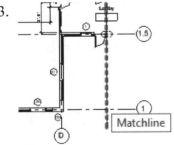 Select the **Green Check** under Mode to finish the matchline.

21. Disable **Crop Region Visible** in the Properties pane.

Extents	
Crop View	☑
Crop Region Visible	☐
Annotation Crop	☐

22. Level 1 - Lobby Detail
Level 1 - West Wing
Level 1- East Wing

Activate the **Level 1 - West Wing** floor plan.

23.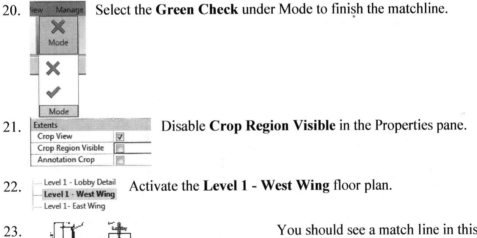

You should see a match line in this view as well.

24. Disable **Crop Region Visible** in the Properties pane.

Extents	
Crop View	☑
Crop Region Visible	☐
Annotation Crop	☐

25.   Add a new **Sheet** using the Sheet Composition panel on the View ribbon.

     Click **OK** to accept the default title block.

26.

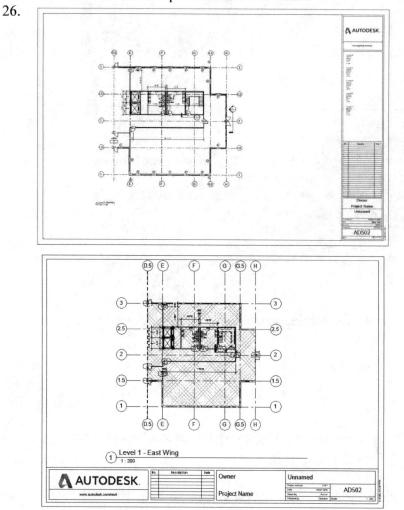

Place the **Level 1-East Wing** floor plan on the sheet.

Adjust the scale so it fills the sheet.

Adjust the grid lines, if needed.

27.

M. Instructor

J. Student

M. Instructor

J. Student

A105

Level 1 - East Wing

Change the Sheet Number to **A105**.
Name the sheet **Level 1 - East Wing**.

Enter in your name and your instructor's name.

28.   Add a new **Sheet** using the Sheet Composition panel on the View ribbon.

     Click **OK** to accept the default title block.

29.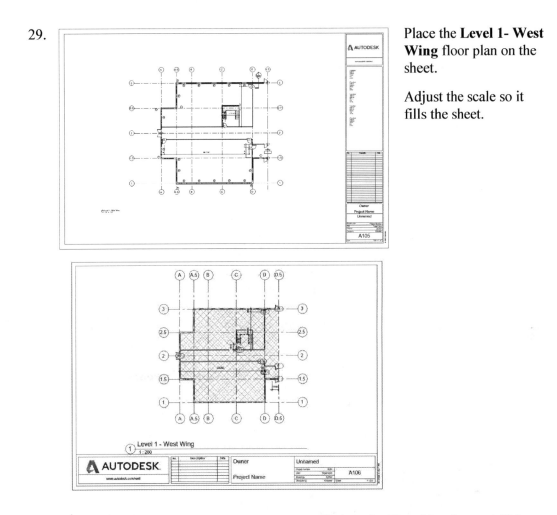

Place the **Level 1- West Wing** floor plan on the sheet.

Adjust the scale so it fills the sheet.

30.

M. Instructor	
J. Student	
M. Instructor	
J. Student	
A106	
Level 1 - West Wing	

Change the Sheet Number to **A106**.

Name the sheet **Level 1 -West Wing**.

Enter in your name and your instructor's name.

31.

(1) Level 1 - West Wing
    3/16" = 1'-0"

Zoom into the view title bar on the sheet.

(1) Level 1 - West Wing
    1 : 200

32.

- Views (all)
  - Floor Plans
    - Level 1
      - Level 1 - East Wing

Activate the **Level 1** floor plan.

33. 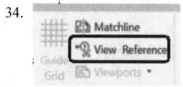 You can see the matchline that was added to the view if you zoom into the D.5 grid line.

34. Activate the View ribbon.

Select the **View Reference** tool on the Sheet Composition panel.

This adds an annotation to a view.

35. 1 / A106   1 / A105

D   D.5   E   Place the note on the right side of the grid line to correspond with the view associated with the right wing.

Cancel out of the command.

36. 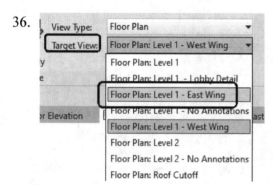 Select the **View Reference** label.

Set the Target View on the ribbon to the **Floor Plan: Level 1 – East Wing.**

1 / A106   1 / A105

D   D.5   E   *The View Reference label updates to the correct sheet number.*

37. 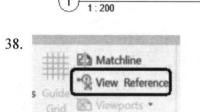 Note that the **A106 – Level 1 – East Wing** sheet view is also labeled 1.

38. Activate the View ribbon.

Select the **View Reference** tool on the Sheet Composition panel.

39.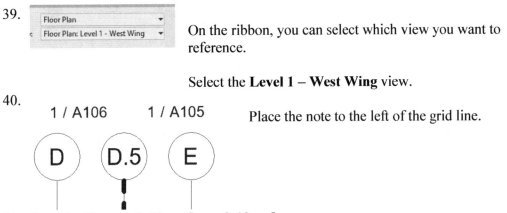

On the ribbon, you can select which view you want to reference.

Select the **Level 1 – West Wing** view.

40.

1 / A106   1 / A105

Place the note to the left of the grid line.

41. Save the file as *ex8-13.rvt [m_ex8-13.rvt]*.

Exercise 8-14
Modifying a Crop Region

Drawing Name: crop_region.rvt [m_ crop_region.rvt]
Estimated Time: 15 minutes

This exercise reinforces the following skills:

□   Crop Region
□   Duplicate View
□   Views

1.   Open *crop_region.rvt [m_ crop_region.rvt]*.

2.   Activate **Level 1 – Interior Plan**.

Views (all)
Floor Plans
Level 1
Level 1 - Lobby Detail
Level 1 - East Wing
Level 1 - Interior Plan
Level 1 - No Annotations
Level 1 - West Wing
Level 2

3.   Turn on the visibility of the crop region.

4.

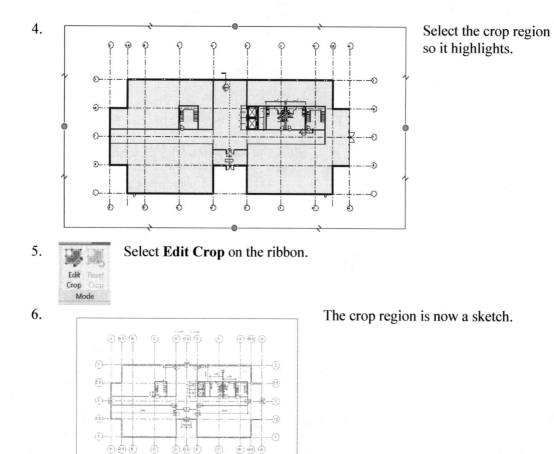

Select the crop region so it highlights.

5.   Select **Edit Crop** on the ribbon.

6.   The crop region is now a sketch.

7.

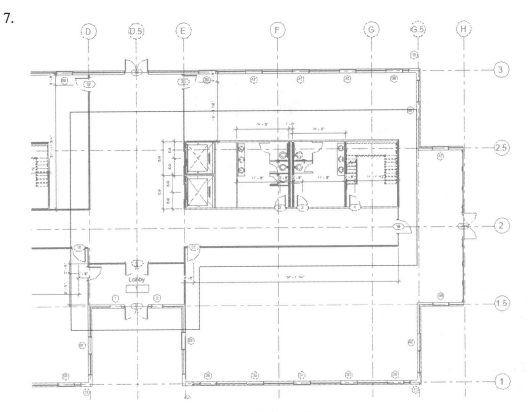

Add lines and trim to enclose the area indicated –
the right stairs, the lavatories, the elevators, and the lobby area.

8. Select the **Green check** to finish.

9. If you see this error, the sketch is not closed or needs to be trimmed.

> **Error - cannot be ignored**
>
> Crop area sketch can either be empty or include one closed, not self-intersecting loop.
>
> Show   More Info   Expand >>

Zoom out to see if there are any stray lines.

Click **Continue** and fix the sketch.

10.

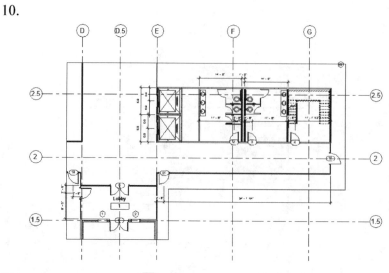

The new view is displayed.

Note that crop region sketches can only use lines, not arcs.

11. 1/8" = 1'-0" ⊠ ⊡ ⌗ ⌗ ⌗ ⌗ ⌗ ♀ ⌗ ◂   Turn off visibility of the crop region.

12. Save as *ex8-14.rvt [m_ex8-14.rvt]*.

Exercise 8-15
Updating a Schedule Using Shared Parameters

Drawing Name: schedule8.rvt [m_schedule8.rvt]
Estimated Time: 5 minutes

This exercise reinforces the following skills:

- ❏ Parameters
- ❏ Schedules

1. 📂   Open *schedule8.rvt [m_schedule8.rvt]*.

2. Locate the Curtain Wall-Double-Glass door in the Project Browser.

Right click and select **Type Properties**.

3. Notice that the Head and Jamb Detail properties updated.

Refer to the Details sheet for the detail view number.

Click **OK**.

4. Schedules/Quantities
    Door Details
    Door Schedule
    Glazing Schedule
    Lobby Keynotes
    Room Schedule

    Activate the **Door Details** schedule.

5. Notice that the schedule has updated.

36" x 84"	3' - 0"	7' - 0"	0' - 2"	HD3	JD4	TD4
36" x 84"	3' - 0"	7' - 0"	0' - 2"	HD3	JD4	TD4
36" x 84"	3' - 0"	7' - 0"	0' - 2"	HD3	JD4	TD4
36" x 84"	3' - 0"	7' - 0"	0' - 2"	HD3	JD4	TD4
12000 x 2290 Opening	2' - 11 1/2"	7' - 6 1/4"	0' - 1 1/4"	HD3	JD3	TD3
12000 x 2290 Opening	2' - 11 1/2"	7' - 6 1/4"	0' - 1 1/4"	HD3	JD3	TD3
Door-Curtain-Wall-Double-Glass	6' - 9"	8' - 0"		HD4	JD5	TD1

Check the detail numbers for the curtain wall.

6. Save as *ex8-15.rvt [m_ ex8-15.rvt]*.

Exercise 8-16
Create a Sheet List

Drawing Name: sheet_list.rvt [m_ sheet_list.rvt]
Estimated Time: 20 minutes

This exercise reinforces the following skills:

- Sheet List Schedule
- Sheets

1. Open *sheet_list.rvt [m_ sheet_list.rvt]*.

2. 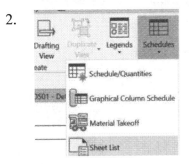 Activate the **View** ribbon.

    Select **Schedules→Sheet List**.

3.

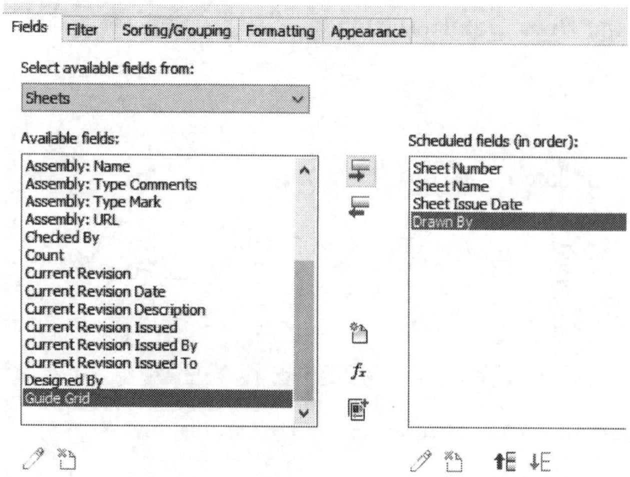

Add the following fields in this order:

- Sheet Number
- Sheet Name
- Sheet Issue Date
- Drawn By

4.

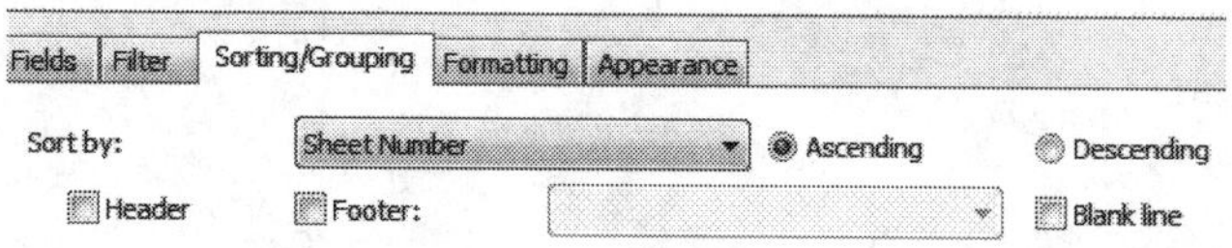

Select the Sorting/Grouping tab.
Sort by **Sheet Number**.

5.

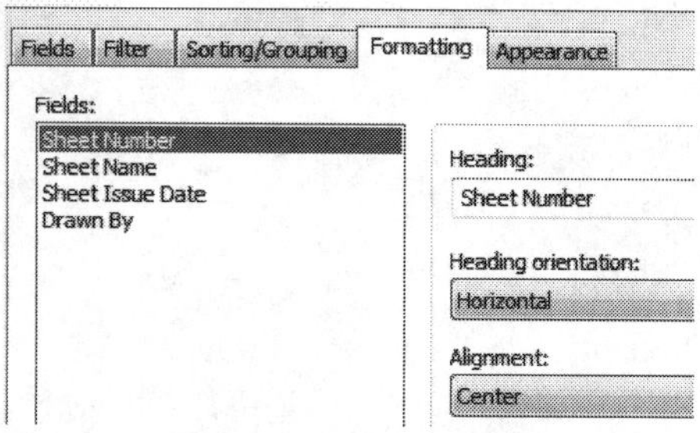

Select the Formatting tab.

Highlight the Sheet Number field.
Change the Alignment to **Center**.

6.

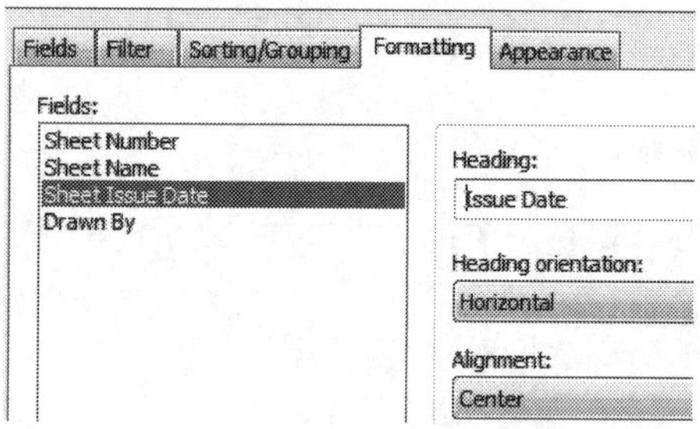

Highlight the Sheet Issue Date.
Change the Heading to **Issue Date**.
Change the Alignment to **Center**.

7.

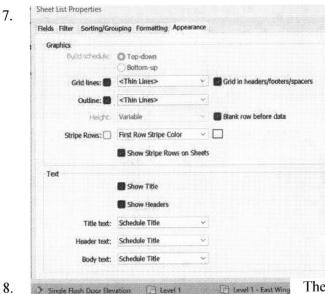

Select the Appearance tab.
Enable Grid lines.
Enable Outline
Enable Grid in headers/
footer/spacers.
Disable blank row before data.
Enable Show Title.
Enable Show Headers.
Change the Title text to
Schedule Title.
Change the Header Text to
Schedule Header.
Change the Body text to
Schedule Cell.
Click **OK**.

8.

A	B	C	D
Sheet Number	Sheet Name	Sheet Issue Date	Drawn By
A101	First Level Floor Plan	01/22/21	J Student
A102	Second Level Floor Plan	01/22/21	J. Student
A103	Lobby Keynotes	01/24/21	J. Student
A104	Level 1 - East Wing	01/28/21	J. Student
A105	Level 1 - West Wing	01/28/21	J Student
A201	Exterior Elevations	01/27/21	J. Student
AD501	Details	01/27/21	J. Student

<Sheet List>

The Sheet List appears.

This is a good way to check if you added your name as the Author to all sheets.

Can you find a sheet without your name?

9.

Glazing Schedule
Sheets (all)
- A101 - First Level Floor Plan
- A102 - Second Level Floor Pla
- A103 - Lobby Keynotes
- A104 - Level 1 - East Wing
- A105 - Level 1 - West Wing
- A201 - Exterior Elevations
- AD501 - Details
- **G101 - Cover Page**

Open the **G101 – Cover Page** sheet.

Sheets (all)
- A101 - Landing Page
- A102 - First Level Floor Plan
- A103 - Second Level Floor Plan
- A104 - LOBBY KEYNOTES
- A105 - Level 1 - East Wing
- A106 - Level 1 - West Wing
- A201 - Exterior Elevations
- AD501 - Details
- **G101 - Cover Page**

10. Drag and drop the sheet list schedule onto the sheet.

Sheet List			
Sheet Number	Sheet Name	Issue Date	Drawn By
A101	First Level Floor Plan	01/22/21	J Student
A102	Second Level Floor Plan	01/22/21	J. Student
A103	Lobby Keynotes	01/24/21	J. Student
A104	Level 1 - East Wing	01/28/21	J. Student
A105	Level 1 - West Wing	01/28/21	J Student
A201	Exterior Elevations	01/27/21	J. Student
AD501	Details	01/27/21	J. Student
G101	Cover Page	01/28/21	J Student

11.

Sheet: Cover Page	∨ 🔠 Edit Type
Approved By	M. Instructor
Designed By	J. Student
Checked By	M. Instructor
Drawn By	J. Student
Sheet Number	G101
Sheet Name	Cover Page
Sheet Issue Date	03/03/22
Appears In Sheet List	☐
Revisions on Sheet	Edit...

Fill in the properties for the sheet. Uncheck **Appears in Sheet List**.

12. Save as *ex8-16.rvt [m_ ex8-16.rvt]*.

Exercise 8-17
Create a PDF Document Set

Drawing Name: print_pdf.rvt
Estimated Time: 10 minutes

This exercise reinforces the following skills:

❑   Printing

1.   Open *print-pdf.rvt*.

2.   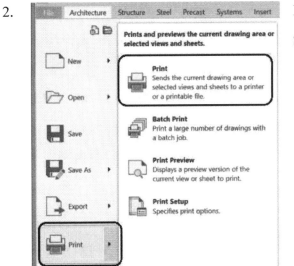   In the Application Menu:

Select **Print→Print**.

3.   Select **Adobe PDF** as the printer.

Enable **combine multiple selected views/ sheets into a single file**.

Enable **Selected views/sheets**.

Click the **Select** button to select which sheets to print.

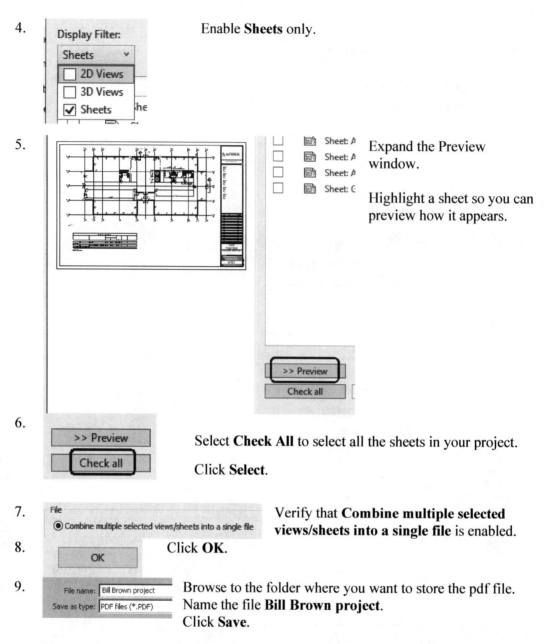

4. Enable **Sheets** only.

5. Expand the Preview window.

    Highlight a sheet so you can preview how it appears.

6. Select **Check All** to select all the sheets in your project.

    Click **Select**.

7. Verify that **Combine multiple selected views/sheets into a single file** is enabled.

8. Click **OK**.

9. Browse to the folder where you want to store the pdf file.
    Name the file **Bill Brown project**.
    Click **Save**.

You may want to save the file with your last name so your instructor can identify your work.

10.

Locate the file and preview it before you turn it in or send it to someone.

Notes:

Additional Projects

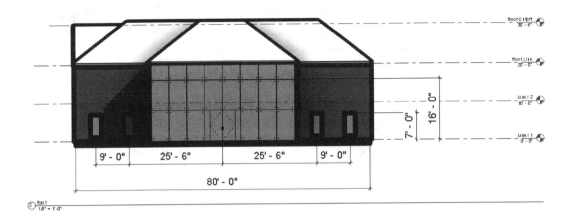

1) Create East & West Elevations view on a sheet.
   Add dimensions to the East Elevation view.

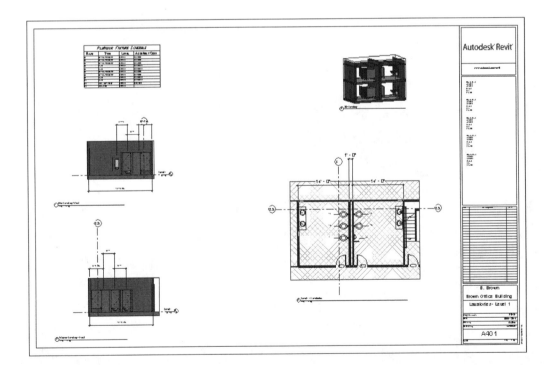

2) Create a sheet for the level 1 Lavatories. The sheet should have a plumbing
   fixture schedule, interior views for the men's and women's lavatories, a cropped
   view of the floor plan and the 3D view as shown.

3) Create Head, Threshold, and Jamb Details for the remaining doors and windows.

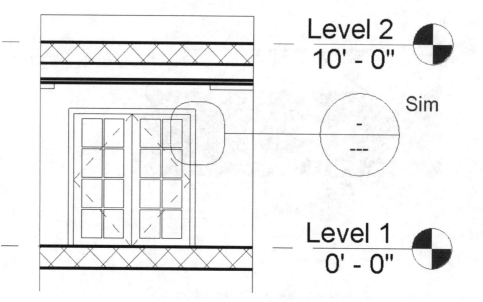

4) Create a sheet with elevations for each door and window.

5) Create a sheet with detail views for the doors and windows.

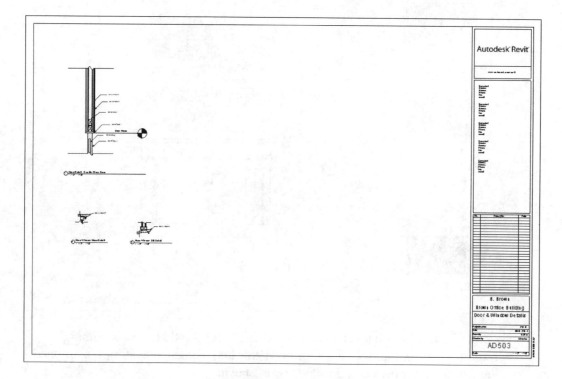

Lesson 8 Quiz

True or False

1. You have to deactivate a view before you can activate a different view.
2. When adding dimensions to an elevation, you add horizontal dimensions, not vertical dimensions.
3. The Linework tool is used to change the line style of a line in a view.
4. If you change the line style of a line in one view, it will automatically update in all views.
5. Double clicking on a blue symbol will automatically activate the view linked to that symbol.

Multiple Choice

6. To add a view to a sheet, you can:

    A. Drag and drop the view name from the browser onto the sheet
    B. Go to View→New→Add View
    C. Select 'Add View' from the View Design Bar
    D. All of the above

7. To add a sheet, you can:

    A. Select the New Sheet tool from the View ribbon
    B. Right click on a sheet and select New Sheet
    C. Highlight Sheets in the Browser, right click and select 'New Sheet'
    D. Go to File→New Sheet

8. To control the depth of an elevation view (visible objects behind objects):

    A. Change to Hidden Line mode
    B. Adjust the location of the elevation clip plane
    C. Adjust the Section Box
    D. Change the View Underlay

9. The keyboard shortcut key for Linework is:

    A. L
    B. LW
    C. LI
    D. LK

10. The text leader option NOT available is:

    A. Leader with no shoulder
    B. Leader with shoulder
    C. Curved Leader
    D. Spline Leader

11. The number 1 in the section symbol shown indicates:

    A. The sheet number
    B. The sheet scale
    C. The elevation number
    D. The detail number on the sheet

12. The values in the section symbol are automatically linked to:

    A. The browser
    B. The sheet where the section view is placed
    C. The text entered by the user
    D. The floor plan

13. To control the line weight of line styles:

    A. Go to Additional Settings→Line Styles
    B. Go to Line Styles→Properties
    C. Pick the Line, right click and select 'Properties'
    D. Go to Tools→Line Weights

14. To re-associate a callout to a new/different detail view:

    A. Delete the callout and create a new callout linked to the new detail view.
    B. Select the callout and use the drop-down list of available drafting views on the Option bar to select a new/different view.
    C. Duplicate the old detail view and then rename it. Name the desired detail view with the name used by the callout.
    D. Activate the desired detail view. In the properties pane, select the Callout to be used for linking.

ANSWERS:

    1) T; 2) F; 3) T; 4) T; 5) T; 6) A & B; 7) A & C; 8) B; 9) B; 10) D; 11) D; 12) B; 13) A; 14) B

Lesson 9
Rendering

Rendering is an acquired skill. It takes practice to set up scenes to get the results you want. It is helpful to become familiar with photography as much of the same lighting and shading theory is applicable. Getting the right scene is a matter of trial and error. It is a good idea to make small adjustments when setting up a scene as you often learn more about the effects of different settings.

The Revit 2024 release replaces the toposurface tool with a toposolid tool. This allows users to add thickness and control the appearance of the topography closer to their desired effect. The toposurface element uses massing features, so if you do not have a good grasp of massing concepts you will need to review that chapter. The toposolid interface was still in beta and some of the tools may change over time.

Exercise 9-1
Create a Toposolid

Drawing Name: toposurface.rvt [m_toposolid.rvt]
Estimated Time: 15 minutes

This exercise reinforces the following skills:

- Site
- Toposolid

Before we can create some nice rendered views, we need to add some background scenery to our model.

1. 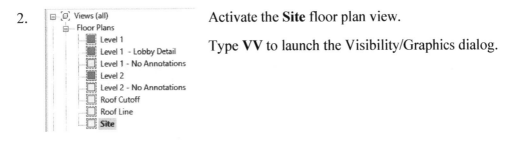 Open *toposurface.rvt*.

2. Activate the **Site** floor plan view.

   Type **VV** to launch the Visibility/Graphics dialog.

3.

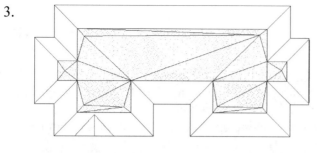

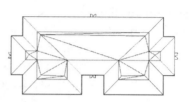

The site plan should just display the roof and building.

4.  Activate the **Massing & Site** ribbon.

5.

Select the **Toposolid→Create from Sketch** tool from the Model Site panel.

6.

Use the rectangle tool to draw a rectangle around the building model.

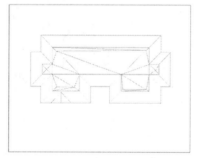

7. Select the **green check** on the Surface panel to **Finish Toposolid**.

8.

An error message will appear.

Close the message box.

9.

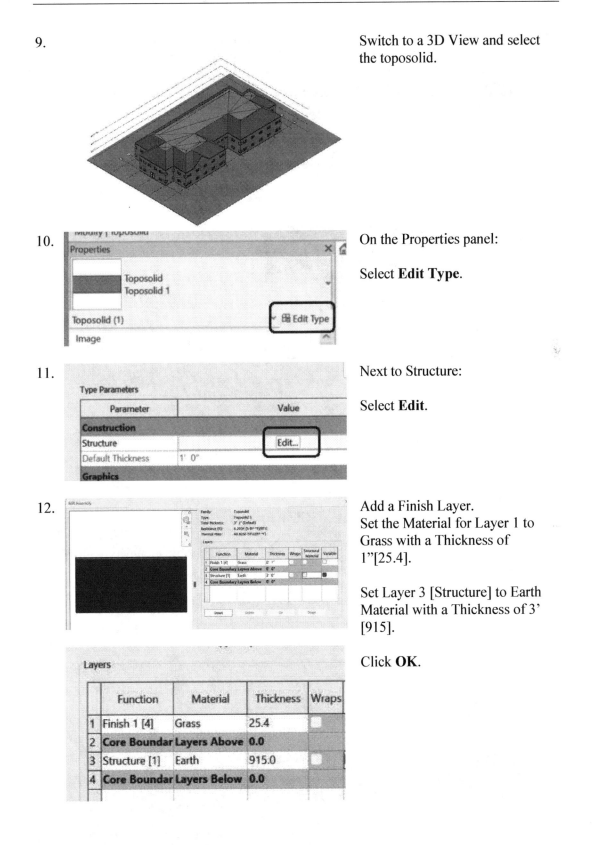

Switch to a 3D View and select the toposolid.

10.

On the Properties panel:

Select **Edit Type**.

11.

Next to Structure:

Select **Edit**.

12.

Add a Finish Layer.
Set the Material for Layer 1 to Grass with a Thickness of 1"[25.4].

Set Layer 3 [Structure] to Earth Material with a Thickness of 3' [915].

Click **OK**.

13.

Click in the **Coarse Scale Fill Pattern** column.

14.

Click **New**.

15.

Enable **Custom**.

Click **Browse.**

Select the *grass2.pat* file included in the exercise files.

Click **OK**.

16.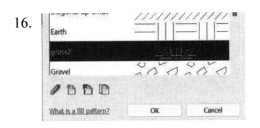

Highlight **grass2**.

Click **OK**.

17.

Graphics	
Coarse Scale Fill Pattern	grass2
Coarse Scale Fill Color	Green
Contour Display	

Change the Coarse Scale Fill Color to **Green**.

Click **OK**.

18.

Constraints	
Level	Level 1
Height Offset From Level	-2' 1"
Room Bounding	☑
Related to Mass	☐
Dimensions	

Select the toposolid.

Set the Height Offset from Level to **-2' 1" [-635]**.

This locates the toposolid below the floor.

19.

Single Flush Door Elevation

— Sections (Wall Section)

   North Exterior Wall

   North Exterior Wall Sketch

Open the **North Exterior Wall** section view.

You don't see the toposolid because it is hidden by the filled regions.

20.

Right click on the **North Exterior Wall** section view.

Select **Duplicate→Duplicate with Detailing**.

21.

Single Flush Door Elevation

— Sections (Wall Section)

   North Exterior Wall

   North Exterior Wall Sketch

   North Exterior Wall with Topo

Rename the section **North Exterior Wall with Topo**.

Delete the two filled regions.

Now you see the floor and the toposolid.

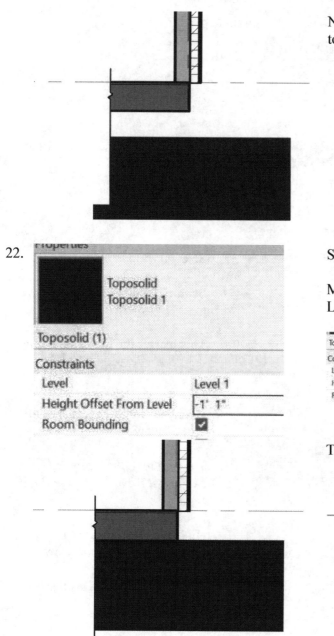

22.

Select the toposolid.

Modify the Height Offset from Level to **-1' 1" [-241]**

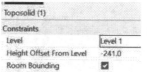

The section view updates.

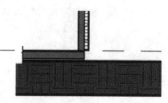

23.   Save as *ex9-1.rvt [m_ex9-1.rvt]*.

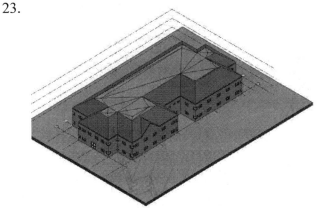

Slab

The structural foundation slab tool creates a foundation element for the model.
Foundation slabs are similar to floors, requiring a boundary sketch and defined structure.
These are system families, which means they are uniquely defined for each project.

Exercise 9-2
Create a Slab

Drawing Name: foundation.rvt [m_foundation.rvt]
Estimated Time: 20 minutes

This exercise reinforces the following skills:

- Site
- Toposolid
- Foundation Slab
- Cut
- Callout

1.   Open *foundation.rvt. [m_foundation.rvt]*

2.   Sections (Wall Section)   Open the **North Exterior Wall with Topo**
              North Exterior Wall   Section view.
              North Exterior Wall Sketch
              North Exterior Wall with Topo

3.

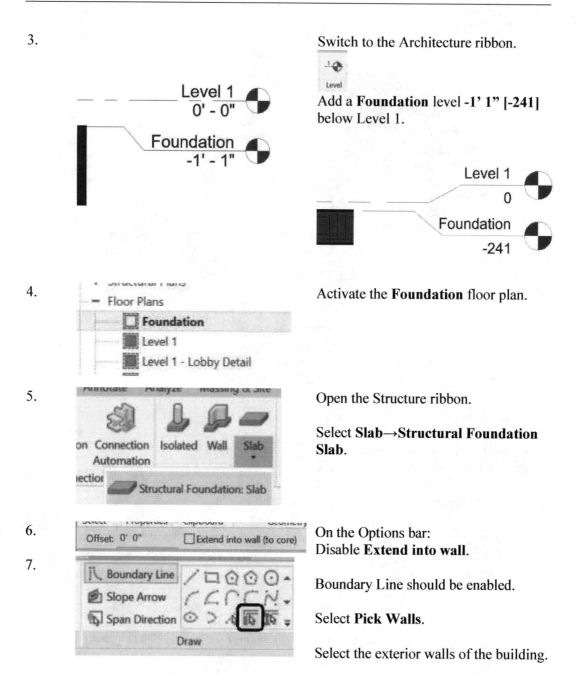

Switch to the Architecture ribbon.

Add a **Foundation** level -1' 1" [-241] below Level 1.

4. Activate the **Foundation** floor plan.

5. Open the Structure ribbon.

Select **Slab→Structural Foundation Slab**.

6. On the Options bar:
Disable **Extend into wall**.

7. Boundary Line should be enabled.

Select **Pick Walls**.

Select the exterior walls of the building.

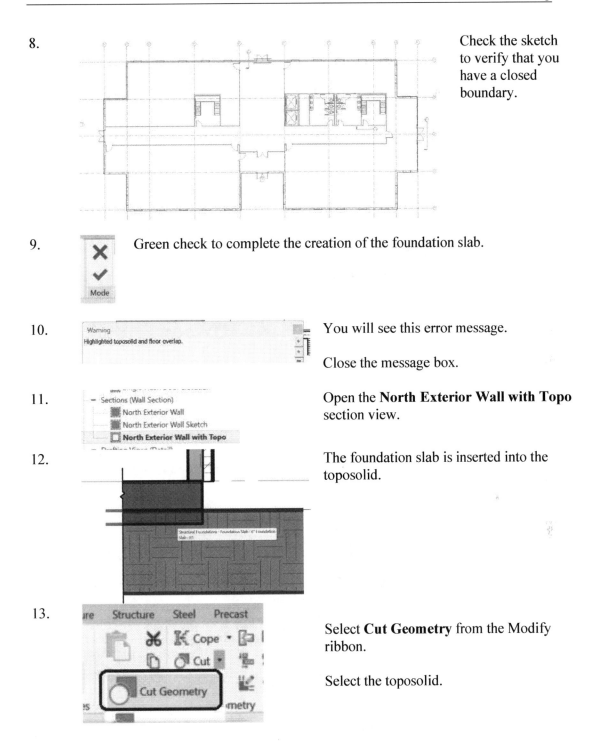

8. Check the sketch to verify that you have a closed boundary.

9. Green check to complete the creation of the foundation slab.

10. You will see this error message.

Close the message box.

11. Open the **North Exterior Wall with Topo** section view.

12. The foundation slab is inserted into the toposolid.

13. Select **Cut Geometry** from the Modify ribbon.

Select the toposolid.

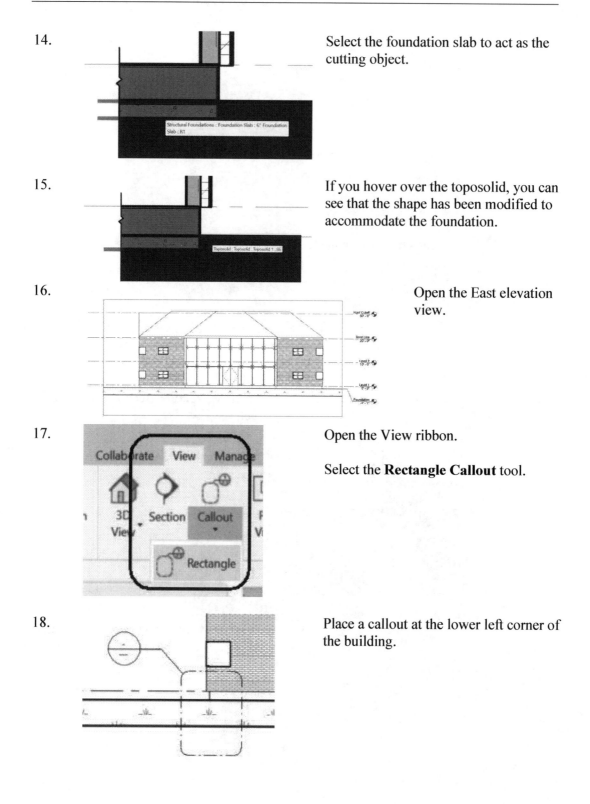

14. Select the foundation slab to act as the cutting object.

15. If you hover over the toposolid, you can see that the shape has been modified to accommodate the foundation.

16. Open the East elevation view.

17. Open the View ribbon.

Select the **Rectangle Callout** tool.

18. Place a callout at the lower left corner of the building.

19.
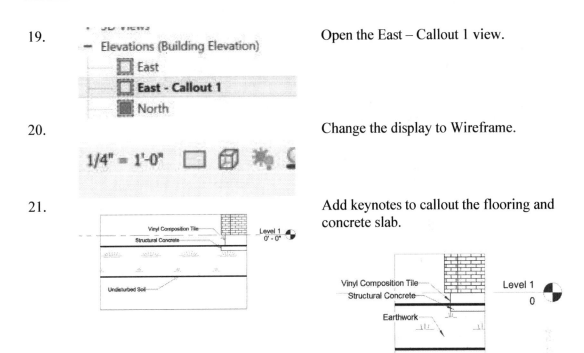
Open the East – Callout 1 view.

20.
Change the display to Wireframe.

21.
Add keynotes to callout the flooring and concrete slab.

22.
Save as *ex9-2.rvt*.

Exercise 9-3

Create a Gravel Fill

Drawing Name: gravel fill.rvt [m_gravel fill.rvt]
Estimated Time: 15 minutes

This exercise reinforces the following skills:

- Site
- Toposolid
- Cut a toposolid

1.   Open *gravel fill.rvt*.

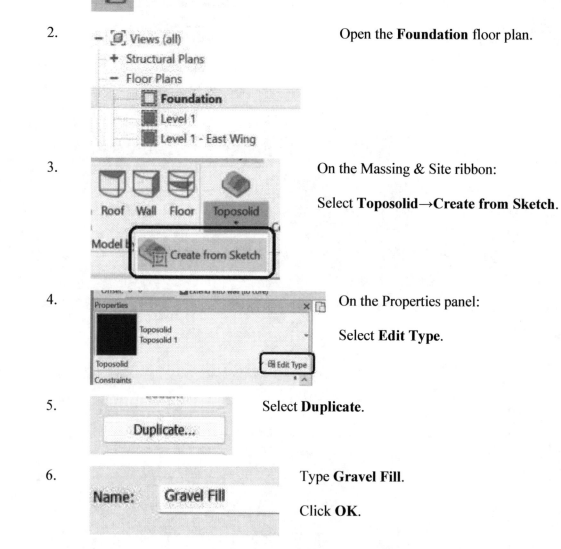

2. Open the **Foundation** floor plan.

3. On the Massing & Site ribbon:

   Select **Toposolid→Create from Sketch**.

4. On the Properties panel:

   Select **Edit Type**.

5. Select **Duplicate**.

6. Type **Gravel Fill**.

   Click **OK**.

7.

Type Parameters	
Parameter	Value
Construction	
Structure	Edit...
Default Thickness	3' 1"

Select **Edit** next to Structure.

8.

Layers

	Function	Material	Thickness	Wr
1	Core Boundar	Layers Above	0' 0"	
2	Structure [1]	Gravel	1' 0"	
3	Core Boundar	Layers Below	0' 0"	

Delete the Finish layer.

Change the Structure layer material to **Gravel**.

Set the Thickness to **1' 0" [305]**.

Click **OK**.

Layers

	Function	Material	Thickness	V
1	Core Boundar	Layers Above	0.0	
2	Structure [1]	Gravel	305.0	
3	Core Boundar	Layers Below	0.0	

9.

Graphics	
Coarse Scale Fill Pattern	Gravel
Coarse Scale Fill Color	RGB 192-192-192
Contour Display	Edit...

Change the Coarse Scale Fill Pattern to **Gravel**.
Set the Coarse Scale Fill Color to **Grey**.
Click **OK**.

10.

Constraints	
Level	Foundation
Height Offset From Level	-0' 6"
Room Bounding	☑
Related to Mass	☐

On the Properties panel:

Change the Height Offset from Level to **-0' 6" [-152]**.

Toposolid
Gravel Fill

Toposolid
Constraints	
Level	Foundation
Height Offset From Level	-152.0
Room Bounding	☑

11.

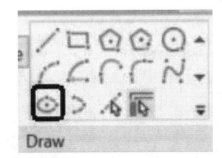

Select the **Ellipse** tool.

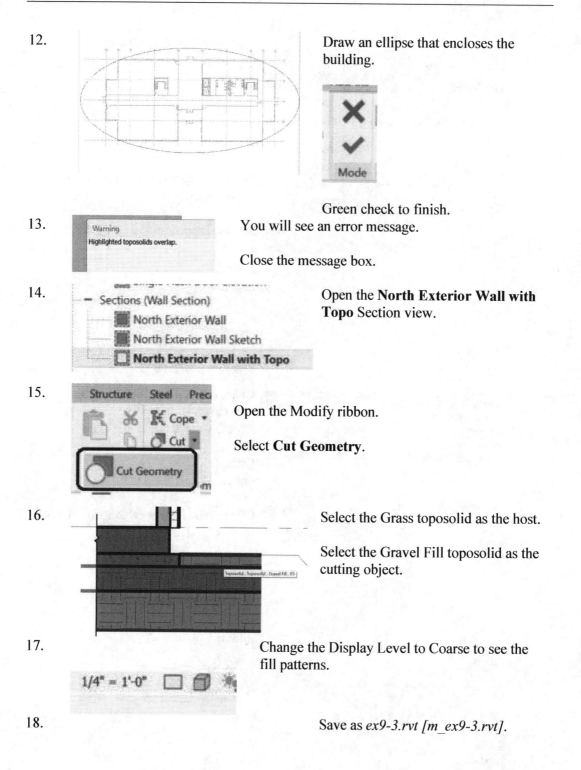

12. Draw an ellipse that encloses the building.

Green check to finish.

13. You will see an error message.

Warning
Highlighted toposolids overlap.

Close the message box.

14. Open the **North Exterior Wall with Topo** Section view.

Sections (Wall Section)
North Exterior Wall
North Exterior Wall Sketch
North Exterior Wall with Topo

15. Open the Modify ribbon.

Select **Cut Geometry**.

16. Select the Grass toposolid as the host.

Select the Gravel Fill toposolid as the cutting object.

17. Change the Display Level to Coarse to see the fill patterns.

1/4" = 1'-0"

18. Save as *ex9-3.rvt [m_ex9-3.rvt]*.

When you select the Site Component tool instead of the Component tool, Revit automatically filters out all loaded components except for Site Components.

Exercise 9-4
Add Site Components

Drawing Name: site_components.rvt [m_site_components.rvt]
Estimated Time: 15 minutes

This exercise reinforces the following skills:

❑   Site
❑   Planting
❑   Entourage

1.   Open *site_components.rvt [m_site_components.rvt].*

2.   Level 2
     Level 2 - No Annotations
     Roof Cutoff
     Roof Line
     Site
                       Activate the **Site** floor plan.

3.   Graphic Display Options...
     Wireframe
     Hidden Line
     Shaded
     Consistent Colors
     Textures
     Realistic
     1" = 20'-0"
                       Switch to a **Hidden Line** display.

4.   Activate the **Massing & Site** ribbon.

5.   Site Component   Select the **Site Component** tool on the Model Site panel.

6.   RPC Tree - Deciduous Red Maple - 30'   Select **Red Maple - 30' [Red Maple - 9 Meters]** from the Properties pane.

7. Place some trees in front of the building.

   To add plants not available in the Type Selector drop-down, load additional families from the Planting folder.

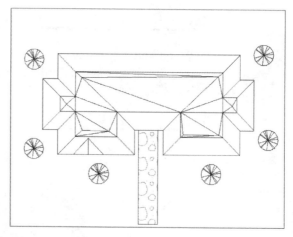

8. Switch to the Architecture ribbon.

   Select the **Place a Component** tool on the Build panel.

9. Select **Load Family** from the Mode panel.

10. Locate the **RPC Female [M_RPC Female.rfa]** file in the downloaded exercise files.

    File name: RPC Female.rfa

    File name: M_RPC Female.rfa

    Click **Open**.

11. Set the Female to **Cathy** on the Properties pane.

    Properties

    RPC Female Cathy

12. 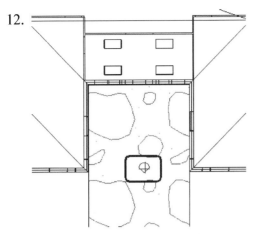 Place the person on the walkway.

Cancel out of the command.

13. If you zoom in on the person, you only see a symbol – you don't see the real person until you perform a Render. You may need to switch to Hidden Line view to see the symbol for the person.

The point indicates the direction the person is facing.

14. Rotate your person so she is facing the building.

15. Save the file as *ex9-4.rvt [m_ex9-4.rvt]*.

Make sure Level 1 or Site is active or your trees could be placed on Level 2 (and be elevated in the air). If you mistakenly placed your trees on the wrong level, you can pick the trees, right click, select Properties, and change the level.

Exercise 9-5
Defining Camera Views

Drawing Name: add_camera_1.rvt [m_add_camera-1.rvt]
Estimated Time: 15 minutes

This exercise reinforces the following skills:

- ❑ Camera
- ❑ Rename View
- ❑ View Properties

1.   Open *add_camera_1.rvt [m_add_camera-1.rvt]*.

2.    Activate **Site** floor plan.

3.   Activate the **View** ribbon.

4.   Select the **3D View→Camera** tool from the Create panel on the View ribbon.

5.    If you move your mouse in the graphics window, you will see a tool tip to pick the location for the camera.

6.

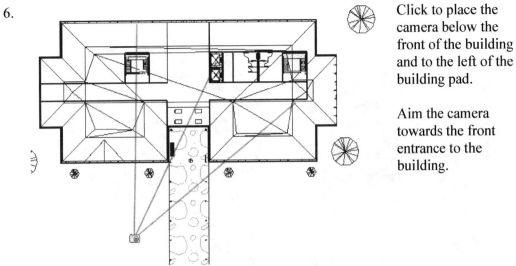

Click to place the camera below the front of the building and to the left of the building pad.

Aim the camera towards the front entrance to the building.

7. A window opens with the camera view of your model.

The person and the tree appear as stick figures because the view is not rendered yet.

8.

Change the Model Graphics Style to **Realistic**.

Graphic Display Options...

Wireframe
Hidden Line
Shaded
Consistent Colors
Textures
Realistic

Perspective

9. Our view changes to a realistic display.

10. | Perspective ☒ 🖥 ✧ ♀ ℃ 🎞 🏠 ⌄ ♀ 🖼 🖼 | Turn ON Sun and Shadows.

11. The sun will not display because Relative to View is selected in Sun Settings. What would you like to do?

    ⇨ Use the specified project location, date, and time instead.
       Sun Settings change to <In-session, Still> which uses the specified project location, date, and time.

    ⇨ Continue with the current settings.
       Sun Settings remain in Lighting mode with Relative to View selected. Sun will not display.

                          [Cancel]

Select Use the specified project location, date and time instead to enable the sun.

12.

    🔒 Save Orientation and Lock View
    🔓 Restore Orientation and Lock View
    🔓 Unlock View

Perspective ☒ 🖥 ✧ ♀ ℃ 🎞 🏠 ⌄ ♀ 🖼 🖼 ◂

If you are happy with the view orientation, select the 3D Lock tool on the bottom of the Display Bar and select **Save Orientation and Lock View**.

13.

Camera	
Rendering Settings	Edit...
Locked Orientation	☑
Projection Mode	Perspective
Eye Elevation	5' 6"
Target Elevation	5' 6"
Camera Position	Explicit

You should see the Locked Orientation enabled in the Properties pane.

14. If you look in the browser, you see that a view has been added to the 3D Views list.

15. Highlight the **3D View 1**.

    Right click and select **Rename**.

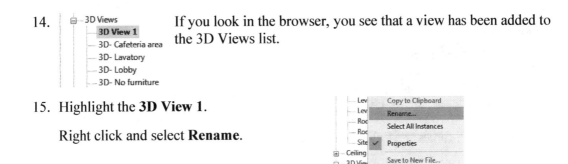

    Rename to **Southeast Perspective**.

The view we have is a perspective view – not an isometric view.
Isometrics are true scale drawings. The Camera View is a perspective view with vanishing points. Isometric views have no vanishing points. Vanishing points are the points at which two parallel lines appear to meet in perspective.

16. Save the file as *ex9-5.rvt*.

Additional material libraries can be added to Revit. You may store your material libraries on a server to allow multiple users access to the same libraries. You must add a path under Settings→Options to point to the location of your material libraries. There are many online sources for photorealistic materials; www.accustudio.com is a good place to start.

Take a minute to add more plants or adjust your people before you proceed.

Exercise 9-6
Rendering Settings

Drawing Name: render_1.rvt
Estimated Time: 20 minutes

This exercise reinforces the following skills:

- ❑ Rendering
- ❑ Settings
- ❑ Save to Project

You can also assign a specific time, date, location, lighting, and environment to your model.

1. 📂 Open *render_1.rvt*.

2. ⊟ 3D Views
        ▢ 3D - Lavatory
        ▢ 3D - Lobby
        ▢ **Southeast Perspective**
        ▢ {3D}

    Activate the **Southeast Perspective** view.

3.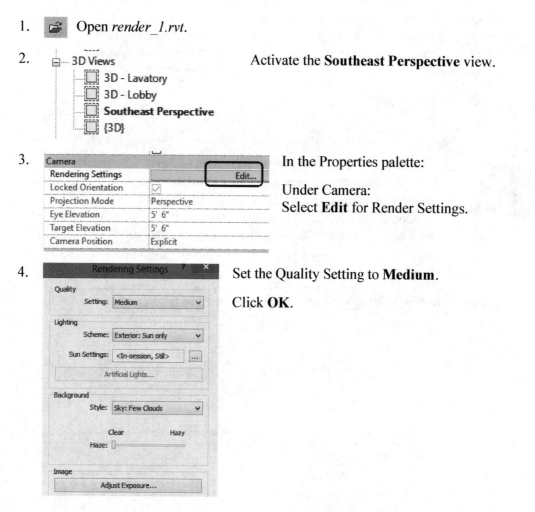

    Camera
    Rendering Settings   Edit...
    Locked Orientation   ☑
    Projection Mode   Perspective
    Eye Elevation   5' 6"
    Target Elevation   5' 6"
    Camera Position   Explicit

    In the Properties palette:

    Under Camera:
    Select **Edit** for Render Settings.

4. Rendering Settings

    Quality
        Setting: Medium

    Lighting
        Scheme: Exterior: Sun only

        Sun Settings: <In-session, Still> ...

        Artificial Lights...

    Background
        Style: Sky: Few Clouds

                Clear   Hazy
        Haze: ▯

    Image
        Adjust Exposure...

    Set the Quality Setting to **Medium**.

    Click **OK**.

5.

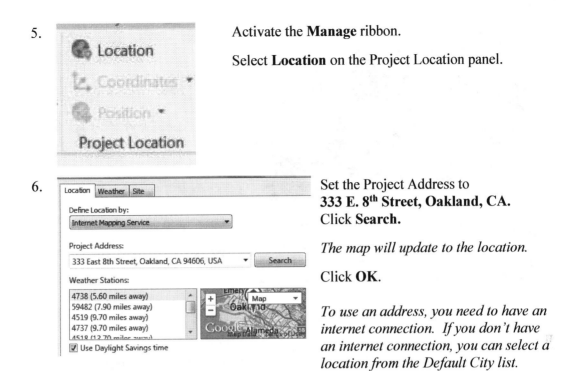

Activate the **Manage** ribbon.

Select **Location** on the Project Location panel.

6.

Set the Project Address to
333 E. 8ᵗʰ Street, Oakland, CA.
Click **Search.**

The map will update to the location.

Click **OK.**

To use an address, you need to have an internet connection. If you don't have an internet connection, you can select a location from the Default City list.

Note that the shadows adjust to the new location.

7.

Select the **Rendering** tool located on the bottom of the screen.

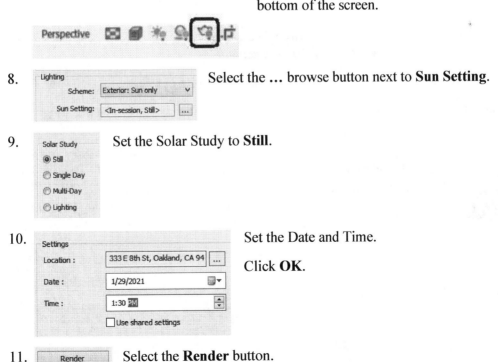

8.

Select the **...** browse button next to **Sun Setting.**

9.

Set the Solar Study to **Still.**

10.

Set the Date and Time.

Click **OK.**

11.

Select the **Render** button.

12.

Your window will be rendered.

13. | Save to Project... | Select **Save to Project**.

14. Rendered images are saved in the Renderings branch of Project Browser.

Name: Southeast Perspective_1

| OK | | Cancel |

Click **OK** to accept the default name.

15. Renderings
  Southeast Perspective_1

Under Renderings, we now have a view called **Southeast Perspective_1**.

You can then drag and drop the image onto a sheet.

16. Display
  Show the model

Select the **Show the Model** button.

Close the Rendering dialog.

Note you can now toggle to Show the Rendering or Show the model.

17. *Our window changes to Realistic – not Rendered mode.*

18. Save the file as *ex9-6.rvt*.

> The point in the person symbol indicates the front of the person. To position the person, rotate the symbol using the Rotate tool.
> Add lights to create shadow and effects in your rendering. By placing lights in strategic locations, you can create a better photorealistic image.

Exercise 9-7
Render Region

Drawing Name: render_2.rvt
Estimated Time: 15 minutes

This exercise reinforces the following skills:

- Components
- Add Camera
- Model Text
- Render Region
- Load Family

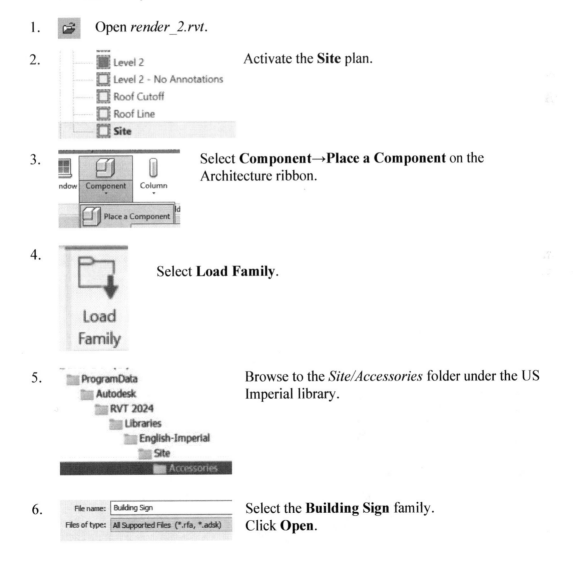

1.   Open *render_2.rvt*.

2.   Activate the **Site** plan.

   Level 2
   Level 2 - No Annotations
   Roof Cutoff
   Roof Line
   Site

3.   Select **Component→Place a Component** on the Architecture ribbon.

4.   Select **Load Family**.

5.   Browse to the *Site/Accessories* folder under the US Imperial library.

   ProgramData
   Autodesk
   RVT 2024
   Libraries
   English-Imperial
   Site
   Accessories

6.   File name: Building Sign
   Files of type: All Supported Files (*.rfa, *.adsk)

   Select the **Building Sign** family.
   Click **Open**.

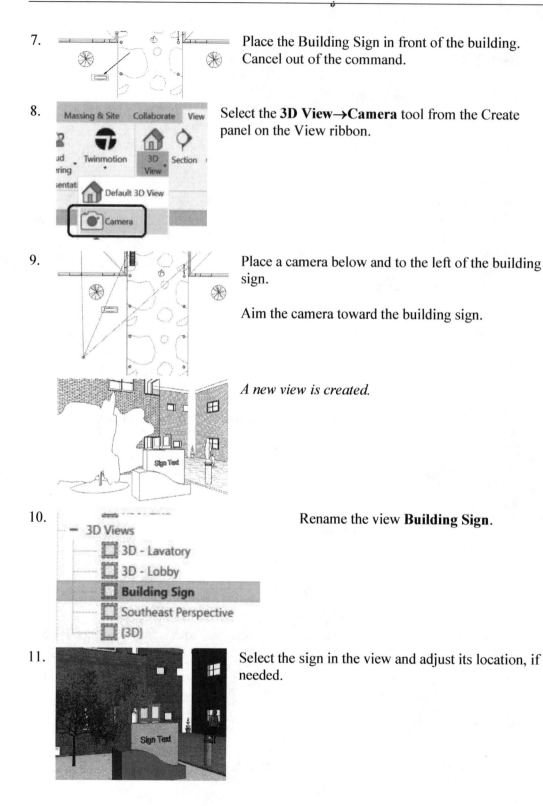

7. Place the Building Sign in front of the building. Cancel out of the command.

8. Select the **3D View→Camera** tool from the Create panel on the View ribbon.

9. Place a camera below and to the left of the building sign.

Aim the camera toward the building sign.

A new view is created.

10. Rename the view **Building Sign**.

11. Select the sign in the view and adjust its location, if needed.

12. With the sign selected:
Select **Edit Type** in the Properties pane.

13. Change the Sign Text to **Autodesk**.
Click **OK**.

14. The sign updates.
The text on the sign was created using Model Text.

15. Turn ON Sun and Shadows.
Set Display to **Realistic**.

16. Select **Use the specified project location, date and time instead** to enable the sun.

The sun will not display because Relative to View is selected in Sun Settings. What would you like to do?

➡ Use the specified project location, date, and time instead.
Sun Settings change to <In-session, Still> which uses the specified project location, date, and time.

➡ Continue with the current settings.
Sun Settings remain in Lighting mode with Relative to View selected. Sun will not display.

Cancel

17. If you are happy with the view orientation, select the 3D Lock tool on the bottom of the Display Bar and select **Save Orientation and Lock View**.

Save Orientation and Lock View
Restore Orientation and Lock View
Unlock View

18. Select the **Render** tool in the task bar.

19. Place a check next to Region.

Render ☑ Region

20.

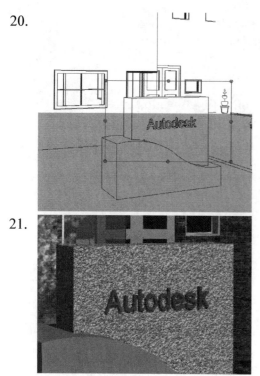

A window will appear in the view. Select the window and use the grips to position the window around the building sign.
Click **Render**.

21.

Just the area inside the rectangle is rendered.

You can use Render Region to check a portion of the view to see how it renders.

Close the Render dialog.

22. Save as ex9-7.rvt.

If you see this error message:

Warnings:

The following render appearance images are missing:
Metals.Metal Fabrications.Metal Stairs.Galvanized.jpg
gravel.PNG

Go to File→Options

Highlight the Rendering category.
Add the lesson folder to the Additional render appearance paths.

I have included material image files in the event your Revit installation doesn't include them.

Space Planning

Drawing Name: space_planning.rvt [m_ space_planning.rvt]
Estimated Time: 20 minutes

This exercise reinforces the following skills:

- Components
- Duplicate View
- View Settings
- Load Family

1. Open *space_planning.rvt [m_ space_planning.rvt]*.

2. Open the **Level 1 – Lobby Space Planning** floor plan.

3. Select the **Component→Place a Component** tool on the Build Panel from the Architecture ribbon.

4. Select **Load** Family from the Mode panel.

5. Locate the *Television-Plasma, Chair-Corbu*, and *Table-Coffee* in the exercise files.

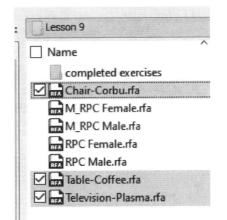

6. Hold down the **CTRL** key to select more than one file.

   Click **Open**.

7.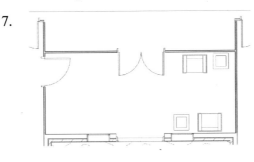

   Place the chair and table in the room.

   You can type in the family you are looking for in the Type Selector so you don't have to spend time scrolling down searching for it.

   If you click the SPACE bar before you click to place you can rotate the object.

8.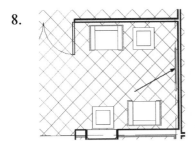

   Place the *Television-Plasma* on the East wall using the **Component** tool.

   Hint: You can use the Search field to quickly locate the desired family.

9.

In the Properties pane:
Set the Offset to **3′ 6″ [1070mm]**.

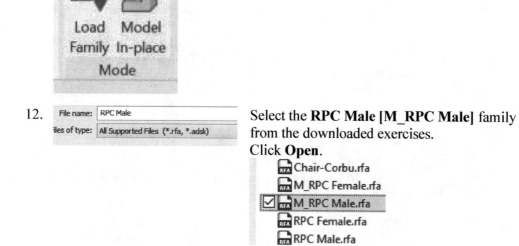

Properties	
Television-Plasma 50″	
Furniture (1)	
Constraints	
Level	Level 1
Elevation from Level	1070.0
Host	Level : Level 1
Offset from Host	1070.0
Moves With Nearby Elem...	☐

This will locate the television 3′ 6″ [1070mm] above Level 1.

10. Select the **Component→Place a Component** tool on the Build Panel from the Architecture ribbon.

11. Select **Load** Family from the Mode panel.

Load Family   Model In-place

Mode

12.

File name: RPC Male

Files of type: All Supported Files (*.rfa, *.adsk)

Select the **RPC Male [M_RPC Male]** family from the downloaded exercises.
Click **Open**.

Chair-Corbu.rfa
M_RPC Female.rfa
☑ M_RPC Male.rfa
RPC Female.rfa
RPC Male.rfa

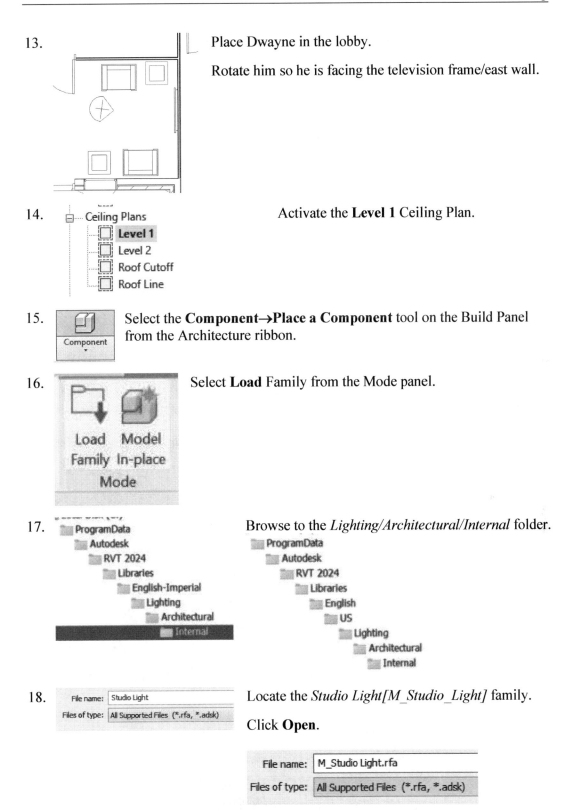

13. Place Dwayne in the lobby.

Rotate him so he is facing the television frame/east wall.

14. Activate the **Level 1** Ceiling Plan.

Ceiling Plans
  Level 1
  Level 2
  Roof Cutoff
  Roof Line

15. Select the **Component→Place a Component** tool on the Build Panel from the Architecture ribbon.

16. Select **Load** Family from the Mode panel.

17. Browse to the *Lighting/Architectural/Internal* folder.

ProgramData
  Autodesk
    RVT 2024
      Libraries
        English-Imperial
          Lighting
            Architectural
              Internal

ProgramData
  Autodesk
    RVT 2024
      Libraries
        English
          US
            Lighting
              Architectural
                Internal

18. Locate the *Studio Light[M_Studio_Light]* family.

Click **Open**.

File name: Studio Light
Files of type: All Supported Files (*.rfa, *.adsk)

File name: M_Studio Light.rfa
Files of type: All Supported Files (*.rfa, *.adsk)

19.

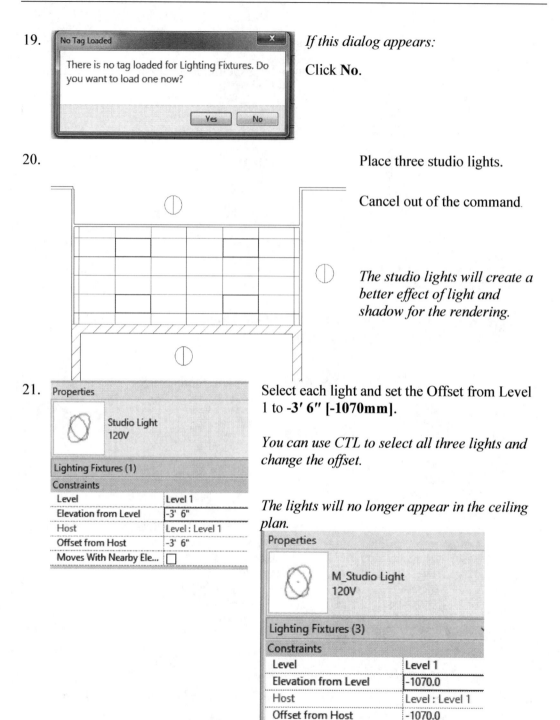

If this dialog appears:

Click **No**.

No Tag Loaded

There is no tag loaded for Lighting Fixtures. Do you want to load one now?

Yes   No

20.

Place three studio lights.

Cancel out of the command.

The studio lights will create a better effect of light and shadow for the rendering.

21.

Properties

Studio Light
120V

Lighting Fixtures (1)

Constraints

Level	Level 1
Elevation from Level	-3' 6"
Host	Level : Level 1
Offset from Host	-3' 6"
Moves With Nearby Ele...	☐

Select each light and set the Offset from Level 1 to **-3' 6" [-1070mm]**.

You can use CTL to select all three lights and change the offset.

The lights will no longer appear in the ceiling plan.

Properties

M_Studio Light
120V

Lighting Fixtures (3)

Constraints

Level	Level 1
Elevation from Level	-1070.0
Host	Level : Level 1
Offset from Host	-1070.0

22. Save as *ex9-8.rvt [m_ex9-8.rvt]*.

Exercise 9-9
Building Sections

Drawing Name: section_view2.rvt [m_ section_view2.rvt]
Estimated Time: 10 minutes

This exercise reinforces the following skills:

- ❑ Section Views
- ❑ Elevations
- ❑ Crop Regions

1. 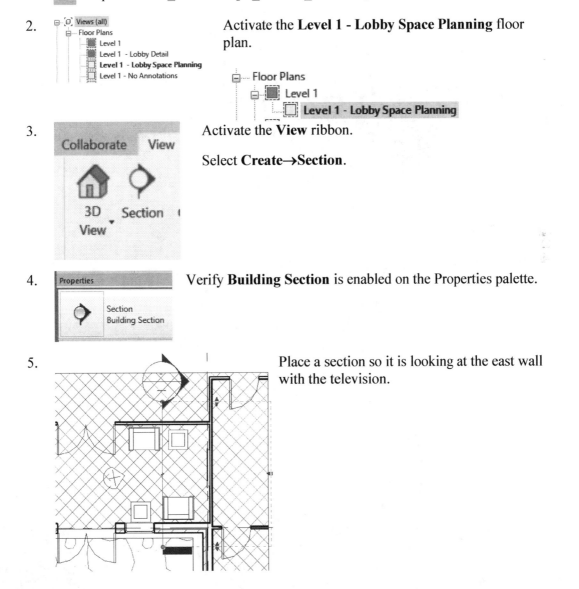 Open *section_view2.rvt [m_ section_view2.rvt]*.

2. Activate the **Level 1 - Lobby Space Planning** floor plan.

3. Activate the **View** ribbon.

   Select **Create→Section**.

4. Verify **Building Section** is enabled on the Properties palette.

5. Place a section so it is looking at the east wall with the television.

6.  Rename the Section **Lobby East Elevation**.

7. Activate the **Lobby East Elevation** section.

8. 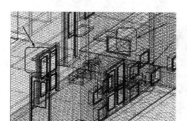 Adjust the crop region so only the lobby area is visible.

Note the position of the studio light. Adjust the elevation of the light so it is positioned on the frame. You can move it simply by selecting the light and moving it up or down.

If you don't see the studio light, check that you are in Wireframe. You may need to adjust your section region in the plan view. You can also switch to a 3D Wireframe view to see where the studio lights are located.

Properties	
Studio Light 120V	
Lighting Fixtures (1)	
Constraints	
Level	Level 1
Elevation from Level	-3' 6"
Host	Level : Level 1
Offset from Host	-3' 6"
Moves With Nearby Elem...	☐

9. Save as *ex9-9.rvt [m_ex9-9.rvt]*.

Exercise 9-10
Decals

Drawing Name: decal1.rvt [m_decal1.rvt]
Estimated Time: 20 minutes

This exercise reinforces the following skills:

- Decal Types
- Decals
- Set Workplane

1.   Open *decal1.rvt [m_decal1.rvt].*

2.   Sections (Building Section)
         Curtain Wall Door Elevation
         Lobby East Elevation
         Single Flush Door Elevation
     Activate the **Lobby East Elevation** Section.

3.   Activate the **Insert** ribbon.

         Select **Decal→Decal Types** from the Link panel.

     Steel   Precast   System
     Link   DWF   Decal
     ography   Markup
     Place Decal
     Decal Types

4.   Select **Create New Decal** from the bottom left of the dialog box.

5.   New Decal
         Name: soccer
     Type a name for your decal.

         The decal will be assigned an image. I have included several images that can be used for this exercise in the downloaded Class Files.

         Click **OK**.

6.

Source

Invalid Image File

Select the Browse button to select the image.

Revit now allows you to use Autodesk Docs to store materials, textures, and decals, if you have a valid subscription. This means you can access these file types from any device that has internet access.

7.

Source soccer.jpg

Brightness 1

Reflectivity: 0%

Browse to where your exercise files are located.

Select *soccer.jpg* or the image file of your choice. Click **Open**.

8.

Settings

Source soccer.jpg

Brightness 1

Reflectivity: 0%

Transparency: 0%

Finish: Matte 0

Luminance (cd/m^2): Dim glow 10

Bump Pattern: Black

Bump Amount: 30%

Cutouts: None

You will see a preview of the image.
Set the Finish to **Matte**.
Set the Luminance **to Dim glow**.

Click **OK**.

9.

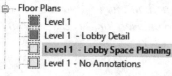

Activate **Level 1 – Lobby Space Planning**.

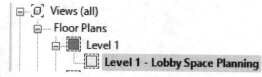

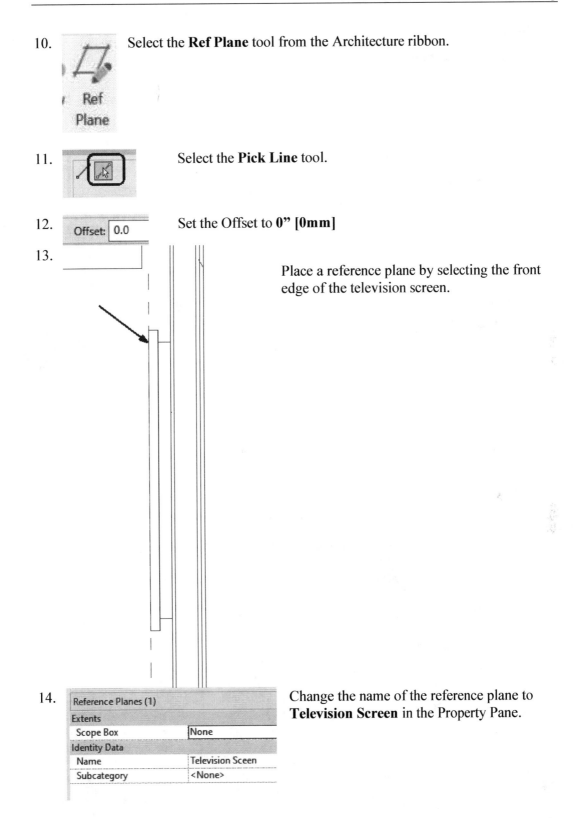

10.　　　　　　　　Select the **Ref Plane** tool from the Architecture ribbon.

Ref Plane

11.　　　　　　　　Select the **Pick Line** tool.

12.　Offset: 0.0　　Set the Offset to **0" [0mm]**

13.　　　　　　　　Place a reference plane by selecting the front edge of the television screen.

14.　Reference Planes (1)　　Change the name of the reference plane to **Television Screen** in the Property Pane.

Extents	
Scope Box	None
Identity Data	
Name	Television Sceen
Subcategory	<None>

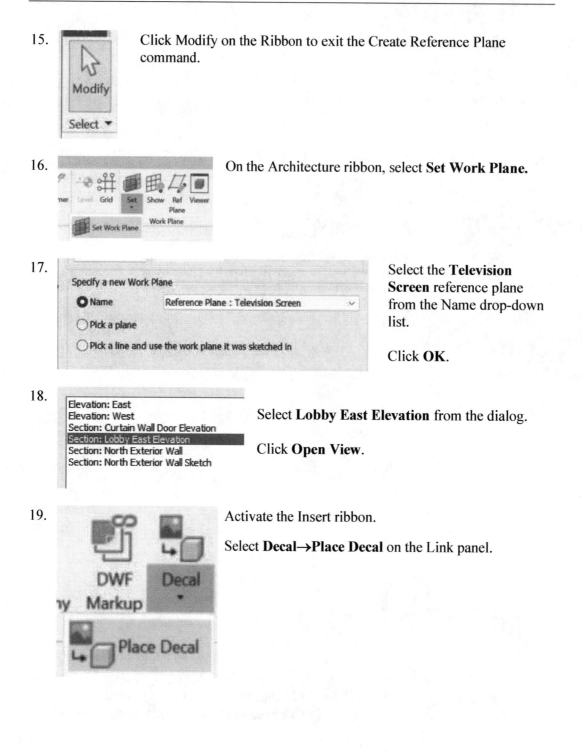

15. Click Modify on the Ribbon to exit the Create Reference Plane command.

16. On the Architecture ribbon, select **Set Work Plane.**

17. Select the **Television Screen** reference plane from the Name drop-down list.

    Click **OK**.

18. Select **Lobby East Elevation** from the dialog.

    Click **Open View**.

19. Activate the Insert ribbon.

    Select **Decal→Place Decal** on the Link panel.

20.

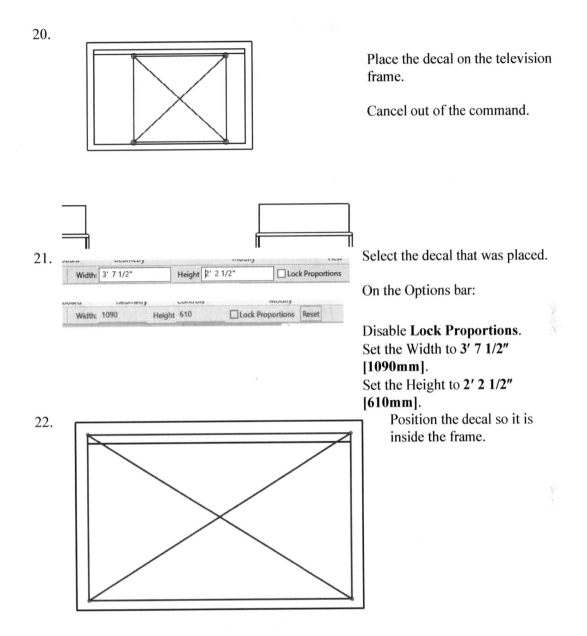

Place the decal on the television frame.

Cancel out of the command.

21.

Width: 3' 7 1/2"   Height 2' 2 1/2"   ☐ Lock Proportions

Width: 1090   Height 610   ☐ Lock Proportions   Reset

Select the decal that was placed.

On the Options bar:

Disable **Lock Proportions**.
Set the Width to **3' 7 1/2"
[1090mm]**.
Set the Height to **2' 2 1/2"
[610mm]**.

22.

Position the decal so it is inside the frame.

23.

Change the display to **Realistic** and you should see the decal inside the television frame.

24. Save as *ex9-10.rvt [m_ ex9-10.rvt]*.

Exercise 9-11
Creating a 3D Camera View (Reprised)

Drawing Name: add_camera2.rvt [m add_camera2.rvt]
Estimated Time: 20 minutes

This exercise reinforces the following skills:

- ❑ 3D Camera
- ❑ Rename View
- ❑ Navigation Wheel
- ❑ Render

1. 📂 Open *add_camera_2.rvt [m add_camera2.rvt]*.

2. 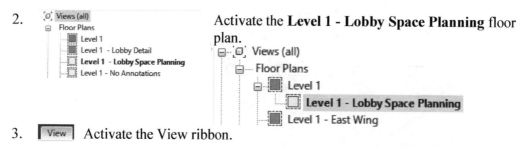 Activate the **Level 1 - Lobby Space Planning** floor plan.

3. View   Activate the View ribbon.

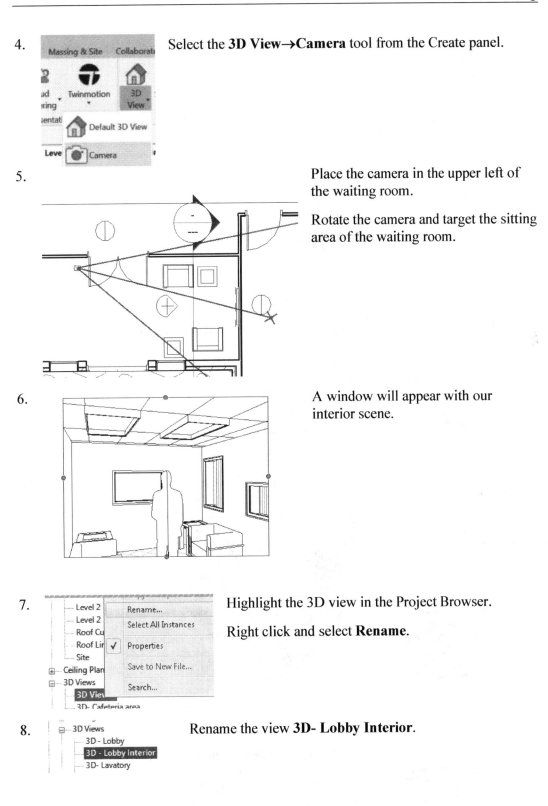

4.   Select the **3D View→Camera** tool from the Create panel.

5.   Place the camera in the upper left of the waiting room.

Rotate the camera and target the sitting area of the waiting room.

6.   A window will appear with our interior scene.

7.   Highlight the 3D view in the Project Browser.

Right click and select **Rename**.

8.   Rename the view **3D- Lobby Interior**.

9. 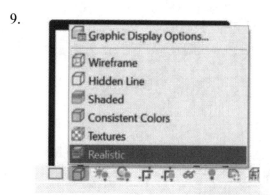 Set the Display to **Realistic**.

10.  Select **Mini Tour Building Wheel** from View toolbar.

Use the Navigation Wheel to adjust your scene. Try the LOOK and WALK buttons.

11.  Adjust the camera view.

You can also use this tool to 'walk through your model'.

12.  Click on the Sun on the Display toolbar. Turn **Sun Path ON**.

13. The sun will not display because Relative to View is selected in Sun Settings. What would you like to do?

> → Use the specified project location, date, and time instead.
> Sun Settings change to <In-session, Still> which uses the specified project location, date, and time.

> → Continue with the current settings.
> Sun Settings remain in Lighting mode with Relative to View selected. Sun will not display.

Select the **Use the specified project location date and time instead** option.

14.

Turn **Shadows On**.

15.

Activate the **Rendering** dialog.

16.

Set the Quality Setting to **Medium**.

Set the Lighting Scheme to **Interior: Sun and Artificial Light**.

Click **Render**.

17. The interior scene is rendered.

A bright light coming from the south window and the door is from the studio light.
Repositioning the light might make the rendering better.
You can also delete the studio lights if you think they interfere with the rendering.

18. [Adjust Exposure...] Select the **Adjust Exposure** button on the Rendering Dialog.

19.

Reset to Default

Settings

Exposure Value: 12
　　　　　　　Brighter　　　　Darker

Highlights: 0.25
　　　　　　　Darker　　　　Brighter

Mid Tones: 1
　　　　　　　Darker　　　　Brighter

Shadows: 0.2
　　　　　　　Lighter　　　　Darker

White Point: 6500
　　　　　　　Cooler　　　　Warmer

Saturation: 1
　　　　　　　Grey　　　　Intense

If necessary:

Adjust the controls and select **Apply**.

Readjust and select **Apply**.

Click **OK**.

20. [Save to Project...] Click **Save to Project**.

21. Rendered images are saved in the Renderings branch of Project Browser.

    Name: 3D Lobby Interior 1

    [OK] [Cancel]

    Click **OK**.

22. Close the Rendering Dialog.

23. Save the file as *ex9-11.rvt [m_ex9-11.rvt]*.

Cameras do not stay visible in plan views. If you need to re-orient the camera location, you can do so by activating your plan view, then right clicking on the name of the 3D view you want to adjust and selecting the 'Show Camera' option.

3D Views
— Southeast Perspective
— Waiting Room View 1
— {3D}
⊕ Elevations
⊟ Sections
— Section 1-East W

Open
Close

Show Camera

Exercise 9-12
Rendering Using Autodesk Cloud

Drawing Name: render3.rvt
Estimated Time: 5 minutes

This exercise reinforces the following skills:

- Autodesk Cloud
- Rendering

If you have an Autodesk account, you can render for free (if you are a student) or a limited number of renderings for free if you are a professional. By rendering using the Cloud, you can continue to work in Revit while the Rendering is being processed. Creating an Autodesk account is free. An internet connection is required to use Autodesk Cloud for rendering.

1.   Open *render3.rvt.*

2. 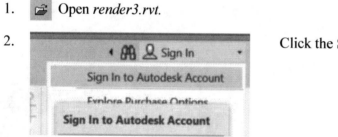   Click the **Sign In** button next to Help.

If you are already signed in, you can skip these steps.

3.   Sign in using your name and password.

4. 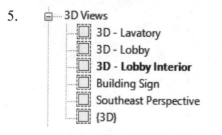   If sign-in is successful, you will see your log-in name in the dialog.

5.   Activate the **3D Lobby Interior** view.

    3D Views
        3D - Lavatory
        3D - Lobby
        3D - Lobby Interior
        Building Sign
        Southeast Perspective
        {3D}

6.

Activate the **View** ribbon.

Select **Render in Cloud**.

7.

Welcome to Rendering in the Cloud for Revit 2021

To begin rendering in the cloud:

Step 1
On the next screen, select 3D views and start rendering.

Step 2
The service will notify you when your images are ready.

Step 3
Select "View > Render Gallery" to view, download, or do more with your completed renderings online.

☐ Don't show this message next time   Continue

Click **Continue.**

8.

Select 3D views to render in the Cloud

3D View	3D - Lobby Interior
Output Type	☐ Render All 3D Views
	☐ (3D)
Render Quality	☐ 3D - Lavatory
	☐ 3D - Lobby
Image Size	☑ 3D - Lobby Interior
	☐ Building Sign
Exposure	☐ Southeast Perspective

Select the **3D – Lobby Interior** view to be rendered using the drop-down list.

9.

Select 3D views to render in the Cloud

3D View	3D - Lobby Interior
Output Type	Still Image
Render Quality	Final
Image Size	Medium (1 Mega Pixel)
Exposure	Advanced

Set the Render Quality to **Final.**
Set the Image Size to **Medium.**
Set the Exposure to **Advanced.**

10.

☐ Email me when complete   Start Rendering

Click **Start Rendering.**

Notice that you can enable email notification for when your rendering is completed.

11.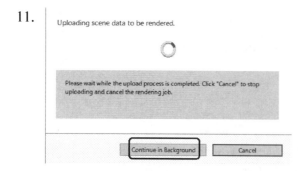

You can continue working while the rendering is being processed.

This is a major advantage for rendering using the Cloud option, plus the quality of the renderings are often better than what your local machine can do.

12.

When the rendering is done, select **Render Gallery** on the View ribbon.

This will bring up a browser and display your rendering.

This only works if you are connected to the internet.

13.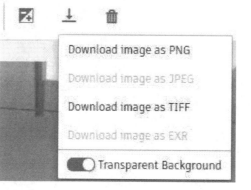

The rendering will be displayed.
To view a larger image, simply click on the thumbnail.

You may need to sign into the Autodesk site in order to view your renderings.

14. Select **Download** from the menu to download the file.

Download image as PNG.

Enable Transparent Background.

Note the editing tools available inside Autodesk for your renderings.

15. Save as *ex9-12.rvt*.

Exercise 9-13

Placing a Rendering on a Sheet

Drawing Name: sheet_1.rvt [m_sheet_1.rvt]
Estimated Time: 5 minutes

This exercise reinforces the following skills:

- View Properties
- View Titles

1.   Open *sheet_1.rvt [m_sheet_1.rvt]*.

2.
```
Sheets (all)
    A101 - First Level Floor Plan
    A102 - Second Level Floor Plan
    A103 - Lobby Keynotes
    A201 - Exterior Elevations
    AD501 - Details
```

In the browser, activate the **Exterior Elevations** sheet.

```
Sheets (all)
    A101 - Landing Page
    A102 - First Level Floor Plan
    A103 - Second Level Floor Plan
    A104 - LOBBY KEYNOTES
    A105 - Level 1 - East Wing
    A106 - Level 1 - West Wing
    A201 - Exterior Elevations
    AD501 - Details
    G101 - Cover Page
```

If you click on the South Elevation, you will see that the view scale property is grayed out.

The scale on the South Elevation is set in the view template.

Properties	
Viewport Viewport 1	
Viewports (1)	
Graphics	
View Scale	1/8" = 1'-0"
Scale Value 1:	96
Display Model	Normal
Detail Level	Medium

Viewport Title w Line	
Viewports (1)	
Graphics	
View Scale	1 : 200
Scale Value 1:	200

3. Select the **View Template** in the Properties panel.

4. Change the View Scale to **1/16″ = 1′-0″ [1:500]**.

Click **Apply** to see how the view changes.

Parameter	Value
View Scale	1 : 500
Scale Value 1:	500

5. Click **Edit** next to **V/G Overrides Annotation**.

Display Model	Normal
Detail Level	Medium
Parts Visibility	Show Original
V/G Overrides Model	Edit
V/G Overrides Annotation	Edit...
V/G Overrides Analytical Model	Edit...
V/G Overrides Import	Edit...
V/G Overrides Filters	Edit...
Model Display	Edit

6. Turn off the visibility of **Reference Planes**.

☑ Railing Tags
☑ Reference Lines
☐ Reference Planes
☑ Reference Points

Click **Apply** and **OK**.

Close the dialog.

7. Right click on the view.

Select **Activate View**.

8. Turn on the visibility of the Crop Region.

Adjust the crop region.

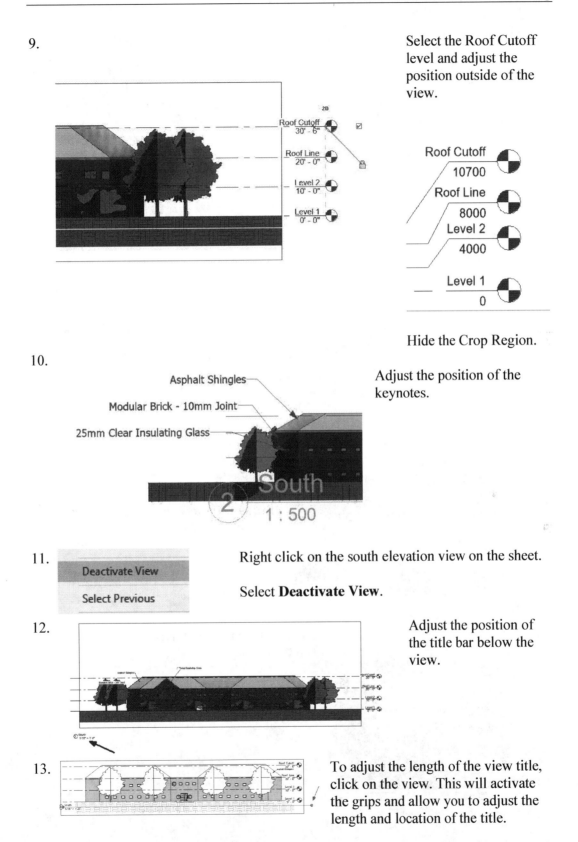

9.   Select the Roof Cutoff level and adjust the position outside of the view.

Hide the Crop Region.

10.   Adjust the position of the keynotes.

Asphalt Shingles

Modular Brick - 10mm Joint

25mm Clear Insulating Glass

South

1 : 500

11.   Right click on the south elevation view on the sheet.

Deactivate View

Select Previous

Select **Deactivate View**.

12.   Adjust the position of the title bar below the view.

13.   To adjust the length of the view title, click on the view. This will activate the grips and allow you to adjust the length and location of the title.

14.

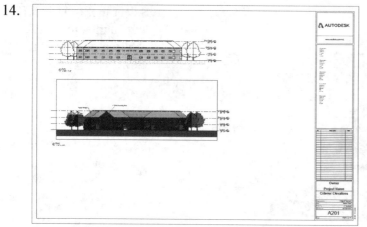

Zoom to Fit.

Shift the elevation views to the left to make room for an additional view.

Note you see a toposurface in the elevation views because one was added for the exterior rendering.

15.

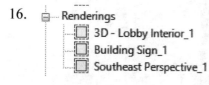

North
1 : 500

Adjust the North elevation view using the same methods as the south elevation view.

16. Renderings
- 3D – Lobby Interior_1
- Building Sign_1
- Southeast Perspective_1

In the browser, locate the **Southeast Perspective** view under *Renderings*.

Drag and drop it onto the sheet.

You see how easy it is to add rendered views to your layouts.

17.

Select the rendering image that was just placed on the sheet.

18. In the Properties pane:

Select **Edit Type**.

19. Note that the current type of Viewport shows the title.

Select **Duplicate**.

20. Name the new Viewport type: **Viewport-No Title No Line**.

Click **OK**.

21. Set View Title to **None**.
Uncheck **Show Extension Line**.

Set Show Title to **No**.

Click **OK**.

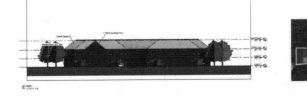

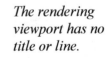

The rendering viewport has no title or line.

The elevation viewport has a title and line.

22.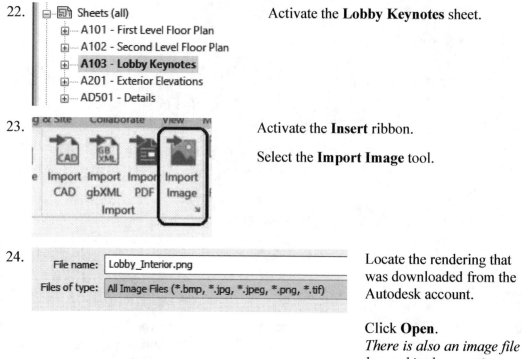
Activate the **Lobby Keynotes** sheet.

23.
Activate the **Insert** ribbon.

Select the **Import Image** tool.

24.

File name:	Lobby_Interior.png
Files of type:	All Image Files (*.bmp, *.jpg, *.jpeg, *.png, *.tif)

Locate the rendering that was downloaded from the Autodesk account.

Click **Open**.
There is also an image file located in the exercise files.

25.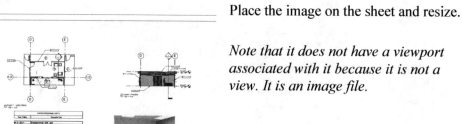
Place the image on the sheet and resize.

Note that it does not have a viewport associated with it because it is not a view. It is an image file.

26. Save the file as *ex9-13.rvt [m_ex9-13.rvt]*.

You can free up memory resources for rendering by closing any windows you are not using. Windows with 3D or renderings take up a considerable amount of memory.

Exercise 9-14
Placing a Path for a Walkthrough

Drawing Name: i_Urban_House.rvt
Estimated Time: 10 minutes

This exercise reinforces the following skills:

    ❏   Creating a Walkthrough view

1. | File name: | i_Urban_House.rvt | Locate the *i_Urban_House.rvt* file.
    This file is included in the Class Files download.
    Select **Open**.

2.   Views (all) — Floor Plans — **FIRST FLOOR** / GROUND FLOOR / ROOF PLAN   Activate the **First Floor** floor plan view.

3.   View   Activate the **View** ribbon.

4.   Twinmotion / 3D View — Default 3D View / Camera / Walkthrough   Select the **Walkthrough** tool under the 3D View drop-down.

5.   ☑ Perspective Scale: 1/8" = 1'-0"   Offset: 5' 6"   From FIRST FLOC ∨

Verify that **Perspective** is enabled in the Status Bar.

This indicates that a perspective view will be created.
The Offset indicates the height of the camera offset from the floor level.

6.   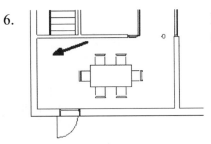   Pick a point to the left of the dining room table as the starting point for your camera's path.

7.

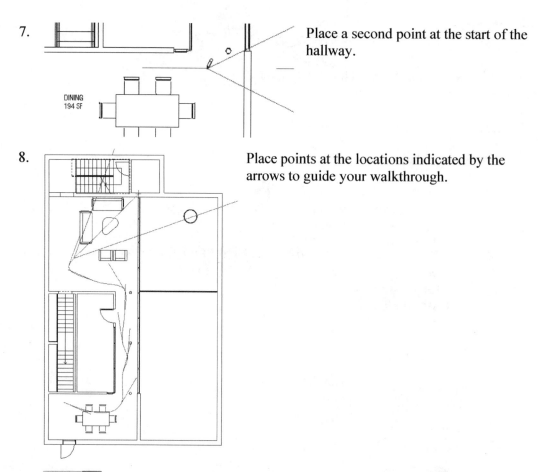

Place a second point at the start of the hallway.

8.

Place points at the locations indicated by the arrows to guide your walkthrough.

9. Select **Walkthrough→Finish Walkthrough** from the ribbon.

10. Walkthroughs
   Walkthrough 1

In the Project Browser, you will now see a Walkthrough listed under Walkthroughs.

11. Save as *ex9-14.rvt*.

Exercise 9-15
Playing the Walkthrough

Drawing Name: walkthrough_1.rvt
Estimated Time: 5 minutes

This exercise reinforces the following skills:

❑ Playing a Walkthrough

1. Open *walkthrough_1.rvt*.

2. Highlight the Walkthrough in the Browser.
Right click and select **Open**.

3. Highlight the Walkthrough in the Browser.
Right click and select **Show Camera**.

4. Select **Edit Walkthrough** on the ribbon.

5. Look at the Options bar.

   Set the Frame number to **1.0**.

6. Activate the **Edit Walkthrough** ribbon.

7.  Click **Play** on the ribbon.

8. Click the button that displays the total frame number. Your frame number value may be different.

9. Change the Total Frames value to **50**.
Click **Apply** and **OK**.

10.  Click **Play**.

Notice that fewer frames speeds up the animation.

11. Save as *ex9-15.rvt*.

Exercise 9-16
Editing the Walkthrough Path

Drawing Name: walk_through_2.rvt
Estimated Time: 15 minutes

This exercise reinforces the following skills:

- ❑ Show Camera
- ❑ Editing a Walkthrough
- ❑ Modifying a Camera View

1. Open *walk_through_2.rvt*.

2. 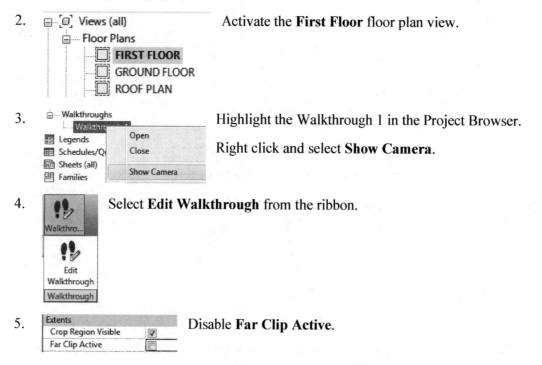 Activate the **First Floor** floor plan view.

3. Highlight the Walkthrough 1 in the Project Browser.

Right click and select **Show Camera**.

4. Select **Edit Walkthrough** from the ribbon.

5. Disable **Far Clip Active**.

6.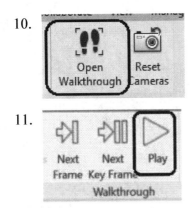

    Select **Edit** next to Graphic Display Options.

7. Set the Style to **Realistic**.
    Enable **Show Edges**.
    Enable **Smooth lines with anti-aliasing**.

    Enable **Cast Shadows**.
    Enable **Show Ambient Shadows**.

8. Set the Lighting to **Interior: Sun and Artificial.**

    Click **Apply** and **OK**.

9. ⊟ Walkthroughs
       Walkthrough 1

    Double left click on **Walkthrough 1** to open the view.

    Right click on the name of the view and select **Show Camera**.

10. Select **Open Walkthrough** from the ribbon.

11. Click **Play**.

12. Save as *ex9-16.rvt*.

The appearance of shaded objects is controlled in the material definition.
To stop playing the walkthrough, click **ESC** at any frame point.

Exercise 9-17
Creating an Animation

Drawing Name: walk_through_3.rvt
Estimated Time: 15 minutes

This exercise reinforces the following skills:

- ❑ Show Camera
- ❑ Editing a Walkthrough
- ❑ Modifying a Camera View

1. Open *walk_through_3.rvt*.

2.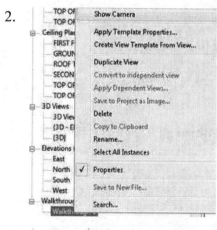

Highlight the Walkthrough 1 in the Project Browser.

Right click and select **Show Camera**.

3. Select **Edit Walkthrough** on the ribbon.

4. Set the Frame number to **1.0**.
Set the total number of frames to **50** by clicking on the total number.

5.

Go to **File→Export→ Images** and **Animations→ Walkthrough**.

6.

Set the Visual Style to **Realistic with Edges**.

Click **OK**.

You can create an animation using Rendering, but it takes a long time to process.

7.

Locate where you want to store your avi file. You can use the Browse button (…) next to the Name field to locate your file.

8. Name your file *Walk_through 1.avi.*

9. Click **Save**.

10.

Select **Microsoft Video 1**.

This will allow you to play the avi on RealPlayer or Microsoft Windows Media Player.
If you select a different compressor, you may not be able to play your avi file.

Click **OK**.

11.

A progress bar will appear in the lower left of your screen to keep you apprised of how long it will take.

12. Walkthrough 1.avi

Locate the file using Explorer.

Double click on the file to play it.

Use Windows Media Player to play the file.

13. Save as *ex9-17.rvt*.

Additional Projects

1) Create a rendering of the cafeteria area.

2) Create a rendering of one of the office cubicles.
   The figure is the Female Entourage family - Cynthia. The publisher's website
   (www.sdcpublications.com/downloads/978-1-63057-600-4) has the images and
   the families for the monitor, phone, keyboard, and CPU. Image files can be
   used in the rendering. *Note: This rendering took my workstation more than a
   day to complete, so I recommend using the cloud.*

3) Add exterior lighting using the bollard lights in the Revit library to both sides of
   the walkway to the building.
   Set the day to December 11 and the time at 7:30 pm.
   Adjust the exposure after the rendering is done.

4) Create a walkthrough avi of one of your rendered scenes.

Lesson 9 Quiz

True or False

1. The Walkthrough tool on the View ribbon allows you to create an avi file.
2. The Environment button on Render Settings dialog allows you to set the time of day.
3. The Sun button on Render Settings dialog allows you to set the time of day.
4. When you create a Rendering, you will be able to see how your people will appear.
5. Rendering is a relatively quick process.
6. Renderings can be added to sheets.
7. You can adjust the number of frames in a walkthrough.
8. You cannot create walkthrough animations that go from one level to another.
9. Once a walkthrough path is placed, it cannot be edited.
10. You cannot modify the view style of a walkthrough from wireframe to shaded.

Multiple Choice

11. The Camera tool is located on the _____ ribbon.

    A. Rendering
    B. Site
    C. Home
    D. View

12. To see the camera in a view:

    A. Launch the Visibility/Graphics dialog and enable Camera in the Model Categories.
    B. Highlight the view in the browser, right click and select Show Camera.
    C. Go to View→Show Camera.
    D. Mouse over the view, right click and select Show Camera.

13. In order to save your rendering, use:

    A. File→Export
    B. Save to Project
    C. Capture Rendering
    D. Export Image

14. When editing a walkthrough, the user can modify the following:

    A. Camera
    B. Path
    C. Add Key Frame
    D. Remove Key Frame
    E. All of the above

15. Decals are visible in the display mode:

    A. Hidden
    B. Wireframe
    C. Realistic
    D. Textiles

16. In order to render using the Cloud, the user needs:

    A. Internet access
    B. An umbrella
    C. A Revision Cloud
    D. A valid ID

ANSWERS:
    1) F; 2) F; 3) T; 4) T; 5) F; 6) T; 7) T; 8) F; 9) F; 10) F; 11) D; 12) C; 13) B; 14) B; 15) C; 16) A;

Lesson 10
Customizing Revit

Exercise 10-1
Creating an Annotation Symbol

File:	north arrow.dwg (located in the Class Files download)
	add_symbol.rft
Estimated Time:	30 minutes

This exercise reinforces the following skills:

- ❑ Import AutoCAD Drawings
- ❑ Full Explode
- ❑ Query
- ❑ Annotation Symbol

Many architects have accumulated hundreds of symbols that they use in their drawings. This exercise shows you how to take your existing AutoCAD symbols and use them in Revit.

1. Locate the *north arrow.dwg* in the Class Files downloaded from the publisher's website.

2. 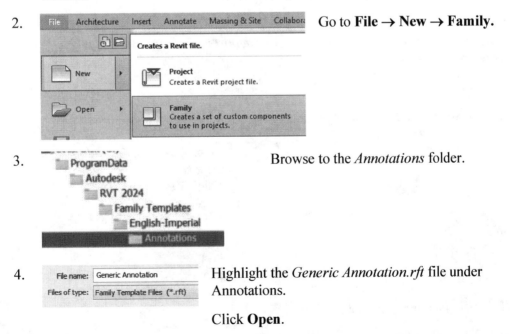 Go to **File → New → Family.**

3. Browse to the *Annotations* folder.

4. Highlight the *Generic Annotation.rft* file under Annotations.

   Click **Open**.

5. Activate the Insert ribbon.

Select **Import CAD**.

6. ☑ North Arrow.dwg — Locate the *north arrow.dwg* file in the Class Files download.

7.

| File name: | North Arrow.dwg |
| Files of type: | DWG Files (*.dwg) |

Colors:	Preserve		Positioning:	Auto - Center to Center
Layers/Levels:	All		Place at:	
Import units:	millimeter	1.000000		☑ Orient to View
	☑ Correct lines that are slightly off axis		Open	Cancel

Under Colors: Enable **Preserve**.
Under Layers: Select **All**.
Under Import Units: Select **millimeter**.
Under Positioning: Select **Auto – Center to Center**.

This allows you to automatically scale your imported data.
Note that we can opt to manually place our imported data or automatically place using the center or origin.

Click **Open**.

8. Import detected no valid elements in the file's Paper space. Do you want to import from the Model space?

The Import tool looks in Paper space and Model space. You will be prompted which data to import.
Click **Yes**.

9. **Zoom In Region** to see the symbol you imported.

10.  Pick the imported symbol.

Select **Query** from the ribbon.

11.

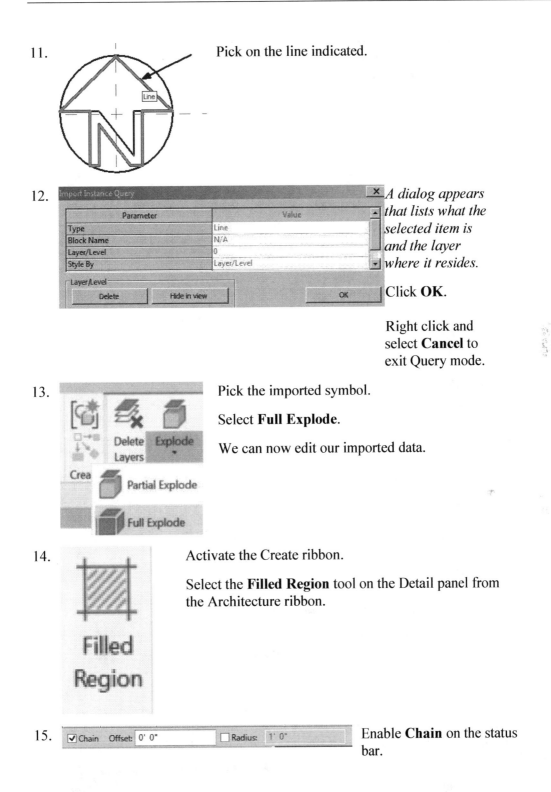

Pick on the line indicated.

12.

A dialog appears that lists what the selected item is and the layer where it resides.

Click **OK**.

Right click and select **Cancel** to exit Query mode.

13.

Pick the imported symbol.

Select **Full Explode**.

We can now edit our imported data.

14.

Activate the Create ribbon.

Select the **Filled Region** tool on the Detail panel from the Architecture ribbon.

15.

Enable **Chain** on the status bar.

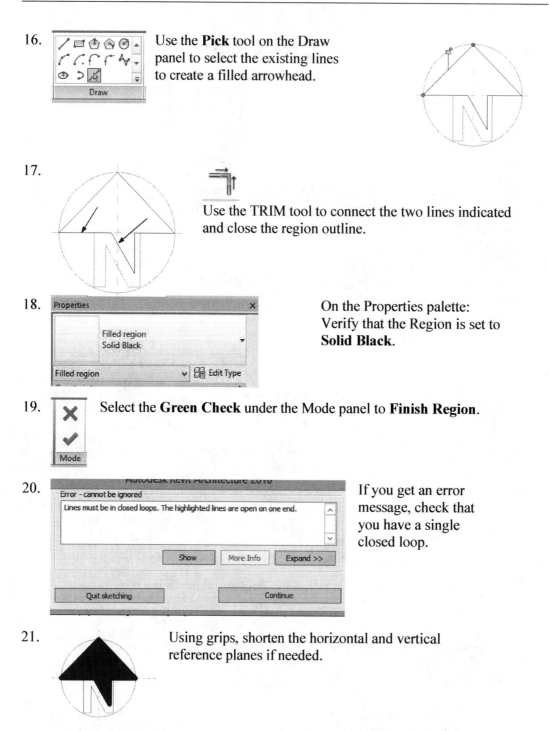

16. Use the **Pick** tool on the Draw panel to select the existing lines to create a filled arrowhead.

17. Use the TRIM tool to connect the two lines indicated and close the region outline.

18. On the Properties palette: Verify that the Region is set to **Solid Black**.

    Properties

    Filled region
    Solid Black

    Filled region   Edit Type

19. Select the **Green Check** under the Mode panel to **Finish Region**.

20. If you get an error message, check that you have a single closed loop.

    Autodesk Revit Architecture 2010
    Error - cannot be ignored
    Lines must be in closed loops. The highlighted lines are open on one end.
    Show   More Info   Expand >>
    Quit sketching   Continue

21. Using grips, shorten the horizontal and vertical reference planes if needed.

You will need to unpin the reference planes before you can adjust the link.

Be sure to pin the reference planes after you have finished any adjustments. Pinning the reference planes ensures they do not move.

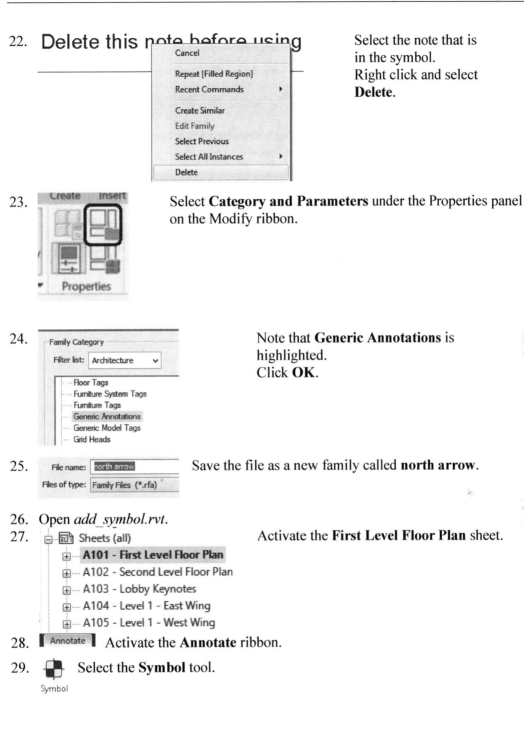

22. Delete this note before using

Select the note that is in the symbol.
Right click and select **Delete**.

23. Select **Category and Parameters** under the Properties panel on the Modify ribbon.

24. Note that **Generic Annotations** is highlighted.
Click **OK**.

25. Save the file as a new family called **north arrow**.

26. Open *add_symbol.rvt*.

27. Activate the **First Level Floor Plan** sheet.

28. Activate the **Annotate** ribbon.

29. Select the **Symbol** tool.

30. 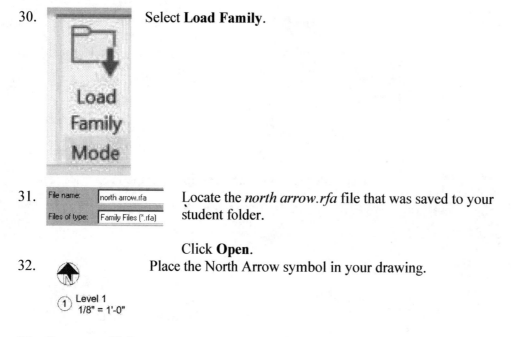 Select **Load Family**.

31. File name: north arrow.rfa   Locate the *north arrow.rfa* file that was saved to your

    Files of type: Family Files [*.rfa]   student folder.

    Click **Open**.

32.   Place the North Arrow symbol in your drawing.

    ① Level 1
       1/8" = 1'-0"

33. Save as *ex10-1.rvt*.

Exercise 10-2
Creating a Custom Title Block

Estimated Time: 60 minutes

This exercise reinforces the following skills:

- ❑ Titleblock
- ❑ Import CAD
- ❑ Labels
- ❑ Text
- ❑ Family Properties
- ❑ Shared Parameters

1. 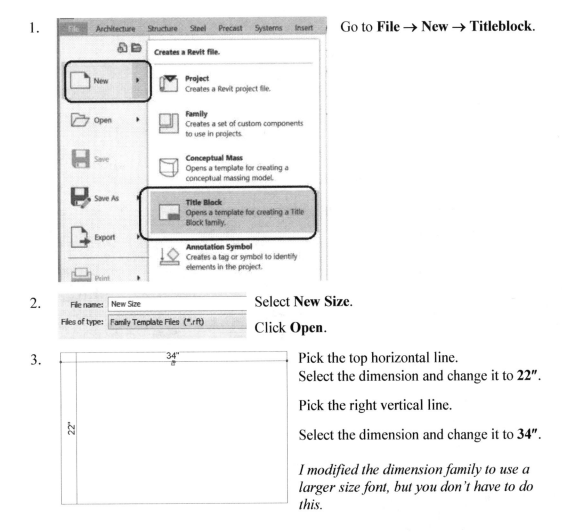 Go to **File → New → Titleblock**.

2. File name: New Size

    Files of type: Family Template Files (*.rft)

    Select **New Size**.

    Click **Open**.

3. Pick the top horizontal line.
    Select the dimension and change it to **22″**.

    Pick the right vertical line.

    Select the dimension and change it to **34″**.

    I modified the dimension family to use a larger size font, but you don't have to do this.

The title block you define includes the sheet size. If you delete the title block, the sheet of paper is also deleted. This means your title block is linked to the paper size you define.

You need to define a title block for each paper size you use.

4.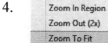

Right click in the graphics window and select **Zoom to Fit**.

You can also double click on the mouse wheel.

5.

Activate the Insert ribbon.
Select **Import CAD**.

6.

| File name: | Architectural Title Block |
| Files of type: | DWG Files (*.dwg) |

Locate the *Architectural Title Block* in the exercise files directory.

7.

| File name: | Architectural Title Block.dwg |
| Files of type: | DWG Files (*.dwg) |

Colors:	Preserve	Positioning:	Auto - Center to Center
Layers/Levels:	All	Place at:	
Import units:	inch 1.000000		☑ Orient to View
	☑ Correct lines that are slightly off axis	Open	Cancel

Set Colors to **Preserve**.
Set Layers to **All**.
Set Import Units to **Inch**.
Set Positioning to: **Auto - Center to Center**.

Click **Open**.

8.

Import detected no valid elements in the file's Paper space. Do you want to import from the Model space?

| Yes | No |

Click **Yes**.

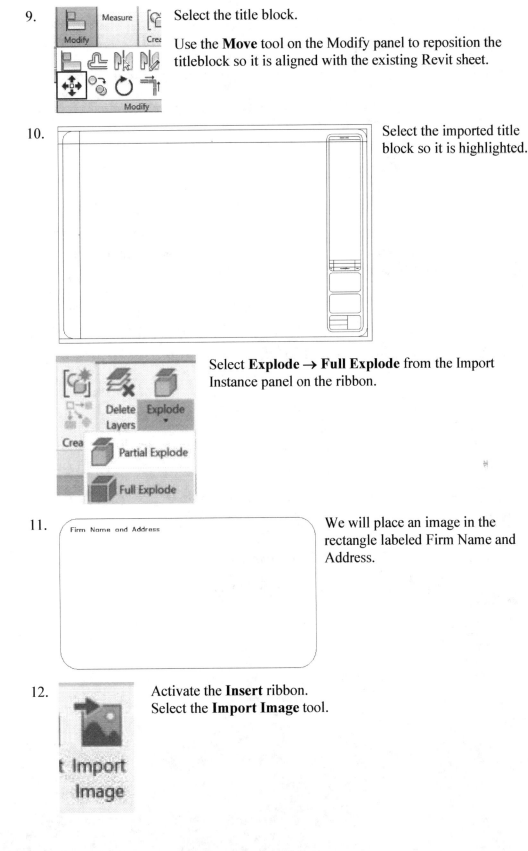

9.   Select the title block.

Use the **Move** tool on the Modify panel to reposition the titleblock so it is aligned with the existing Revit sheet.

10.   Select the imported title block so it is highlighted.

Select **Explode → Full Explode** from the Import Instance panel on the ribbon.

11.   We will place an image in the rectangle labeled Firm Name and Address.

Firm Name and Address

12.   Activate the **Insert** ribbon.
Select the **Import Image** tool.

13.

| File name: | Company_Logo |
| Files of type: | All Image Files (*.bmp, *.jpg, *.jpeg, *.png, *.tif) |

Open the *Company_Logo.jpg* file from the downloaded Class Files.

14.

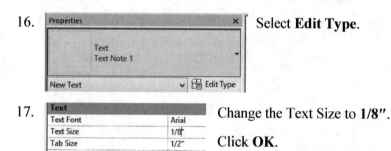

Place the logo on the right side of the box.

Scale and move it into position.

To scale, just select one of the corners and drag it to the correct size.

Revit can import jpg, jpeg, or bmp files. Imported images can be resized, rotated, and flipped using activated grips.

15. **A** Select the **Text** tool from the **Create** Ribbon.

Text

16.

Properties	✕
Text Text Note 1	
New Text	☑ Edit Type

Select **Edit Type**.

17.

Text	
Text Font	Arial
Text Size	1/8"
Tab Size	1/2"

Change the Text Size to **1/8"**.

Click **OK**.

Revit can use any font that is available in your Windows font folder.

18.

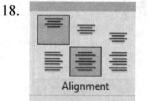
Alignment

Select **CENTER** on the Format panel.

19.

Type in the name of your school or company next to the logo.

 The data entered in the Value field for labels will be the default value used in the title block. To save edit time, enter the value that you will probably use.

---

20. Select the **Label** tool from the Create ribbon.

Labels are similar to attributes. They are linked to file properties.

21. Left pick in the Project Name and Address box.

22. Click **OK**.
Left click to release the selection.
Pick to place when the dashed line appears.

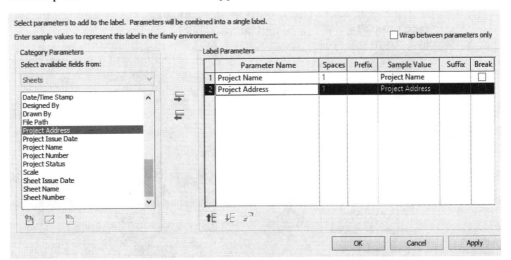

Select **Project Name**.
Use the **Add** button to move it into the Label Parameters list.
Add **Project Address**.

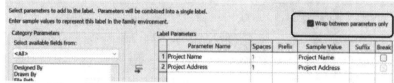

23. Enable **Wrap between parameters**.
Click **OK**.

24.

Cancel out of the Label command.

Position the label.
Use the grips to expand the label.

Project Name
Project Address

25.

Select the label.

In the Properties panel:

Verify that the Horizontal Align is set to
Center.

*The Wrap between parameters only will
appear in the Properties palette if you
have enabled it when the properties were
selected.*

26.

Project Name
Project Address

Use Modify→Move to adjust the position of the project
name and address label.

27.

Select the **Label** tool.

28.

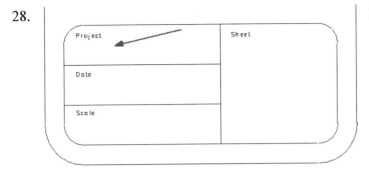

Left click in the Project box.

29. Locate **Project Number** in the Parameter list.
Move it to the right pane.

	Parameter Name	Spaces	Prefix	Sample Value	Suffix	Break
1	Project Number	1		A201		

Add a sample value, if you like.

Click **OK**.

30.

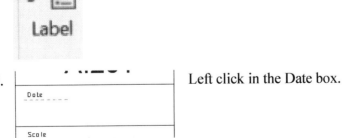

Left click to complete placing the label.

Select and reposition as needed.

31. Select the **Label** tool from the Create ribbon.

32. Left click in the Date box.

33.

	Parameter Name	Spaces	Prefix	Sample Value
1	Project Issue Date	1		02/26/23

Label Parameters

Highlight **Project Issue Date**.
Click the Add button.
In the Sample Value field, enter the default date to use.

Click **OK**.

34.

Position the date in the date field.

35.

Select the **Label** tool.

Left click in the Scale box.

36.

Select parameters to add to the label. Parameters will be combined into a single label.

Enter sample values to represent this label in the family environment.

☐ Wrap between parameters only

Category Parameters

Select available fields from:

Sheets

Date/Time Stamp
Designed By
Drawn By
File Path
Project Address
Project Issue Date
Project Name
Project Number
Project Status
Scale
Sheet Issue Date
Sheet Name
Sheet Number

Label Parameters

	Parameter Name	Spaces	Prefix	Sample Value	Suffix	Break
1	Scale	1		1/8" = 1'-0"		

Highlight **Scale**.

Click the Add button.

Click **OK**.

37.

Properties

Tag Label
Tag 1

Title Blocks (1)   ☰ Edit Type

Graphics
Sample Text   1/8" = 1'-0"...
Label   Edit...
Wrap between parameters only   ☐

Highlight the scale label that was just placed and select **Edit Type** on the Properties panel.

38.

Type Parameters	
Parameter	
Graphics	
Color	Black
Line Weight	1
Background	Transparent
Show Border	☐
Leader/Border Offset	5/64"

Set the Background to **Transparent**.

Click **OK**.

39.

Scale
1/8" = 1'-0"

Note that the border outline and the block text are no longer hidden by the label background.

40.

Edit Type

Select the scale label.

On the Properties panel, select **Edit Type**.

41. Duplicate...

Select **Duplicate**.

42.

Name: Tag - Small

Rename **Tag – Small**.

Click **OK**.

43.

Text	
Text Font	Arial
Text Size	1/8"
Tab Size	1/2"
Bold	☐
Italic	☐
Underline	☐
Width Factor	1.000000

Set the Text Size to **1/8"**.

Click **OK**.

44.

Project A201
Date 02/26/23
Scale 1/8" = 1'-0"

Adjust the lines, text and labels so the Sheet label looks clean.

45.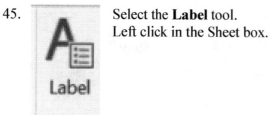

Select the **Label** tool.
Left click in the Sheet box.

46.

mbined into a single label.

ronment.

	Parameter Name	Spaces	Prefix	Sample Value	Suffix	Break
1	Sheet Number	1		A101		☑
2	Sheet Name	1		Level 1		☐

☑ Wrap between parameters only

Label Parameters

Highlight **Sheet Number**.

Click the **Add** button.

Highlight **Sheet Name**.

Click the **Add** button.
Type in a sample value for the sheet name.
Enable **Wrap between parameters only** in the upper right corner of the dialog.

Click **OK**.

47.

Project	**A201**	Sheet	
Date	**02/26/23**	A101	
		Level 1	
Scale	1/8" = 1'-0"		

Position the Sheet label.

Use the Tag-Small for the new label.

48.

Go to the Manage ribbon.

Select **Shared Parameters**.

49.

If you need to, locate the shared parameters file that was created earlier.

The file is located in the Lesson 6 folder.

Switch to the **General** parameter group.

Click **New** parameter.

50.

Set the Name to **View Type**.
Set the Discipline to **Common**.
Set the Data Type to **Text**.

Click **OK**.

Click **OK** to close the Shared Parameters dialog.

Name:

View Type

Discipline:

Common

Data Type:

Text

51.

Switch to the **Create** ribbon.

Select the **Label** tool.
Left click in the Sheet box.

52.

Category Parameters

Select available fields from:

<All>

Approved By
Assembly: Assembly Code
Assembly: Assembly Description
Assembly: Cost
Assembly: Description
Assembly: Keynote
Assembly: Manufacturer
Assembly: Model
Assembly: Name
Assembly: Type Comments
Assembly: Type Mark
Assembly: URL
Author
Building Name

Select **New Parameter**.

53.

Parameter Properties

Parameter Type

Shared parameter

(Can be shared by multiple projects and families, exported to ODBC, and appear in schedules and tags)

Select...   Export...

Click **Select** under Shared parameter.

54.

Choose a parameter group, and a parameter.

Parameter group:

General

Parameters:

Hardware Gorup
View Type

Switch to the General parameter group.

Highlight **View Type**.

Click **OK**.

55.

Parameter Data

Name:

View Type

Discipline:

Common

Data Type:

Text

Click **OK**.

56.

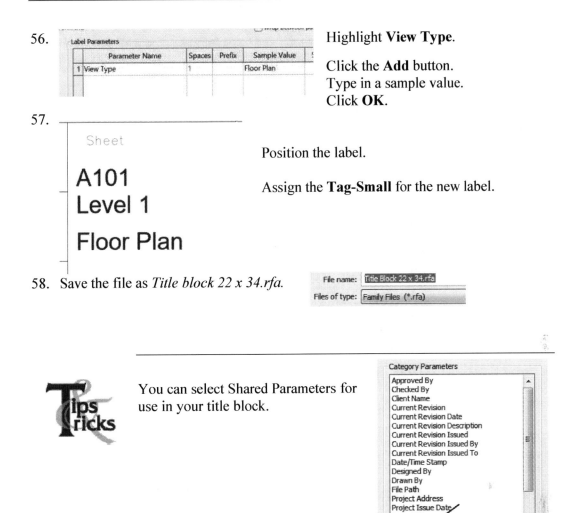

Highlight **View Type**.

Click the **Add** button.
Type in a sample value.
Click **OK**.

57.

Position the label.

Assign the **Tag-Small** for the new label.

58. Save the file as *Title block 22 x 34.rfa*.

You can select Shared Parameters for use in your title block.

Exercise 10-3
Using a Custom Title Block

File: title_block.rvt
Estimated Time: 10 minutes

This exercise reinforces the following skills:

- ❑ Title block
- ❑ Type Properties
- ❑ Project Parameters

1.   Open *title_block.rvt*.

2.    Activate the Insert ribbon.

Select **Load Family** in the Load from Library panel.

3.   File name: Title block 22 x 34
     Files of type: All Supported Files (*.rfa, *.adsk)

Browse to the folder where you saved the custom title block.

Select it and click **Open**.

4.   Sheets (all)
      A101 - First Level Floor Plan
      A102 - Second Level Floor Plan
      A103 - Lobby Keynotes
      A104 - Level 1 - East Wing
      A105 - Level 1 - West Wing

Activate the **First Level Floor Plan** sheet.

5.   Properties

     D 22 x 34 Horizontal

     Title Blocks (1)   Edit Type

Select the title block so it is highlighted.

You will see the name of the title block in the Properties pane.

6.

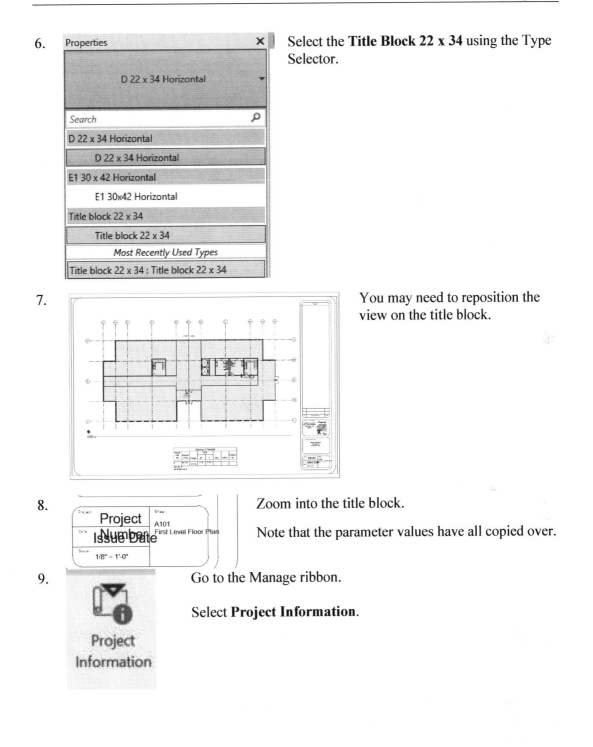

Select the **Title Block 22 x 34** using the Type Selector.

7.

You may need to reposition the view on the title block.

8.

Zoom into the title block.

Note that the parameter values have all copied over.

9.

Go to the Manage ribbon.

Select **Project Information**.

10.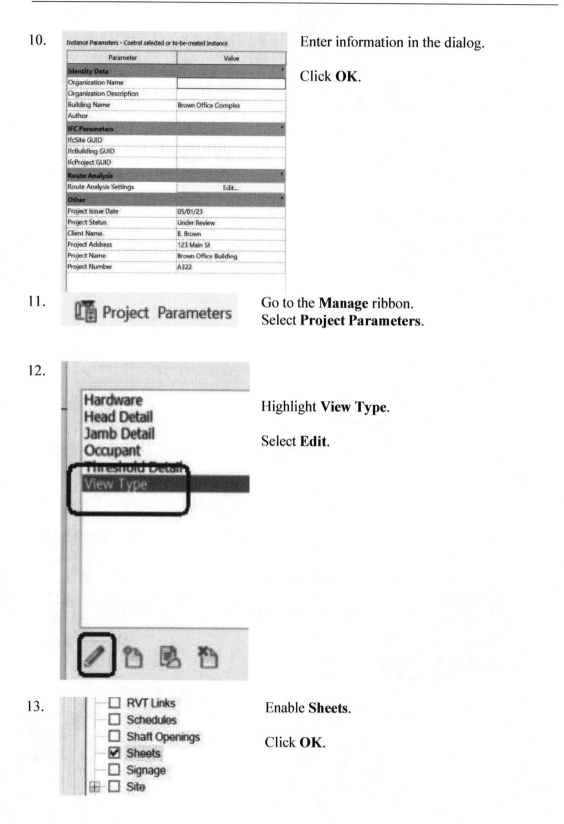

Enter information in the dialog.

Click **OK**.

11. Go to the **Manage** ribbon.
Select **Project Parameters**.

12. Highlight **View Type**.

Select **Edit**.

13. Enable **Sheets**.

Click **OK**.

14.

Select the title block.

In the Properties panel:

Type **Floor Plan** in the View Type field.

The title block updates.

Project	A322	Sheet	A101
Date	05/01/23		First Level
			Floor Plan
Scale	1/8" = 1'-0"		

15. Save the file as *ex10-3.rvt*.

It is a good idea to save all your custom templates, families, and annotations to a directory on the server where everyone in your team can access them.

Exercise 10-4
Creating a Line Style

Drawing Name: line_style_1.rvt
Estimated Time: 10 minutes

This exercise reinforces the following skills:

❑   Line Styles

1.   Open *line_style_1.rvt*.

2.   Activate the **Site** Floor plan.

> Level 2 - No Annotations
> Roof Line
> **Site**

3.   Activate the **Manage** ribbon.

Go to **Additional Settings→Line Styles**.

4. Select **New** under Modify Subcategories.

    Modify Subcategories

    | New | Delete |

5. Enter **Sewer** in the Name field.

    Name:
    Sewer

    Click **OK**.

    Subcategory of:
    Lines

    | OK | Cancel |

6.

Lines	1	RGB 000-166-000	Solid
Medium Lines	3	Black	Solid
Sewer	1	Black	Solid
Thin Lines	1	Black	Solid
Wide Lines	12	Black	Solid

    Sewer appears in the list.

    Note Revit automatically alphabetizes any new line styles.

7.

<Space Separation>	1	Black	
<Thin Lines>	1	Black	Solid
<Wide Lines>	5	Black	Solid
Sewer	3	Blue	Dash dot dot

    Set the Line Weight to **3**.
    Set the Color to **Blue**.
    Set the Line Pattern to **Dash Dot Dot**.
    Click **OK**.

8. Annotate   Activate the **Annotate** ribbon.

9.  Select the **Detail Line** tool under Detail.

10. Line Style:
    Sewer

    <Beyond>
    <Centerline>
    <Demolished>
    <Hidden Lines>
    <Hidden>
    <Lines>
    <Medium Lines>
    <Overhead>
    <Path of Travel Lines>
    <Thin Lines>
    <Wide Lines>
    Sewer

    Select **sewer** from the drop-down list in the Properties palette.

    You can also set the Line Style from the ribbon drop-down list.

Graphics	
Line Style	Sewer
Detail Line	☑

11.

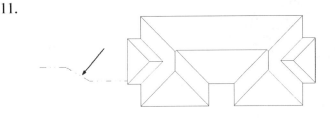

Draw a line from the building to the toposurface edge.

12. Save the file as *ex10-4.rvt*.

Revit has three line tools: *Lines*, *Detail Lines*, and *Linework*. All of them use the same line styles, but for different applications. The *Lines* command draws model lines on a specified plane and can be seen in multiple views (i.e., score joints on an exterior elevation). The *Detail Lines* command draws lines that are view specific and will only appear on the view where they were placed (i.e., details). The *Linework* tool is used to change the appearance of model-generated lines in a view.

Exercise 10-5
Defining Keyboard Shortcuts

Estimated Time: 5 minutes

This exercise reinforces the following skills:

❑ Keyboard Shortcuts

1.

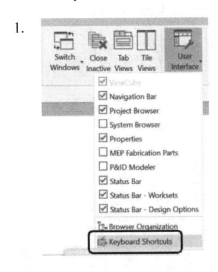

You do not need to have any files open for this exercise.

Activate the View ribbon.

Select **User Interface→Keyboard Shortcuts** from the Windows panel.

2.

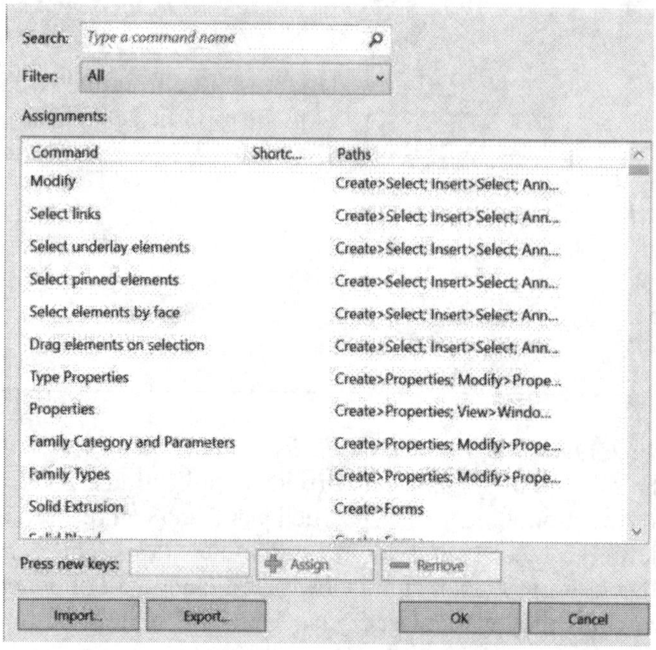

A dialog opens which allows you to assign shortcut keys to different Revit commands.

3.

Search: mea

Filter: All

Type **mea** in the Search box.

4.

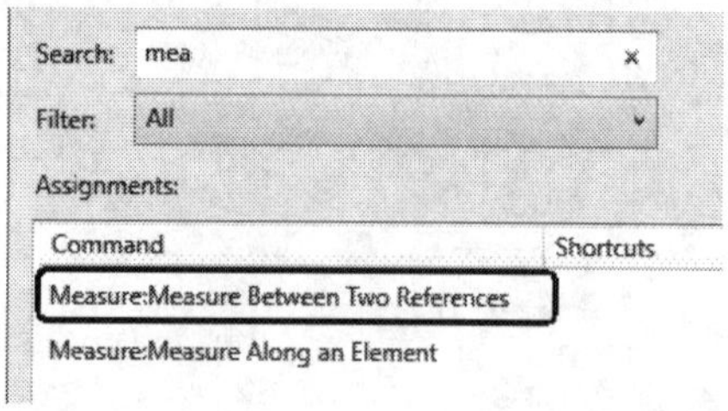

Search: mea

Filter: All

Assignments:

Command	Shortcuts
Measure:Measure Between Two References	
Measure:Measure Along an Element	

Highlight the first command to **Measure between two references.**

5.

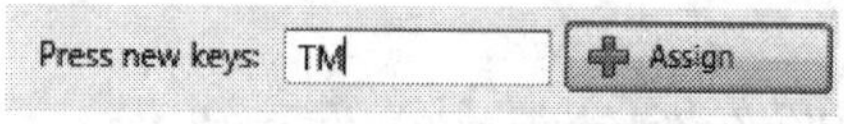

Press new keys: TM   Assign

Type **TM** in the Click new keys text box.

Do not hold down the Caps Lock or shift key – Revit will automatically capitalize the letters.

Click **Assign**.

6.

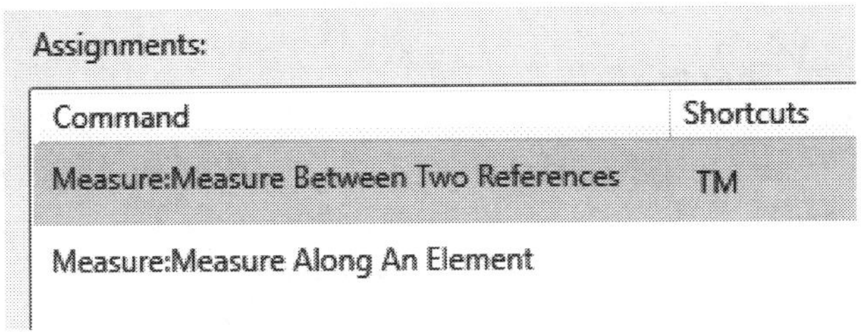

Assignments:	
Command	**Shortcuts**
Measure:Measure Between Two References	TM
Measure:Measure Along An Element	

The shortcut appears next to the command.

Click **OK** to close the dialog.

7. To test, you need to open a file and type **TM**.

Total Length appears on the Options bar, and your cursor is in Measure mode.

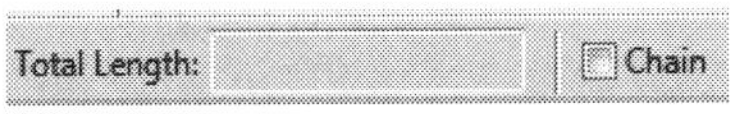

Total Length:		☐ Chain

Creating Custom Families

One of the greatest tools inside of Revit is the ability to create custom families for common components, such as windows and doors. These are more powerful than blocks because they are completely parametric. You can create one set of geometry controlled by different sets of dimensions.

The steps to create a family are the same, regardless of whether you are creating a door, window, furniture, etc.

Step 1:
Select the appropriate family template to use.

Step 2:
Define sub-categories for the family.
Sub-categories determine how the object will appear in different views.

Step 3:
Lay out reference planes.

Step 4:
Dimension planes to control the parametric geometry.

Step 5:
Label dimensions to become type or instance parameters.

Step 6:
Create your types using the 'FamilyTypes' tool.

Step 7:
Activate different types and verify that the reference planes shift correctly with the dimensions assigned.

Step 8:
Label your reference planes.

Step 9:
Create your geometry and constrain to your reference planes using Lock and Align.

Step 10:
Activate the different family types to see if the geometry reacts correctly.

Step 11:
Save family and load into a project to see how it performs within the project environment.

Exercise 10-6
Defining Reference Plane Object Styles

Estimated Time: 15 minutes

This exercise reinforces the following skills:

❑ Reference Planes
❑ Object Styles

Users have the ability to create different styles of reference planes. This is especially useful when creating families as reference planes are used to control family geometry.

1. [File] Architecture Structure Steel Precast Systems Insert

    Creates a Revit file.

    New ► Project — Creates a Revit project file.

    Open ► Family — Creates a set of custom components to use in projects.

    Start a New family.

2. File name: Generic Model

    Files of type: Family Template Files (*.rft)

    Select the *Generic Model* template.

    Note that there are already two reference planes in the file. The intersection of the two planes represents the insertion point for the family. Both reference planes are pinned in place so they cannot be moved or deleted.

3. Reference Planes (2)   Edit Type

    Construction
    Wall Closure   ☐
    Extents
    Scope Box   None
    Identity Data
    Name   <varies>
    Subcategory   <None>
    Other
    Is Reference   <varies>
    Defines Origin   ☑

    Hold down the CTRL key and select the two reference planes.

    Note in the Properties pane **Defines Origin** is enabled.

    This is how you can control the insertion point for the family. The intersection of any two reference planes may be used to define the origin.

4.

    Object Styles

    Release the selection,
    Activate the Manage ribbon.

    Select **Object Styles** from the Settings panel.

5. Select the **Annotation Objects** tab.

Highlight **Reference Planes**.

6. Select **New** under Modify Subcategories.

7. Type **Table Outline** in the Name field.

Note that this is a subcategory of Reference Planes.

Click **OK**.

8. Change the Color to **Magenta**.
Set the Line Pattern to **Aligning Line.**

9. Select **New** under Modify Subcategories.

10. Type **Leg Locations** in the Name field.

Note that this is a subcategory of Reference Planes.

Click **OK**.

11. Change the Color to **Green**.
Set the Line Pattern to **Aligning Line.**

12. Click **OK** to close the dialog.

13. Save as *ex10-6.rfa.*

Exercise 10-7
Creating a Furniture Family

Estimated Time: 120 minutes
File: Coffee Table_1.rfa

This exercise reinforces the following skills:

- Standard Component Families
- Templates
- Reference Planes
- Align
- Dimensions
- Parameters
- Sketch Tools
- Solid extrusion
- Materials
- Types

1. Select the **Reference Plane** tool from the Datum panel on the Create ribbon.

   Reference
   Plane

2. On the ribbon, set the Subcategory to **Table Outline**.

   Subcategory:
   Table Outline

3. Draw four reference planes: two horizontal and two vertical on each side of the existing reference planes.

   These will act as the outside edges of the tabletop.

   Note they are a different color because they are using the reference plane style you defined earlier.

4. Add a continuous aligned dimension horizontally and vertically so they can be set equal.

   Measure

5.

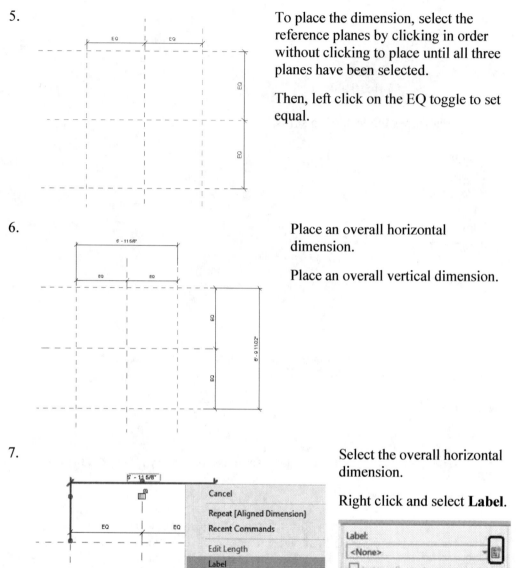

To place the dimension, select the reference planes by clicking in order without clicking to place until all three planes have been selected.

Then, left click on the EQ toggle to set equal.

6.

Place an overall horizontal dimension.

Place an overall vertical dimension.

7.

Select the overall horizontal dimension.

Right click and select **Label**.

You can also select Label from the Options bar.

8.

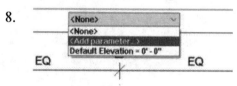

Select **Add Parameter**...

9.

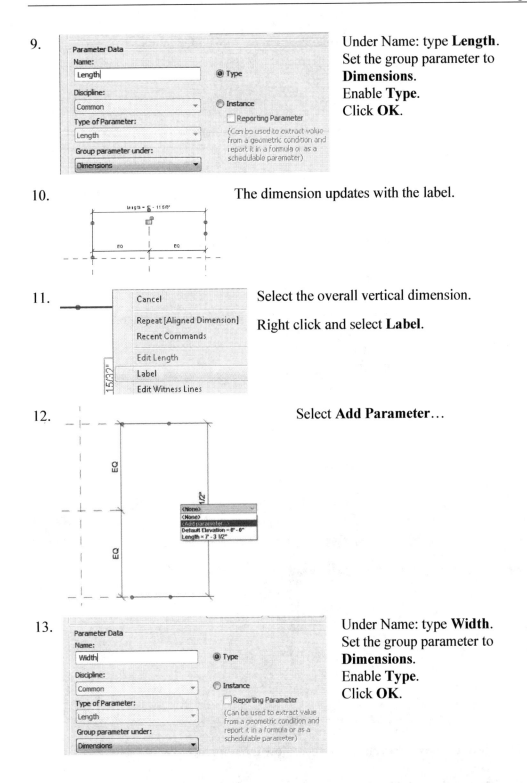

Under Name: type **Length**.
Set the group parameter to
Dimensions.
Enable **Type**.
Click **OK**.

10.

The dimension updates with the label.

11.

Select the overall vertical dimension.

Right click and select **Label**.

12.

Select **Add Parameter**…

13.

Under Name: type **Width**.
Set the group parameter to
Dimensions.
Enable **Type**.
Click **OK**.

14. 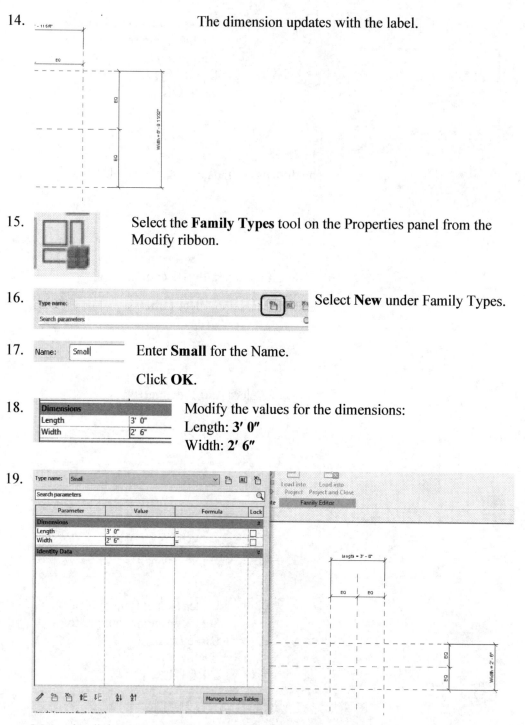 The dimension updates with the label.

15. Select the **Family Types** tool on the Properties panel from the Modify ribbon.

16. Select **New** under Family Types.

17. Name: Small Enter **Small** for the Name.

Click **OK**.

18. Modify the values for the dimensions:
Length: **3′ 0″**
Width: **2′ 6″**

19.

Move the dialog over so you can observe how the reference planes react.
Click **Apply**.
Note how the dimensions update.
This is called *flexing* the model.

20. 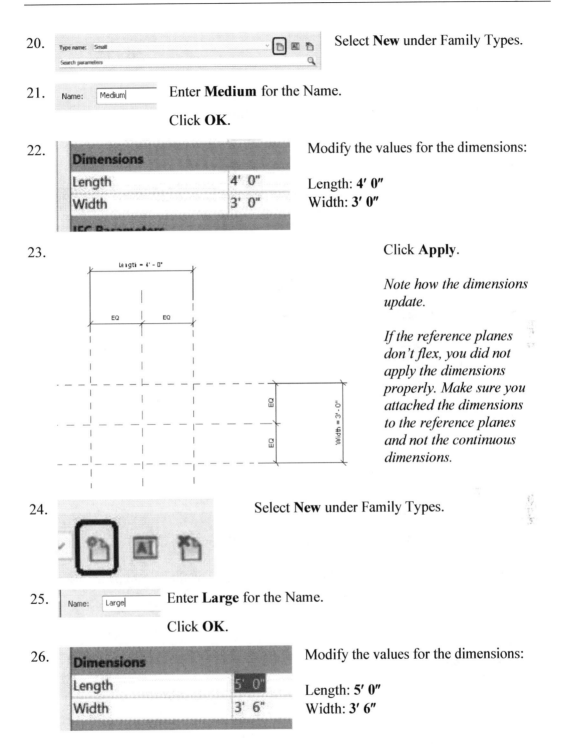 Select **New** under Family Types.

21. Enter **Medium** for the Name.

Click **OK**.

22. Modify the values for the dimensions:

Length: **4' 0"**
Width: **3' 0"**

23. Click **Apply**.

Note how the dimensions update.

If the reference planes don't flex, you did not apply the dimensions properly. Make sure you attached the dimensions to the reference planes and not the continuous dimensions.

24. Select **New** under Family Types.

25. Enter **Large** for the Name.

Click **OK**.

26. Modify the values for the dimensions:

Length: **5' 0"**
Width: **3' 6"**

27.

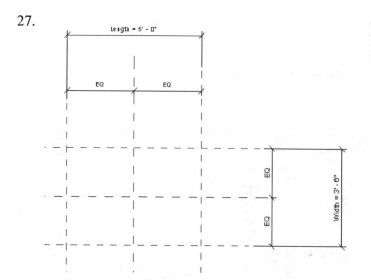

Click **Apply**.
Note how the dimensions update.

28. Left click on the drop-down arrow on the Name bar.
Switch between the different family names and click Apply to see the dimensions change.

Click **OK** to close the dialog.

29. Elevations (Elevation 1)
  - Back
  - **Front**
  - Left
  - Right

Activate the **Front** elevation.

Zoom In Region
Zoom Out (2x)
Zoom To Fit

Perform a **Zoom to Fit** to see the entire view.

30. Select the **Reference Plane** tool on the Datum panel from the Create ribbon.

31. Subcategory:
Table Outline

On the ribbon:

Set the Subcategory to **Table Outline**.

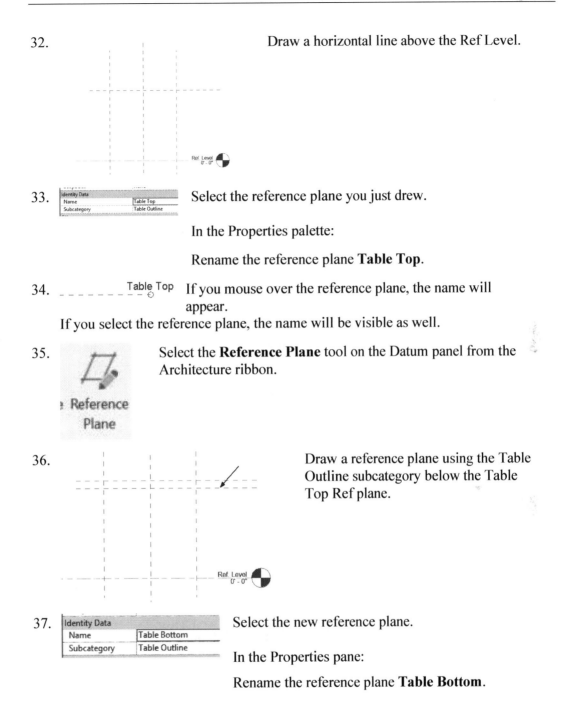

32.   Draw a horizontal line above the Ref Level.

33.   Select the reference plane you just drew.

In the Properties palette:

Rename the reference plane **Table Top**.

34.   If you mouse over the reference plane, the name will appear.

If you select the reference plane, the name will be visible as well.

35.   Select the **Reference Plane** tool on the Datum panel from the Architecture ribbon.

36.   Draw a reference plane using the Table Outline subcategory below the Table Top Ref plane.

37.   Select the new reference plane.

In the Properties pane:

Rename the reference plane **Table Bottom**.

38.

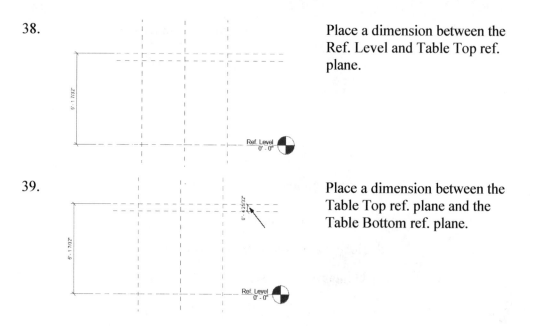

Place a dimension between the Ref. Level and Table Top ref. plane.

39.

Place a dimension between the Table Top ref. plane and the Table Bottom ref. plane.

40. Select the table height dimension, the dimension between the Ref. Level and Top ref. plane.

Right click and select **Label**.

Select *Add Parameter* from the **Label** drop-down on the Option bar.

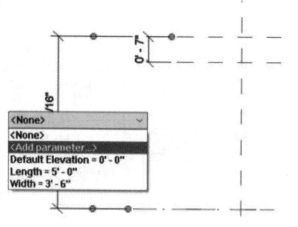

41.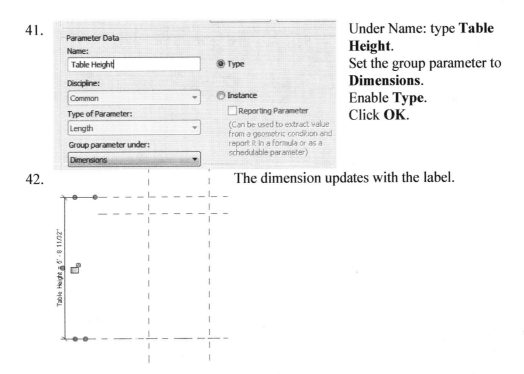

Under Name: type **Table Height**.
Set the group parameter to **Dimensions**.
Enable **Type**.
Click **OK**.

42.

The dimension updates with the label.

43. Select the table thickness dimension, the dimension between the Table Top ref. plane and the Table Bottom ref. plane below it.

Select *Add Parameter* from the **Label** drop-down on the Option bar.

44.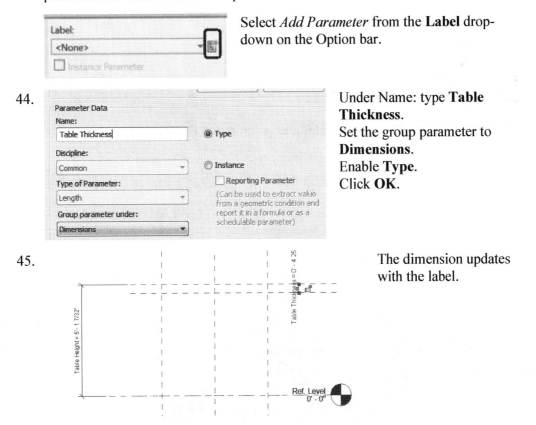

Under Name: type **Table Thickness**.
Set the group parameter to **Dimensions**.
Enable **Type**.
Click **OK**.

45.

The dimension updates with the label.

46. 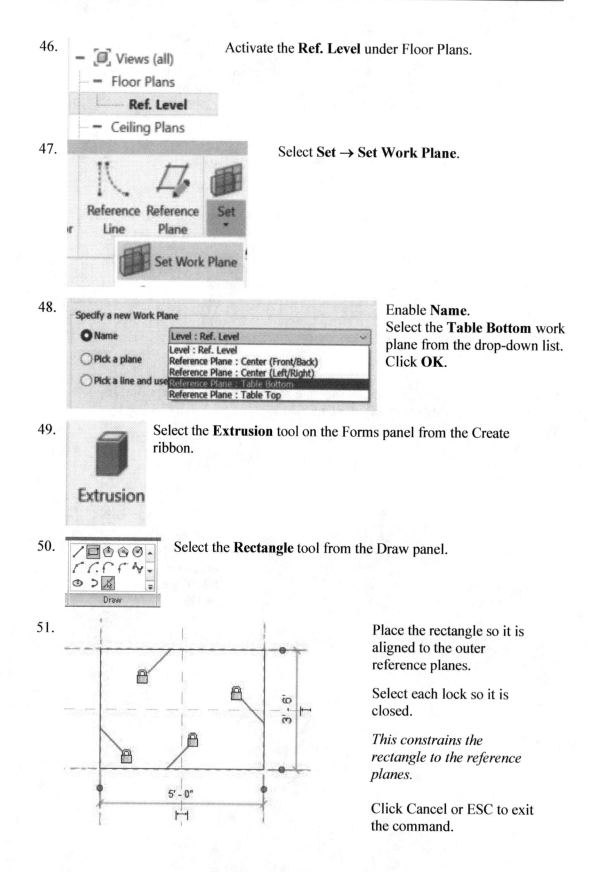 Activate the **Ref. Level** under Floor Plans.

47. Select **Set → Set Work Plane**.

48. Enable **Name**.
Select the **Table Bottom** work plane from the drop-down list.
Click **OK**.

49. Select the **Extrusion** tool on the Forms panel from the Create ribbon.

50. Select the **Rectangle** tool from the Draw panel.

51. Place the rectangle so it is aligned to the outer reference planes.

Select each lock so it is closed.

This constrains the rectangle to the reference planes.

Click Cancel or ESC to exit the command.

52.  Select the **Family Types** tool from the Properties pane.

53. 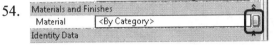 Select each type and click **Apply**.

Verify that the rectangle flexes properly with each size.
If it does not flex properly, check that the side has been locked to the correct reference line using the Align tool.

Click **OK** to close the dialog.

54. | Materials and Finishes | |
    | --- | --- |
    | Material | \<By Category\> |
    | Identity Data | |

    Select the small button in the far right of the Material column.

    This allows you to link the material to a parameter.

55. | | |
    | --- | --- |
    | Family parameter: | Material |
    | Parameter type: | Material |

    Existing family parameters of compatible type:

    Search parameters

    \<none\>

    Select **New parameter**.

56. **Parameter Data**

    Name:

    Table Top

    Discipline:

    Common

    Type of Parameter:

    Material

    Group parameter under:

    Materials and Finishes

    ● Type
    ○ Instance
    ☐ Reporting Parameter
    (Can be used to extract value from a geometric condition and report it in a formula or as a schedulable parameter)

    Type **Table Top** for the name.

    Enable **Type**.
    Set the Group Parameter under **Materials and Finishes**.

    Click **OK**.

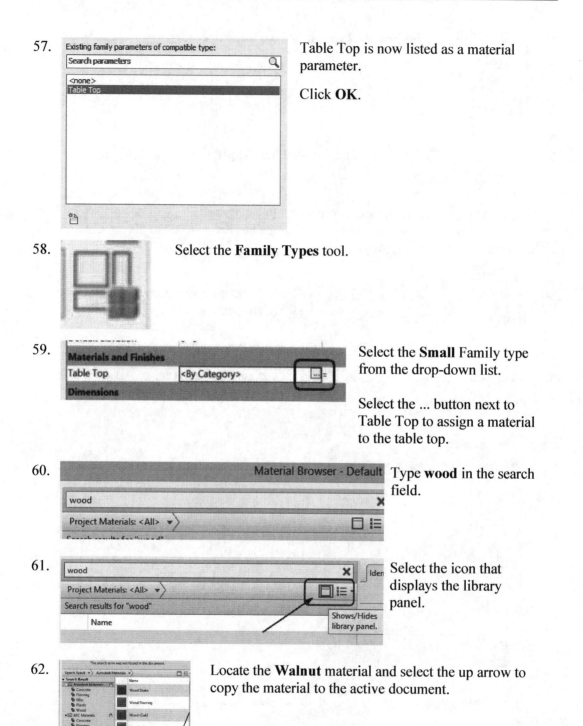

57. Table Top is now listed as a material parameter.

Click **OK**.

58. Select the **Family Types** tool.

59. Select the **Small** Family type from the drop-down list.

Select the ... button next to Table Top to assign a material to the table top.

60. Type **wood** in the search field.

61. Select the icon that displays the library panel.

62. Locate the **Walnut** material and select the up arrow to copy the material to the active document.

63. 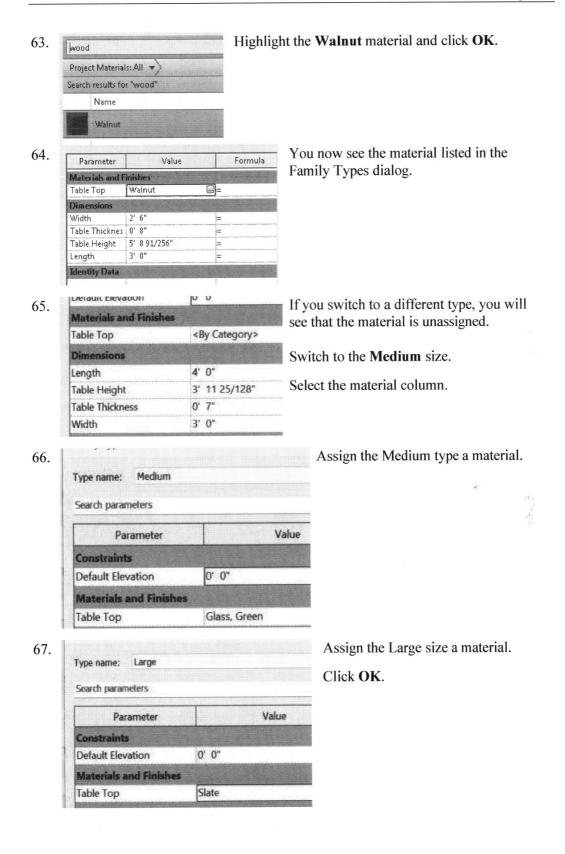 Highlight the **Walnut** material and click **OK**.

64. You now see the material listed in the Family Types dialog.

65. If you switch to a different type, you will see that the material is unassigned.

Switch to the **Medium** size.

Select the material column.

66. Assign the Medium type a material.

67. Assign the Large size a material.

Click **OK**.

68.

Click the button in the Extrusion End row.

69.

Select **Table Thickness** from the parameters list.

Click **OK**.

70.

Select the **Family Types** tool from the Properties pane.

71. Table Thickness has now been added as a parameter to the Family Types.

Change the value of the Table Thickness to **1/8″** for each type.

Click **Apply** before switching to the next type.

Click **OK**.

72.

Select the **green check** under Mode to **Finish Extrusion**.

73.

Activate the **Front** Elevation.

74.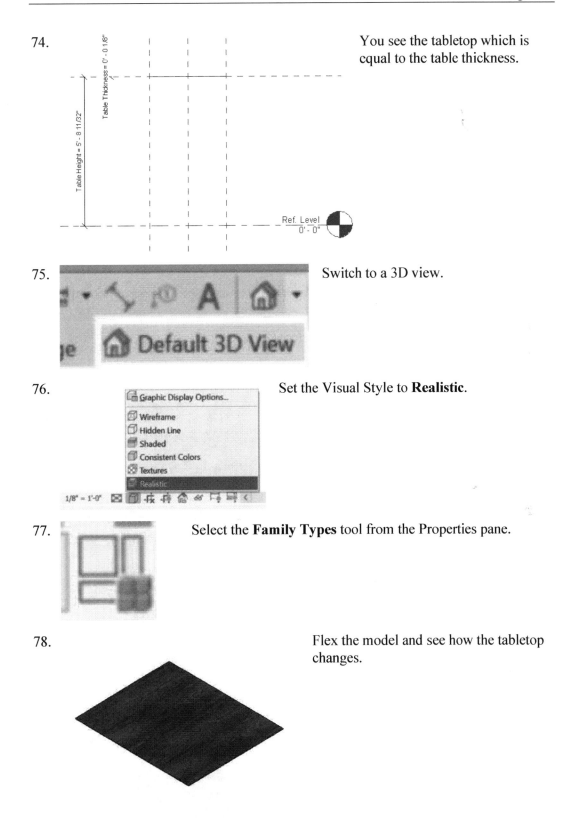

You see the tabletop which is equal to the table thickness.

75. Switch to a 3D view.

76. Set the Visual Style to **Realistic**.

77. Select the **Family Types** tool from the Properties pane.

78. Flex the model and see how the tabletop changes.

79.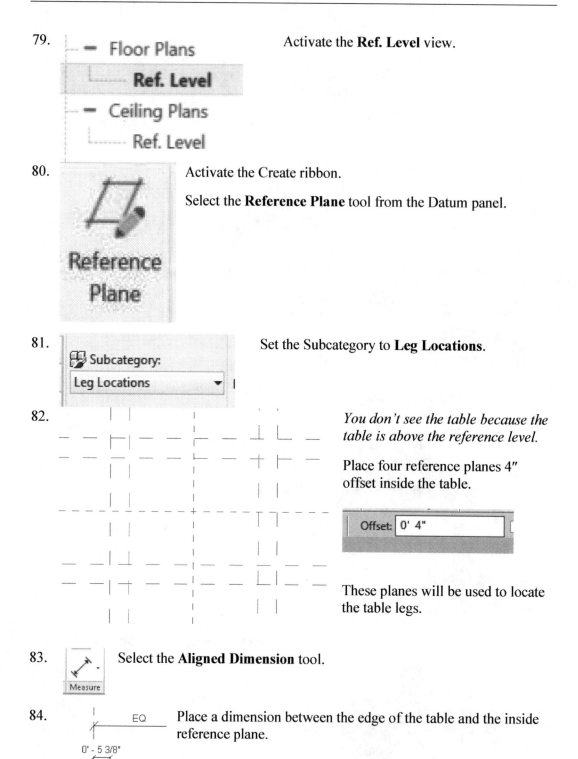
    Floor Plans
    Ref. Level
    Ceiling Plans
    Ref. Level

Activate the **Ref. Level** view.

80. Activate the Create ribbon.

Select the **Reference Plane** tool from the Datum panel.

Reference Plane

81. Subcategory:
    Leg Locations

Set the Subcategory to **Leg Locations**.

82. *You don't see the table because the table is above the reference level.*

Place four reference planes 4″ offset inside the table.

Offset: 0' 4"

These planes will be used to locate the table legs.

83. Select the **Aligned Dimension** tool.

Measure

84. EQ

0' - 5 3/8"

Place a dimension between the edge of the table and the inside reference plane.

85.

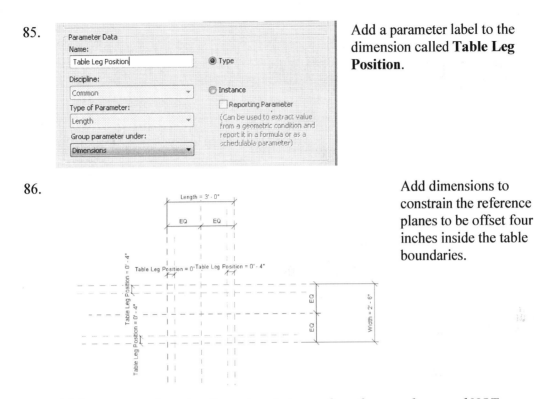

Add a parameter label to the dimension called **Table Leg Position**.

86.

Add dimensions to constrain the reference planes to be offset four inches inside the table boundaries.

Make sure you place the dimensions between the reference planes and NOT between the table extrude edges and the reference planes. Do not lock any of the dimensions or reference planes or they won't shift when the table size changes.

87. Select each dimension and use the label pull-down on the Options bar to assign the Table Leg position parameter to the dimension.

88.

Select the **Family and Types** tool.

89.

Parameter	Value
Materials and Finishes	
Table Top	Walnut
Dimensions	
Length	3' 0"
Table Height	5' 1 29/128"
Table Leg Position	0' 4"
Table Thickness	0' 0 1/8"
Width	2' 6"

Type name: Small

Search parameters

Modify the Table Leg Position parameter to **4"**.

Click **Apply** and verify that the reference planes flex.

Repeat for each family name and flex the model to verify that the reference planes adjust.

Click **OK**.

90. 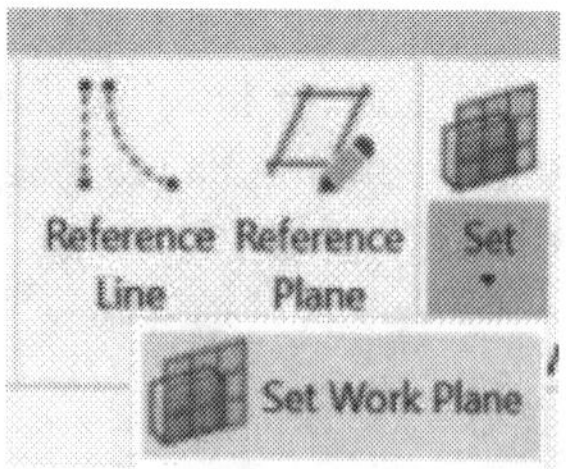
Reference Line
Reference Plane
Set
Set Work Plane

Select the **Set Work Plane** tool.

91. 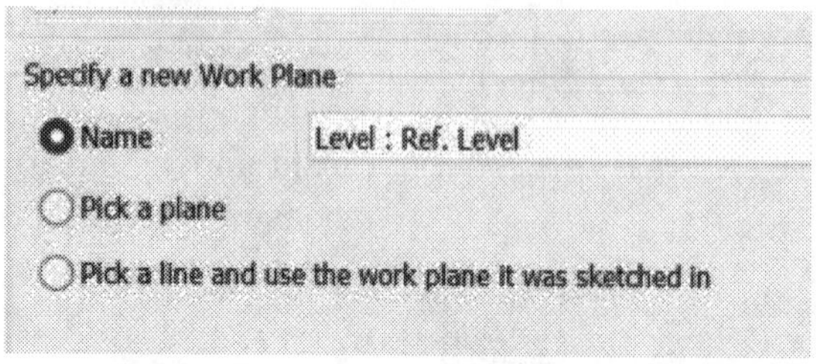
Specify a new Work Plane
Name
Level : Ref. Level
Pick a plane
Pick a line and use the work plane it was sketched in

Select the **Ref. Level** for the sketch.

Click **OK**.

92. 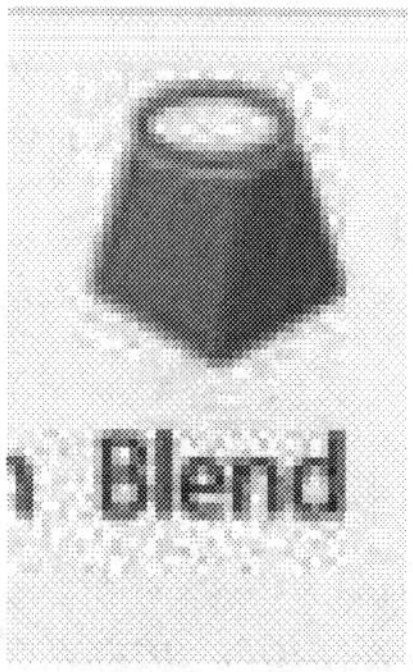
Blend

Select **Blend** from the Create ribbon.

93. 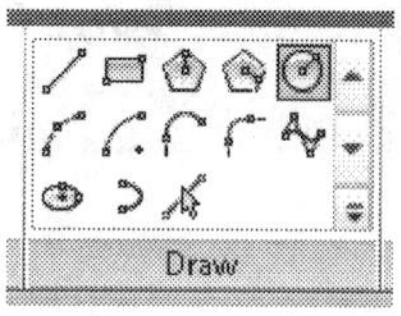
Draw

Select the **Circle** tool from the Draw panel.

94. 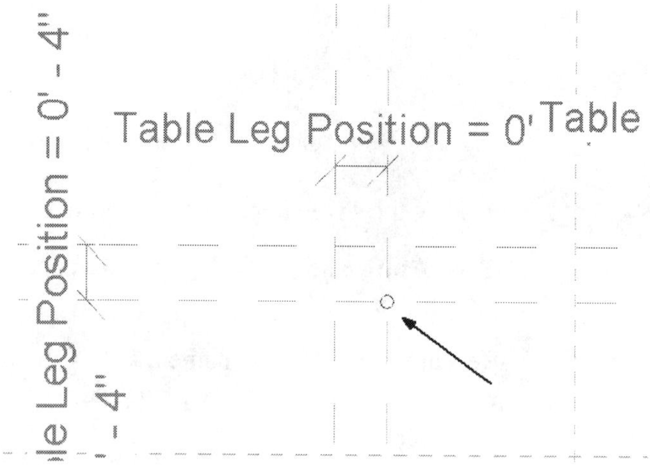
Table Leg Position = 0' - 4"
Table Leg Position = 0' Table

Draw a circle with a 1/2″ radius so the center is at the intersection of the two offset reference planes.

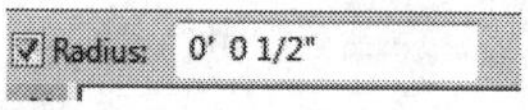
Radius: 0' 0 1/2"

You can set the radius in the Options bar.

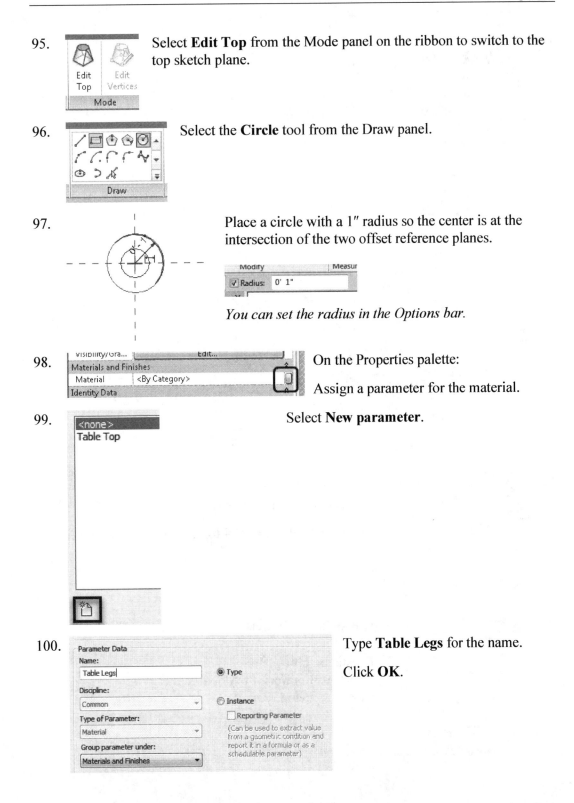

95. Select **Edit Top** from the Mode panel on the ribbon to switch to the top sketch plane.

96. Select the **Circle** tool from the Draw panel.

97. Place a circle with a 1″ radius so the center is at the intersection of the two offset reference planes.

You can set the radius in the Options bar.

98. On the Properties palette:

Assign a parameter for the material.

99. Select **New parameter**.

100. Type **Table Legs** for the name.

Click **OK**.

101. 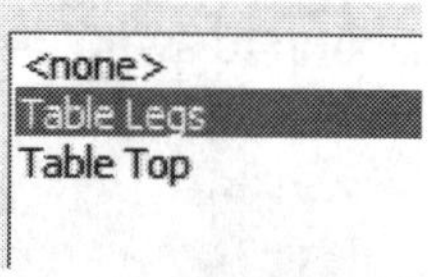
<none>
Table Legs
Table Top

Highlight **Table Legs.**

Click **OK.**

102. 

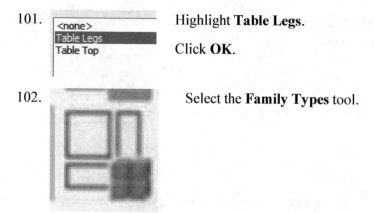

Select the **Family Types** tool.

103. *Table Legs is now a material to be assigned for each type.*

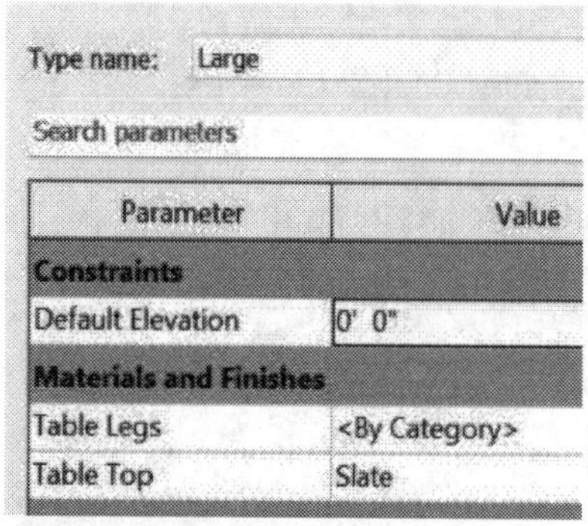
Type name: Large
Search parameters
Parameter
Value
Constraints
Default Elevation
0' 0"
Materials and Finishes
Table Legs
<By Category>
Table Top
Slate

Select the **...** button in the Table Legs Value column.

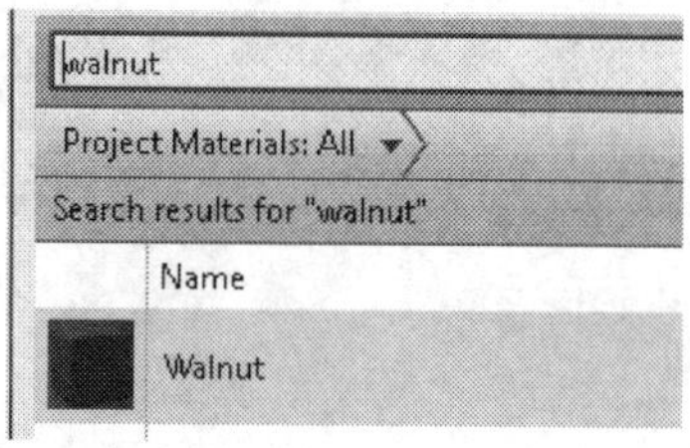
walnut
Project Materials: All
Search results for "walnut"
Name
Walnut

104. 
concrete

Type **concrete** in the search field.

105. 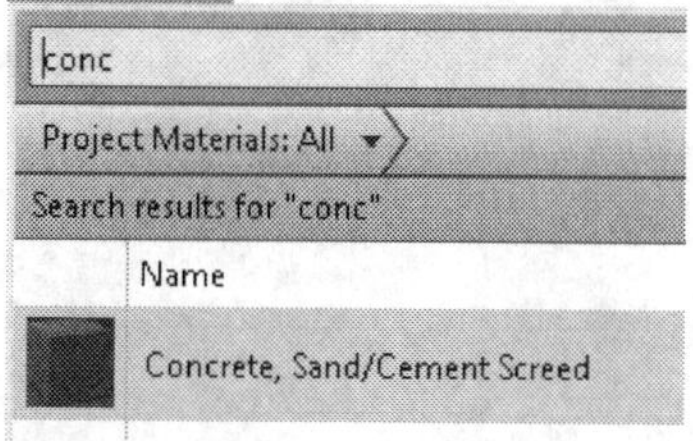
conc
Project Materials: All
Search results for "conc"
Name
Concrete, Sand/Cement Screed

Locate a concrete material.

Select **Add material to document**.

106. 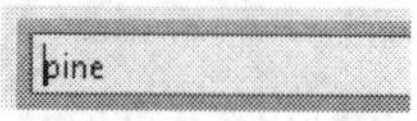
pine

Type **pine** in the search field.

107. 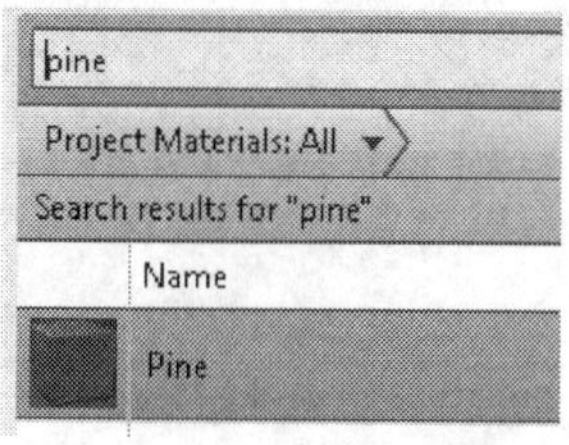
pine
Project Materials: All
Search results for "pine"
Name
Pine

Locate the **pine** material.
Select **Add material to document.**
Close the dialog.

108.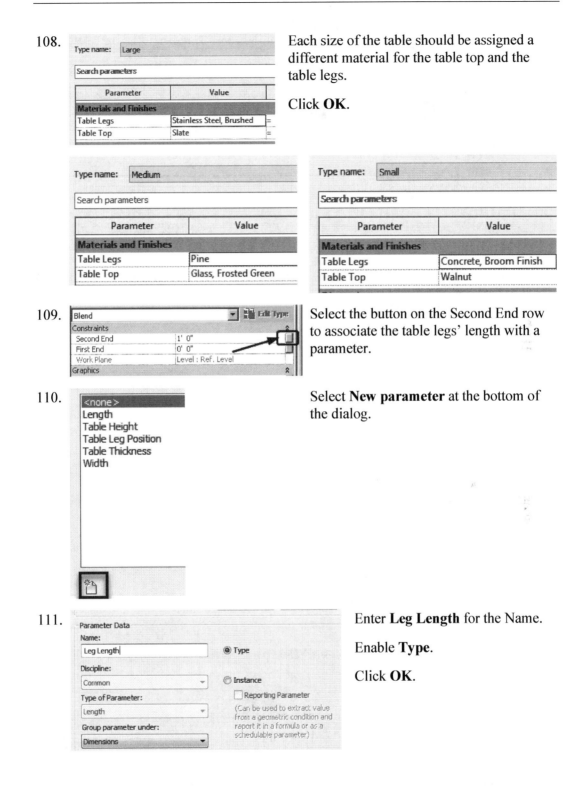

Each size of the table should be assigned a different material for the table top and the table legs.

Click **OK**.

109.

Select the button on the Second End row to associate the table legs' length with a parameter.

110.

Select **New parameter** at the bottom of the dialog.

111.

Enter **Leg Length** for the Name.

Enable **Type**.

Click **OK**.

112.

Highlight **Leg Length** in the parameter list.

Click **OK**.

Search parameters

<none>
Default Elevation
Leg Length
Length
Table Height
Table Leg Position
Table Thickness
Width

113.

Select **Family Types** from Properties pane on the Create ribbon.

114.

Dimensions		
Leg Length	5' 5 89/128"	= Table Height - Table Thickness
Length	3' 0"	=
Table Height	5' 5 105/128"	=
Table Leg Position	0' 4"	=
Table Thickness	0' 0 1/8"	=
Width	2' 6"	=

In the Formula column, enter **Table Height – Table Thickness**.

Click **OK**.

Note, you only have to enter the formula once and it works for all three sizes.

The variable names must match exactly! If you have a spelling error or use an abbreviation, you have to copy it.

115.

Constraints	
Second End	3' 9 73/128"
First End	0' 0"
Work Plane	Level : Ref. Level

The value for Second End will update.

116.

Select the **green check** on the Mode panel to finish the extrusion.

117.

{3D}
Elevations (Elevation 1)
— Back
— **Front**
— Left
— Right

Activate the **Front** Elevation view.

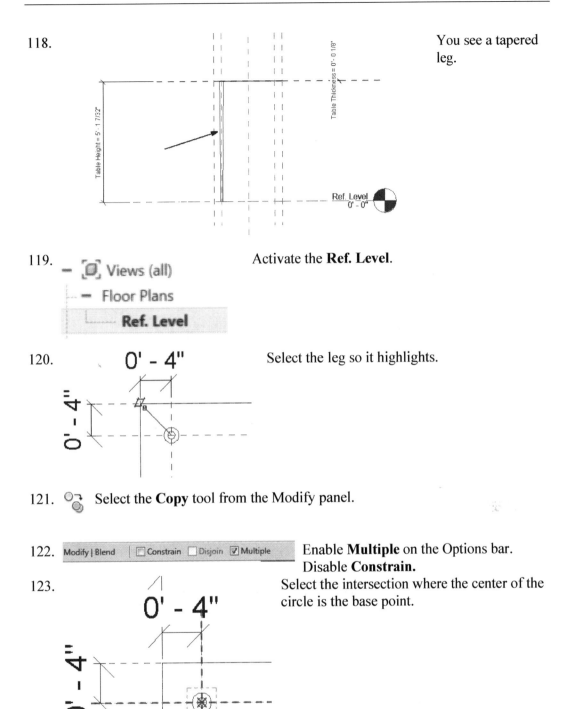

118.   You see a tapered leg.

119.   Activate the **Ref. Level**.

— 🗗 Views (all)
    — Floor Plans
        Ref. Level

120.   Select the leg so it highlights.

121.   Select the **Copy** tool from the Modify panel.

122.   Modify | Blend   ☐ Constrain ☐ Disjoin ☑ Multiple   Enable **Multiple** on the Options bar.
        Disable **Constrain.**

123.   Select the intersection where the center of the circle is the base point.

124.

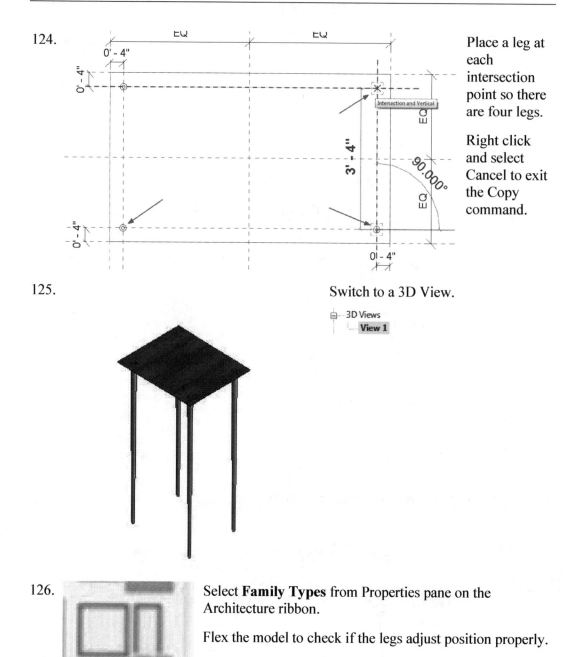

Place a leg at each intersection point so there are four legs.

Right click and select Cancel to exit the Copy command.

125.　Switch to a 3D View.

⊟ 3D Views
　└ **View 1**

126.　Select **Family Types** from Properties pane on the Architecture ribbon.

Flex the model to check if the legs adjust position properly.

127. Change the value for the Table Height and Table Thickness for each size to see what happens to the model.

Type name: Small

Search parameters

Parameter	Value	Formula
Materials and Finishes		
Table Legs	Concrete, Broom Finish	=
Table Top	Walnut	=
Dimensions		
Leg Length	1' 8 7/8"	= Table Height - Table Thickness
Length	3' 0"	=
Table Height	1' 9"	=
Table Leg Position	0' 4"	=
Table Thickness	0' 0 1/8"	=
Width	2' 6"	=
Identity Data		

Type name: Medium

Search parameters

Parameter	Value	Formula
Materials and Finishes		
Table Legs	Pine	=
Table Top	Glass, Frosted Green	=
Dimensions		
Leg Length	1' 11 7/8"	= Table Height - Table Thickness
Length	4' 0"	=
Table Height	2' 0"	=
Table Leg Position	0' 4"	=
Table Thickness	0' 0 1/8"	=
Width	3' 0"	=
Identity Data		

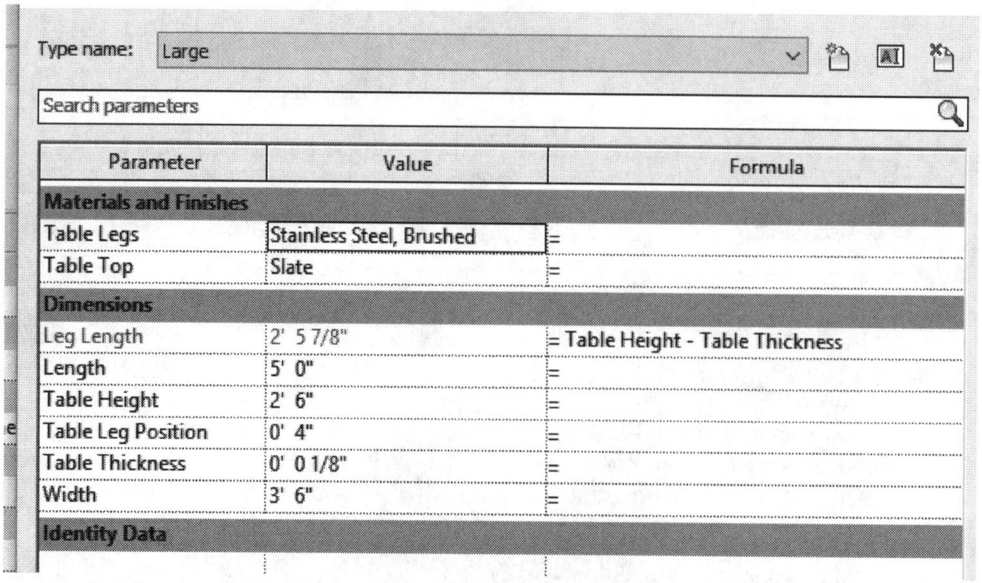

Parameter	Value	Formula
Materials and Finishes		
Table Legs	Stainless Steel, Brushed	=
Table Top	Slate	=
Dimensions		
Leg Length	2' 5 7/8"	= Table Height - Table Thickness
Length	5' 0"	=
Table Height	2' 6"	=
Table Leg Position	0' 4"	=
Table Thickness	0' 0 1/8"	=
Width	3' 6"	=
Identity Data		

Type name: Large

Search parameters

128. Save the family as *Coffee Table.rfa*.

129. Close the file.

Exercise 10-8
Modifying a Family

Estimated Time: 10 minutes
File: Office_2.rvt

This exercise reinforces the following skills:

- Standard Component Families
- Types

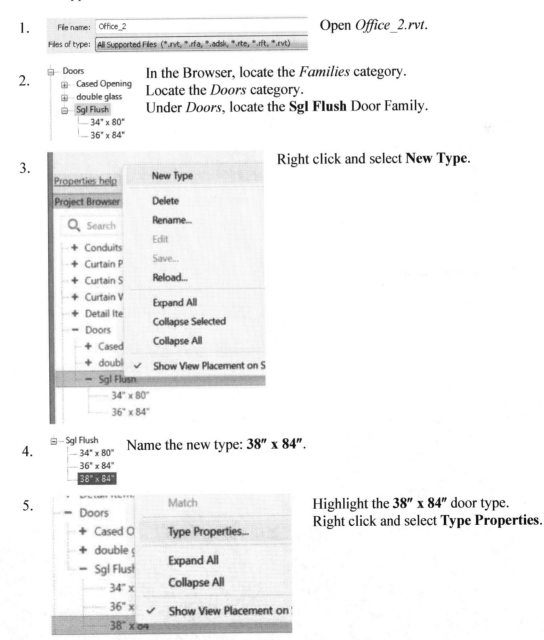

1.   Open *Office_2.rvt*.

2.   In the Browser, locate the *Families* category.
Locate the *Doors* category.
Under *Doors*, locate the **Sgl Flush** Door Family.

3.   Right click and select **New Type**.

4.   Name the new type: **38″ x 84″**.

5.   Highlight the **38″ x 84″** door type.
Right click and select **Type Properties**.

6.

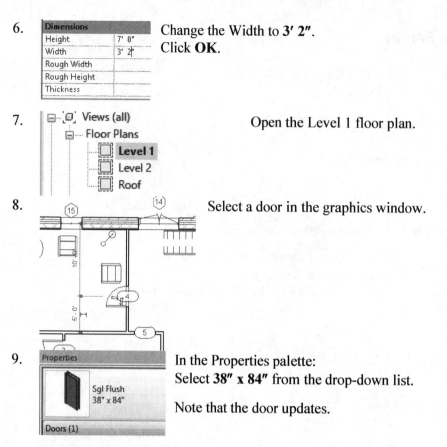

Dimensions	
Height	7' 0"
Width	3' 2"
Rough Width	
Rough Height	
Thickness	

Change the Width to **3' 2"**.
Click **OK**.

7.

⊟ ⌐Oᵀ Views (all)
   ⊟ Floor Plans
       Level 1
       Level 2
       Roof

Open the Level 1 floor plan.

8.

Select a door in the graphics window.

9.

Properties
Sgl Flush 38" x 84"
Doors (1)

In the Properties palette:
Select **38" x 84"** from the drop-down list.

Note that the door updates.

10. Close without saving.

Exercise 10-9
Adding a Shared Parameter to a View Label

Estimated Time: 60 minutes
File: basic_project.rvt

This exercise reinforces the following skills:

- □ View Labels
- □ Shared Parameters
- □ View Properties

1. Open *basic_project.rvt*.

2. Activate the Manage ribbon.

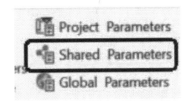

Select **Shared Parameters**.

If you cannot locate the text file, browse for custom parameters.txt in the exercise files.

3. Shared parameter file:

    C:\Revit 2024 Basics\Revit 2024 Basics exx   [Browse...]

    Select the **Browse** button.

4. File name: [custom parameters.txt]

    Files of type: Shared Parameter Files (*.txt)

    Open the *custom paramters.txt*.

5. Parameter group:
    [General]

    Select the **General** parameter group from the drop-down.

6. Parameters
    [New...]

    Select **New** under Parameters.

7. **Parameter Properties** [x]

    Name:
    [View Type]

    Discipline:
    [Common ▼]

    Type of Parameter:
    [Text ▼]

    Tooltip Description:
    <No tooltip description. Edit this parameter to write a custom...

    [Edit Tooltip...]

    Type **View Type** in the Name field.
    Set the Type of Parameter to **Text**.

    Click **OK**.

8.

Shared parameter file:

C:\Users\emoss\SkyDrive\Revit Basics 20   Browse

Parameter group:

General

Parameters:

Hardware Group
View Type

The parameter is now listed.

Click **OK** to close the dialog box.

9.

File   Architecture   Structure   Steel   Precast   Systems   Insert

Creates a Revit file.

New

Project
Creates a Revit project file.

Open

Family
Creates a set of custom components
to use in projects.

On the Applications menu:

Select **New** → **Family**.

10.

ProgramData

Autodesk

RVT 2024

Family Templates

English-Imperial

Annotations

Browse to the *Annotations* folder under *Family Templates*.

11.

File name: View Title

Files of type: Family Template Files (*.rft)

Select the View Title template.

Click **Open**.

12.

before using.

Zoom into the intersection of the two reference planes.

The template contains a line and two reference planes.

13.

Label

Select the **Label** tool on the ribbon.

Left click to place the label above the line.

14.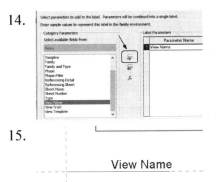

Highlight **View Name** from the list on the left pane.

Select add parameter to label (middle icon) to add the parameter to the right panel.

Click **OK**.

15. Locate the View Name above the line.

View Name

16. Select the label.

In the Properties panel:

Set the Horizontal Alignment for the label to **Left.**

You must have the label selected in order to see its properties.

Click ESC to release the selection.

View Titles (1)		Edit Type
Graphics		⦙
Sample Text	View Name...	
Label	Edit...	
Wrap between parameters...	☐	
Horizontal Align	Left	
Vertical Align	Middle	
Keep Readable	☑	
Visible	☑	

17. Select the **Label** tool on the Create ribbon.

Left click to place the label.

18.

	Parameter Name	Spaces	Prefix	Sample Value
1	View Scale	1	Scale:	1/8" = 1'-0"

Select **View Scale** to add the parameter to the label.

Type **Scale:** in the prefix field.
Click **OK**.

19. Place the Scale label below the view name.

View Name
Scale:1/8" = 1'-0"

20. In the Properties panel:

    Set the Horizontal Alignment for the label to Left.

    You must have the label selected in order to see its properties.

21.

    Select the **Label** tool on the Create ribbon.

    Left click to place the label.

22. Select the **New Parameter** tool at the bottom left of the dialog.

23. Click **Select**.

24. Select the **General** parameter group.

    Select the **View Type**.

    Click **OK**.

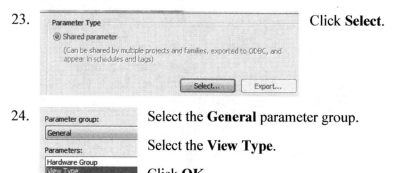

25. Click **OK**.

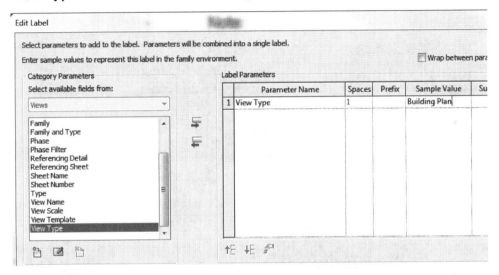

26. View Type is now listed as an available parameter.

Add it to the Label Parameters box.
Type a Sample Value, like Building Plan.

Click **OK**.

27. Set the Horizontal Align to **Left**.

28. This is what you should have so far.

Select the label for the View Type.

Building Plan

View Name

Scale: 1/8" = 1'-0"

29.

Properties	✕ ▣
	Tag Label
	1/8"
View Titles (1)	⊞ Edit Type
Graphics	⌃

Click **Edit Type**.

30. Duplicate...   Click **Duplicate**.

31. Name: 1/8" Technic Italic   Rename **1/8″ Tahoma Italic**.

     OK   Click **OK**.

32.

Text	
Text Font	Tahoma
Text Size	1/8"
Tab Size	1/2"
Bold	☐
Italic	☑
Underline	☐
Width Factor	1.000000

Change the Text Font to **Tahoma**.

Enable **Bold**.
Enable **Italic**.
Click **OK**.

33.

Building Plan

View Name

Scale: 1/8" = 1'-0"

Position the labels so they are aligned.

34.

Note:
View Title system line terminates at
intersection of r

Dummy line sh
Title line.

Delete this note
before using.

Cancel
Repeat [Label]
Recent Commands ▸
Create Similar
Edit Family
Select Previous
Select All Instances ▸
Delete
Find Referring Views

Select the note.
Right click and select **Delete**.

35. Extend the line so that it is past the scale label.

BUILDING PLAN

View Name

Scale: 1/8" = 1'-0"

36. Select **Family Types**.

37. Type name: Select **New** under Family Type.

38. Name: View Type Visible

Type **View Type Visible**.
Click **OK**.

OK

39. Type name: Select New under Family Type.

40. Name: View Type Hidden

Type **View Type Hidden**.
Click **OK**.

OK

41. Select **New Parameter**.

42. Parameter Data

Name:
View Type Label

Discipline:
Common

Type of Parameter:
Yes/No

Group parameter under:
Visibility

○ Type
○ Instance
☐ Reporting Parameter
(Can be used to extract value
from a geometric condition and
report it in a formula or as a
schedulable parameter)

Type **View Type Label** under Name.
Enable **Type**.
Set Type of Parameter to **Yes/No**.
Group under **Visibility**.
Click **OK**.

43. Close the Family Types dialog.

44. Select the view type label.

BUILDING PLAN

View Name

Scale: 1/8" = 1'-0"

45.

View Titles (1)	⌄	🔲 Edit Type
Graphics		⌃
Sample Text	Building Plan...	
Label	Edit...	
Wrap between parameters...	☐	
Horizontal Align	Left	
Vertical Align	Middle	
Keep Readable	☑	
Visible	☑	

On the Visible field, select the small button to the right.

46.

Family parameter: Visible
Parameter type: Yes/No

Existing family parameters of compatible type:

<none>
View Type Label

Add parameter...

OK

Highlight **View Type Label**.

Click **OK**.

47.

Select **Family Types**.

48.

Name: View Type Visible

Parameter	Value
Visibility	
View Type Label	☑
Other	
Name	
Elevation	

Select the **View Type Visible** type.
Enable the **View Type Label**.

49.

Name: View Type Hidden

Parameter	Value
Visibility	
View Type Label	☐
Other	
Name	
Elevation	

Select the **View Type Hidden** type.
Disable the **View Type Label**.

Click **OK**.

50.

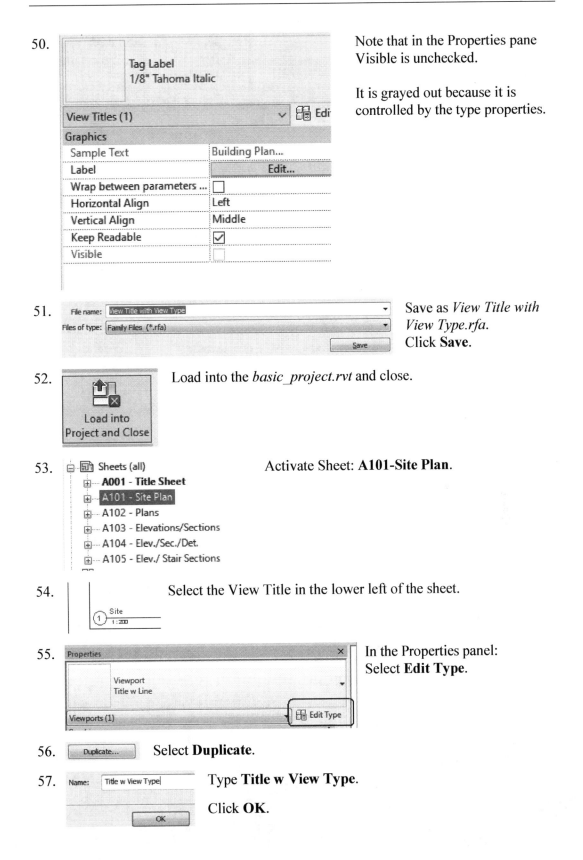

Note that in the Properties pane Visible is unchecked.

It is grayed out because it is controlled by the type properties.

51. Save as *View Title with View Type.rfa*. Click **Save**.

52. Load into the *basic_project.rvt* and close.

53. Activate Sheet: **A101-Site Plan**.

54. Select the View Title in the lower left of the sheet.

55. In the Properties panel: Select **Edit Type**.

56. Select **Duplicate**.

57. Type **Title w View Type**.

Click **OK**.

58.

Type Parameters	
Parameter	Value
Graphics	☆
Title	M_View Title
Show Title	\<none\>
Show Extension Line	M_View Title
Line Weight	View Title with View Type : View Type Hidden
Color	View Title with View Type : View Type Visible
Line Pattern	Solid

Select **View Title with View Type Visible** from the Title drop-down.

Click **OK**.

59.

Site
_____

Scale: 1 : 200
_____

The view title updates, but the view type is not visible because we have not associated the parameter with a view yet.

60. Systems   Insert   Annota

Project Parameters

Shared Parameters

ters

Global Parameters

Select **Project Parameters** on the Manage ribbon.

61. Parameter Name Search: [   ]

▶ Filter

Parameters available to elements in this project:

Occupant
Recycled Content

Select **New Parameter**.

62.

Parameter Type

◯ Project parameter

(Can appear in schedules but not in tags)

◉ Shared parameter

(Can be shared by multiple projects and families, exported to ODBC, and appear in schedules and tags)

[Select...]   [Export...]

Enable **Shared parameter**.

Click **Select**.

63.

Parameter group:

General

Parameters:

Hardware Group
View Type

Select **General**.
Highlight **View Type**.
Click **OK**.

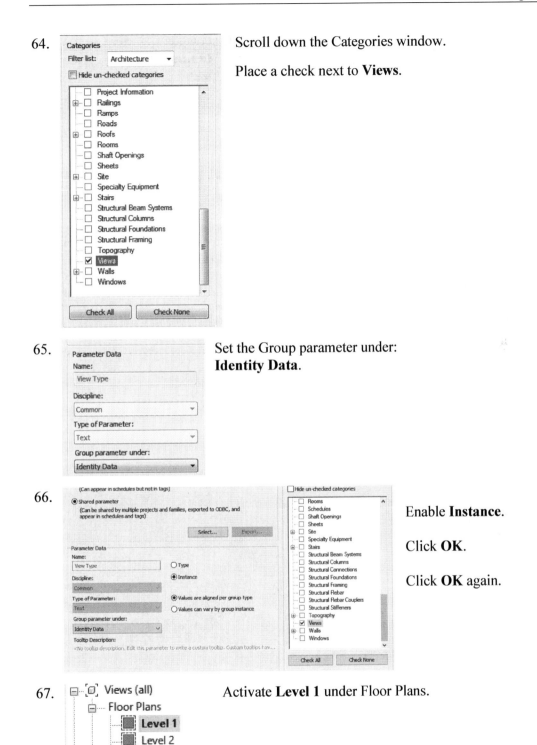

64. Scroll down the Categories window.

Place a check next to **Views**.

65. Set the Group parameter under:
Identity Data.

66. Enable **Instance**.

Click **OK**.

Click **OK** again.

67. Activate **Level 1** under Floor Plans.

68.

Sun Path	☐
Identity Data	☆
View Template	<None>
View Name	Level 1
Dependency	Independent
Title on Sheet	
Sheet Number	A102
Sheet Name	Plans
Referencing Sheet	A103
Referencing Detail	1
View Type	Building Plan

Scroll down to **Identity Data**.
Type **Building Plan** in the View Type field.

69.

Floor Plans
    Level 1
    Level 2
    Site

Activate **Level 2** under Floor Plans.

70.

Sun Path	☐
Identity Data	☆
View Template	<None>
View Name	Level 2
Dependency	Independent
Title on Sheet	
Sheet Number	A102
Sheet Name	Plans
Referencing Sheet	A103
Referencing Detail	1
View Type	Building Plan
Extents	☆

Scroll down to **Identity Data**.
Type **Building Plan** in the View Type field.

Note you can use the drop arrow to select a value.

71.

Views (all)
    Floor Plans
        Level 1
        Level 2
        Site

Activate **Site** under Floor Plans.

72.

Identity Data	
View Template	<None>
View Name	Site
Dependency	Independent
Title on Sheet	
Sheet Number	A101
Sheet Name	Site Plan
Referencing Sheet	A103
Referencing Detail	1
View Type	Survey Plan

Scroll down to **Identity Data**.
Type **Survey Plan** in the View Type field.

73.

Planting Schedule
Sheets (all)
    A001 - Title Sheet
    A101 - Site Plan
    A102 - Plans
    A103 - Elevations/Sections
    A104 - Elev./Sec./Det.
    A105 - Elev./ Stair Sections
Families

Activate Sheet: **A101 - Site Plan**.

74.

Survey Plan

Site
_____

Scale: 1 : 200

The view title has updated.

If you don't see the view type, activate the view and verify that the View Type is set to Survey Plan.

75. 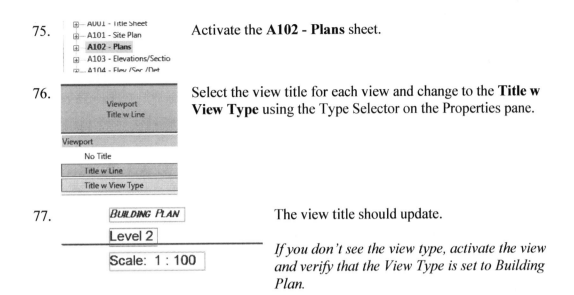 Activate the **A102 - Plans** sheet.

76. Select the view title for each view and change to the **Title w View Type** using the Type Selector on the Properties pane.

77. The view title should update.

If you don't see the view type, activate the view and verify that the View Type is set to Building Plan.

Exercise 10-10
Managing Family Subcategories

Estimated Time: 20 minutes
File: Coffee Table_2.rfa

This exercise reinforces the following skills:

- Families
- OmniClass Number
- Family Subcategories
- Managing Family Visibility

1. Open the *coffee table_2.rfa*

2.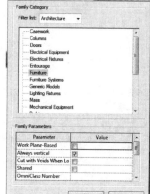

    Activate the Create ribbon.

    Select **Family Categories.**

3.

    Highlight **Furniture**.

4.

    Select the Browse button next to OmniClass Number.

5. Locate the number for coffee tables.

   Click **OK**.

   OmniClass Number and Title
   - No classification
   - 23.40.00.00 - Equipment and Furnishings
     - 23.40.10.00 - Exterior Equipment and Furnishings
     - 23.40.20.00 - General Furniture and Specialties
       - 23.40.20.11 - Wardrobe and Closet Specialties
       - 23.40.20.14 - Residential Furniture
         - 23.40.20.14.14 - Residential Seating
         - 23.40.20.14.17 - Residential Tables and Cabinets
           - 23.40.20.14.17.11 - Dining Room Tables
           - 23.40.20.14.17.14 - China Cabinets
           - 23.40.20.14.17.17 - Sideboards
           - 23.40.20.14.17.21 - End Tables
           - 23.40.20.14.17.24 - Coffee Tables

The OmniClass Title will automatically fill in.

Click **OK**.

Family Parameters

Parameter	Value
Cut with Voids When Loaded	
Shared	
OmniClass Number	23.40.20.14.17.24
OmniClass Title	Coffee Tables
Room Calculation Point	

6. Select the table top.

   Note in the Properties panel there is no subcategory assigned.

   Subcategories can be used to manage level of detail (LOD).

   Other (1) — Edit Type

   Constraints
Extrusion End	0' 0 1/8"
Extrusion Start	0' 0"
Work Plane	Reference Plane : Table Bottom

   Graphics
Visible	☑
Visibility/Graphics Overrides	Edit...

   Materials and Finishes
Material	Walnut

   Identity Data
Subcategory	None
Solid/Void	Solid

7. Activate the **Manage** ribbon.
   Select **Object Styles**.

   Create   Insert   Annota

   Materials   Object Styles   Snaps

8. Select **New** under Modify Subcategories.

   Modify Subcategories
   New   Delete

9. Type **Table Top**.

   Click **OK**.

   New Subcategory

   Name:
   Table Top

   Subcategory of:
   Furniture

   OK   Cancel

10. Select **New** under Modify Subcategories.

11. Type **Table Legs**.

12.

Category	Line Weight		Line Color	Line Pattern
	Projection	Cut		
Furniture	1		Black	
Hidden Lines	1		Black	Dash
Table Legs	1		Black	Hidden
Table Top	3		Blue	Solid

Change the Lineweight of the Table Top to **3**.

Change the Line pattern of the Table Legs to **Hidden**.
Change the Line Color of the Table Top to **Blue**.

13.

Category	Line Weight		Line Color	Line Pattern	Material
	Projection	Cut			
Furniture	1		Black		
Hidden Lines	1		Black	Dash	
Table Legs	1		Black	Hidden	Walnut
Table Top	3		Blue	Solid	Walnut

Assign the Walnut Material to both subcategories.

14.

Modify the Walnut material to have a surface pattern and a cut pattern.

Click **OK**.

Close the Object Styles dialog.

15.

Material	Walnut
Identity Data	
Subcategory	Table Top
Solid/Void	Solid

Select the Table Top and assign the Table Top subcategory.

16.

Identity Data	
Subcategory	Table Legs
Solid/Void	Solid

Select the Table Legs and assign the Table Legs category.

17.
- Views (all)
  - Floor Plans
    - **Ref. Level**

Switch to the **Ref Level** view.

18.

Switch to different displays and observe how the model appears.

Notice how the subcategories control the display of the two different element types.

Graphic Display Options...
- Wireframe
- Hidden Line
- Shaded
- Consistent Colors
- Textures
- Realistic

19. Save the family as *Coffee Table 3.rfa*.

The following families are not cuttable and are always shown in projection in views:
- Balusters
- Detail Items
- Electrical Equipment
- Electrical Fixtures
- Entourage
- Furniture
- Furniture Systems
- Lighting Fixtures
- Mechanical Equipment
- Parking
- Planting
- Plumbing Fixtures
- Specialty Equipment

Families which are cuttable, where you can control the line style, color, and visibility in plan, RCP, section, and elevation views:

- Casework
- Columns
- Doors
- Site
- Structural Columns
- Structural Foundations
- Structural Framing
- Topography
- Walls
- Window

Lesson 10 Quiz

True or False

1. Family subcategories control the visibility of family geometry.
2. A model family can be loaded using the Insert ribbon.
3. The Symbol tool lets you place 2D annotation symbols.
4. Dimensions in a family should be placed on sketches.
5. Shared parameters cannot be added to loadable families.
6. Shared parameters cannot be used in View Titles.
7. 2D and 3D geometry can be used to create families.
8. Geometry in families should be constrained to reference planes.
9. The ALIGN tool can be used to constrain geometry to a reference plane in the Family Editor.
10. Automatic dimensions are not displayed by default.

Multiple Choice

11. The following families are not cuttable (meaning the edges are always shown as solid):

    A. Furniture
    B. Walls
    C. Entourage
    D. Lighting Fixtures

12. The following families are cuttable (meaning line style, color and visibility can be controlled):

    A. Furniture
    B. Walls
    C. Entourage
    D. Doors

13. The intersection of the two default reference planes in a family template indicates:

    A. The elevation
    B. X marks the spot
    C. The insertion point
    D. The origin

14. A model family has the file extension:

    A. rvt
    B. rft
    C. rfa
    D. mdl

15. The graphical editing mode in Revit that allows users to create families is called:

    A. The Family Editor
    B. Family Types
    C. Family Properties
    D. Revit Project
    E. Draw

16. OMNI Class Numbers are assigned in this dialog:

    A. Type Properties
    B. Properties panel
    C. Family Types
    D. Family Categories

17. Use a wall-based template to create a model family if the component is to be placed:

    A. in or on a wall
    B. in or on a roof
    C. in or on a reference plane
    D. in a project

18. To insert an image into a title block family:

    A. Load a family
    B. Use the Insert ribbon
    C. Use the Image tool
    D. Use the Import tool

ANSWERS:
1) T; 2) T; 3) T; 4) F; 5) F; 6) F; 7) T; 8) T; 9) T; 10) T; 11) A, C, & D; 12) B & D; 13) C & D; 14) C; 15) A; 16) D; 17) A; 18) C

Revit Hot Keys

3F	Fly Mode		EC	Check Electrical Circuits
3O	Object Mode		EE	Electrical Equipment
3W	Walk		EG	Edit Group
32	2D Mode		EH	Hide Element
AA	Adjust Analytical Model		EL	Spot Elevation
AC	Activate Controls/Dimensions		EOD	Override Graphics in View
AD	Attach Detail Group		EOG	Graphic Override by Element
AL	Align			
AP	Add to Group		EOH	Graphic Override by Element
AR	Array			
AT	Air Terminal		EOT	Graphic Override by Element
BM	Structural Beam			
BR	Structural Brace		EP	Edit Part
BS	Space Type Settings		ER	Editing Requests
BX	Selection Box		ES	Electrical Settings
CC	Copy		EU	Unhide Element
CFG	Configuration		EW	Arc Wire
CG	Cancel		EX	Exclude
CL	Structural Column		FD	Flex Duct
CM	Component		FG	Finish
CN	Conduit		FP	Flexible Pipe
CO	Copy		FR	Find/Replace Text
CP	Apply Coping		FS	Fabrication Settings
CS	Create Similar		FT	Foundation Wall
CT	Cable Tray		GD	Graphic Display Options
CV	Convert to Flex Duct		GL	Global Parameters
CX	Reveal Constraints Toggle		GP	Model Group
			GR	Grid
DA	Duct Accessory		HC	Hide/Isolate Category
DC	Check Duct Systems		HH	Hide/Isolate Objects
DE	Delete		HI	Hide/Isolate Objects
DI	Aligned Dimension		HL	Hidden Line
DF	Duct Fitting		HR	Reset Temporary Hide/Isolate
DL	Detail Lines			
DM	Mirror - Draw Axis		HT	Help Tooltip
DR	Door		IC	Isolate Category
DT	Duct		KS	Keyboard Shortcuts
			LC	Lose Changes

LD	Loads	RE	Resize\Scale
LF	Lighting Fixture	RF	Reinforcement
LG	Link	RG	Render Gallery
LI	Model Line	RH	Reveal Hidden Elements
LL	Level	RM	Room
LO	Heating and Cooling Loads	RN	Reinforcement Numbers
		RO	Rotate
LW	Linework	RL	Reload Latest Worksets
MA	Match Properties	RP	Reference Plane
MD	Modify	RR	Render
ME	Mechanical Equipment	RT	Room Tag
MM	Mirror - Pick Axis	RW	Reload Latest Worksets
MP	Move to Project	RY	Ray Trace
MR	Multipoint Routing	S	Split Walls and Lines
MS	Mechanical Settings	SA	Select All Instances in Entire Project
MV	Move		
NF	Conduit Fitting	SB	Structural Floor
OF	Offset	SC	Snap to Center
PA	Pipe Accessory	SD	Shading with Edges On
PB	Fabrication Part	SE	Snap to Endpoint
PC	Check Pipe Systems	SF	Split Face
PI	Pipe	SH	Snap to horizontal/vertical
PE	Plumbing Equipment		
PF	Pipe Fitting	SI	Snap to Intersection
PK	Pick a Plane	SK	Sprinkler
PL	Split	SL	Split Element
PN	Pin	SM	Snap to Midpoint
PP or VP or Ctl+1	Properties	SN	Snap to Nearest
		SO	Snaps OFF
PP	Pin Position	SP	Snap to Perpendicular
PR	Properties	SQ	Snap to Quadrants
PS	Panel Schedules	SR	Snap to Remote Objects
PT	Paint	SS	Turn Snap Override Off
PX	Plumbing Fixture	ST	Snap to Tangent
R3	Define a new center of rotation	SU	Sun Settings
		SW	Snap to Workplane Grid
RA	Reset Analytical Model	SX	Snap to Points
RB	Restore Excluded Member	SZ	System Zone
		TF	Cable Tray Fitting
RC	Remove Coping	TG	Tag by Category
RD	Render in Cloud		

TL	Thin Lines		//	Divide Surface
TR	Trim/Extend		Alt+F4	Close Revit
TW	Tab Window		Alt+Backspace	Undo
TX	Text		Ctl++	Schedule View Zoom In
UN	Project Units		Ctl+-	Schedule View Zoom Out
UG	Ungroup			
UP	Unpin		Ctl+0	Schedule View Zoom Restore
VH	Category Invisible			
VI	View Invisible Categories		Ctl+F4	Close Project file
			Ctl+1	Properties
VG	Visibility/Graphics		Ctl+	Activate Contextual Tab
VOG	Graphic Override by Category		Ctl+=	Subscript
			Ctl+B	Bold
VOH	Graphic Override by Category		Ctl+C	Copy to Clipboard
			Ctl+D	Toggle Home
VOT	Graphic Override by Category		Ctl+F	Project Browser Search
			Ctl+I	Italic
VP	View Properties		Ctl+N	New Project
VR	View Range		Ctl+O	Open a Project file
VU	Unhide Category		Ctl+P	Print
VV	Visibility/Graphics		Ctl+Q	Close Text Editor
WA	Wall		Ctl+S	Save a file
WC	Window Cascade		Ctl+U	Underline
WF	Wire Frame		Ctl+V	Paste from Clipboard
WN	Window		Ctl+W	Close window
WT	Window Tile		Ctl+X	Cut to Clipboard
ZA	Zoom to Fit		Ctl-Y	Redo
ZC	Previous Zoom		Ctl-Z	Undo
ZE	Zoom to Fit		Ctl+Shift+=	Superscript
ZF	Zoom to Fit		Ctl+Shift+A	All Caps
ZN	Zoom Next		Ctl+Shift+Z	Undo
ZO	Zoom Out (2X)		Ctl+Ins	Copy to Clipboard
ZP	Zoom Previous		Shift+W	Dynamic View
ZR	Zoom in region (window)		F1	Revit Help
			F5	Refresh Screen
ZS	Zoom to Sheet Size (limits)		F7	Spelling
			F8	Dynamic View
ZV	Zoom Out (2X)		F9	System Browser
ZX	Zoom to fit		F11	Status Bar
ZZ	Zoom in region (window)			

Notes:

The Revit Beta software did not have access to hot keys. This should be fixed in the release software.

About the Author
Autodesk
Certified Instructor

Elise Moss has worked for the past thirty years as a mechanical designer in Silicon Valley, primarily creating sheet metal designs. She has written articles for Autodesk's Toplines magazine, engineering.com, AUGI's PaperSpace, DigitalCAD.com and Tenlinks.com. She is President of Moss Designs, creating custom applications and designs for corporate clients. She has taught CAD classes at Laney College, DeAnza College, Silicon Valley College, and for Autodesk resellers. Autodesk has named her as a Faculty of Distinction for the curriculum she has developed for Autodesk products and she is a Certified Autodesk Instructor. She holds a baccalaureate degree in mechanical engineering from San Jose State.

She is married with two sons. Her older son, Benjamin, is an electrical engineer. Her younger son, Daniel, works with AutoCAD Architecture in the construction industry. Her husband, Ari, has a distinguished career in software development.

Elise is a third-generation engineer. Her father, Robert Moss, was a metallurgical engineer in the aerospace industry. Her grandfather, Solomon Kupperman, was a civil engineer for the City of Chicago.

She can be contacted via email at elise_moss@mossdesigns.com.

More information about the author and her work can be found on her website at www.mossdesigns.com.

Other books by Elise Moss

AutoCAD Architecture 2023 Fundamentals
Autodesk Revit Architecture 2023 Basics

Notes:

PGIL2023USA